REFRIGERATION AND AIR CONDITIONING

McGraw-Hill Series in Mechanical Engineering

Jack P. Holman, *Southern Methodist University*
Consulting Editor

Anderson: *Modern Compressible Flow: With Historical Perspective*
Barron: *Cryogenic Systems*
Eckert: *Introduction to Heat and Mass Transfer*
Eckert and Drake: *Analysis of Heat and Mass Transfer*
Eckert and Drake: *Heat and Mass Transfer*
Ham, Crane, and Rogers: *Mechanics of Machinery*
Hartenberg and Denavit: *Kinematic Synthesis of Linkages*
Hinze: *Turbulence*
Hutton: *Applied Mechanical Vibrations*
Jacobsen and Ayre: *Engineering Vibrations*
Juvinall: *Engineering Considerations of Stress, Strain, and Strength*
Kays and Crawford: *Convective Heat and Mass Transfer*
Lichty: *Combustion Engine Processes*
Martin: *Kinematics and Dynamics of Machines*
Phelan: *Dynamics of Machinery*
Phelan: *Fundamentals of Mechanical Design*
Pierce: *Acoustics: An Introduction to Its Physical Principles and Applications*
Raven: *Automatic Control Engineering*
Schenck: *Theories of Engineering Experimentation*
Schlichting: *Boundary-Layer Theory*
Shames: *Mechanics of Fluids*
Shigley: *Dynamic Analysis of Machines*
Shigley: *Kinematic Analysis of Mechanisms*
Shigley: *Mechanical Engineering Design*
Shigley: *Simulation of Mechanical Systems*
Shigley and Uicker: *Theory of Machines and Mechanisms*
Stoecker and Jones: *Refrigeration and Air Conditioning*

REFRIGERATION AND AIR CONDITIONING

Second Edition

W. F. Stoecker

Professor of Mechanical Engineering
University of Illinois at Urbana-Champaign

J. W. Jones

Associate Professor of Mechanical Engineering
University of Texas at Austin

McGraw-Hill, Inc.
New York St. Louis San Francisco Auckland Bogotá
Caracas Lisbon London Madrid Mexico City Milan
Montreal New Delhi San Juan Singapore
Sydney Tokyo Toronto

This book was set in Press Roman by Jay's Publishers Services, Inc.
The editors were Diane D. Heiberg and David A. Damstra;
the production supervisor was Diane Renda.
The drawings were done by George Morris, Scientific Illustrators.

REFRIGERATION AND AIR CONDITIONING

15 16 17 18 19 20 BRBBRB 00 9 8 7 6 5 4 3 2 1 0

ISBN 0-07-061619-1

Library of Congress Cataloging in Publication Data

Stoecker, W. F. (Wilbert F.), date
Refrigeration and air conditioning.

(McGraw-Hill series in mechanical engineering)
Includes bibliographical references and indexes.
1. Refrigeration and refrigerating machinery.
2. Air conditioning. I. Jones, J. W. (Jerold W.)
II. Title. III. Series.
TP492.S8 1982 621.5′6 81-19385
ISBN 0-07-061619-1 AACR2

CONTENTS

PREFACE

There are conventional as well as special reasons for writing a second edition of "Refrigeration and Air Conditioning." The conventional reason is that in the 24 years since the appearance of the first edition some of the equipment and systems have slipped to lesser importance while new products and concepts have emerged. An updating is therefore timely. A special reason for a new edition is the impact that energy effectiveness now exerts on heating and cooling systems. Energy consciousness has perceptibly changed the equipment and design concepts in refrigeration and air conditioning. Also, since most engineers predict that the days of low-cost energy will never return, energy-conservative strategies are now permanent.

The second edition differs from the first in the following major ways. A heavier presentation of air-conditioning *systems* is provided, but the technical emphasis on vapor-compression systems is not only maintained but strengthened. The new material directed toward the air conditioning of buildings replaces several chapters (cryogenics, steam jet, and air-cycle) which are of admitted importance, but the engineering of these systems in professional practice is normally handled by other engineers than those working in comfort air conditioning and industrial refrigeration and air conditioning. The digital computer is now a standard tool of engineers, and where appropriate calculations and problems use the computer.

Some sections and particularly the style of the first edition have been maintained. The first edition achieved a certain degree of acceptance throughout the world and apparently meets the need of those seeking a technical book that goes beyond pure descriptions into quantitative treatment of the subjects. At the same time the second edition attempts to maintain an emphasis on qualitative evaluations and trends, yet introducing no additional complexity unless there is a payoff of improved understanding.

The authors are indebted to numerous professional associates whose views have influenced the topic selection, emphases, and technical presentations in this book. Special thanks are due the following colleagues: Larry G. Berglund, John B. Pierce Foundation; John C. Chato, University of Illinois at Urbana-Champaign; Arthur M. Clausing, University of Illinois at Urbana-Champaign; James E. Shahan, Transco, Inc.; Gary C. Vliet, University of Texas at Austin; and James E. Woods, Iowa State University.

W. F. Stoecker
J. W. Jones

CHAPTER
ONE
APPLICATIONS OF REFRIGERATION AND AIR CONDITIONING

1-1 Major uses The fields of refrigeration and air conditioning are interconnected, but each also has its own province. The interrelationship and independence can be schematized as in Fig. 1-1. The largest application of refrigeration, which is the process of cooling, is for air conditioning. In addition, refrigeration embraces *industrial refrigeration*, including the processing and preservation of food; removing heat from substances in chemical, petroleum, and petrochemical plants; and numerous special applications such as those in the manufacturing and construction industries.

In a similar manner, air conditioning embraces more than cooling. The definition[1]† of comfort air conditioning is "the process of treating air to control simultaneously its temperature, humidity, cleanliness, and distribution to meet the comfort requirements of the occupants of the conditioned space." Air conditioning therefore includes the entire heating operation (which does not involve refrigeration except for heat pumps) as well as the regulation of velocity, thermal radiation, and the quality of air, including removal of foreign particles and vapors.

Engineers are employed in the research, development, and application of products serving these fields, and other engineers have the responsibility of designing systems using the products. While there is no rigid barrier that prevents engineers from moving freely through the various provinces of Fig. 1-1, the concentration of interest of commercial firms, and thus the emphasis of their engineers, lies either in air conditioning or industrial refrigeration. The temperature range of interest in industrial refrigeration probably extends down to about -60°C. Another field, cryogenics,[2] which covers temperatures still lower, includes the industrial-gas industry (separation of air into nitrogen and oxygen), liquefied natural gas, and the pursuit of temperatures approaching absolute zero.

This chapter seeks to convey some of the breadth and diversity of the uses of refrigeration and air conditioning. The categories of some of the air-conditioning applications described include medium-sized and large buildings, industrial, residential, and vehicular. For the industrial-refrigeration field examples will also be cited in the food

† Numbered references are given at the end of the chapter.

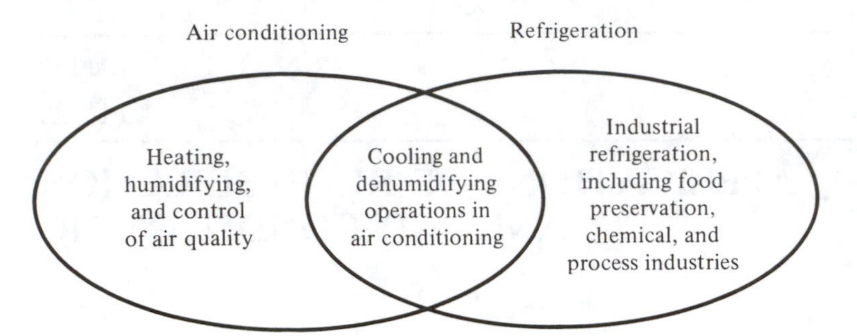

Figure 1-1 Relationship of the refrigeration and air-conditioning fields.

and the thermal-processing industries. While the cooling capacity associated with special uses of refrigeration is small relative to air conditioning or food refrigeration, many of these special applications are intriguing and technically challenging.

1-2 Air conditioning of medium-sized and large buildings Most of the air-conditioning units in service provide *comfort air conditioning*, the purpose of which is to supply comfortable conditions for people. Summer cooling systems have become a standard utility in large buildings throughout the world. Even in climates where summer temperatures are not high, large buildings may have to be cooled in order to remove the heat generated internally by people, lights, and other electrical equipment. In hot climates, the existence of summer cooling systems is the difference between workers performing effectively and not. Some form of central system usually serves large buildings. It may consist of one or more water-chilling plants and a water heater (tradition-

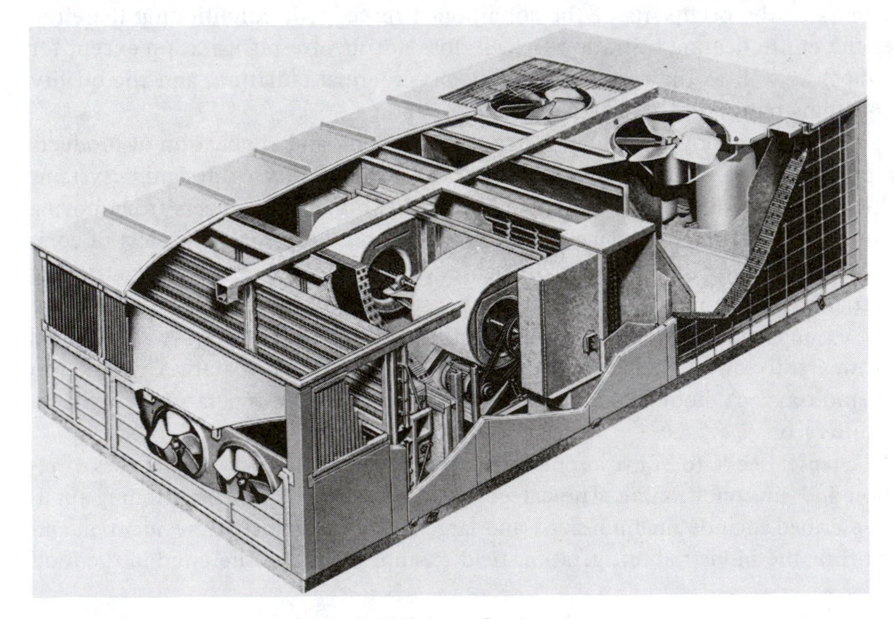

Figure 1-2 A rooftop unit. *(Lennox Industries, Inc.)*

ally referred to as a boiler) located in a machine room. The conditioned spaces are served by one or more air-supply and return systems, as explained in Chap. 5, or the hot or chilled water may be piped to heat exchangers in the conditioned space.

Single-story commercial buildings such as stores and factories are often served by rooftop units (Fig. 1-2) mounted on the roof and providing conditioned air to the space below. The unit in Fig. 1-2 is a heat pump (discussed in Chap. 18), which is capable of providing either heated or cooled air to the conditioned space.

Another important class of air-conditioning installation is for hospitals[3] and other medical buildings. Many of the same requirements that prevail in office-building air conditioning apply to hospitals, but there are a number of additional concerns as well. Ventilation requirements often specify the use of 100 percent outdoor air, and humidity limits may be more severe in operating rooms to avoid static electricity. The design of an energy-efficient system for a hospital that also meets the special requirements poses an engineering challenge.

1-3 Industrial air conditioning The term *industrial air conditioning* will refer here to providing at least a partial measure of comfort for workers in hostile environments but also to controlling air conditions so that they are favorable to processing some object or material.

Spot heating During cold weather it may be more practical to warm a confined zone where a worker is located. One such approach is through the use of an infrared heater.[4] When its surfaces are heated to a high temperature by means of a burner or by electricity, they radiate heat to the affected area.

Spot cooling It may be impractical to cool an entire steel mill, but conditions may be kept tolerable for workers by directing a stream of cool air[5] onto occupied areas.

Environmental laboratories The role of air conditioning varies from one environmental laboratory to another.[6] In one a temperature of -40°C must be maintained to test engines at low temperatures, and in another a high temperature and humidity may be maintained to study the behavior of animals[7] in tropical climates.

Printing Control of humidity is one of the primary reasons for air conditioning printing plants. In some printing processes the paper is run through several different presses, and the air conditioning must be maintained to provide proper registration. Other troubles caused by improper humidity are static electricity, curling or buckling of the paper, or failure of the ink to dry.

Textiles Like paper, textiles are sensitive to changes in humidity and to a lesser extent changes in temperature. The yarn in modern textile plants moves at tremendous speeds, and changes in the flexibility and strength of the textile or generation of static electricity must be prevented.

Precision parts and clean rooms For manufacturers of precision metal parts air conditioning performs three services: keeping the temperatures uniform so that the metal will not expand and contract, maintaining a humidity so that rust is prevented, and filtering the air to minimize dust. A technology for *clean rooms*[8] (Fig. 1-3) has developed for the design and construction of such enclosures for manufacturing electronic components and other materials.

Photographic products The photographic-products industry is a large user of air con-

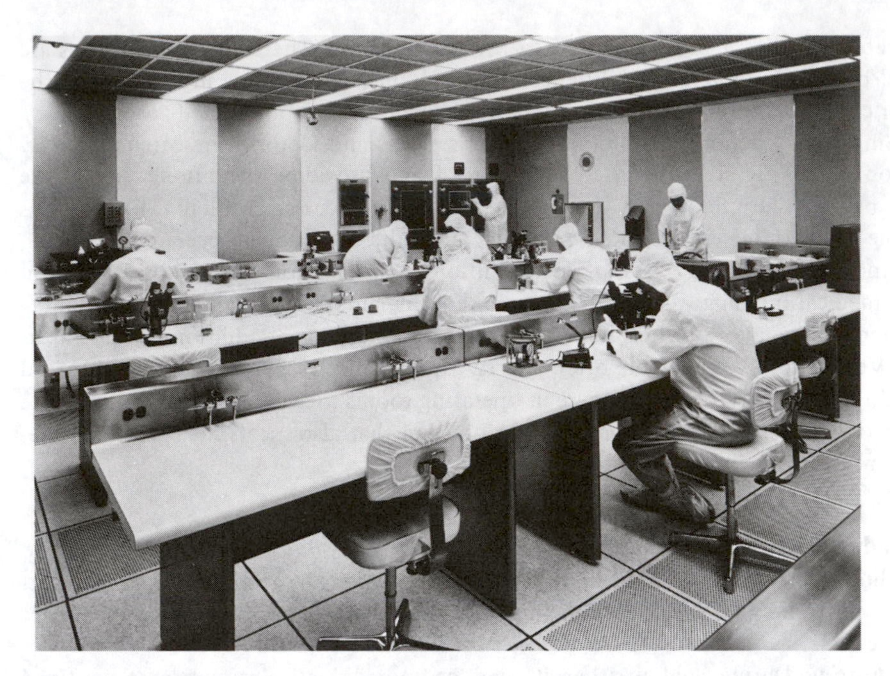

Figure 1-3 A clean room. *(Weber Technical Products, subsidiary of Walter Kidde & Co., Inc.)*

ditioning and refrigeration. Raw photographic material deteriorates rapidly in high temperatures and humidities, and other materials used in coating film require careful control of temperature.

Computer rooms The air-conditioning system for computer rooms should control the temperature, humidity, and cleanness of the air. Some electronic components operate in a faulty manner if they become too hot, and one means of preventing such localized high temperature is to maintain the air temperature in the computer room in the range of 20 to 23°C. The electronic components in the computer function favorably at even lower temperatures, but this temperature is a compromise with the lowest comfortable temperature for occupants. A relative humidity between 30 and 45 percent is desirable for handling cards, since too high a relative humidity can result in improper feeding of cards and too low a relative humidity could cause static electricity in card decks. For long-term storage of magnetic tape the uniformity of temperature is an important requirement. Well-filtered air is necessary to facilitate low-maintenance operation of printers, tape drives, and card readers.

Power plants Traditionally steam power plants were kept tolerable for workers by ventilating with outdoor air. The compactness and the increase in heat-flow intensities now no longer leave adequate space for air ducts. To provide cooling for many confined spaces in modern power plants air-cooling coils are supplied with refrigerated water that is conveyed through pipes much smaller than the conventional air ducts.

1-4 Residential air conditioning In the United States approximately 5 million room air conditioners are sold each year, and most of them are used for residential service.

In another type of residential air conditioner, the *central* or *unitary system*, a condensing unit consisting of the compressor and condenser is located out of doors and the evaporator coil in the interior air duct. Annual sales of this class of air conditioner are usually between 3 and 5 million units.

During the past several decades in the United States there has been a shift in population to the sun belt in the southern part of the country. Leaders of the air-conditioning industry point out that this demographic shift would probably not have occurred had it not been for the widespread use of air conditioning in homes and places of work, business, and entertainment.

Another system that is growing in importance for the combined heating and cooling of residences is the heat pump. It first appeared on the market in the 1950s amid optimistic predictions of how it would make competing equipment obsolete. This entry into the market floundered, however, primarily because of the mechanical failure rate of the heat pumps of that era. Improved design and quality of manufacture led to a resurgence of heat pumps in the 1980s such that annual sales in the United States now run between 0.5 and 1 million units.

1-5 Air conditioning of vehicles The most air-conditioned vehicle is the automobile,[9] for which between 5 and 10 million systems are sold annually. But many other conveyances are air-conditioned as well, including buses, trains, trucks (Fig. 1-4), recrea-

Figure 1-4 A truck air conditioner. *(Kysor Manufacturing Co.)*

tional vehicles, tractors, crane cabs, aircraft, and ships. The major contributor to the cooling load in many of these vehicles is heat from solar radiation, and, in the case of public transportation, heat from people. The loads are also characterized by rapid changes and by a high intensity per unit volume in comparison to building air conditioning.

1-6 Food storage and distribution Many meats, fish, fruits, and vegetables are perishable, and their storage life can be extended by refrigeration. Fruits, many vegetables, and processed meat, such as sausages, are stored at temperatures just slightly above freezing to prolong their life. Other meats, fish, vegetables, and fruits are frozen and stored many months at low temperatures until they are defrosted and cooked by the consumer.

The frozen-food chain typically consists of the following links: freezing, storage in refrigerated warehouses, display in a refrigerated case at food markets, and finally storage in the home freezer or frozen-food compartment of a domestic refrigerator.

Freezing Early attempts to freeze food resulted in a product laced with ice crystals until it was discovered that the temperature must be plunged rapidly through the freezing zone. Approaches[10] to freezing food include air-blast freezing, where air at approximately -30°C is blown with high velocity over packages of food stacked on fork-lift pallets; contact freezing, where the food is placed between metal plates and surfaces; immersion freezing, where the food is placed in a low-temperature brine; fluidized-bed freezing, where the individual particles are carried along a conveyor belt and kept in suspension by an upward-directed stream of cold air (Fig.1-5); and freezing with a cryogenic substance such as nitrogen or carbon dioxide.

Figure 1-5 Freezing peas on a fluidized bed conveyor belt. *(Lewis Refrigeration Company.)*

Figure 1-6 A refrigerated warehouse. *(International Association of Refrigerated Warehouses.)*

Storage Fruits and vegetables should be frozen quickly after harvesting and meats frozen quickly after slaughter to maintain high quality. Truckload and railcar-load lots are then moved to refrigerated warehouses (Fig. 1-6), where they are stored at -20 to -23°C, perhaps for many months. To maintain a high quality in fish, the storage temperature is even lower.

Distribution Food moves from the refrigerated warehouses to food markets as needed to replenish the stock there. In the market the food is kept refrigerated in display cases held at 3 to 5°C for dairy products and unfrozen fruits and vegetables at approximately -20°C for frozen foods and ice cream. In the United States about 100,000 refrigerated display cases are sold each year.

The consumer finally stores the food in a domestic refrigerator or freezer until used. Five million domestic refrigerators are sold each year in the United States, and for several decades styling and first cost were paramount considerations in the design and manufacture of domestic refrigerators. The need for energy conservation,[11] however, has brought back the engineering challenge in designing these appliances.

1-7 Food processing Some foods need operations in addition to freezing and refrigerated storage, and these processes entail refrigeration as well.

Dairy products The chief dairy products are milk, ice cream, and cheese. To pasteurize milk the temperature is elevated to approximately 73°C and held for about 20 s. From that process the milk is cooled and ultimately refrigerated to 3 or 4°C for storage. In manufacturing ice cream[12] the ingredients are first pasteurized and thoroughly mixed. Then, refrigeration equipment cools the mix to about 6°C, whereupon it enters a freezer. The freezer drops the temperature to -5°C, at which temperature the mix stiffens but remains fluid enough to flow into a container. From this point on the ice cream is stored below freezing temperatures.

There are hundreds of varieties of cheese, each prepared by a different process, but typical steps include bringing the temperature of milk to about 30°C and then adding several substances, including a cheese starter and sometimes rennet. Part of the mixture solidifies into the curds, from which the liquid whey is drained. A curing period in refrigerated rooms follows for most cheeses at temperature of the order of 10°C.

Beverages Refrigeration is essential in the production of such beverages as concentrated fruit juice, beer, and wine. The taste of many drinks can be improved by serving them cold.

Juice concentrates are popular because of their high quality and reasonable cost. It is less expensive to concentrate the juice close to the orchards and ship it in its frozen state than to ship the raw fruit. To preserve the taste of juice, its water must be boiled off at a low temperature, requiring the entire process to be carried out at pressures much below atmospheric.

In the brewing industry refrigeration controls the fermentation reaction and preserves some of the intermediate and final products. A key process in the production of alcohol is fermentation, an exothermic reaction. For producing a lager-type beer, fermentation should proceed at a temperature between 8 and 12°C, which is maintained by refrigeration. From this point on in the process the beer is stored in bulk and ultimately bottled or kegged (Fig. 1-7) in refrigerated spaces.

The major reason for refrigerating bakery products is to provide a better match between production and demand and thus prevent waste. Many breads and pastries are frozen following baking to provide a longer shelf life before being sold to the consumer. A practice that provides freshly baked products (and the enticing aroma as well) in individual supermarkets but achieves some of the advantages of high production is to prepare the dough in a central location, freeze it, and then transport it to the supermarket, where it is baked as needed.

Some biological and food products are preserved by freeze drying, in which the product is frozen and then the water is removed by sublimation (direct transition from ice to water vapor). The process takes place in a vacuum while heat is carefully applied to the product to provide the heat of sublimation. Some manufacturers of instant coffee use the freeze-drying process.

1-8 Chemical and process industries The chemical and process industries include the manufacturers of chemicals, petroleum refiners, petrochemical plants, paper and pulp industries, etc. These industries require good engineering for their refrigeration since

Figure 1-7 Refrigeration is essential in such beverage industries as breweries. *(Anheuser Busch Company, Inc.)*

almost every installation is different and the cost of each installation is so high. Some important functions served by refrigeration[13] in the chemical and process industries are (1) separation of gases, (2) condensation of gases, (3) solidification of one substance in a mixture to separate it from others, (4) maintenance of a low temperature of stored liquid so that the pressure will not be excessive, and (5) removal of heat of reaction.

A mixture of hydrocarbon gases can be separated into its constituents by cooling the mixture so that the substance with the high-temperature boiling point condenses and can be physically removed from the remaining gas. Sometimes in petrochemical plants (Fig. 1-8) hydrocarbons, such as propane, are used as the refrigerant. Propane is relatively low in cost compared with other refrigerants, and the plant is completely equipped to handle flammable substances. In other applications separate refrigeration units, such as the large packaged unit shown in Fig. 1-9, provide refrigeration for the process.

1-9 Special applications of refrigeration Other uses of refrigeration and air conditioning span sizes and capacities from small appliances to the large industrial scale.

Figure 1-8 Refrigeration facility in a petrochemical plant. The building contains refrigeration compressors requiring 6500 kW. Associated refrigeration equipment is immediately to the right of the compressor house. *(U.S. Industrial Chemicals Company, Division of National Distillers and Chemical Corp.)*

Figure 1-9 Two-stage packaged refrigeration unit for condensing CO_2 at −23°C. *(Refrigeration Engineering Corporation).*

Drinking fountains Small refrigeration units chill drinking water for storage and use as needed.

Dehumidifiers An appliance to dehumidify air in homes and buildings uses a refrigeration unit by first passing the air to be dehumidified through the cold evaporator coil of the system, where the air is both cooled and dehumidified. Then this cool air flows over the condenser and is discharged to the room.

Ice makers The production of ice may take place in domestic refrigerators, ice makers serving restaurants and motels, and large industrial ice makers serving food-processing and chemical plants.

Ice-skating rinks Skaters, hockey players, and curlers cannot rely upon the weather to provide the cold temperatures necessary to freeze the water in their ice rinks. Pipes carrying cold refrigerant or brine are therefore embedded in a fill of sand or sawdust, over which water is poured and frozen.[14]

Construction Refrigeration is sometimes used to freeze soil to facilitate excavations. A further use of refrigeration is in cooling huge masses of concrete[15] (the chemical reaction which occurs during hardening gives off heat, which must be removed so that it cannot cause expansion and stress the concrete). Concrete may be cooled by chilling the sand, gravel, water, and cement before mixing, as in Fig. 1-10, and by embedding chilled-water pipes in the concrete.

Desalting of seawater One of the methods available for desalination of seawater[16] is to freeze relatively salt-free ice from the seawater, separate the ice, and remelt it to redeem fresh water.

Figure 1-10 Precooling materials for concrete in a dam. *(Sulzer Brothers, Inc.)*

1-10 Conclusion The refrigeration and air-conditioning industry is characterized by steady growth. It is a stable industry in which replacement markets join with new applications to contribute to its health.

The high cost of energy since the 1970s has been a significant factor in stimulating technical challenges for the individual engineer. Innovative approaches to improving efficiency which once were considered impractical now receive serious consideration and often prove to be economically justified. An example is the recovery of low-temperature heat by elevating the temperature level of this energy with a heat pump (which is a refrigeration system). The days of designing the system of lowest first cost with little or no consideration of the operating cost now seem to be past.

REFERENCES

1. "ASHRAE Handbook, Fundamentals Volume," American Society of Heating, Refrigerating, and Air-Conditioning Engineers, Atlanta, Ga.,† 1981.
2. G. Haselden: "Cryogenic Fundamentals," Academic, New York, 1971.
3. H. H. Stroeh and J. E. Woods: Development of a Hospital Energy Management Index, *ASHRAE Trans.* vol. 84, pt. 2, 1978.
4. J. E. Janssen: Field Evaluation of High Temperature Infrared Space Heating Systems, *ASHRAE Trans.*, vol. 82, pt. 1, pp. 31–37, 1976.
5. Environmental Control in Industrial Plants, Symp. PH-79-1, *ASHRAE Trans.*, vol. 85, pt. 1, pp. 307–333, 1979.
6. Environmental Control for the Research Laboratory, Symp. AT-78-3, *ASHRAE Trans.*, vol. 84, pt. 1, pp. 511–560, 1978.
7. Environmental Considerations for Laboratory Animals, Symp. BO-75-8, *ASHRAE Trans.*, vol. 81, pt. 2, 1975.
8. Clean Rooms, *ASHRAE Symp.* DE-67-3, American Society of Heating, Refrigerating, and Air-Conditioning Engineers, Atlanta, Ga., 1967.
9. D. W. Ruth: Simulation Modeling of Automobile Comfort Cooling Requirements, *ASHRAE J.*, vol. 17, no. 5, pp. 53–55, May 1975.
10. "Handbook and Product Directory, Applications Volume," chap. 25, American Society of Heating, Refrigerating, and Air-Conditioning Engineers, Atlanta, Ga., 1978.
11. What's Up with Watts Down on Refrigerators and Air Conditioners, Symp CH-77-13, *ASHRAE Trans.*, vol. 83, pt. 1, pp. 793–838, 1977.
12. "Handbook and Product Directory, Applications Volume," chap. 31, American Society of Heating, Refrigerating, and Air-Conditioning Engineers, Atlanta, Ga., 1978.
13. R. N. Shreve and J. A. Brink, Jr.: "Chemical Process Industries," 4th ed., McGraw-Hill, New York, 1977.
14. "Handbook and Product Directory, Applications Volume," chap. 55, American Society of Heating, Refrigerating, and Air-Conditioning Engineers, Atlanta, Ga., 1978.
15. E. Casanova: Concrete Cooling on Dam Construction for World's Largest Hydroelectric Power Station, *Sulzer Tech. Rev.*, vol. 61, no. 1, pp. 3–19, 1979.
16. W. E. Johnson: Survey of Desalination by Freezing–Its Status and Potential, *Natl. Water Supply Improv. Assoc. J.*, vol. 4, no. 2, pp. 1–14, July 1977.

†Before 1981 the actual place of publication for all ASHRAE material was New York, but the present address is given for the convenience of readers who may wish to order from the society.

CHAPTER

TWO

THERMAL PRINCIPLES

2-1 Roots of refrigeration and air conditioning Since a course in air conditioning and refrigeration might easily be titled Applications of Thermodynamics and Heat Transfer, it is desirable to begin the technical portion of this text with a brief review of the basic elements of these subjects. This chapter extracts some of the fundamental principles that are important for calculations used in the design and analysis of thermal systems for buildings and industrial processes. The presentation of these principles is intended to serve a very specific purpose and makes no attempt to cover the full range of applications of thermodynamics and heat transfer. Readers who feel the need of a more formal review are directed to basic texts in these subjects.[1-4]

This chapter does, however, attempt to present the material in a manner which establishes a pattern of analysis that will be applied repeatedly throughout the remainder of the text. This process involves the identification of the essential elements of the problem or design, the use of simplifications or idealizations to model the system to be designed or analyzed, and the application of the appropriate physical laws to obtain the necessary result.

2-2 Concepts, models, and laws Thermodynamics and heat transfer have developed from a general set of concepts, based on observations of the physical world, the specific models, and laws necessary to solve problems and design systems. Mass and energy are two of the basic concepts from which engineering science grows. From our own experience we all have some idea what each of these is but would probably find it difficult to provide a simple, concise, one-paragraph definition of either mass or energy. However, we are well enough acquainted with these concepts to realize that they are essential elements in our description of the physical world in which we live.

As the physical world is extremely complex, it is virtually impossible to describe it precisely. Even if it were, such detailed descriptions would be much too cumbersome for engineering purposes. One of the most significant accomplishments of engineering science has been the development of models of physical phenomena which, although they are approximations, provide both a sufficiently accurate description and a tractable means of solution. Newton's model of the relationship of force to mass and acceleration is an example. Although it cannot be applied universally, within its range of application it is accurate and extremely useful.

Models in and of themselves, however, are of little value unless they can be expressed in appropriate mathematical terms. The mathematical expressions of models provide the basic equations, or laws, which allow engineering science to explain or predict natural phenomena. The first and second laws of thermodynamics and the heat-transfer rate equations provide pertinent examples here. In this text we shall be discussing the use of these concepts, models, and laws in the description, design, and analysis of thermal systems in buildings and the process industries.

2-3 Thermodynamic properties Another essential element in the analysis of thermal systems is the identification of the pertinent thermodynamic properties. A property is any characteristic or attribute of matter which can be evaluated quantitatively. Temperature, pressure, and density are all properties. Work and heat transfer can be evaluated in terms of changes in properties, but they are not properties themselves. A property is something matter "has." Work and heat transfer are things that are "done" to a system to change its properties. Work and heat can be measured only at the boundary of the system, and the amount of energy transferred depends on how a given change takes place.

As thermodynamics centers on energy, all thermodynamic properties are related to energy. The thermodynamic state or condition of a system is defined by the values of its properties. In our considerations we shall examine equilibrium states and find that for a simple substance two intensive thermodynamic properties define the state. For a mixture of substances, e.g., dry air and water vapor, it is necessary to define three thermodynamic properties to specify the state. Once the state of the substance has been determined, all the other thermodynamic properties can be found since they are not all independently variable.

The thermodynamic properties of primary interest in this text are temperature, pressure, density and specific volume, specific heat, enthalpy, entropy, and the liquid-vapor property of state.

Temperature The temperature t of a substance indicates its thermal state and its ability to exchange energy with a substance in contact with it. Thus, a substance with a higher temperature passes energy to one with a lower temperature. Reference points on the Celsius scale are the freezing point of water (0°C) and the boiling point of water (100°C).

Absolute temperature T is the number of degrees above absolute zero expressed in kelvins (K); thus $T = t\ ^\circ\mathrm{C} + 273$. Since temperature intervals on the two scales are identical, differences between Celsius temperatures are stated in kelvins.

Pressure Pressure p is the normal (perpendicular) force exerted by a fluid per unit area against which the force is exerted. *Absolute pressure* is the measure of pressure above zero; *gauge pressure* is measured above existing atmospheric pressure.

The unit used for pressure is newtons per square meter (N/m^2), also called a *pascal* (Pa). The newton is a unit of force.

Standard atmospheric pressure is 101,325 Pa = 101.3 kPa.

Pressures are measured by such instruments as pressure gauges or manometers, shown schematically installed in the air duct of Fig. 2-1. Because one end of the manometer is open to the atmosphere, the deflection of water in the manometer indicates gauge pressure, just as the pressure gauge does.

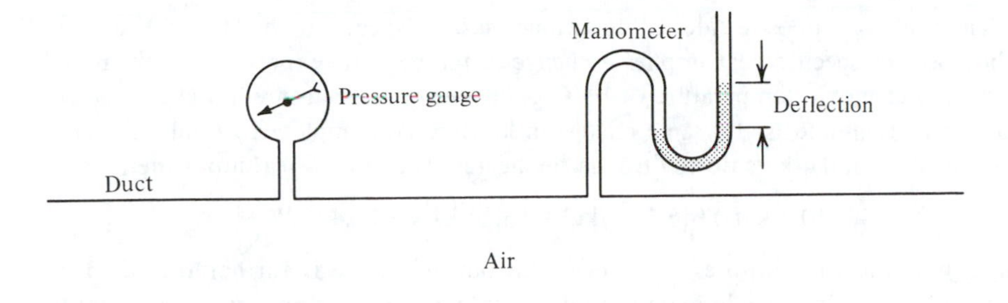

Figure 2-1 Indicating the gauge pressure of air in a duct with a pressure gauge and a manometer.

Density and specific volume The density ρ of a fluid is the mass occupying a unit volume; the specific volume v is the volume occupied by a unit mass. The density and specific volumes are reciprocals of each other. The density of air at standard atmospheric pressure and 25°C is approximately 1.2 kg/m^3.

Example 2-1 What is the mass of air contained in a room of dimensions 4 by 6 by 3 m if the specific volume of the air is 0.83 m^3/kg?

Solution The volume of the room is 72 m^3, and so the mass of air in the room is

$$\frac{72 \text{ m}^3}{0.83 \text{ m}^3/\text{kg}} = 86.7 \text{ kg}$$

Specific heat The specific heat of a substance is the quantity of energy required to raise the temperature of a unit mass by 1 K. Since the magnitude of this quantity is influenced by how the process is carried out, how the heat is added or removed must be described. The two most common descriptions are specific heat at constant volume c_v and specific heat at constant pressure c_p. The second is the more useful to us because it applies to most of the heating and cooling processes experienced in air conditioning and refrigeration.

The approximate specific heats of several important substances are

$$c_p = \begin{cases} 1.0 \text{ kJ/kg}\cdot\text{K} & \text{dry air} \\ 4.19 \text{ kJ/kg}\cdot\text{K} & \text{liquid water} \\ 1.88 \text{ kJ/kg}\cdot\text{K} & \text{water vapor} \end{cases}$$

where J symbolizes the unit of energy, the joule.

Example 2-2 What is the rate of heat input to a water heater if 0.4 kg/s of water enters at 82°C and leaves at 93°C?

Solution The pressure of water remains essentially constant as it flows through the heater, so c_p is applicable. The amount of energy in the form of heat added to each kilogram is

$$(4.19 \text{ kJ/kg}\cdot\text{K})(93 - 82°\text{C}) = 46.1 \text{ kJ/kg}$$

The units on opposite sides of equations must balance, but the °C and K do cancel because the specific heat implies a *change* in temperature expressed in kelvins and 93 - 82 is a change in temperature of 11°C. A change of temperature in Celsius degrees of a given magnitude is the same change in kelvins. To complete Example 2-2, consider the fact that 0.4 kg/s flows through the heater. The rate of heat input then is

$$(0.4 \text{ kg/s})(46.1 \text{ kJ/kg}) = 18.44 \text{ kJ/s} = 18.44 \text{ kW}$$

Enthalpy If the constant-pressure process introduced above is further restricted by permitting no work to be done on the substance, e.g., by a compressor, the amount of heat added or removed per unit mass is the change in enthalpy of the substance. Tables and charts of enthalpy h are available for many substances. These enthalpy values are always based on some arbitrarily chosen datum plane. For example, the datum plane for water and steam is an enthalpy value of zero for liquid water at 0°C. Based on that datum plane, the enthalpy of liquid water at 100°C is 419.06 kJ/kg and of water vapor (steam) at 100°C is 2676 kJ/kg.

Since the change in enthalpy is that amount of heat added or removed per unit mass in a constant-pressure process, the change in enthalpy of the water in Example 2-2 is 46.1 kJ/kg. The enthalpy property can also express the rates of heat transfer for processes where there is vaporization or condensation, e.g., in a water boiler or an air-heating coil where steam condenses.

Example 2-3 A flow rate of 0.06 kg/s of water enters a boiler at 90°C, at which temperature the enthalpy is 376.9 kJ/kg. The water leaves as steam at 100°C. What is the rate of heat added by the boiler?

Solution The change in enthalpy in this constant-pressure process is

$$\Delta h = 2676 - 377 \text{ kJ/kg} = 2299 \text{ kJ/kg}$$

The rate of heat transfer to the water in converting it to steam is

$$(0.06 \text{ kg/s})(2299 \text{ kJ/kg}) = 137.9 \text{ kW}$$

Entropy Although entropy s has important technical and philosophical connotations, we shall use this property in a specific and limited manner. Entropy does appear in many charts and tables of properties and is mentioned here so that it will not be unfamiliar. The following are two implications of this property:

1. If a gas or vapor is compressed or expanded frictionlessly without adding or removing heat during the process, the entropy of the substance remains constant.
2. In the process described in implication 1, the change in enthalpy represents the amount of work per unit mass required by the compression or delivered by the expansion.

Possibly the greatest practical use we shall have for entropy is to read lines of constant entropy on graphs in computing the work of compression in vapor-compression refrigeration cycles.

Liquid-vapor properties Most heating and cooling systems use substances that pass between liquid and vapor states in their cycle. Steam and refrigerants are prime examples of these substances. Since the pressures, temperatures, and enthalpies are key properties during these changes, the relationships of these properties are listed in tables or displayed on charts, e.g., the pressure-enthalpy diagram for water shown in Fig. 2-2.

The three major regions on the chart are (1) the subcooled-liquid region to the left, (2) the liquid-vapor region in the center, and (3) the superheated-vapor region on the right. In region 1 only liquid exists, in region 3 only vapor exists, and in region 2 both liquid and vapor exist simultaneously. Separating region 2 and region 3 is the saturated-vapor line. As we move to the right along a horizontal line at constant pressure from the saturated-liquid line to the saturated-vapor line, the mixture of liquid and vapor changes from 100 percent liquid to 100 percent vapor.

Three lines of constant temperature are shown in Fig. 2-2, $t = 50°C$, $t = 100°C$, and $t = 150°C$. Corresponding to our experience, water boils at a higher temperature when the pressure is higher. If the pressure is 12.3 kPa, water boils at 50°C, but at standard atmospheric pressure of 101 kPa it boils at 100°C.

Also shown in the superheated vapor region are two lines of constant entropy.

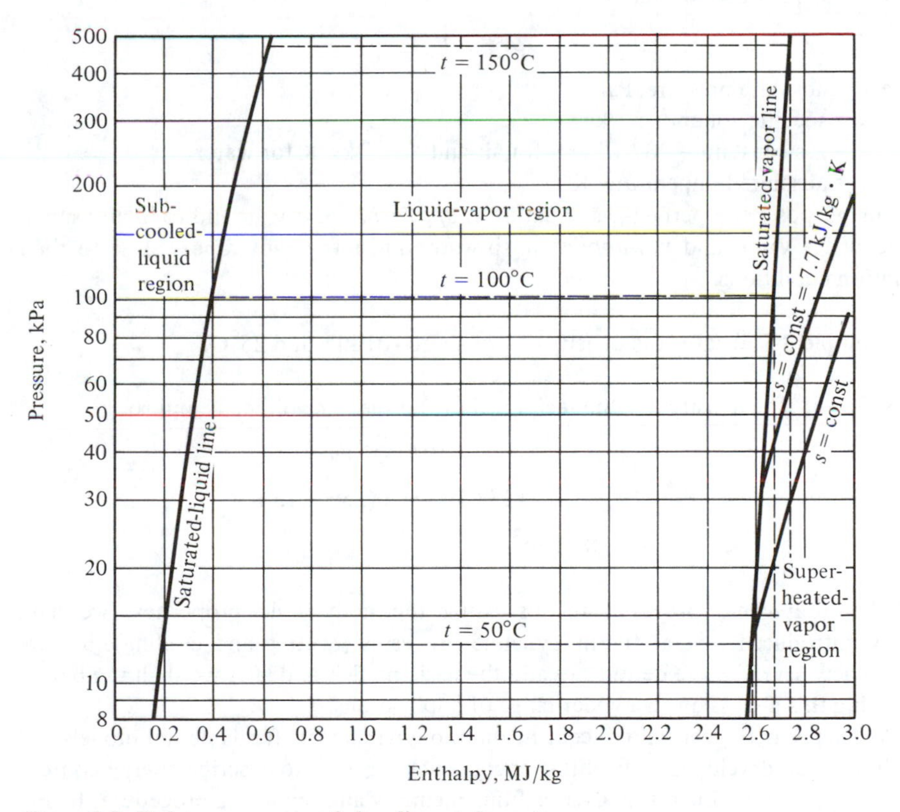

Figure 2-2 Skeleton pressure-enthalpy diagram for water.

Example 2-4 If 9 kg/s of liquid water at 50°C flows into a boiler, is heated, boiled, and superheated to a temperature of 150°C and the entire process takes place at standard atmospheric pressure, what is the rate of heat transfer to the water?

Solution The process consists of three distinct parts: (1) bringing the temperature of the subcooled water up to its saturation temperature, (2) converting liquid at 100°C into vapor at 100°C, and (3) superheating the vapor from 100 to 150°C. The rate of heat transfer is the product of the mass rate of flow multiplied by the change in enthalpy. The enthalpy of entering water at 50°C and 101 kPa is 209 kJ/kg, which can be read approximately from Fig. 2-2 or determined more precisely from Appendix Table A-1. The enthalpy of superheated steam at 150°C and 101 kPa is 2745 kJ/kg. The rate of heat transfer is

$$q = (9 \text{ kg/s})(2745 - 209 \text{ kJ/kg}) = 22{,}824 \text{ kW}$$

Perfect-gas law As noted previously, the thermodynamic properties of a substance are not all independently variable but are fixed by the state of a substance. The idealized model of gas behavior which relates the pressure, temperature, and specific volume of a *perfect gas* provides an example

$$pv = RT$$

where p = absolute pressure, Pa
v = specific volume, m^3/kg
R = gas constant = 287 J/kg·K for air and 462 J/kg·K for water
T = absolute temperature, K

For our purposes the perfect-gas equation is *applicable* to dry air and to highly superheated water vapor and *not applicable* to water and refrigerant vapors close to their saturation conditions.

Example 2-5 What is the density of dry air at 101 kPa and 25°C?

Solution The density ρ is the reciprocal of the specific volume v, and so

$$\rho = \frac{1}{v} = \frac{p}{RT} = \frac{101{,}000 \text{ Pa}}{(287 \text{ J/kg·K})(25 + 273 \text{ K})}$$

$$\rho = 1.18 \text{ kg/m}^3$$

2-4 Thermodynamic processes In discussing thermodynamic properties, we have already introduced several thermodynamic processes (heating and cooling), but we must review several more definitions and the basic models and laws we shall use before expanding this discussion to a wider range of applications.

As energy is the central concept in thermodynamics, its fundamental models and laws have been developed to facilitate energy analyses, e.g., to describe energy content and energy transfer. Energy analysis is fundamentally an accounting procedure. In any accounting procedure whatever it is that is under consideration must be clearly identified. In this text we use the term *system* to designate the object or objects considered in the analysis or discussion. A system may be as simple as a specified volume of a homogeneous fluid or as complex as the entire thermal-distribution network in a large building. In most cases we shall define a system in terms of a specified region in space (sometimes referred to as a *control volume*) and entirely enclosed by a closed surface, referred to as the *system boundary* (or *control surface*). The size of the system and the shape of the system boundary are arbitrary and are specified for each problem so that they simplify accounting for the changes in energy storage within the system or energy transfers across a system boundary. Whatever is not included in the system is called the *environment*.

Consider the simple flow system shown in Fig. 2-3, where mass is transferred from the environment to the system at point 1 and from the system to the environment at point 2. Such a system could be used to analyze something as simple as a pump or as complex as an entire building. The definition of the system provides the framework for the models used to describe the real objects considered in thermodynamic analysis.

The next step in the analysis is to formulate the basic laws so that they are applicable to the system defined. The laws of conservation of mass and conservation of energy provide excellent examples, as we shall be applying them repeatedly in every aspect of air-conditioning and refrigeration design.

2-5 Conservation of mass Mass is a fundamental concept and thus is not simply defined. A definition is often presented by reference to Newton's law

$$\text{Force} = ma = m\frac{dV}{d\theta}$$

where m = mass, kg
a = acceleration, m/s^2
V = velocity, m/s
θ = time, s

An object subjected to an unbalanced force accelerates at a rate dependent upon the magnitude of the force. In this context the mass of an object is conceived of as being characteristic of its resistance to change in velocity. Two objects which undergo the same acceleration under action of identical forces have the same mass. Further, our

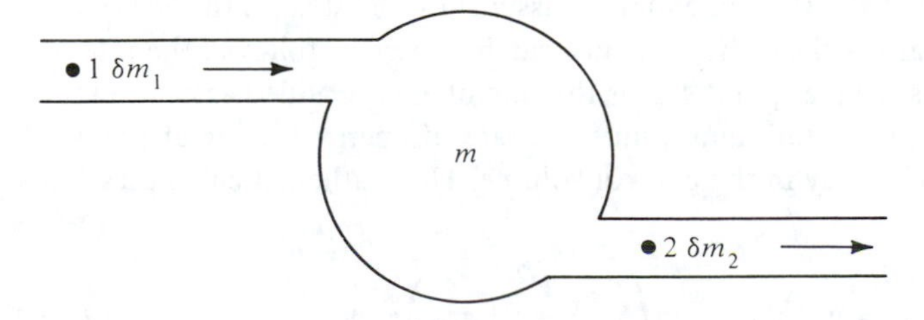

Figure 2-3 Conservation of mass in a simple flow system.

concept of mass holds that the mass of two objects taken together is the sum of their individual masses and that cutting a homogeneous body into two identical parts produces two identical masses, each half of the original mass. This idea is the equivalent of the law of conservation of mass.

In the present context the principle of conservation of mass states that mass is neither created nor destroyed in the processes analyzed. It may be stored within a system or transferred between a system and its environment, but it must be accounted for in any analysis procedure. Consider Fig. 2-3 again. The mass in the system may change over time as mass flows into or out of the system. Assume that during a time increment $d\theta$ of mass δm_1 enters the system and an increment δm_2 leaves. If the mass in the system at time θ is m_θ and that at time $\theta + \delta\theta$ is $m_{\theta + \delta\theta}$, conservation of mass requires that

$$m_\theta + \delta m_1 = m_{\theta + \delta\theta} + \delta m_2$$

Dividing by $\delta\theta$ gives

$$\frac{m_{\theta + \delta\theta} - m_\theta}{\delta\theta} + \frac{\delta m_2}{\delta\theta} - \frac{\delta m_1}{\delta\theta} = 0$$

If we express the mass flux as

$$\dot{m} = \frac{\delta m}{\delta\theta}$$

we can write the rate of change at any instant as

$$\frac{dm}{d\theta} + \dot{m}_2 - \dot{m}_1 = 0$$

If the rate of change of mass within the system is zero,

$$\frac{dm}{d\theta} = 0 \quad \text{and} \quad \dot{m}_1 = \dot{m}_2$$

and we have *steady flow*. Steady flow will be encountered frequently in our analysis.

2-6 Steady-flow energy equation In most air-conditioning and refrigeration systems the mass flow rates do not change from one instant to the next (or if they do, the rate of change is small); therefore the flow rate may be assumed to be steady. In the system shown symbolically in Fig. 2-4 the energy balance can be stated as follows: the rate of energy entering with the stream at point 1 plus the rate of energy added as heat minus the rate of energy performing work and minus the rate of energy leaving at point 2 equals the rate of change of energy in the control volume. The mathematical expression for the energy balance is

$$\dot{m}\left(h_1 + \frac{V_1^2}{2} + gz_1\right) + q - \dot{m}\left(h_2 + \frac{V_2^2}{2} + gz_2\right) - W = \frac{dE}{d\theta} \qquad (2\text{-}1)$$

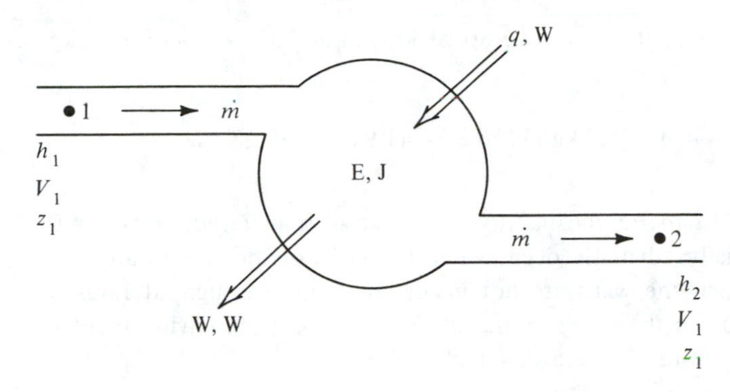

Figure 2-4 Energy balance on a control volume experiencing steady flow rates.

where $\dot{m}$ = mass rate of flow, kg/s
h = enthalpy, J/kg
V = velocity, m/s
z = elevation, m
g = gravitational acceleration = 9.81 m/s^2
q = rate of energy transfer in form of heat, W
W = rate of energy transfer in form of work, W
E = energy in system, J

Because we are limiting consideration to steady-flow processes, there is no change of E with respect to time; the $dE/d\theta$ term is therefore zero, and the usual form of the steady-flow energy equation appears:

$$\dot{m}\left(h_1 + \frac{V_1^2}{2} + gz_1\right) + q = \dot{m}\left(h_2 + \frac{V_2^2}{2} + gz_2\right) + W \tag{2-2}$$

This form of the energy equation will be frequently used in the following chapters. Some applications of Eq. (2-2) will be considered at this point.

2-7 Heating and cooling In many heating and cooling processes, e.g., the water heater in Example 2-2 and the boiler in Example 2-3, the changes in certain of the energy terms are negligible. Often the magnitude of change in the kinetic-energy term $V^2/2$ and the potential-energy term $9.81z$ from one point to another is negligible compared with the magnitude of change of enthalpy, the work done, or heat transferred. If no work is done by a pump, compressor, or engine in the process, $W = 0$. The energy equation then reduces to

$$q + \dot{m}h_1 = \dot{m}h_2 \qquad \text{or} \qquad q = \dot{m}(h_2 - h_1)$$

i.e, the rate of heat transfer equals the mass rate of flow multiplied by the change in enthalpy, as assumed in Examples 2-2 and 2-3.

Example 2-6 Water flowing at a steady rate of 1.2 kg/s is to be chilled from 10 to 4°C to supply a cooling coil in an air-conditioning system. Determine the necessary rate of heat transfer.

Solution From Table A-1, at 4°C $h = 16.80$ kJ/kg and at 10°C $h = 41.99$ kJ/kg. Then

$$q = \dot{m}(h_2 - h_1) = (1.2 \text{ kg/s}) (16.80 - 41.99) = -30.23 \text{ kW}$$

2-8 Adiabatic processes *Adiabatic* means that no heat is transferred; thus $q = 0$. Processes that are essentially adiabatic occur when the walls of the system are thermally insulated. Even when the walls are not insulated, if the throughput rates of energy are large in relation to the energy transmitted to or from the environment in the form of heat, the process may be considered adiabatic.

2-9 Compression work An example of a process which can be modeled as adiabatic is the compression of a gas. The change in kinetic and potential energies and the heat-transfer rate are usually negligible. After dropping out the kinetic- and potential-energy terms and the heat-transfer rate q from Eq. (2-2) the result is

$$W = \dot{m}(h_1 - h_2)$$

The power requirement equals the mass rate of flow multiplied by the change in enthalpy. The W term is negative for a compressor and positive for an engine.

2-10 Isentropic compression Another tool is available to predict the change in enthalpy during a compression. If the compression is adiabatic and without friction, the compression occurs at constant entropy. On the skeleton pressure-enthalpy diagram of Fig. 2-5 such an ideal compression occurs along the constant-entropy line from 1 to 2. The usefulness of this property is that if the entering condition to a compression (point 1) and the leaving pressure are known, point 2 can be located and the power predicted by computing $\dot{m}(h_1 - h_2)$. The actual compression usually takes place along a path to the right of the constant-entropy line (shown by the dashed line to point 2′ in Fig. 2-5), indicating slightly greater power than for the ideal compression.

Example 2-7 Compute the power required to compress 1.5 kg/s of saturated water vapor from a pressure of 34 kPa to one of 150 kPa.

Solution From Fig. 2-2, at $p_1 = 34$ kPa and saturation

$$h_1 = 2630 \text{ kJ/kg} \qquad \text{and} \qquad s_1 = 7.7 \text{ kJ/kg} \cdot \text{K}$$

At $p_2 = 150$ kPa and $s_2 = s_1$

$$h_2 = 2930 \text{ kJ/kg}$$

Then

$$W = (1.5 \text{ kg/s}) (2630 - 2930 \text{ kJ/kg}) = -450 \text{ kW}$$

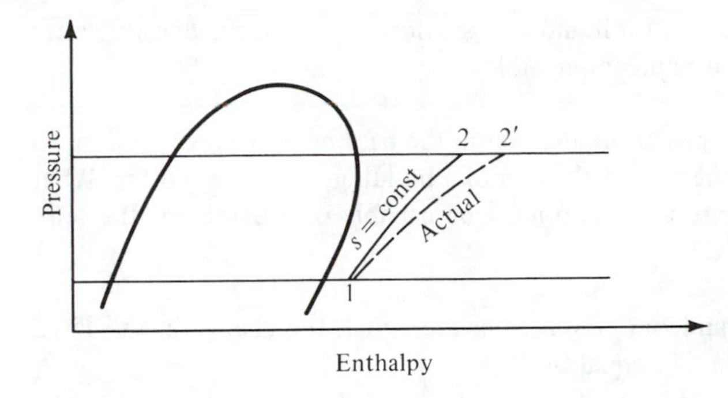

Figure 2-5 Pressure-enthalpy diagram showing a line of constant entropy.

2-11 Bernoulli's equation Bernoulli's equation is often derived from the mechanical behavior of fluids, but it is also derivable as a special case of the energy equation through second-law considerations. It can be shown that

$$T\,ds = du + p\,dv \tag{2-3}$$

where u is the internal energy in joules per kilogram. This expression is referred to as the *Gibb's equation*. For an adiabatic process $q = 0$, and with no mechanical work $W = 0$. Such a process might occur when fluid flows in a pipe or duct. Equation (2-2) then requires that

$$h + \frac{V^2}{2} + gz = \text{const}$$

Differentiation yields

$$dh + V\,dV + g\,dz = 0 \tag{2-4}$$

The definition of enthalpy $h = u + pv$ can be differentiated to yield

$$dh = du + p\,dv + v\,dp \tag{2-5}$$

Combining Eqs. (2-3) and (2-5) results in

$$T\,ds = dh - v\,dp \tag{2-6}$$

Applying Eq. (2-6) to an isentropic process it follows that, since $ds = 0$,

$$dh = v\,dp = \frac{1}{\rho}\,dp$$

Substituting this expression for dh into Eq. (2-4) for isentropic flow gives

$$\frac{dp}{\rho} + V\,dV + g\,dz = 0 \tag{2-7}$$

For constant density Eq. (2-7) integrates to the Bernoulli equation

$$\frac{p}{\rho} + \frac{V^2}{2} + gz = \text{const} \tag{2-8}$$

We shall use the Bernoulli equation for liquid and gas flows in which the density varies only slightly and may be treated as incompressible.

Example 2-8 Water is pumped from a chiller in the basement, where $z_1 = 0$ m, to a cooling coil located on the twentieth floor of a building, where $z_2 = 80$ m. What is the minimum pressure rise the pump must be capable of providing if the temperature of the water is 4°C?

Solution Since the inlet and outlet velocities are equal, the change in the $V^2/2$ term is zero; so from Bernoulli's equation

$$\frac{p_1}{\rho} + gz_1 = \frac{p_2}{\rho} + gz_2$$

The density $\rho = 1000 \text{ kg/m}^3$, and $g = 9.81 \text{ m/s}^2$, therefore

$$p_1 - p_2 = (1000 \text{ kg/m}^3)(9.81 \text{ m/s}^2)(80 \text{ m}) = 785 \text{ kPa}$$

2-12 Heat transfer Heat-transfer analysis is developed from the thermodynamic laws of conservation of mass and energy, the second law of thermodynamics, and three rate equations describing conduction, radiation, and convection. The rate equations were developed from the observation of the physical phenomena of energy exchange. They are the mathematical descriptions of the models derived to describe the observed phenomena.

Heat transfer through a solid material, referred to as *conduction*, involves energy exchange at the molecular level. *Radiation*, on the other hand, is a process that transports energy by way of photon propagation from one surface to another. Radiation can transmit energy across a vacuum and does not depend on any intervening medium to provide a link between the two surfaces. *Convection* heat transfer depends upon conduction from a solid surface to an adjacent fluid and the movement of the fluid along the surface or away from it. Thus each heat-transfer mechanism is quite distinct from the others; however, they all have common characteristics, as each depends on temperature and the physical dimensions of the objects considered.

2-13 Conduction Observation of the physical phenomena and a series of reasoned steps establishes the rate equation for conduction. Consider the energy flux arising from conduction heat transfer along a solid rod. It is proportional to the temperature difference and the cross-sectional area and inversely proportional to the length. These observations can be verified by a series of simple experiments. Fourier provided a mathematical model for this process. In a one-dimensional problem

$$q = -kA \frac{\Delta t}{L} \tag{2-9}$$

where A = cross-sectional area, m^2

Δt = temperature difference, K

Table 2-1 Thermal conductivity of some materials

Material	Temperature, °C	Density, kg/m^3	Conductivity, W/m·K
Aluminum (pure)	20	2707	204
Copper (pure)	20	8954	386
Face brick	20	2000	1.32
Glass (window)	20	2700	0.78
Water	21	997	0.604
Wood (yellow pine)	23	640	0.147
Air	27	1.177	0.026

L = length, m
k = thermal conductivity, W/m·K

The thermal conductivity is a characteristic of the material, and the ratio k/L is referred to as the *conductance*.

The thermal conductivity, and thus the rate of conductive heat transfer, is related to the molecular structure of materials. The more closely packed, well-ordered molecules of a metal transfer energy more readily than the random and perhaps widely spaced molecules of nonmetallic materials. The free electrons in metals also contribute to a higher thermal conductivity. Thus good electric conductors usually have a high thermal conductivity. The thermal conductivity of less ordered inorganic solids is lower than that of metals. Organic and fibrous materials, such as wood, have still lower thermal conductivity. The thermal conductivities of nonmetallic liquids are generally lower than those of solids, and those of gases at atmospheric pressure are lower still. The reduction in conductivity is attributable to the absence of strong intermolecular bonding and the existence of widely spaced molecules in fluids. Table 2-1 indicates the order of magnitude of the thermal conductivity of several classes of materials.

The rate equation for conduction heat transfer is normally expressed in differential form

$$q = -kA \frac{dt}{dx}$$

2-14 Radiation As noted previously, radiant-energy transfer results when photons emitted from one surface travel to other surfaces. Upon reaching the other surfaces radiated photons are either absorbed, reflected, or transmitted through the surface.

The energy radiated from a surface is defined in terms of its emissive power. It can be shown from thermodynamic reasoning that the emissive power is proportional to the fourth power of the absolute temperature. For a perfect radiator, generally referred to as a *blackbody*, the emissive power E_b W/m^2 is

$$E_b = \sigma T^4$$

where σ = Stefan-Boltzmann constant = 5.669×10^{-8} W/m^2 · K^4
T = absolute temperature, K

Since real bodies are not "black," they radiate less energy than a blackbody at the same temperature. The ratio of the actual emissive power E W/m^2 to the blackbody emissive power is the *emissivity* ϵ, where

$$\epsilon = \frac{E}{E_b}$$

In many real materials the emissivity and absorptivity may be assumed to be approximately equal. These materials are referred to as *gray bodies,* and

$$\epsilon = \alpha$$

where α is the absorptivity (dimensionless).

Another important feature of radiant-energy exchange is that radiation leaving a surface is distributed uniformly in all directions. Therefore the geometric relationship between two surfaces affects the radiant-energy exchange between them. For example, the radiant heat transferred between two black parallel plates 1 by 1 m separated by a distance of 1 m with temperatures of 1000 and 300 K, respectively, would be 1.13 kW. If the same two plates were moved 2 m apart, the heat transfer would be 0.39 kW. If they were set perpendicular to each other with a common edge, the rate of heat transfer would again be 1.13 kW. The geometric relationship can be determined and accounted for in terms of a *shape factor* F_A.

The optical characteristics of the surfaces, i.e., their emissivities, absorptivities, reflectivities, and transmissivities, also influence the rate of radiant heat transfer. If these effects are expressed by a factor F_ϵ, the radiant-energy exchange can be expressed as

$$q_{1-2} = \sigma A F_\epsilon F_A (T_1^4 - T_2^4) \tag{2-10}$$

Procedures for evaluating F_ϵ and F_A can be found in heat-transfer texts and handbooks.

2-15 Convection The rate equation for convective heat transfer was originally proposed by Newton in 1701, again from observation of physical phenomena,

$$q = h_c A (t_s - t_f) \tag{2-11}$$

where h_c = convection coefficient, W/m^2 · K
t_s = surface temperature, °C
t_f = fluid temperature, °C

This equation is widely used in engineering even though it is more a definition of h_c than a phenomenological law for convection. In fact, the essence of convective heat-transfer analysis is the evaluation of h_c. Experiments have demonstrated that the convection coefficient for flows over plane surfaces, in pipes and ducts, and across tubes can be correlated with flow velocity, fluid properties, and the geometry of the solid surface. Extensive theory has been developed to support the experimentally observed correlations and to extend them to predict behavior in untested flow configurations. It is the correlations rather than the theories, however, which are used in practical engineering analysis. Dimensionless parameters provide the basis of most of the pertinent correlations. These parameters were identified by applying *dimensional analysis* in grouping the variables influencing convective heat transfer. The proper selection of the variables to be considered, of course, depends on understanding the physical phenomena involved and on the ability to construct reasonable models for basic flow configurations. A detailed presentation of these techniques is beyond the scope of this chapter, and the interested reader is referred to heat-transfer texts. For our present purposes it will be sufficient to identify the pertinent parameters and the forms of the correlations which will often be used in evaluating the convection coefficient h_c:

$$\text{Reynolds number Re} = \frac{\rho V D}{\mu}$$

$$\text{Prandtl number Pr} = \frac{\mu c_p}{k}$$

$$\text{Nusselt number Nu} = \frac{h_c D}{k}$$

Expressions have been developed for particular flow configurations so that the relationship between the Nusselt number and the Reynolds and Prandtl numbers can be expressed as

$$\text{Nu} = C(\text{Re}^n)(\text{Pr}^m)$$

with values of the constant C and the exponents n and m being determined experimentally. Alternatively these relationships can be illustrated graphically, as in Fig. 2-6,

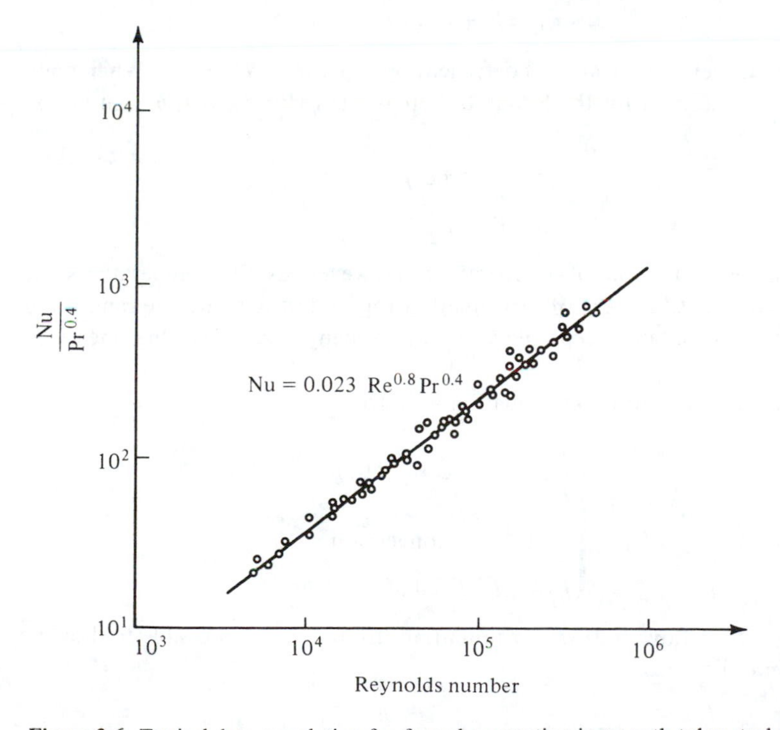

Figure 2-6 Typical data correlation for forced convection in smooth tubes, turbulent flow.

Table 2-2 Typical range of values of convective, boiling, and condensing heat-transfer coefficients[5]

Process	h_c, W/m^2 · K
Free convection, air	5–25
Free convection, water	20–100
Forced convection, air	10–200
Forced convection, water	50–10,000
Boiling water	3000–100,000
Condensing water	5000–100,000

giving magnitudes applicable to turbulent flow in a smooth tube. Table 2-2 provides typical values of h_c for convective heat transfer with water and air and for boiling and condensing of water.

2-16 Thermal resistance It is of interest that the rate equations for both conduction and convection are linear in the pertinent variables—conductance, area, and temperature difference. The radiation rate equation, however, is nonlinear in temperature. Heat-transfer calculations in which conduction, radiation, and convection all occur could be simplified greatly if the radiation heat transfer could be expressed in terms of a *radiant conductance* such that

$$q = h_r A \, \Delta t$$

where h_r is the equivalent heat-transfer coefficient by radiation, W/m^2 · K. When comparing the above equation with the Stefan-Boltzmann law, Eq. (2-10), h_r can be expressed as

$$h_r = \frac{\sigma F_\epsilon F_A (T_1^4 - T_2^4)}{T_1 - T_2}$$

which is a nonlinear function of temperature. However, as the temperatures are absolute, it turns out that h_r does not vary greatly over modest temperature ranges and it is indeed possible to obtain acceptable accuracy in many cases using the linearized equation.

With the linearized radiation rate equation we have

$$q = \begin{cases} \dfrac{k}{L} A \, \Delta t & \text{conduction} \\ h_c A \, \Delta t & \text{convection} \\ h_r A \, \Delta t & \text{radiation} \end{cases}$$

Noting that q is a heat flow and Δt is a potential difference, it is possible to draw an analogy with Ohm's law

$$E = IR \qquad \text{or} \qquad I = \frac{E}{R}$$

where E = potential difference
I = current
R = resistance

When the heat-transfer equation is written according to the electrical analogy,

$$q = \frac{\Delta t}{R_T^*}$$

where R_T^* is thermal resistance. For the three modes of heat transfer

$$R_T^* = \begin{cases} \dfrac{L}{kA} & \text{conduction} \\[2ex] \dfrac{1}{h_c A} & \text{convection} \\[2ex] \dfrac{1}{h_r A} & \text{radiation} \end{cases}$$

With these definitions of thermal resistance it is possible by analogy to apply certain concepts from circuit theory to heat transfer. Recall that the conductance C is the reciprocal of the resistance, $C = 1/R^*$, and that in series circuits the resistances sum but for parallel circuits the conductances sum. In the transfer of energy from one room to another through a solid wall (Fig. 2-7) assume that both the gas and the other walls of

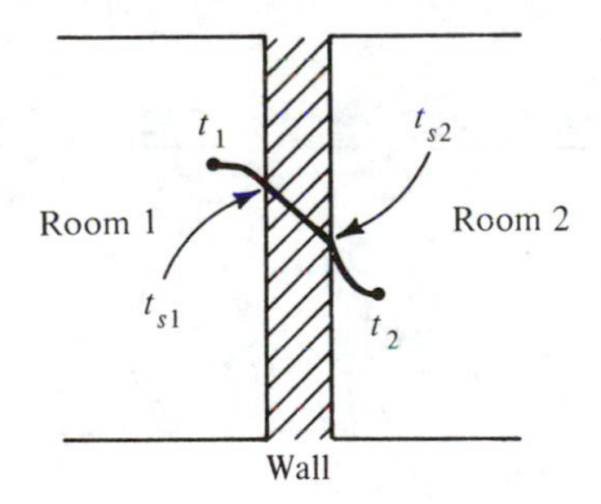

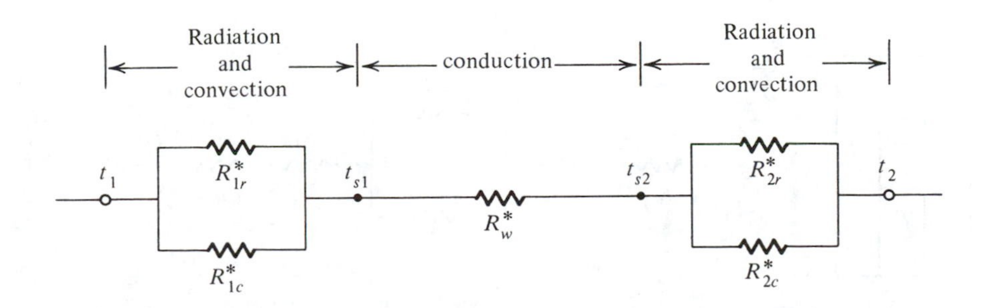

Figure 2-7 Heat transfer from one room to another across a solid wall.

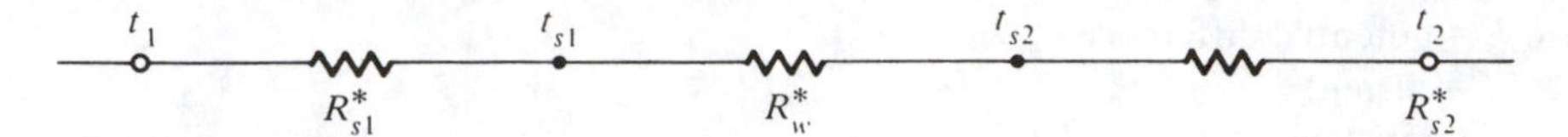

Figure 2-8 Heat-transfer circuit when the convective and radiative resistances are combined into a single surface resistance.

room 1 are at t_1 and those in room 2 are at t_2. In this application the total resistance is

$$R^*_{\text{tot}} = \frac{1}{C_{1r} + C_{1c}} + R^*_w + \frac{1}{C_{2r} + C_{2c}}$$

or

$$R^*_{\text{tot}} = \frac{1}{1/R^*_{1r} + 1/R^*_{1c}} + R^*_w + \frac{1}{1/R^*_{2r} + 1/R^*_{2c}}$$

where the subscripts 1 and 2 refer to rooms 1 and 2, respectively, and the subscripts r, c, and w refer to radiation, convection, and wall, respectively.

Convection and radiation occur simultaneously at a surface, and the convective and radiative conductances can be combined into a single conductance, a usual practice in heating- and cooling-load calculations followed in Chap. 4. The combined surface conductance becomes $(h_c + h_r)A$. In the heat-transfer situation shown in Fig. 2-7, the circuit reduces to a series of resistances, as shown in Fig. 2-8, where $R^*_s = 1/(h_c + h_r)A$.

It is frequently necessary to determine the temperature of the surface knowing the temperatures in the space. Since the heat flux from one room to the other is constant at steady-state conditions,

$$q = \frac{t_1 - t_{s,1}}{R^*_{r-c,1}} = \frac{t_{s,1} - t_{s,2}}{R^*_w} = \frac{t_{s,2} - t_2}{R^*_{r-c,2}} = \frac{t_1 - t_2}{R^*_{\text{tot}}}$$

If $t_{s,1}$ is sought, for example,

$$t_{s,1} = t_1 - \frac{R^*_{r-c,1}}{R^*_{\text{tot}}}(t_1 - t_2)$$

Another instance of parallel heat flow arises when structural elements are present in wall sections. In Fig. 2-9, for example, element C may be a structural member and

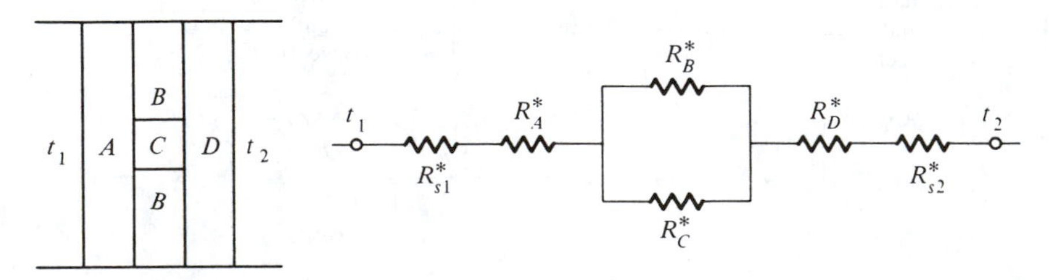

Figure 2-9 Heat transfer through parallel paths.

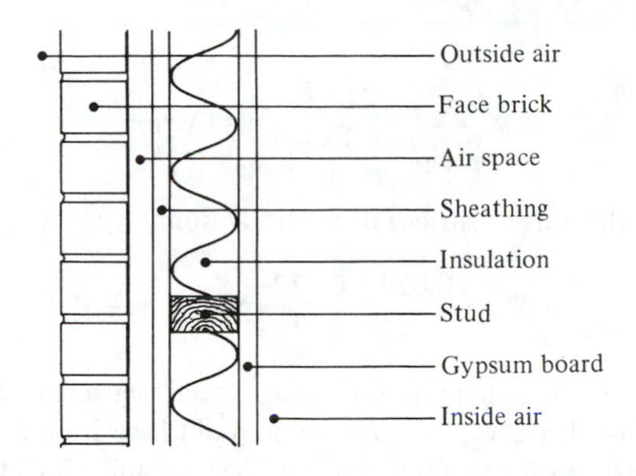

	L, m	k, W/m · K	A, m^2	R^*_{s1}	R^*_A	R^*_B	R^*_C	R^*_D	R^*_{s2}
Outside air			1.0	.029					
Face brick	.09	1.30	1.0		.070				
Air space			1.0		.170				
Sheathing	.013	0.056	1.0		.232				
Insulation	.09	0.038	0.8			2.96			
Stud	.09	0.14	0.2				3.2		
Gypsum board	.013	0.16	1.0					.08	
Inside air			1.0						.125
		Subtotal		.029	.472	2.96	3.2	.08	.125

Figure 2-10 Wall section in Example 2-9.

the space between structural members is occupied by a different material, perhaps insulation. The total resistance between air on one side of the wall and air on the other is

$$R^*_{tot} = R^*_{s,1} + R^*_A + \frac{R^*_B R^*_C}{R^*_B + R^*_C} + R^*_D + R^*_{s,2}$$

Example 2-9 Using the data given in Fig. 2-10, determine the heat transfer in watts per square meter through the wall section and the temperature of the outside surface of the insulation if $t_o = 0°C$ and $t_i = 21°C$. In the insulated portion of the wall assume that 20 percent of the space is taken up with the structural elements, which are wood studs.

Solution From the data given, for each 1 m^2 of surface

$$R^*_{tot} = 0.029 + 0.472 + \frac{2.96(3.2)}{2.96 + 3.2} + 0.08 + 0.125$$

$$= 2.24 \text{ K/W}$$

Thus

$$q = \frac{t_i - t_o}{R^*_{tot}} = \frac{21 - 0}{2.24} = 9.37\ \mathrm{W/m^2}$$

The temperature at the outside surface of the insulation is

$$t_{ins,o} = 0°\mathrm{C} + \frac{0.029 + 0.472}{2.24}(21 - 0°\mathrm{C}) = 4.7°\mathrm{C}$$

In Example 2-9 if the structural element has been neglected, the total resistance calculated would have been R_{tot} = 3.67 and q would have been 5.73 W, which indicates that the presence of structural elements has a significant effect on the heat-transfer calculation.

The heat-transfer equation

$$q = \frac{t_i - t_o}{R^*_{tot}} \tag{2-12}$$

frequently appears in the form

$$q = UA(t_i - t_o)$$

where U = overall heat-transfer coefficient, W/m^2 · K
A = surface area, m^2

Comparing this equation with Eq. (2-12), we see that

$$UA = \frac{1}{R^*_{tot}}$$

and thus

$$U = \frac{1}{(R^*_{tot})A}$$

Many construction materials are available in standard thicknesses so that it is possible to present the resistance of the material directly without having to calculate the L/kA term for the material. An additional convention is that resistances are expressed on the basis of 1 m^2. The relationship of these resistances R, which have units of m^2 · K/W, to the R^* resistances we have dealt with so far is

$$R^*A = \begin{cases} \dfrac{L}{k} = R_w & \text{conduction} \\[2ex] \dfrac{1}{h_c + h_r} = R_s & \text{surface} \end{cases}$$

For a plane wall for which A is the same for all surfaces,

$$U = \frac{1}{R_{s,1} + R_w + R_{s,2}} = \frac{1}{R_{tot}}$$

Resistances presented in Chap. 4 (for example, in Table 4-3) will be the unstarred variety and have units of $m^2 \cdot K/W$.

2-17 Cylindrical cross section The previous discussion applies to plane geometries, but when heat is transferred through circular pipes, geometries are cylindrical. The area through which the heat flows is not constant, and a new expression for the resistance is required.

$$R_{cyl} = \frac{\ln (r_o/r_i)}{2\pi kl} \tag{2-13}$$

where r_o = outside radius, m
r_i = inside radius, m
l = length, m

2-18 Heat exchangers Heat exchangers are used extensively in air conditioning and refrigeration. A heat exchanger is a device in which energy is transferred from one fluid stream to another across a solid surface. Heat exchangers thus incorporate both convection and conduction heat transfer. The resistance concepts discussed in the previous sections prove useful in the analysis of heat exchangers as the first fluid, the solid wall, and the second fluid form a series thermal circuit (Fig. 2-11).

$$q = \frac{\Delta t}{R_{tot}}$$

where

$$R_{tot} = \frac{1}{h_1 A_1} + \frac{\ln (r_o/r_i)}{2\pi kl} + \frac{1}{h_2 A_2} \tag{2-14}$$

The subscripts 1 and 2 refer to fluids 1 and 2.

At a particular point in the heat exchanger the heat flux can be expressed by the thermal resistance and the temperature difference between the fluids. However, since

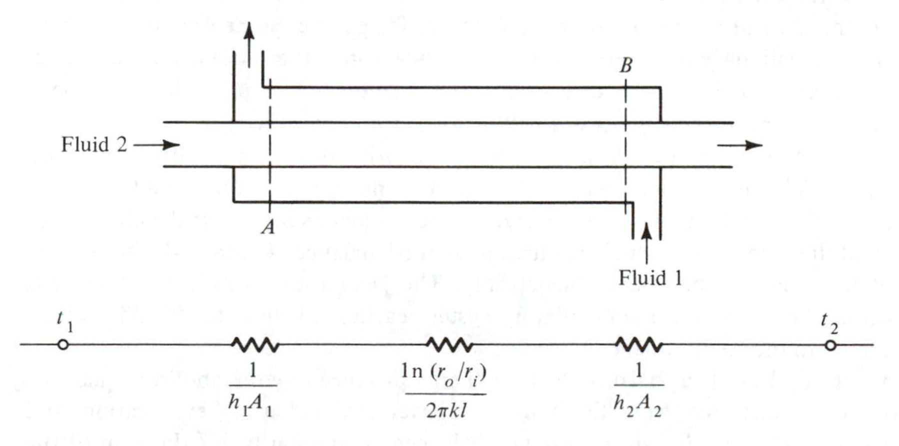

Figure 2-11 Counterflow heat exchanger.

the temperature of one or both fluids may vary as they flow through the heat exchanger, analysis is difficult unless a mean temperature difference can be determined which will characterize the overall performance of the heat exchanger. The usual practice is to use the logarithmic-mean temperature difference (LMTD) and a configuration factor which depends upon the flow arrangement through the heat exchanger. The LMTD is defined as

$$\text{LMTD} = \frac{\Delta t_A - \Delta t_B}{\ln (\Delta t_A / \Delta t_B)} \tag{2-15}$$

where Δt_A = temperature difference between two fluids at position A, K
Δt_B = temperature difference between two fluids at position B, K

The analysis of heat exchangers will be examined at greater length in Chap. 12.

Example 2-10 Determine the heat-transfer rate for the heat exchanger shown in Fig. 2-11, given the following data: $h_1 = 50\ \text{W/m}^2 \cdot \text{K}$, $h_2 = 80\ \text{W/m}^2 \cdot \text{K}$, $t_{1,\text{in}} = 60°\text{C}$, $t_{1,\text{out}} = 40°\text{C}$, $t_{2,\text{in}} = 20°\text{C}$, $t_{2,\text{out}} = 30°\text{C}$, $r_o = 11$ mm, $r_i = 10$ mm, length = 1 m, and for the metal $k = 386\ \text{W/m} \cdot \text{K}$.

Solution $A_1 = 2\pi r_o l = 0.069\ \text{m}^2 \quad A_2 = 0.063\ \text{m}^2$

$$R_{\text{tot}} = \frac{1}{0.069(50)} + \frac{\ln \frac{11}{10}}{2\pi(1)(386)} + \frac{1}{0.063(80)} = 0.487\ \text{W/K}$$

$$\text{LMTD} = \frac{(60 - 30) - (40 - 20)}{\ln \frac{30}{20}} = 24.7°\text{C}$$

$$q = \frac{24.7}{0.487} = 50.7\ \text{W}$$

2-19 Heat-transfer processes used by the human body The primary objective of air conditioning is to provide comfortable conditions for people. Some principles of thermal comfort will be explained in Chap. 4. Since some thermodynamic and heat-transfer processes such as those discussed in this chapter help explain the phenomena presented in Chap. 4, these processes will be explored now.[6] From the thermal standpoint the body is an inefficient machine but a remarkably good regulator of its own temperature. The human body receives fuel in the form of food, converts a fraction of the energy in the fuel into work, and rejects the remainder as heat. It is the continuous process of heat rejection which requires a thermal balance. Figure 2-12 shows the thermal functions of the body schematically. The generation of heat occurs in cells throughout the body, and the circulatory system carries this heat to the skin, where it is released to the environment.

In a steady-state heat balance the heat energy produced by metabolism equals the rate of heat transferred from the body by convection, radiation, evaporation, and respiration. If the metabolism rate is not balanced momentarily by the sum of the transfer of heat, the body temperature will change slightly, providing thermal storage

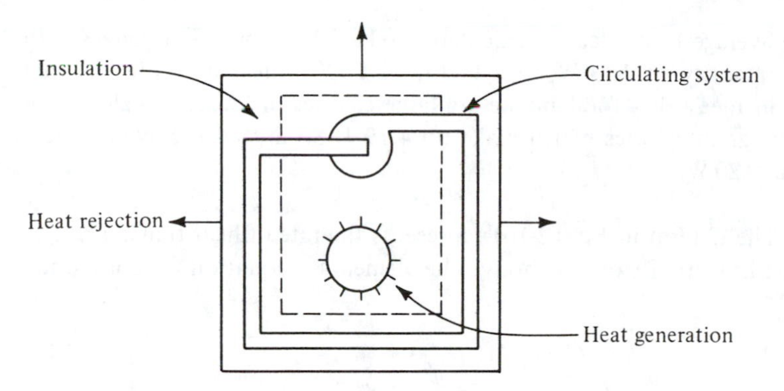

Figure 2-12 The body as a heat generator and rejector.

in the body. The complete equation for the heat balance becomes

$$M = \mathcal{E} \pm \mathcal{R} \pm C + B \pm S \tag{2-16}$$

where M = metabolism rate, W
$\mathcal{E}$ = heat loss by evaporation, W
$\mathcal{R}$ = rate of heat transfer by radiation, W
C = rate of heat transfer by convection, W
B = heat loss by respiration, W
S = rate of change of heat storage in body, W

Several of the terms on the right side of Eq. (2-16) always represent losses of heat from the body, while the radiation, convection, and thermal-storage terms can either be plus or minus. In other words, the body in different situations may either gain or lose heat by convection or radiation.

2-20 Metabolism Metabolism is the process which the body uses to convert energy in food into heat and work. Look at the body again as a heat machine: a person can convert food energy into work with an efficiency as high as 15 to 20 percent for short periods. In nonindustrial applications, particularly during light activity, the efficiency of conversion into work is of the order of 1 percent. The *basal metabolic rate* is the average possible rate, which occurs when the body is at rest, but not asleep.

We have several interests in the metabolism rate: (1) it is the M term in the heat-balance equation (2-16) that must be rejected from the body through the various mechanisms; (2) this heat contributes to the cooling load of the air-conditioning system. The heat-rejection rate by an occupant in a conditioned space may vary from 120 W for sedentary activities to more than 440 W for vigorous activity. The heat input by occupants is particularly important in the design of the environmental system for schoolrooms, conference rooms, theaters, and other enclosures where there is a concentration of people.

Example 2-11 For a rough estimate, consider the body as a heat machine assuming an intake in the form of food of 2400 cal/d (1 cal = 4.19 J). If all intake is oxidized and is rejected in the form of heat, what is the average heat release in watts?

Solution The average heat release is calculated to be 0.12 W, which disagrees with the expected quantity of 120 W by a factor of 1000. The explanation is that calories used in measuring food intake are large calories or kilogram calories, in contrast to the gram calories which make up 4.19 J. So indeed the average heat release is about 120 W.

2-21 Convection The C term in Eq. (2-16) represents the rate of heat transfer due to air flow convecting heat to or from the body. The elementary equation for convection applies

$$C = h_c A(t_s - t_a) \tag{2-17}$$

where A = body surface area, m^2
t_s = skin or clothing temperature, °C
t_a = air temperature, °C

The surface area of the human body is usually in the range of 1.5 to 2.5 m^2, depending upon the size of the person. The heat-transfer coefficient h_c depends upon the air velocity across the body and consequently also upon the position of the person and orientation to the air current. An approximate value of h_c during forced convection can be computed from

$$h_c = 13.5 V^{0.6} \tag{2-18}$$

where V is the air velocity in meters per second.

The skin temperature is controllable to a certain extent by the temperature-regulating mechanism of the body and generally ranges between 31 and 33°C for those parts of the body covered by clothing. The clothing temperature will normally lie somewhere between the skin and the air temperature, unless lowered because it is wet and is evaporating moisture.

2-22 Radiation The equation for heat transfer between the human body and its surroundings has already appeared as Eq. (2-10). Not all parts of the body radiate to the surroundings; some radiate to other parts of the body. The effective area of the body for radiation is consequently less than the total surface area, usually about 70 percent of the total.

The emissivity of the skin and clothing is very close to that of a blackbody and thus has a value of nearly 1.0. The temperature to which the body radiates is often referred to as the *mean radiant temperature,* a fictitious uniform temperature of the complete enclosure that duplicates the rate of radiant heat flow of the actual enclosure. The mean radiant temperature will usually be close to that of the air temperature except for influences of outside walls, windows, and inside surfaces that are affected by solar radiation.

2-23 Evaporation Removal of heat from the body by evaporation of water from the skin is a major means of heat rejection. The transfer of heat between the body and the environment by convection and radiation may be either toward the body or away from it, depending upon the ambient conditions. Evaporation, on the other hand, always constitutes a rejection of heat from the body. In hot environments, sweating provides the dominant method of heat removal from the body.

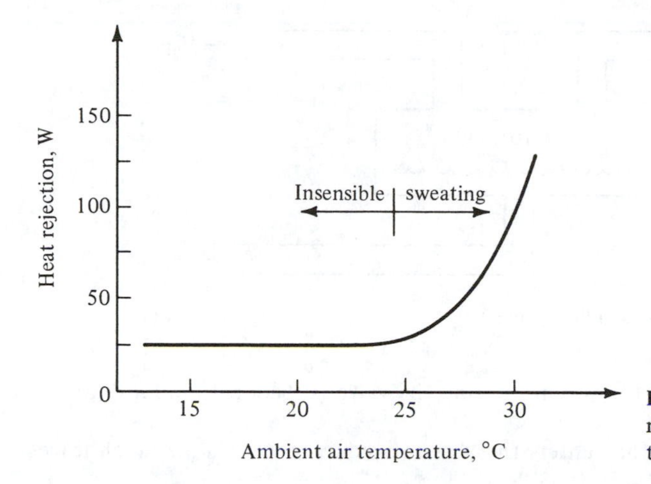

Figure 2-13 Heat rejection by nonsweating (insensible) evaporation and by sweating.

There are two modes by which the body wets the skin, diffusion and sweating. Diffusion, or insensible evaporation, is a constant process, while sweating is controlled by the thermoregulatory system. Typical magnitudes of heat rejection are shown in Fig. 2-13. The rate of heat transfer by insensible evaporation is controlled by the resistance of the deep layers of the epidermis to the diffusion of water from beneath the skin surface to the ambient air. The rate of this transfer is given

$$q_{ins} = h_{fg} A C_{diff} (p_s - p_a) \tag{2-19}$$

where q_{ins} = rate of heat transfer by insensible evaporation, W

h_{fg} = latent heat of water, J/kg

A = area of body, m^2

C_{diff} = coefficient of diffusion, $kg/Pa \cdot s \cdot m^2$

p_s = vapor pressure of water at skin temperature, Pa

p_a = vapor pressure of water vapor in ambient air, Pa

The dominant mechanism for rejecting large rates of heat from the body is by sensible sweating and subsequent evaporation of this sweat. Upon a rise in deep body temperature, the thermoregulatory system activates the sweat glands. Maximum rates of sweating, at least for short periods of time, are of the order of 0.3 g/s, so if all this sweat evaporates and removes 2430 kJ/kg, the potential heat removal rate by sweating is approximately 700 to 800 W.

Heat generation and the heat-transfer processes of convection, radiation, and evaporation all influence the design of air-conditioning systems in order to maintain comfortable conditions for the occupants.

PROBLEMS

2-1 Water at 120°C and a pressure of 250 kPa passes through a pressure-reducing valve and then flows to a separating tank at standard atmospheric pressure of 101.3 kPa, as shown in Fig. 2-14.

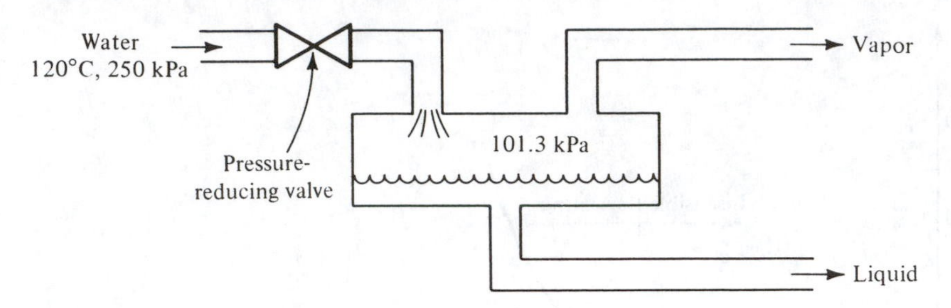

Figure 2-14 Pressure-reducing valve in Problem 2-1.

(*a*) What is the state of the water entering the valve (subcooled liquid, saturated liquid, or vapor)?

(*b*) For each kilogram that enters the pressure-reducing valve, how much leaves the separating tank as vapor? *Ans.* 0.0375

2-2 Air flowing at a rate of 2.5 kg/s is heated in a heat exchanger from −10 to 30°C. What is the rate of heat transfer? *Ans.* 100 kW

2-3 One instrument for measuring the rate of airflow is a venturi, as shown in Fig. 2-15, where the cross-sectional area is reduced and the pressure difference between positions *A* and *B* measured. The flow rate of air having a density of 1.15 kg/m^3 is to be measured in a venturi where the area at position *A* is 0.5 m^2 and the area at *B* is 0.4 m^2. The deflection of water (density = 1000 kg/m^3) in a manometer is 20 mm. The flow between *A* and *B* can be considered to be frictionless so that Bernoulli's equation applies.

(*a*) What is the pressure difference between positions *A* and *B*?

(*b*) What is the airflow rate? *Ans.* 12.32 m^3/s

2-4 Use the perfect-gas equation with R = 462 J/kg · K to compute the specific volume of saturated vapor water at 20°C. Compare with data of Table A-1. *Ans.* Deviation = 0.19%

2-5 Using the relationship shown on Fig. 2-6 for heat transfer when a fluid flows inside a tube, what is the percentage increase or decrease in the convection heat-transfer coefficient h_c if the viscosity of the fluid is decreased 10 percent? *Ans.* 4.3% increase

2-6 What is the order of magnitude of heat release by convection from a human body when the air velocity is 0.25 m/s and its temperature is 24°C? *Ans.* 60 W

2-7 What is the order of magnitude of radiant heat transfer from a human body in a comfort air-conditioning situation? *Ans.* 40 W

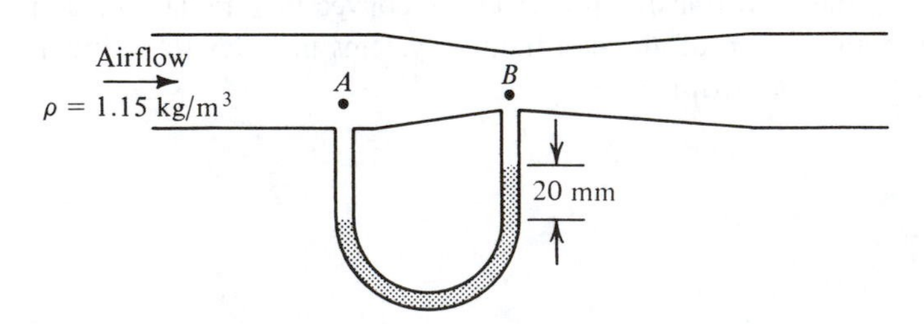

Figure 2-15 A venturi for measuring air flow.

2-8 What is the approximate rate of heat loss due to insensible evaporation if the skin temperature is 32°C, the vapor pressure is 4750 Pa, and the vapor pressure of air is 1700 Pa? The latent heat of water is 2.43 MJ/kg; $C_{\text{diff}} = 1.2 \times 10^{-9}$ kg/Pa · s · m^2. *Ans.* 18 W

REFERENCES

1. G. J. Van Wylen and R. E. Sonntag: "Fundamentals of Classical Thermodynamics," Wiley, New York, 1978.
2. W. D. Reynolds and H. C. Perkins: "Engineering Thermodynamics," McGraw-Hill, New York, 1970.
3. K. Wark: "Thermodynamics," 2d ed., McGraw-Hill, New York, 1976.
4. J. P. Holman: "Heat Transfer," 4th ed., McGraw-Hill, New York, 1976.
5. F. Kreith and W. Z. Black: "Basic Heat Transfer," Harper & Row, New York, 1980.
6. "ASHRAE Handbook, Fundamentals Volume," chap. 8, American Society of Heating, Refrigerating, and Air Conditioning Engineers, Atlanta, Ga., 1981.

CHAPTER

THREE

PSYCHROMETRY AND WETTED-SURFACE HEAT TRANSFER

3-1 Importance Psychrometry is the study of the properties of mixtures of air and water vapor. The subject is important in air-conditioning practice because atmospheric air is not completely dry but a mixture of air and water vapor. In some air-conditioning processes water is removed from the air-water-vapor mixture, and in others water is added. Psychrometric principles are applied in later chapters in this book, e.g., to load calculations, air-conditioning systems, cooling and dehumidifying coils, cooling towers, and evaporative condensers.

In some equipment there is a heat- and mass-transfer process between air and a wetted surface. Examples include some types of humidifiers, dehumidifying and cooling coils, and water-spray equipment such as cooling towers and evaporative condensers. Some convenient relations can be developed to express the rates of heat and mass transfer using *enthalpy potential*, discussed later in this chapter. But first the psychrometric chart is explored, property by property, followed by a discussion of the most common air-conditioning processes.

3-2 Psychrometric chart Since charts showing psychrometric properties are readily available (Fig. 3-1), why should we concern ourselves with the development of a chart? Two reasons are to become aware of the bases of the chart and to be able to calculate properties at new sets of conditions, e.g., nonstandard barometric pressure.

The step-by-step development of the psychrometric chart that follows will make use of a few simplifying assumptions. They will be pointed out along the way with recommendations for making a more accurate calculation. The chart that can be devel-

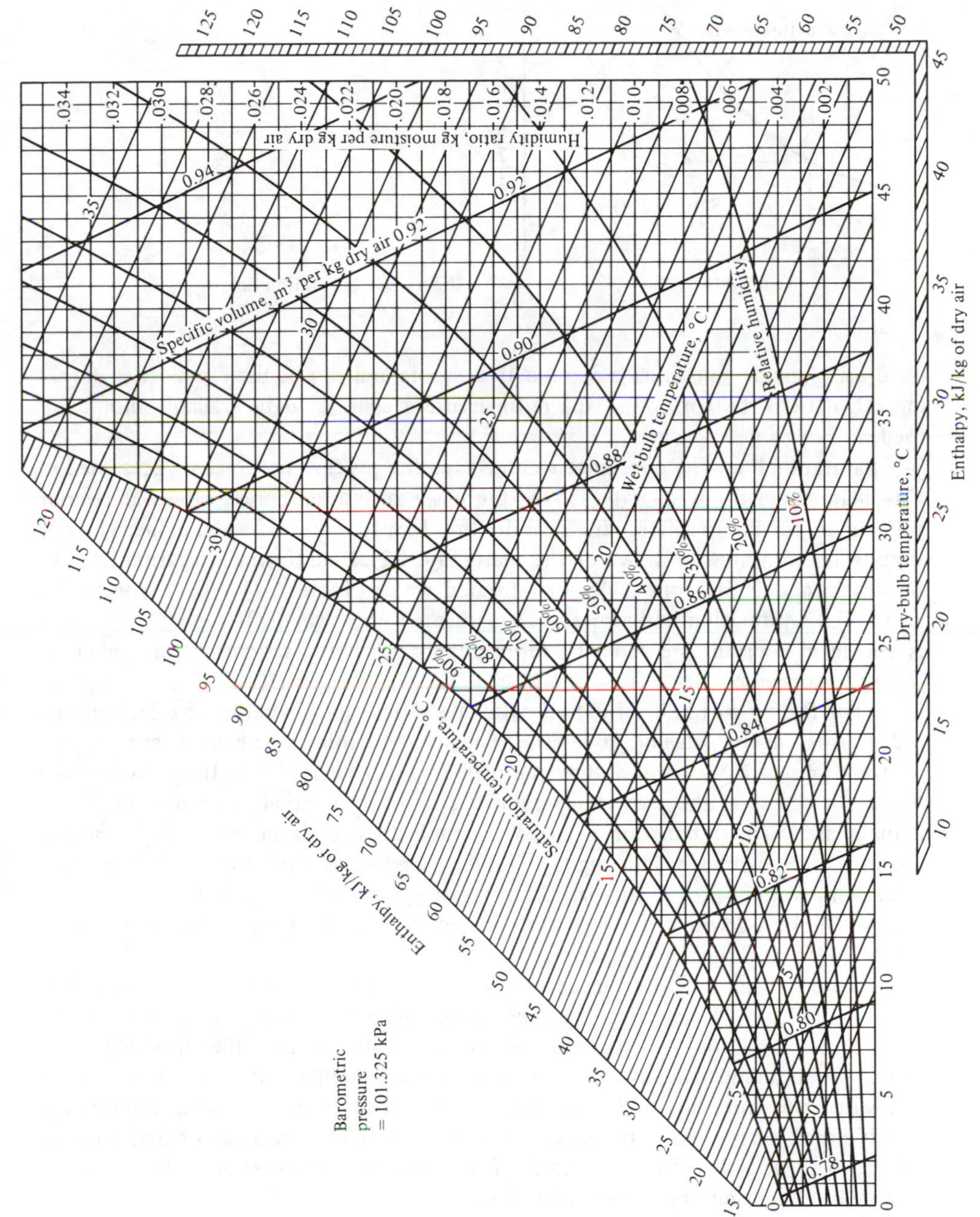

Figure 3-1 Psychrometric chart.

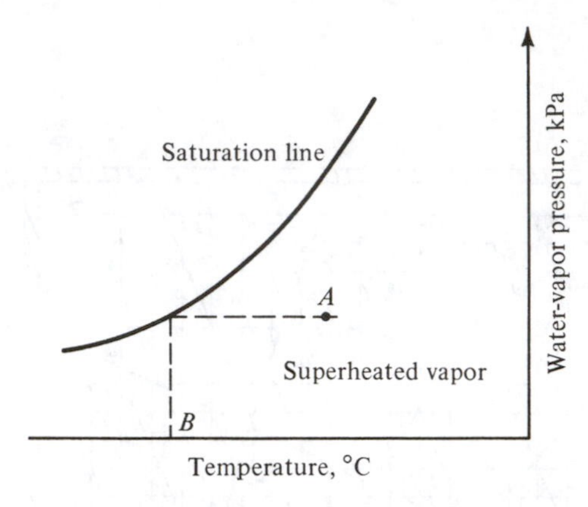

Figure 3-2 Saturation line.

oped using simple equations is reasonably accurate and can be used in most engineering calculations, but of course the most accurate chart or data available should be used.

3-3 Saturation line The coordinates chosen for the psychrometric chart presented in this chapter are the temperature t for the abscissa and temporarily the water-vapor pressure p_s for the ordinate. First consider the chart to represent water alone. The saturation line can now be drawn on the chart (Fig. 3-2). Data for the saturation line can be obtained directly from tables of saturated water (Table A-1). The region to the right of the saturation line represents superheated water vapor. If superheated vapor is cooled at constant pressure, it will eventually reach the saturation line, where it begins to condense.

Thus far, no air has been present with the water vapor. What is the effect on Fig. 3-2 if air is present? Ideally, none. The water vapor continues to behave as though no air were present. At a given water-vapor pressure, which is now a partial pressure, condensation occurs at the same temperature as it would if no air were present. There actually is a slight interaction between the molecules of air and water vapor, which changes the steam-table data slightly. Table A-2 presents the properties of air saturated with water vapor. A comparison of vapor pressures of the water in the air mixture of Table A-2 with that of pure water shown in Table A-1 reveals practically no difference in pressure at a given temperature.

Figure 3-2 can now be considered applicable to an air–water-vapor mixture. The portion of the chart now of significance is bounded by the saturation line and the axes. If the condition of the mixture lies on the saturation line, the air is said to be *saturated*, meaning that any decrease in temperature will result in condensation of the water vapor into liquid. To the right of the saturation line the air is unsaturated.

If point A represents the condition of the air, the temperature of that mixture will have to be reduced to temperature B in order for condensation to begin. Air at A is said to have a *dew-point temperature* of B.

3-4 Relative humidity The relative humidity ϕ is defined as the ratio of the mole fraction of water vapor in moist air to mole fraction of water vapor in saturated air at

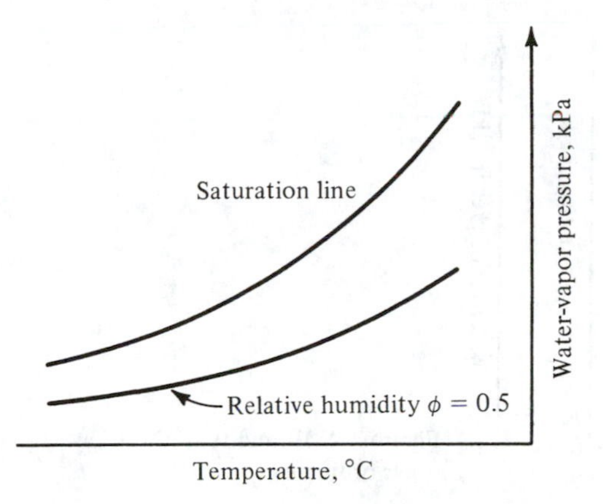

Figure 3-3 Relative-humidity line.

the same temperature and pressure. From perfect-gas relationships another expression for ϕ is

$$\phi = \frac{\text{existing partial pressure of water vapor}}{\text{saturation pressure of pure water at same temperature}}$$

Lines of constant relative humidity can be added to the chart, as in Fig. 3-3, by marking off vertical distances between the saturation line and the base of the chart. The relative humidity of 0.50, for example, has an ordinate equal to one-half that of the saturation line at that temperature.

3-5 Humidity ratio The humidity ratio W is the mass of water interspersed in each kilogram of dry air. The humidity ratio, like the next several properties to be studied—enthalpy and specific volume—is based on 1 kg of dry air. The perfect-gas equation can be summoned to solve for the humidity ratio. Both water vapor and air may be assumed to be perfect gases (obey the equation $pv = RT$ and have constant specific heats) in the usual air-conditioning applications. Air is assumed to be a perfect gas because its temperature is high relative to its saturation temperature, and water vapor is assumed to be a perfect gas because its pressure is low relative to its saturation pressure

$$W = \frac{\text{kg of water vapor}}{\text{kg of dry air}} = \frac{p_s V/R_s T}{p_a V/R_a T} = \frac{p_s/R_s}{(p_t - p_s)/R_a} \tag{3-1}$$

where W = humidity ratio, (kg of water vapor)/(kg of dry air)
V = arbitrary volume of air-vapor mixture, m^3
p_t = atmospheric pressure = $p_a + p_s$, Pa
p_a = partial pressure of dry air, Pa
R_a = gas constant of dry air = 287 J/kg·K
R_s = gas constant of water vapor = 461.5 J/kg·K
T = absolute temperature of air-vapor mixture, K

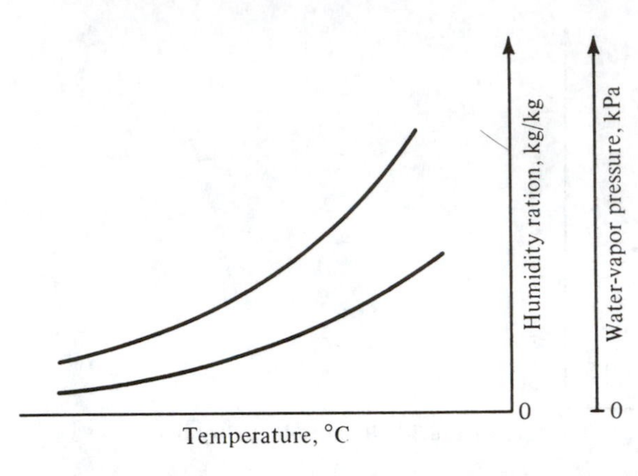

Figure 3-4 Humidity ratio W as another ordinate.

Substituting the numerical values of R_a and R_s into Eq. (3-1) gives

$$W = \frac{287}{461.5} \frac{p_s}{p_t - P_s} = 0.622 \frac{p_s}{p_t - P_s} \tag{3-2}$$

The atmospheric pressure p_t has now appeared on the scene, and from this point on in the development of the psychrometric chart the chart will be unique to a given barometric pressure. Equation (3-2) shows the relationship between the humidity ratio and the water-vapor pressure, so that companion scales can be shown as ordinates of the psychrometric chart, as illustrated in Fig. 3-4. As Eq. (3-2) shows, the relation between W and p_s is not perfectly linear. In Fig. 3-1 and in most psychrometric charts the W scale is divided linearly, which makes the p_s scale slightly nonlinear.

Example 3-1 Compute the humidity ratio of air at 60 percent relative humidity when the temperature is 30°C. The barometric pressure is the standard value of 101.3 kPa.

Solution The water-vapor pressure of saturated air at 30°C is 4.241 kPa from Table A-1. Since the relative humidity is 60 percent, the water-vapor pressure of the air is 0.60 (4.241 kPa) = 2.545 kPa. From Eq. (3-2)

$$W = 0.622 \frac{2.545}{101.3 - 2.545} = 0.0160 \text{ kg/kg}$$

This result checks the value read from Fig. 3-1.

3-6 Enthalpy The enthalpy of the mixture of dry air and water vapor is the sum of the enthalpy of the dry air and the enthalpy of the water vapor. Enthalpy values are always based on some datum plane, and the zero value of the dry air is chosen as air at 0°C. The zero value of the water vapor is saturated liquid water at 0°C, the same datum plane that is used for tables of steam. An equation for the enthalpy is

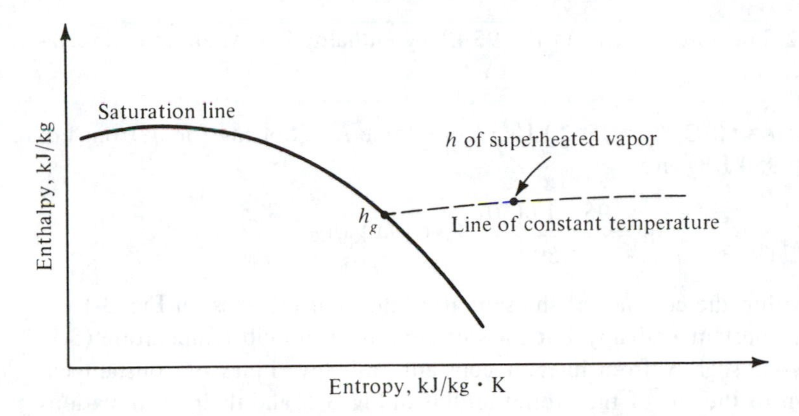

Figure 3-5 Line of constant temperature shows that the enthalpy of superheated water vapor is approximately equal to the enthalpy of saturated vapor at the same temperature.

$$h = c_p t + W h_g \qquad \text{kJ/kg dry air} \tag{3-3}$$

where c_p = specific heat of dry air at constant pressure = 1.0 kJ/kg·K
t = temperature of air-vapor mixture, °C
h_g = enthalpy of saturated steam at temperature of air-vapor mixture, kJ/kg

Equation (3-3) gives quite accurate results, although several refinements can be made. The specific heat c_p actually varies from 1.006 at 0°C to 1.009 at 50°C. The enthalpy of water vapor h_g is for saturated steam, but the water vapor in the air-vapor mixture is likely to be superheated. No appreciable error results, however, because of the fortunate relationship of enthalpy and temperature shown on the Mollier diagram of Fig. 3-5.

A line of constant enthalpy can now be added to the psychrometric chart, as in Fig. 3-6. Suppose, for example, that the 95 kJ/kg enthalpy line is to be constructed. Several arbitrary temperatures can be chosen and the humidity ratio computed at 95 kJ/kg using Eq. (3-3). The humidity ratio thus computed and the temperature locate one point on the line of constant enthalpy.

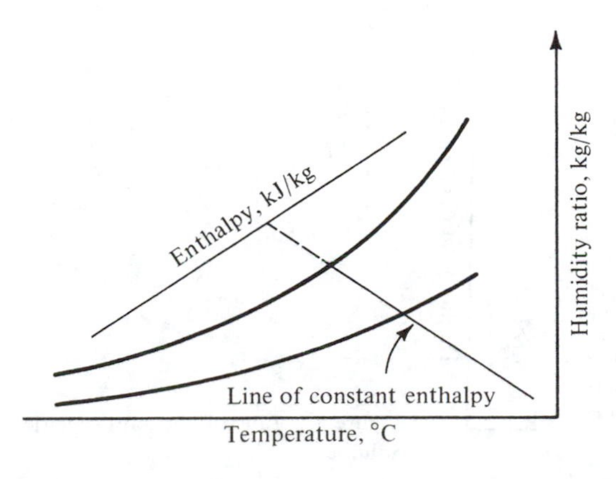

Figure 3-6 Line of constant enthalpy.

Example 3-2 Locate the point on the 95 kJ/kg enthalpy line where the temperature is 50°C.

Solution At $t = 50°C$, $h_g = 2592$ kJ/kg from Table A-1. Solving for W from Eq. (3-3) for $h = 95$ kJ/kg gives

$$W = \frac{95 - 1.0(50)}{2592} = 0.0174 \text{ kg/kg}$$

The lines within the confines of the saturation line and the axes on Fig. 3-1 are not the lines of constant enthalpy but lines of constant wet-bulb temperature (Sec. 3-9), which deviate slightly from lines of constant enthalpy. Lines of constant enthalpy are shown to the left of the saturation line in Fig. 3-1, and their continuations are shown at the right and bottom borders of the chart. The procedure for reading enthalpy values off the psychrometric chart will be explained later.

3-7 Specific volume The perfect-gas equation is used to calculate the specific volume of the air-vapor mixture. The specific volume is the number of cubic meters of mixture per kilogram of dry air. It could just as well be the cubic meters of dry air or the cubic meters of mixture per kilogram of dry air, since the volumes occupied by the individual substances are the same.

From the perfect-gas equation, the specific volume v is

$$v = \frac{R_a T}{p_a} = \frac{R_a T}{p_t - p_s} \quad \text{m}^3\text{/kg dry air} \tag{3-4}$$

To establish points on a line of constant specific volume, 0.90 m^3/kg for example, substitute 0.9 for v, the barometric pressure for p_t, and at arbitrary values of T solve for p_s. The pairs of p_s and t values then describe the line of constant v, as in Fig. 3-7.

Example 3-3 What is the specific volume of an air-water-vapor mixture having a temperature of 24°C and a relative humidity of 20 percent at standard barometric pressure?

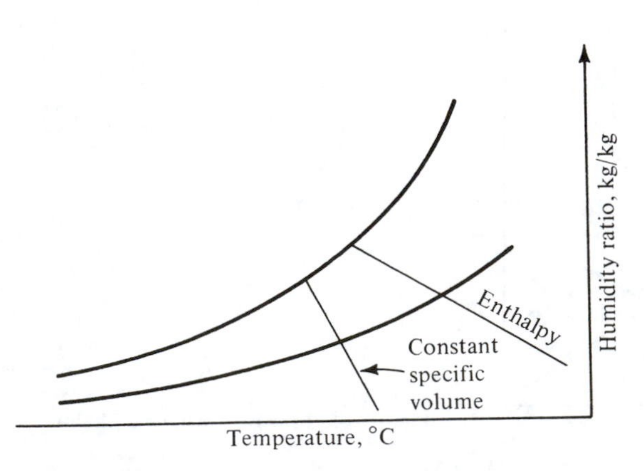

Figure 3-7 Line of constant specific volume.

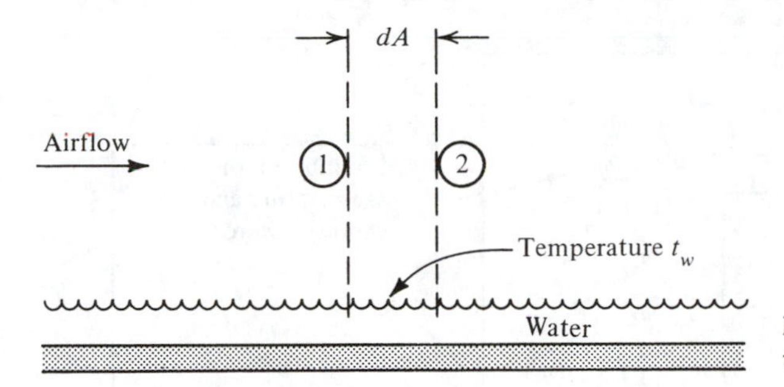

Figure 3-8 Air passing over a wetted surface.

Solution The water-vapor pressure of saturated air at 24°C is, from Table A-1, 2.982 kPa; so the vapor pressure with a relative humidity of 20 percent is 0.2(2.982) = 0.5964 kPa = 596.4 Pa. Applying Eq. (3-4), we get

$$v = \frac{287(24 + 273.15)}{101{,}300 - 596} = 0.85 \text{ m}^3/\text{kg dry air}$$

This result checks the value from Fig. 3-1.

3-8 Combined heat and mass transfer; the straight-line law The final psychrometric property to be considered is the wet-bulb temperature, but in order to improve our understanding of this property a short detour will be made. It leads into the combined process of heat and mass transfer and proposes the *straight-line law*. This law states that when air is transferring heat and mass (water) to or from a wetted surface, the condition of the air shown on a psychrometric chart drives toward the saturation line at the temperature of the wetted surface. If air flows over a wetted surface, as in Fig. 3-8, the condition of air passing over differential area dA changes from condition 1 to condition 2 on the psychrometric chart, Fig. 3-9. The straight-line law asserts that point 2 lies on a straight line drawn between point 1 and the saturation curve at the wetted-surface temperature.

It is no surprise that the warm air at 1 drops in temperature when in contact with water at temperature t_w. It is also to be expected that the air at 1, having a higher

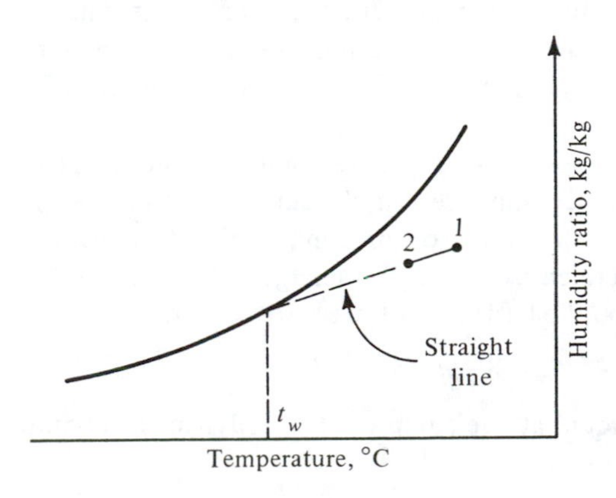

Figure 3-9 Condition of air drives toward saturation line at temperature of wetted surface.

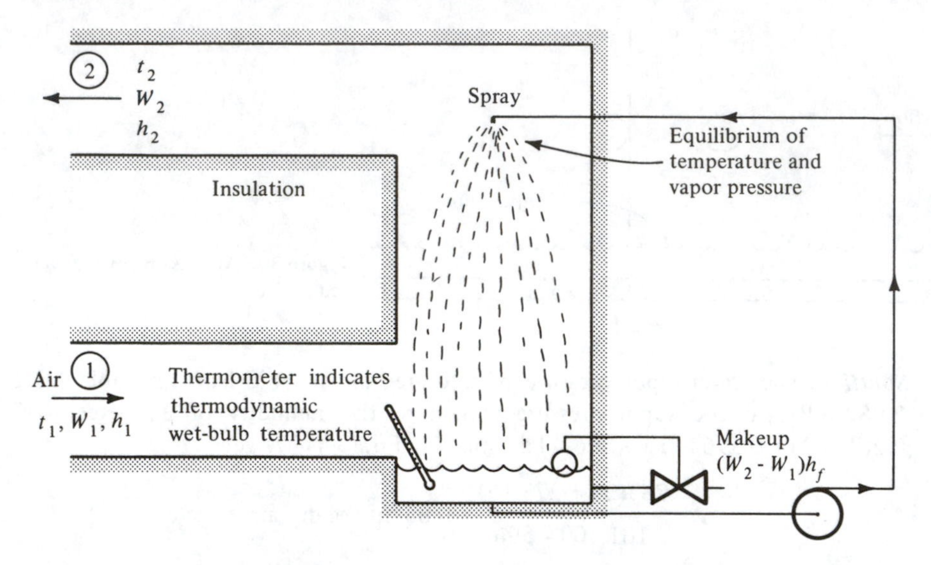

Figure 3-10 Adiabatic saturation.

vapor pressure than the liquid at temperature t_w, will transfer mass by condensing some water vapor and dropping the humidity ratio of the air. What is unique is that the rates of heat and mass transfer are so related that the path is a straight line driving toward the saturation line at the wetted-surface temperature. This special property is due to the value of unity of the Lewis relation, a dimensionless group that will be explained in Sec. 3-14.

3-9 Adiabatic saturation and thermodynamic wet-bulb temperature An *adiabatic saturator* (Fig. 3-10) is a device in which air flows through a spray of water. The water circulates continuously, and the spray provides so much surface area that the air leaves the spray chamber in equilibrium with the water, with respect to both temperature and vapor pressure. The device is adiabatic in that the walls of the saturator are insulated, and no heat is added to, or extracted from, the water line that circulates the water from the sump back to the sprays. In order to perpetuate the process it is necessary to provide makeup water to compensate for the amount of water evaporated into the air. The temperature of this makeup water is controlled so that it is the same as that in the sump.

After the adiabatic saturator has achieved a steady-state condition, the temperature indicated by an accurate thermometer immersed in the sump is the *thermodynamic wet-bulb temperature*. Certain combinations of air conditions will result in a given sump temperature and can be defined by writing an energy balance about the saturator. This energy balance, written on the basis of unit mass flow of air, is

$$h_1 = h_2 - (W_2 - W_1)h_f \tag{3-5}$$

where h_f is the enthalpy of saturated liquid at the sump or thermodynamic wet-bulb temperature.

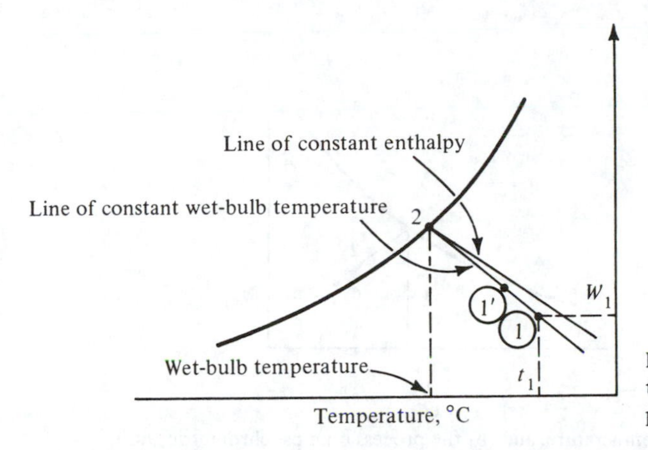

Figure 3-11 Line of constant thermodynamic wet-bulb temperature.

On the psychrometric chart in Fig. 3-11, point 1 lies below the line of constant enthalpy that passes through point 2. Any other condition of air that results in the same sump temperature, such as point 1′, has the same wet-bulb temperature. This line is straight because of the straight-line law, which states that the entering air at point 1 drives toward the saturation line at the wetted-surface temperature. The straight line between points 1 and 2 represents the path of the air as it passes through the saturator.

Lines of constant wet-bulb temperature are shown on psychrometric charts, as in Fig. 3-1, but lines of constant enthalpy are rarely shown. The enthalpy scale to the left of the saturation line applies to air that is saturated. For unsaturated air the enthalpy scale on the left must be combined with the enthalpy scale shown at the right and bottom borders of the chart. The deviation between the enthalpy and wet-bulb-temperature lines will be explained next.

3-10 Deviation between enthalpy and wet-bulb lines As Fig. 3-11 indicates, readings of enthalpy obtained by following the wet-bulb line to the saturation curve specify values of enthalpy that are too high. The psychrometric chart, Fig. 3-1, shows lines of constant thermodynamic wet-bulb temperature and not lines of constant enthalpy. The enthalpy scale shown at the left applies only to the conditions on the saturation line, and both the scale at the left and the scales at the right and bottom borders should be used for more precise determinations of enthalpy.

To check an enthalpy deviation, compare the chart reading with a calculation for air having a dry-bulb temperature of 40°C and a relative humidity of 41 percent. The wet-bulb temperature of air at this condition is 28°C.

In Fig. 3-1 a straightedge can be set at 40°C dry-bulb temperature and 41 percent relative humidity and pivoted about that point until the enthalpy values on the left and right enthalpy scales match. That value is 89 kJ/kg.

Equation (3-5) permits calculation of the enthalpy of the point in question, h_1, by correcting the value of h_2 (the enthalpy of saturated air at the same wet-bulb temperature)

$$h_1 = 89.7 \text{ kJ/kg} - (W_2 - W_1)h_f$$

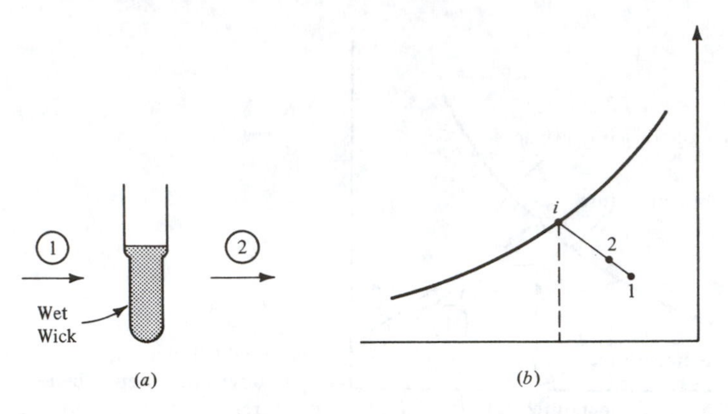

Figure 3-12 (a) The wet-bulb temperature, and (b) the process on a psychrometric chart.

where $W_1 = 0.019$ kg/kg
$W_2 = 0.0241$ kg/kg
$h_f = h_f$ at 28°C = 117.3 kJ/kg
$h_1 = 89.7 - 117.3(0.0241 - 0.019) = 89.1$ kJ/kg

3-11 Wet-bulb thermometer Although the adiabatic saturator of Fig. 3-10 is not a convenient device for routine measurements, a thermometer having a wetted wick, as in Fig. 3-12, would be convenient. We must therefore determine whether the wet-bulb thermometer truly indicates the thermodynamic wet-bulb temperature. The wetted area of the wick is finite, rather than infinite like the saturator in Fig. 3-10, so the change in state of air passing over the wetted bulb can be represented by process 1-2 in Fig. 3-12*b*. Since the energy balance about the bulb is

$$h_1 + W_1 h_f = h_2 + W_2 h_f$$

points 1 and 2 lie on the same thermodynamic wet-bulb line. The important question, however, is: What is the temperature of the water on the wick? The answer, which comes from the application of the straight-line law, is that the condition of the air starting at point 1 has been driving toward the saturation line at the temperature of the wetted surface in order to reach point 2. Had more wetted surface been available, the state of the air would continue to drive along the straight line toward the saturation curve.

Carrier,[1] in his pioneer paper on psychrometry, assumed that the temperature of water on a wet-bulb thermometer was the same as that in an adiabatic saturator. Lewis[2] in 1922 grouped the terms that bear his name and concluded that a value of unity of this dimensionless group results in identical temperatures of a wetted wick and adiabatic spray. In 1933 Lewis[3] demonstrated that in atmospheres other than air and water vapor the reading of a wet-bulb thermometer and the saturated spray are different. We shall hereafter consider the temperature of the wet-bulb thermometer and the adiabatic spray to be the same and drop the qualification "thermodynamic" on the wet-bulb temperature, simply calling it the wet-bulb temperature.

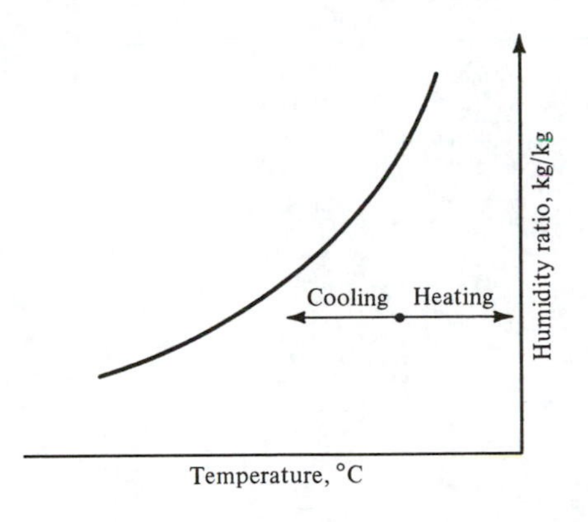

Figure 3-13 Sensible heating or cooling.

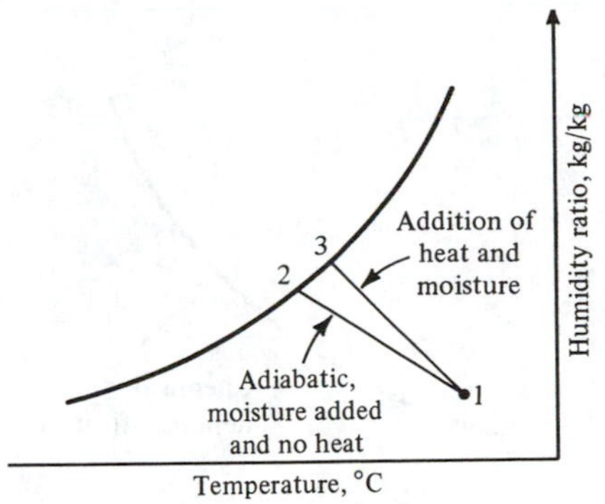

Figure 3-14 Humidification.

3-12 Processes Processes performed on air can be plotted on the psychrometric chart for quick visualization. Of even more importance is the fact that the chart can be used to determine changes in such significant properties as temperature, humidity ratio, and enthalpy for the processes. Some of the basic processes will now be shown, including (1) sensible heating or cooling, (2) humidification, adiabatic and nonadiabatic, (3) cooling and dehumidification, (4) chemical dehumidification, and (5) mixing.

1. Sensible heating or cooling refers to a rate of heat transfer attributable only to a change in dry-bulb temperature of the air. Figure 3-13 shows a change in dry-bulb temperature with no change in humidity ratio.
2. Humidification, as shown in Fig. 3-14, may be adiabatic, as shown in process 1-2, or with addition of heat, as in process 1-3.
3. Cooling and dehumidification results in a reduction of both the dry-bulb temperature and the humidity ratio (Fig. 3-15). A cooling and dehumidifying coil performs such a process. The refrigeration capacity in kilowatts during a cooling and dehumidifying process is given by

$$\text{Refrigeration capacity} = w(h_1 - h_2)$$

where w is in kilograms per second and h_1 and h_2 in kilojoules per kilogram.

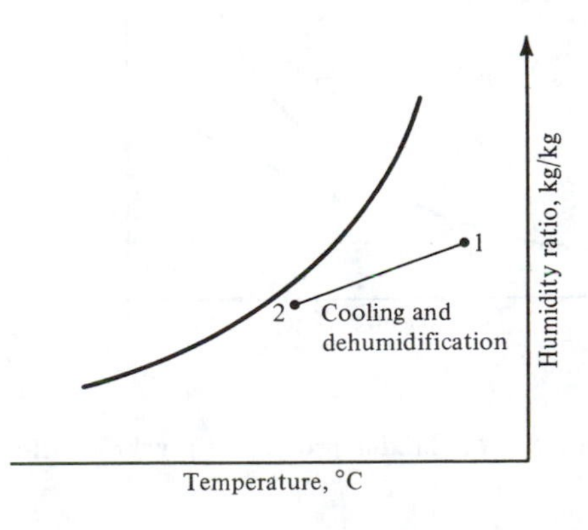

Figure 3-15 Cooling and dehumidification.

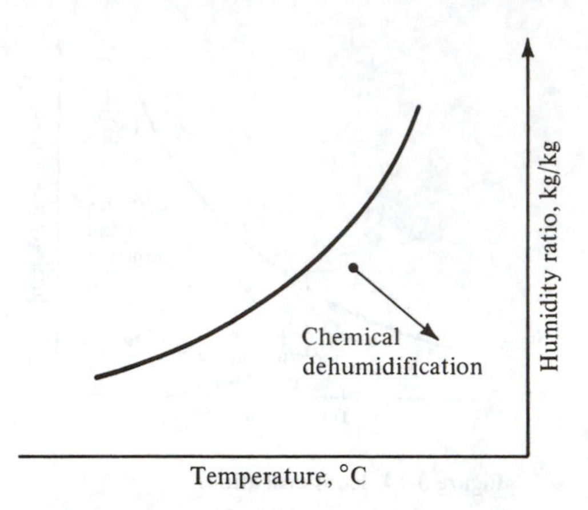

Figure 3-16 Chemical dehumidification.

4. In the process of chemical dehumidification (Fig. 3-16) the water vapor from the air is absorbed or adsorbed by a hygroscopic material. Since the process, if thermally isolated, is essentially one of constant enthalpy, and since the humidity ratio decreases, the temperature of the air must increase.
5. Mixing of two streams of air is a common process in air conditioning. Figure 3-17*a* shows the mixing of w_1 kg/s of air at condition 1 with w_2 kg/s of air at condition 2. The result is condition 3, shown on the psychrometric chart in Fig. 3-17*b*. The fundamental equations applicable to the mixing process are an energy balance and a mass balance. The energy balance is

$$w_1 h_1 + w_2 h_2 = (w_1 + w_2) h_3 \tag{3-6}$$

and the balance of the mass of water is

$$w_1 W_1 + w_2 W_2 = (w_1 + w_2) W_3 \tag{3-7}$$

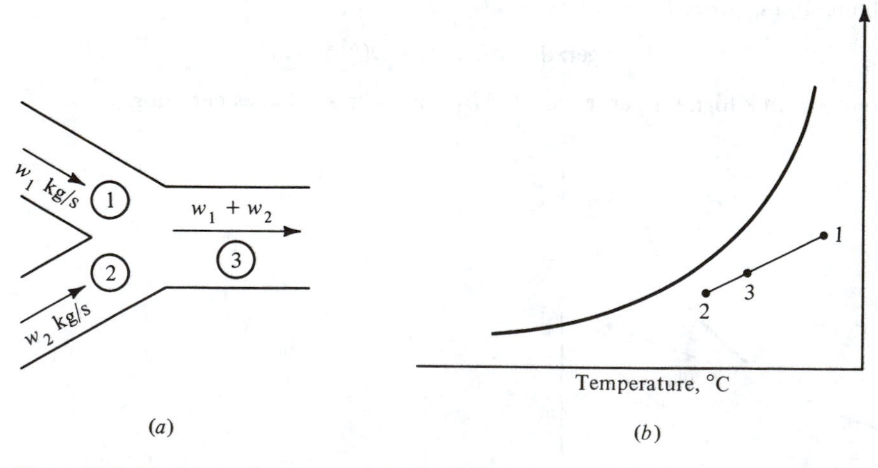

Figure 3-17 (*a*) Schematic arrangement of mixing process. (*b*) Mixing process on psychrometric chart.

Equations (3-6) and (3-7) show that the final enthalpy and humidity ratio are weighted averages of the entering enthalpies and humidity ratios. An approximation used by many engineers is that the final *temperature* and humidity ratio are weighted averages of the entering values. With that approximation, the point on the psychrometric chart representing the result of a mixing process lies on a straight line connecting the points representing the entering conditions. Furthermore, the ratio of distances on the line, (1-3)/(2-3), equals the ratio of the flow rates, w_2/w_1. The error of this approximation is caused by the variation in specific heats of moist air and is usually less than 1 percent.

3-13 Comment on the basis of 1 kg of dry air The enthalpy, humidity ratio, and specific volume are all based on 1 kg of *dry* air, and it may seem strange to speak of, say, the mass of water in 1 kg of dry air. (Correctly the humidity ratio should be expressed as the mass of water *associated* with 1 kg of dry air.) A review of some of the processes presented in Sec. 3-12, however, shows the usefulness of the basis of dry air. In the processes shown in Figs. 3-14 to 3-16 the total mass changes throughout the process because of the addition or extraction of water. If total mass of mixture were used as the basis, it would be necessary to recalculate the mass flow rate after each of these processes. The flow rate of dry air, however, remains constant through the processes.

3-14 Transfer of sensible and latent heat with a wetted surface When air flows past a wetted surface, as shown in Fig. 3-18, there is a likelihood of transfer of both sensible and latent heat. If there is a difference in temperature between the air t_a and the wetted surface t_i, heat will be transferred. If there is a difference in the partial pressure of water vapor in the air $p_{s,a}$ and that of the water $p_{s,i}$, there will be a transfer of mass (water). This transfer of mass causes a thermal-energy transfer as well, because if vapor condenses from the air, the latent heat must be removed at the water. Conversely, if some liquid evaporates from the water layer, the latent heat of this vaporized water must be supplied to the water.

The rate of sensible-heat transfer from the water surface to the air q_s can be calculated by the convection equation

$$dq_s = h_c \, dA \, (t_i - t_a) \tag{3-8}$$

where q_s = rate of sensible-heat transfer, W
h_c = convection coefficient, $W/m^2 \cdot K$
A = area, m^2

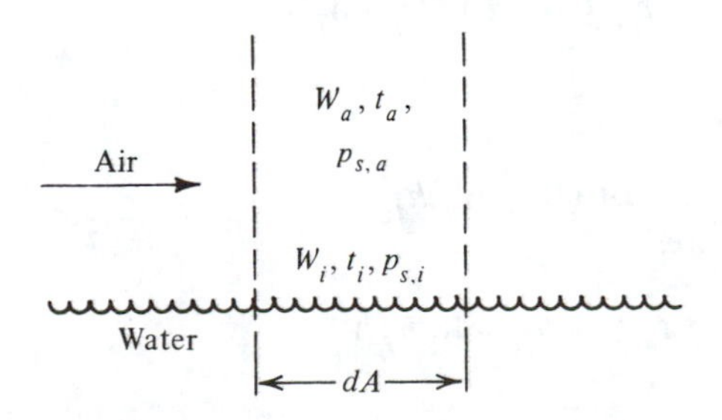

Figure 3-18 Heat and mass transfer between air and a wetted surface.

The rate of mass transfer from the water surface to the air is proportional to the pressure difference, $p_{s,i} - p_{s,a}$. Section 3-5 showed that the humidity ratio is approximately proportional to the vapor pressure, and so a corresponding proportionality can be established

$$\text{rate of mass transfer} = h_D\, dA(W_i - W_a) \quad \text{kg/s}$$

where h_D = proportionality constant, kg/m^2

W_i = humidity ratio of saturated air at wetted-surface temperature

Since the mass transferred to or from the water causes a transfer of heat due to the condensation or evaporation,

$$dq_L = h_D\, dA(W_i - W_a)h_{fg} \tag{3-9}$$

where q_L = rate of latent-heat transfer, W

h_{fg} = latent heat of water at t_i, J/kg

Difficult though it may seem to determine accurate values of the convection coefficient h_c for a particular situation, many more data are available on convection coefficients for sensible-heat transfer than for the mass-transfer proportionality constant h_D. Fortunately the transport mechanism at the water surface that controls the rate of sensible-heat transfer is the same one that controls the rate of mass transfer. There should, and indeed does, exist a proportional relation between h_D and h_c. Some of the details of the boundary-layer analysis are given in Ref. 4. This proportionality is expressed by

$$h_D = \frac{h_c}{c_{pm}} \tag{3-10}$$

where c_{pm} is the specific heat of moist air, J/kg·K.

The specific heat of moist air is based on 1 kg of dry air and is thus the sum of the specific heat of dry air and that of the water vapor is

$$c_{pm} = c_p + W_a c_{ps} \tag{3-11}$$

3-15 Enthalpy potential The concept of *enthalpy potential* is a useful one in quantifying the transfer of total heat (sensible plus latent) in those processes and components where there is direct contact between air and water. The expression for transfer of total heat dq_t through a differential area dA is available from a combination of Eqs. (3-8) and (3-9)

$$dq_t = dq_s + dq_L = h_c\, dA\,(t_i - t_a) + h_D\, dA\,(W_i - W_a)h_{fg}$$

Applying the expression for h_D from Eq. (3-10) gives

$$dq_t = h_c\, dA\,(t_i - t_a) + \frac{h_c}{c_{pm}}\, dA\,(W_i - W_a)h_{fg}$$

or

$$dq_t = \frac{h_c\, dA}{c_{pm}}(c_{pm}t_i - c_{pm}t_a + W_i h_{fg} - W_a h_{fg})$$

Substituting Eq. (3-11), we get

$$dq_t = \frac{h_c\,dA}{c_{pm}} [(c_p t_i + W_i h_{fg}) - (c_p t_a + W_a c_{ps} t_a - W_a c_{ps} t_i + W_a h_{fg})] \qquad (3\text{-}12)$$

The final approximation is to add to Eq. (3-12) an expression that is almost negligible compared with the other terms. That expression is $W_i h_f - W_a h_f$, where h_f is the enthalpy of saturated liquid water at temperature t_i. Equation (3-12) then becomes

$$dq_t = \frac{h_c dA}{c_{pm}} \{[c_p t_i + W_i(h_f + h_{fg})] - [c_p t_a + W_a(h_f + h_{fg} + c_{ps} t_a - c_{ps} t_i)]\} \qquad (3\text{-}13)$$

The expression in the first set of brackets of Eq. (3-13) is precisely the enthalpy of saturated air at the wetted-surface temperature, and the expression in the second set of brackets is precisely the enthalpy of the air in the free stream. The units of both enthalpies are kilojoules per kilogram of dry air. Thus

$$dq_t = \frac{h_c\,dA}{c_{pm}} (h_i - h_a) \qquad (3\text{-}14)$$

The name enthalpy potential originates from Eq. (3-14) because the potential for the transfer of the sum of sensible and latent heats is the difference between the enthalpy of saturated air at the wetted-surface temperature h_i and the enthalpy of air in the free stream h_a.

The specific heat of moist air c_{pm} is expressed by Eq. (3-11), but for states of air near those of normal room conditions a value of 1.02 kJ/kg·K may be used. For example, with air at 25°C and 50 percent relative humidity, $c_p = 1.00$ kJ/kg·K, $W = 0.011$ kg/kg, $c_{ps} = 1.88$, and $c_{pm} = 1.0207$ kJ/kg·K.

3-16 Insights provided by enthalpy potential In addition to helping quantify the calculations of heat and mass transfer in cooling and dehumidifying coils, sprayed coils, evaporative condensers, and cooling towers, the enthalpy potential provides a qualitative indication of the direction of total heat flow. Three different cases are illustrated in Figs. 3-19 to 3-21. Air at condition *a* is in contact with water at three different temperatures in cases 1, 2, and 3. In case 1

$$\left.\begin{matrix} dq_s \\ dq_L \\ dq_t \end{matrix}\right\} \text{is from the air to the water since} \left\{\begin{matrix} t_a > t_i \\ W_a > W_i \\ h_a > h_i \end{matrix}\right.$$

and since both dq_s and dq_L are from the air to the water.

In case 2

dq_s is from the air to the water since $t_a > t_i$
dq_L is from the water to the air since $W_a < W_i$
dq_t is from the air to the water since $h_a > h_i$

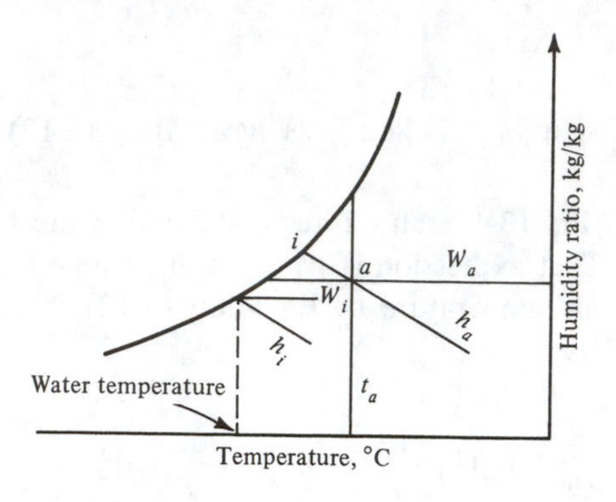

Figure 3-19 Case 1, q_t from air to water.

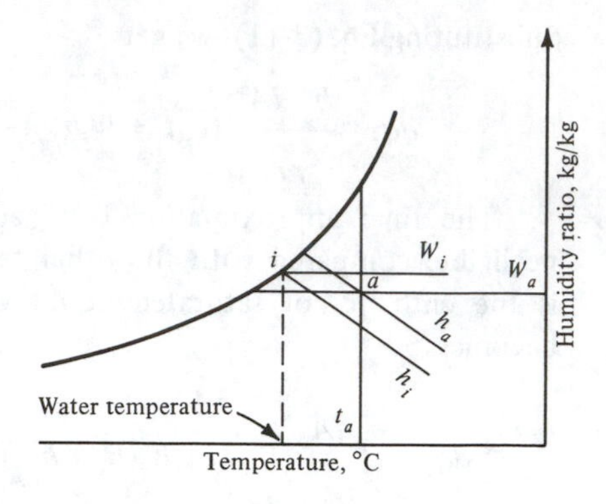

Figure 3-20 Case 2, q_t from air to water.

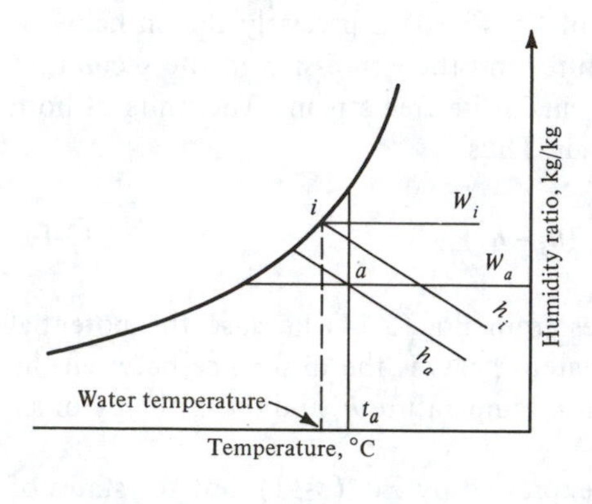

Figure 3-21 Case 3, q_t from the water to air.

Before the concept of enthalpy potential was developed, we were unable to determine immediately which way dq_t was flowing because we did not know the relative magnitudes of dq_s and dq_L. Now the relative values of h_a and h_i provide the clue.

In case 3

dq_s is from the air to the water since $t_a > t_i$
dq_L is from the water to the air since $W_a < W_i$
dq_t is from the water to the air since $h_a < h_i$

An interesting situation occurs in Fig. 3-21, where heat flows from the low-temperature water to the high-temperature air. The second law of thermodynamics is not violated, however, because the transfer due to the difference in partial pressure of the water vapor must also be considered.

PROBLEMS

3-1 Calculate the specific volume of an air-vapor mixture in cubic meters per kilogram of dry air when the following conditions prevail: $t = 30°C$, $W = 0.015$ kg/kg, and $p_t =$ 90 kPa. *Ans.* 0.99 m^3/kg.

3-2 A sample of air has a dry-bulb temperature of 30°C and a wet-bulb temperature of 25°C. The barometric pressure is 101 kPa. Using steam tables and Eqs. (3-2), (3-3), and (3-5), calculate (*a*) the humidity ratio if this air is adiabatically saturated, (*b*) the enthalpy of the air if it is adiabatically saturated, (*c*) the humidity ratio of the sample using Eq. (3-5), (*d*) the partial pressure of water vapor in the sample, and (*e*) the relative humidity. *Ans.* (*a*) 0.0201 kg/kg, (*b*) 76.2 kJ/kg, (*c*) 0.0180 kg/kg, (*d*) 2840 Pa, (*e*) 67%.

3-3 Using humidity ratios from the psychrometric chart, calculate the error in considering the wet-bulb line to be the line of constant enthalpy at the point of 35°C dry-bulb temperature and 50 percent relative humidity.

3-4 An air-vapor mixture has a dry-bulb temperature of 30°C and a humidity ratio of 0.015. Calculate at two different barometric pressures, 85 and 101 kPa, (*a*) the enthalpy and (*b*) the dew-point temperature. *Ans.* (*a*) 68.3 and 68.3 kJ/kg, (*b*) 17.5 and 20.3°C.

3-5 A cooling tower is a device that cools a spray of water by passing it through a stream of air. If 15 m^3/s of air at 35°C dry-bulb and 24°C wet-bulb temperature and an atmospheric pressure of 101 kPa enters the tower and the air leaves saturated at 31°C, (*a*) to what temperature can this airstream cool a spray of water entering at 38°C with a flow rate of 20 kg/s and (*b*) how many kilograms per second of makeup water must be added to compensate for the water that is evaporated? *Ans.* (*a*) 31.3°C, (*b*) 0.245 kg/s.

3-6 In an air-conditioning unit 3.5 m^3/s of air at 27°C dry-bulb temperature, 50 percent relative humidity, and standard atmospheric pressure enters the unit. The leaving condition of the air is 13°C dry-bulb temperature and 90 percent relative humidity. Using properties from the psychrometric chart, (*a*) calculate the refrigerating capacity in kilowatts and (*b*) determine the rate of water removal from the air. *Ans.* (*a*) 88 kW, (*b*) 0.0113 kg/s.

3-7 A stream of outdoor air is mixed with a stream of return air in an air-conditioning system that operates at 101 kPa pressure. The flow rate of outdoor air is 2 kg/s, and its condition is 35°C dry-bulb temperature and 25°C wet-bulb temperature. The flow rate of return air is 3 kg/s, and its condition is 24°C and 50 percent relative humidity. Determine (*a*) the enthalpy of the mixture, (*b*) the humidity ratio of the mixture, (*c*) the dry-bulb temperature of the mixture from the properties determined in parts (*a*) and (*b*), and (*d*) the dry-bulb temperature by weighted average of the dry-bulb temperatures of the entering streams. *Ans.* (*a*) 59.1 kJ/kg, (*b*) 0.01198 kg/kg, (*c*) 28.6°C, (*d*) 28.4°C.

3-8 The air conditions at the intake of an air compressor are 28°C, 50 percent relative humidity, and 101 kPa. The air is compressed to 400 kPa, then sent to an intercooler. If condensation of water vapor from the compressed air is to be prevented, what is the minimum temperature to which the air can be cooled in the intercooler? *Ans.* 40.3°C.

3-9 A winter air-conditioning system adds for humidification 0.0025 kg/s of saturated steam at 101 kPa pressure to an airflow of 0.36 kg/s. The air is initially at a temperature of 15°C with a relative humidity of 20 percent. What are the dry- and wet-bulb temperatures of the air leaving the humidifier? *Ans.* 16.0 and 13.8°C.

3-10 Determine for the three cases listed below the magnitude in watts and the direction of transfer of sensible heat [using Eq. (3-8)], latent heat [using Eq. (3-9)], and total heat [using Eq. (3-14)]. The area is 0.15 m^2 and $h_c = 30\ W/m^2 \cdot K$. Air at 30°C and 50 percent relative humidity is in contact with water that is at a temperature of (*a*) 13°C, (*b*) 20°C, and (*c*) 28°C. *Ans.* (*a*) –76.5, –42.3, –120.4 W; (*b*) –45.0, 15.1, –29.6 W; (*c*) –9.0, 116.5, 113.8 W.

REFERENCES

1. Carrier, W. H.: Rational Psychrometric Formulae, *ASME Trans.*, vol. 33, p. 1005, 1911.
2. Lewis, W. K.: The Evaporation of a Liquid into a Gas, *Trans. ASME*, vol. 44, p. 325, 1922.
3. Lewis, W. K.: The Evaporation of a Liquid into a Gas–A Correction, *Mech. Eng.*, vol. 55, p. 1567, September 1933.
4. Stoecker, W. F.: "Principles for Air Conditioning Practice," Industrial Press, Inc., New York, 1968.

CHAPTER

FOUR

HEATING- AND COOLING-LOAD CALCULATIONS

4-1 Introduction Buildings are built to provide a safe and comfortable internal environment despite variations in external conditions. The extent to which the desired interior conditions can be economically maintained is one important measure of the success of a building design. Although control of inside conditions is usually attributed to the active heating and cooling system, the design of heating, ventilating, and air conditioning (HVAC) must start with an examination of the thermal characteristics of the envelope. They influence both the equipment capacity and the energy required for its operation.

The primary intent of this chapter is to examine procedures for evaluating the impact of the thermal characteristics of the building envelope on the design of the HVAC systems used to maintain comfort. As the objective of the system is to provide comfort, however, it is advisable to begin with a brief discussion of the factors which influence comfort.

4-2 Health and comfort criteria The human body is an amazingly adaptable organism. With long-term conditioning the body can function under quite extreme thermal conditions. Variations in outdoor temperature and humidity, however, often go beyond the normal limits of adaptability, and it becomes necessary to provide modified conditions indoors in order to maintain a healthy, comfortable environment.

4-3 Thermal comfort Figure 4-1 illustrates the factors that influence thermal comfort. First, body heat is generated by metabolic processes to maintain body temperature. Metabolic processes are influenced by such factors as age, health, and level of

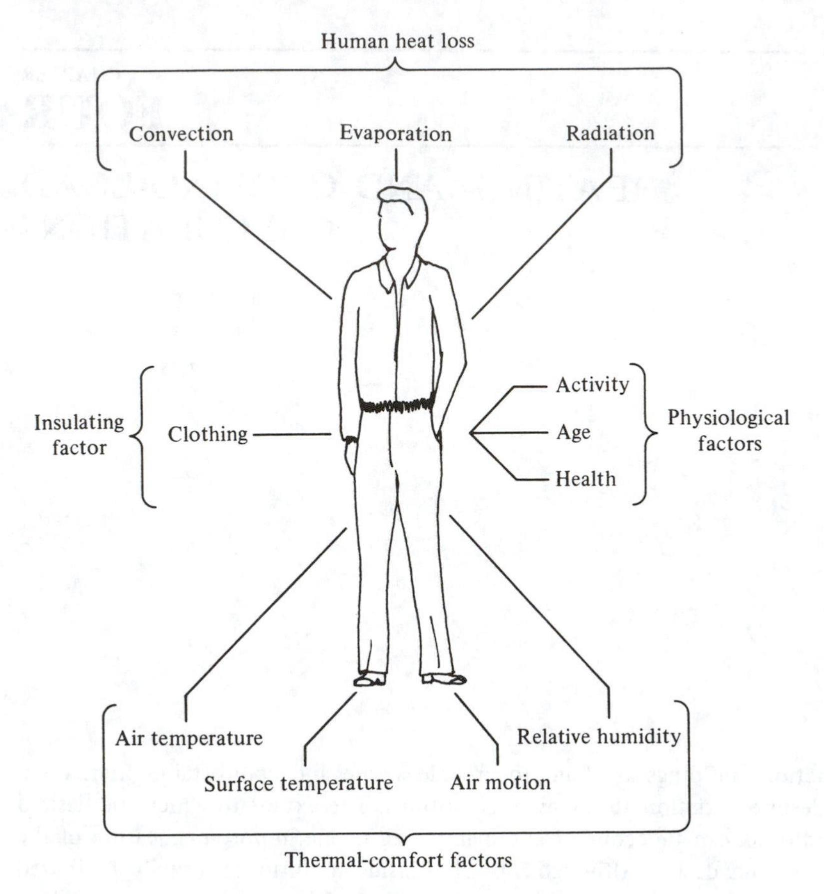

Figure 4-1 Factors influencing thermal comfort.

activity. For example, a given range of environmental conditions might be quite acceptable in a space occupied by a healthy person but unacceptable for one who is ill. When people are willing to adjust their dress habits with the changing seasons, they find that they are comfortable over a broader range of environmental conditions than they would expect.

The body is continuously generating heat, which must be dissipated to maintain a constant body temperature. The various mechanisms by which temperature control is accomplished were described in Sec. 2-19 and are shown in Fig. 4-1. For a person at rest or doing light work in a conditioned space, the body dissipates heat primarily by convection (carried away by the surrounding air) and radiation (to surrounding surfaces that are at a lower temperature than the body surface). Each of these components of heat dissipation accounts for approximately 30 percent of the heat loss. Evaporation, from both respiration and perspiration, accounts for the remaining 40 percent. As environmental conditions or levels of activity change, these percentages will vary. For example, if a person is doing strenuous work, the primary heat-dissipation mechanism will be evaporation.

Four environmental factors influence the body's ability to dissipate heat: air temperature, the temperature of the surrounding surfaces, humidity, and air velocity. The amount and type of clothing and the activity levels of the occupants interact with these factors. In designing an air-conditioning system we turn our attention to the control of these four factors. If a person is wearing appropriate clothing, the following ranges should usually be acceptable:[1]

Operative temperature. 20 to 26°C
Humidity. A dew-point temperature of 2 to 17°C
Average air velocity. Up to 0.25 m/s

The operative temperature is approximately the average of the air dry bulb temperature and the mean radiant temperature as long as the mean radiant temperature is less than 50°C and the average air velocity is less than 0.4 m/s. The mean radiant temperature is the uniform surface temperature of an imaginary black enclosure with which an occupant would have the same radiant energy exchange as in the actual nonuniform space. A person wearing heavy clothing may be comfortable at lower temperatures; conversely, lighter clothing and higher air velocity may provide comfort despite higher temperatures. The temperatures of surrounding surfaces have an influence on comfort as great as that of the air temperature and can not be neglected.

4-4 Air quality Air quality must also be maintained to provide a healthy, comfortable indoor environment. Sources of pollution exist in both the internal and external environment. Indoor air quality is controlled by removal of the contaminant or by dilution. Ventilation plays an important role in both processes. Ventilation is defined as supplying air by natural or mechanical means to a space. Normally, ventilation air is made up of outdoor air and recirculated air. The outdoor air is provided for dilution. In most cases odor and irritation of the upper respiratory tract or eyes are the reason for ventilation rather than the presence of health-threatening contaminants. The possibility of contaminants cannot be overlooked, however.

Reference 2 prescribes both necessary quantities of ventilation for various types of occupancies and methods of determining the proportions of outside air and recirculated air. If the level of contaminants in outdoor air exceeds that for minimum air-quality standards, extraordinary measures beyond the scope of this text must be used. For the present discussion it will be presumed that outdoor-air quality is satisfactory for dilution purposes. Table 4-1 presents outdoor-air requirements for ventilation for three occupancy types listed in the standard. As noted in the table, much larger quantities of air are required for dilution in areas where smoking is permitted.

Ventilation imposes a significant load on heating and cooling equipment and thus is a major contribution to energy use. Space occupancies and the choice of ventilation rates should be considered carefully. For example, if smoking is permitted in part of a building but restricted in another part of the building, ventilation rates for smoking should not be assumed uniformly. Also, the prospect of filtering and cleaning air for recirculation must be examined carefully. The use of recirculated air will conserve

Table 4-1 Outdoor-air requirements for ventilation

Function	Estimated occupancy per 100 m^2 floor area	Outdoor-air requirements per person, L/s	
		Smoking	Nonsmoking
Offices	7	10	2.5
Meeting and waiting spaces	60	17.5	3.5
Lobbies	30	7.5	2.5

energy whenever the outdoor-air temperature is extremely high or low. The ASHRAE Standard[2] provides the following procedure for determining the allowable rate for recirculation

$$\dot{V} = \dot{V}_r + \dot{V}_m$$

where $\dot{V}$ = rate of supply air for ventilation purposes, L/s
$\dot{V}_r$ = recirculation air rate, L/s
$\dot{V}_m$ = minimum outdoor-air rate for specified occupancy, for example the nonsmoking value from Table 4-1, but never less than 2.5 L/s per person

also

$$\dot{V}_r = \frac{\dot{V}_o - \dot{V}_m}{E}$$

where $\dot{V}_o$ = outdoor-air rate from Table 4-1 for specified occupancy (smoking or nonsmoking, as appropriate), L/s
E = efficiency of contaminant removal by air-cleaning device. The efficiency must be determined relative to the contaminant to be removed. Table 4-2 provides values appropriate for removal of 1-μm particles

Example 4-1 Determine the ventilation rate, outdoor-air rate, and recirculated-air rate for an office-building meeting room if smoking is permitted. An air-cleaning device with E = 60 percent for removal of tobacco smoke is available.

Table 4-2 ASHRAE dust spot efficiencies (1-μm particles)[3]

Filter type	Efficiency range, %	Application
Viscous impingement	5–25	Dust and lint removal
Dry media:		
Glass fiber, multi-ply cellulose, wool felt	25–40	Same as above and for some industrial applications
Mats of 3- to 10-μm fiber 6 to 20 mm thick	40–80	Building recirculated- and fresh-air systems
Mats of 0.5- to 4-μm fiber (usually glass)	80–98	Hospital surgeries, clean rooms, special applications
Electrostatic (depending on type)	20–90	Pollen and airborne particles

Solution Table 4-1 indicates that 17.5 L/s of outdoor air per person would be required to ventilate the space without any recirculation and air cleaning. The table also indicates that 3.5 L/s per person is the required outdoor-air rate for nonsmoking spaces and may be assumed, for purposes of this example, to be the minimum rate. There are two possible solutions for this design problem: (*a*) supply 17.5 L/s of outdoor air per person or (*b*) calculate the allowable recirculation rate and corresponding required ventilation rate as follows:

$$\dot{V}_r = \frac{\dot{V}_o - \dot{V}_m}{E} = \frac{17.5 - 3.5}{60/100} = 23.3 \text{ L/s}$$

Then $\dot{V} = 23.3 + 3.5 = 26.8$ L/s per person.

Although the total ventilation rate is higher for the second approach in Example 4-1, the energy requirements may be less due to the reduced outside-air flow rate.

If contamination such as tobacco smoke, body odor, moisture, or a high CO_2 content is the result of occupancy, ventilation is not required when the space is not occupied. If other sources of contamination exist, however, such as equipment or processes, outgassing from materials, or naturally occurring production of radon, an appropriate level of ventilation must be maintained even if the space is unoccupied.

Each of the factors influencing comfort must be kept in mind in the design of an air-conditioning system. These factors have an impact on system capacity, system control, and the design and placement of the duct system or terminal units. For example, placing heating units under a window or along an exterior wall may offset the effects of the lower temperature of those surfaces.

4-5 Estimating heat loss and heat gain Heat transfer through a building envelope is influenced by the materials used; by geometric factors such as size, shape, and orientation; by the existence of internal heat sources; and by climatic factors. System design requires each of these factors to be examined and the impact of their interactions to be carefully evaluated.

The primary function of heat-loss and heat-gain calculations is to estimate the capacity that will be required for the various heating and air-conditioning components necessary to maintain comfort within a space. These calculations are therefore based on peak-load conditions for heating and cooling and correspond to environmental conditions which are near the extremes normally encountered. Standard outside design values of temperature, humidity, and solar intensities are usually available from handbooks.

A number of load-calculation procedures have been developed over the years. Those developed by ASHRAE[4] will be used here. Although other procedures differ in some respects, they are all based on a systematic evaluation of the components of heat loss and heat gain. Loads are generally divided into the following four categories (Fig. 4-2):

Transmission. Heat loss or heat gain due to a temperature difference across a building element

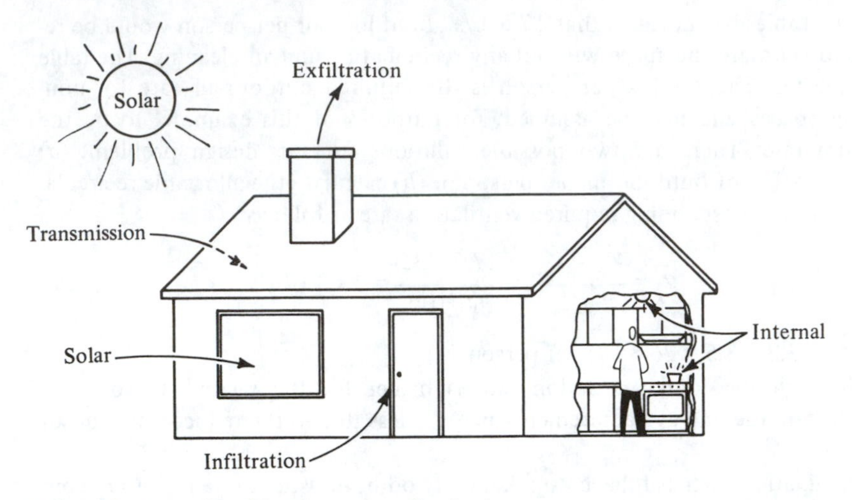

Figure 4-2 Categories of heating and cooling loads.

Solar. Heat gain due to transmission of solar energy through a transparent building component or absorption by an opaque building component

Infiltration. Heat loss or heat gain due to the infiltration of outside air into a conditioned space

Internal. Heat gain due to the release of energy within a space (lights, people, equipment, etc.)

In response to these loads the temperature in the space will change or the heating or cooling equipment will operate to maintain a desired temperature. In the following paragraphs we outline procedures for evaluating each of these load components. A more detailed presentation is available in Ref. 4.

4-6 Design conditions The design conditions usually specified for estimating heating loads are the inside and outside dry-bulb temperatures. For heating operation an indoor temperature of 20 to 22°C is generally assumed, and for cooling operation 24 to 26°C is typical. A minimum relative humidity of 30% in the winter and a maximum of 60% in the summer is also assumed. For heating operation the 97.5 percent value of the outside temperature is usually chosen. This means that on a long-term basis the outside dry-bulb temperature equals or exceeds this value for 97.5 percent of the hours during the coldest months of the year. At the 97.5 percent outdoor temperature the air is assumed to be saturated.

The set of conditions specified for cooling-load estimates is more complex and includes dry-bulb temperature, humidity, and solar intensity. Peak-load conditions during the cooling season usually correspond to the maximum solar conditions rather than to the peak outdoor-air temperature. Thus, it is often necessary to make several calculations at different times of the day or times of the year to fix the appropriate maximum-cooling-capacity requirements. When the cooling-load calculation is made will depend on the geographic location and on the orientation of the space being considered. For example, peak solar loading on an east-facing room may occur at 8 A.M., while for a west room the maximum load may occur at 4 P.M. Peak solar loads for south-facing rooms will occur during the winter rather than the summer. Of course, when a cooling system serves several spaces with different orientations, the peak system load may occur at a time other than the peak for any of the several spaces. Fortunately, after making a number of such calculations one begins to recognize likely choices for times when the peak load may occur.

Table 4-3 provides outdoor design temperature data for a number of locations. The table provides the 97.5 percent dry-bulb temperature for winter and the 2.5 percent dry-bulb and coincident wet-bulb temperature for summer. The 2.5 percent dry-

Table 4-3 Design temperature data

	Winter	Summer	
City	97.5% dry bulb, °C	2.5% dry bulb/ coincident wet bulb, °C	August daily average, °C
Albuquerque, N. Mex.	- 9	33/16	24
Atlanta, Ga.	- 6	33/23	26
Boise, Idaho	-12	34/18	22
Boston, Mass.	-13	31/22	22
Chicago, Ill.	-18	33/23	23
Columbus, Ohio	-15	32/23	23
Dallas, Tex.	- 6	36/24	29
Denver, Colo.	-17	33/15	22
El Paso, Tex.	- 5	37/18	27
Great Falls, Mont.	-26	31/16	19
Houston, Tex.	0	34/25	28
Las Vegas, Nev.	- 2	41/18	31
Los Angeles, Calif.	4	32/21	21
Memphis, Tenn.	- 8	35/24	27
Miami, Fla.	8	32/25	28
Minneapolis, Minn.	-24	37/23	22
New Orleans, La.	- 4	33/26	28
New York, N.Y.	- 9	32/23	24
Phoenix, Ariz.	1	42/22	32
Pittsburgh, Pa.	-14	31/22	22
Portland, Oreg.	- 4	30/20	20
Sacramento, Calif.	0	37/21	26
Salt Lake City, Utah	-13	35/17	24
San Francisco, Calif.	4	22/17	17
Seattle, Wash.	- 3	28/19	18
Spokane, Wash.	-17	32/17	20
St. Louis, Mo.	-13	34/24	25
Washington, D.C.	- 8	33/23	25

bulb temperature is the temperature exceeded by 2.5 percent of the hours during June to September. The mean coincident wet-bulb temperature is the mean wet-bulb temperature occurring at that 2.5 percent dry-bulb temperature. Tables 4-10 to 4-12, which provide additional data relative to the solar load, will be discussed when the solar load for windows and the thermal transmission for walls and roof are studied.

Example 4-2 Select outside and inside design temperatures for a building to be constructed in Denver, Colorado.

Solution From Table 4-3 the summer design conditions are given as

Summer design dry-bulb temperature = 33°C
Coincident wet-bulb temperature = 15°C

Assuming that no special requirements exist, an inside design temperature of 25°C and 60% relative humidity is chosen. The winter design outside temperature is -17°C from Table 4-3, and if no special requirements exist, an inside design temperature of 20°C and 30% relative humidity is chosen.

It should be noted that the inside design temperature only limits the conditions that can be maintained in extreme weather. During heating operation, when the outside temperature is above the outside design value, an inside temperature greater than 20°C can be maintained if desired.

4-7 Thermal transmission The general procedure for calculating heat loss or heat gain by thermal transmission is to apply Eq. (2-12)

$$q = \frac{\Delta t}{R^*_{tot}} = \frac{A\,\Delta t}{R_{tot}} = UA(t_o - t_i)$$

where $UA = 1/R^*_{tot}$, W/K
R^*_{tot} = total thermal resistance, K/W
U = overall heat-transfer coefficient W/m^2 · K
A = surface area, m^2
$t_o - t_i$ = outside-inside temperature difference, K

For heating-load estimates the temperature difference is simply the 97.5 percent outside value minus the inside design value.

As discussed in Chap. 2, the overall heat-transfer coefficient U is a function of the thermal resistances. Table 4-4 (p. 68) provides values of thermal resistance applicable to 1 m^2 of surface for commonly used building materials, enclosed air spaces, and building envelope boundaries. Example 4-3 illustrates determination of the U value of a typical wall cross section. The areas used in the transmission calculations are nominal inside areas of the spaces.

Example 4-3 Determine the total thermal resistance of a unit area of the wall section shown in Fig. 4-3.

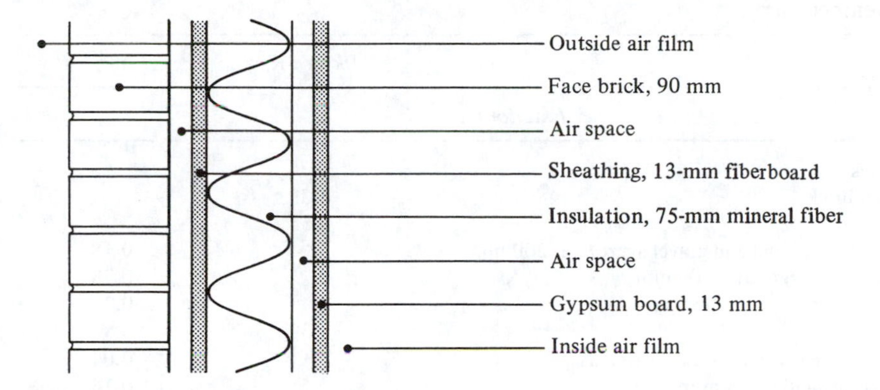

Figure 4-3 Wall section in Example 4-3.

If below-grade spaces are not conditioned, the heat loss through below-grade surfaces is often neglected. Heating loads are included in such cases based on an estimate of the temperature of these unconditioned spaces and transmission through the floor above them. If the below-grade spaces are to be conditioned, transmission heat losses are based on the wall and floor thermal resistance, the inside temperature to be maintained, and an estimate of the ground temperature adjacent to the surface.

For slab-on-grade construction the heat loss is more nearly proportional to the length of the perimeter of the slab (in meters) than its area. Thus

$$q_{\text{slab}} = F(\text{perimeter})\,(t_o - t_i) \qquad \text{where } F = \text{const}$$

Little information is available on which to base values of F for large slabs. Values for residential-scale slabs are given[4] as $F = 1.4$ W/m · K for an uninsulated edge and $F = 0.9$ W/m · K for a slab with 2.5 cm of insulation at the edge. These values must be viewed as approximate and are generally considered too high.

4-8 Infiltration and ventilation loads The entry of outside air into the space influences both the air temperature and the humidity level in the space. Usually a distinction is made between the two effects, referring to the temperature effect as *sensible load* and the humidity effect as *latent load.* This terminology applies to the other load components as well. For example, transmission and solar loads are sensible, as they affect only temperature, while internal loads arising from occupancy have both sensible and latent components. Heat loss or heat gain due to the entry of outside air is then expressed as

$$q_{is} = 1.23\dot{Q}(t_o - t_i) \qquad q_{il} = 3000\dot{Q}(W_o - W_i)$$

where $\dot{Q}$ = volumetric flow rate of outside air, L/s

W = humidity ratio, water to air, kg/kg

Infiltration, defined as the uncontrolled entry of unconditioned outside air directly into the building, results from natural forces, e.g., wind and buoyancy due to the temperature difference between inside and outside. For our purposes we define ventilation as air intentionally brought into the building by mechanical means. Of

Table 4-4 Thermal resistance of unit areas of selected building materials at 24°C mean temperature

	$1/k$, m · K/W	R, m^2 · K/W
Exterior material		
Face brick	0.76	
Common brick	1.39	
Stone	0.55	
Concrete block, sand and gravel aggregate, 200 mm		0.18
Lightweight aggregate, 200 mm		0.38
" " 150 mm		0.29
Stucco	1.39	
Siding, asbestos-cement, 6 mm, lapped		0.04
Asphalt insulating, 13 mm		0.14
Wood plywood, 10 mm		0.10
Aluminum or steel, backed with insulating board, 10 mm		0.32
Sheathing		
Asbestos-cement	1.73	
Plywood	8.66	
Fiberboard, regular density, 13 mm		0.23
Hardboard, medium density	9.49	
Particle board, medium density	7.35	
Roofing		
Asphalt shingles		0.08
Built-up roofing, 10 mm		0.06
Concrete		
Sand and gravel aggregate	0.55	
Lightweight aggregate	1.94	
Insulating materials		
Blanket and batt, mineral fiber, 75-90 mm		1.94
135-165 mm		3.35
Board and slab, glass fiber, organic bond	27.7	
Expanded polystyrene, extruded	27.7	
Cellular polyurethane	43.8	
Loose fill, mineral fiber, 160 mm		3.35
Cellulosic	21.7-25.6	
Interior materials		
Gypsum or plaster board, 15 mm		0.08
16 mm		0.10
Plaster materials, cement plaster	1.39	
Gypsum plaster, lightweight, 16 mm		0.066
Wood, soft (fir, pine, etc.)	8.66	
Hardwood (maple, oak, etc.)	6.31	

Table 4-4 Thermal resistance of unit areas of selected building materials at 24°C mean temperature (cont.)

	$1/k$, m · K/W	R, m² · K/W
Air resistance		
Surface, still air (surface emissivity of 0.9) horizontal, heat flow up		0.11
Horizontal heat flow down		0.16
Vertical, heat flow horizontal		0.12
Surface, moving air, heating season, 6.7 m/s		0.029
" " " cooling season, 3.4 m/s		0.044
Air space, surface emissivity of 0.8, horizontal		0.14
Vertical		0.17
Surface emissivity of 0.2, horizontal		0.24
Vertical		0.36

Flat glass

	U, W/m² · K†	
	Summer	Winter
Single glass	5.9	6.2
Double glass, 6-mm air space	3.5	3.3
13-mm air space	3.2	2.8
Triple glass, 6-mm air spaces	2.5	2.2
13-mm air spaces	2.2	1.8
Storm windows, 25 to 100-mm air space	2.8	2.3

† Includes inside and outside air film resistance

Solution The following resistances are obtained from Table 4-4:

Outside air film	0.029 m² · K/W
Face brick, 90 mm	0.068
Air space	0.170
Sheathing, 13-mm fiberboard	0.232
Insulation, 75-mm mineral fiber	1.940
Air space	0.170
Gypsum board, 13 mm	0.080
Inside air film	0.120
R_{tot}	2.809 m² · K/W

course, the air entering must also leave by natural means, i.e., exfiltration, or be exhausted by mechanical means.

In commercial and institutional buildings it is considered advisable to control the entry of outside air to assure proper ventilation and minimize energy use. As infiltration is uncontrolled, these buildings are designed and constructed to limit it as much as possible. This is done by sealing the building envelope where possible, using vestibules or revolving doors, or maintaining a pressure within the building slightly in excess of

Table 4-5 Infiltration constants for infiltration in Eq. (4-1)

Quality of construction	*a*	*b*	*c*
Tight	0.15	0.010	0.007
Average	0.20	0.015	0.014
Loose	0.25	0.020	0.022

that outside. However, if a building does not have mechanical ventilation, or if the fans in the system are not operating, infiltration will occur. The volumetric flow rate of infiltration air is rather difficult to determine with any measure of precision. It will vary with the quality of construction, wind speed and direction, indoor-outdoor temperature difference, and internal pressure in the building. One procedure that is often used in load calculations is to estimate the infiltration in terms of the number of air changes per hour. One air change per hour would be a volumetric flow rate numerically equal to the internal volume of the space. The number of air changes per hour for a smaller building with no internal pressurization can be estimated as a function[5] of wind velocity and temperature difference

$$\text{Number of air changes} = a + bV + c(t_o - t_i) \tag{4-1}$$

where a, b, c = experimentally determined constants
V = wind velocity, m/s

Typical values of a, b, and c are presented in Table 4-5. For nonresidential buildings it is customary to use estimates of infiltration for load calculations only when the fans in the ventilation system are not operating. An example would be for sizing a heating system to maintain a minimum temperature during an unoccupied nighttime period.

The volumetric flow rate of outside air for ventilation is computed from Table 4-1 and the methods outlined in Section 4-4. The slightly positive pressure in the building is maintained by sizing the exhaust fans to handle less air than brought in from the outside by the ventilation system. Also, exhaust fans are generally located in restrooms, mechanical rooms, or kitchens to ensure that air and odors from these spaces will not be recirculated throughout the building. Chapters 5 and 6 will provide more information on the design of the air-distribution system. It should be noted at this point, however, that although the outdoor component of ventilation imposes a load on heating and cooling equipment, the load occurs at the point where air is conditioned rather than in the space. It is therefore necessary to distinguish between equipment loads and the space loads used to determine the airflow required for the building spaces.

4-9 Summary of procedure for estimating heating loads In estimating the heating loads for a building, it is important to use an organized, step-by-step procedure. The necessary steps can be outlined as follows:

1. Select design values for outdoor winter design (97.5 percent value) from Table 4-3.

2. Select an indoor design temperature appropriate to the activities to be carried out in the space and a minimum acceptable relative humidity.
3. Determine whether any special conditions will exist, such as adjacent unconditioned spaces. Estimate temperatures in the unconditioned spaces as necessary.
4. On the basis of building plans and specifications, calculate heat-transfer coefficients and areas for the building components in each enclosing surface. Any surfaces connecting with spaces to be maintained at the same temperature may be omitted, i.e., interior walls.
5. On the basis of building components, system design and operation, wind velocity, and indoor-outdoor temperature difference, estimate the rate of infiltration and/or ventilation outside air. Note that the latent component of the infiltration and/or ventilation load is included only if the conditioned air is to be humidified to maintain a specified minimum indoor humidity level. Humidification is often omitted from air-conditioning systems in mild climates.
6. Using the above design data, compute transmission heat losses for each surface of the building envelope and the heat loss from infiltration and/or ventilation. Sum these values to determine the total estimated heat loss and the required capacity of the heating equipment.
7. Consider any special circumstances that might influence equipment sizing. Three circumstances may influence equipment capacity.
   a. If a building and its heating system are designed to take advantage of passive solar gain and thermal storage, heating capacities should be based on dynamic rather than static heat-loss analysis (see Chap. 20).
   b. In a building that has an appreciable steady internal load (heat release) at the time of the maximum transmission and ventilation heat loss, heating-equipment capacity may be reduced by the amount of the internal heat release. An example would be a hospital or industrial building which operates on a 24-h basis.
   c. A building that does not operate on a continuous basis and indoor temperatures are allowed to drop over a lengthy unoccupied period, additional capacity may be required to bring the air temperature and building indoor surface temperatures back to an acceptable level in a short time. An alternative to the additional capacity is to bring the heating system into operation earlier and allow the building to heat more gradually.

4-10 Components of the cooling load Estimating the cooling load is more complex than estimating the heating load. Additional consideration must be given to internal loads, latent loads, and solar loads.

4-11 Internal loads The primary sources of internal heat gain are lights, occupants, and equipment operating within the space. Internal loads are a major factor in most nonresidential buildings. The amount of heat gain in the space due to lighting depends on the wattage of the lamps and the type of fixture. When fluorescent lighting is used, the energy dissipated by the ballast must also be included in the internal load. As lighting is often the largest single component of the internal load, care must be exercised in its evaluation. The portion of the heat emanating from lighting which is in the form of radiant energy is not an instantaneous load on the air-conditioning system. The radiant energy from the lights is first absorbed by the walls, floor, and furnishings of the space, and their temperatures then increase at a rate dependent on their mass. As the surface temperature of these objects rises above the air temperature, heat is convected from the surfaces and finally becomes a load on the cooling system. Thus because of the mass of the objects absorbing the radiation there is a delay between turning the light on and the energy from the lights having an effect on the load. The cooling load from the lighting persists after the lights are turned off for the same reason. To accommodate these circumstances the following format has been developed for estimating the internal heat gain from lights:[4]

$$q = (\text{lamp rating in watts})\,(F_u)\,(F_b)\,(\text{CLF})$$

where F_u = utilization factor or fraction of installed lamps in use
F_b = ballast factor for fluorescent lamps = 1.2 for most common fluorescent fixtures
CLF = cooling-load factor from Table 4-6

Table 4-6 provides cooling-load factors for two common fixture arrangements and

Table 4-6 Cooling-load factors for lighting[4]

No. of hours after lights are turned on	Fixture X†, hours of operation		Fixture Y†, hours of operation	
	10	16	10	16
0	0.08	0.19	0.01	0.05
1	0.62	0.72	0.76	0.79
2	0.66	0.75	0.81	0.83
3	0.69	0.77	0.84	0.87
4	0.73	0.80	0.88	0.89
5	0.75	0.82	0.90	0.91
6	0.78	0.84	0.92	0.93
7	0.80	0.85	0.93	0.94
8	0.82	0.87	0.95	0.95
9	0.84	0.88	0.96	0.96
10	0.85	0.89	0.97	0.97
11	0.32	0.90	0.22	0.98
12	0.29	0.91	0.18	0.98
13	0.26	0.92	0.14	0.98
14	0.23	0.93	0.12	0.99
15	0.21	0.94	0.09	0.99
16	0.19	0.94	0.08	0.99
17	0.17	0.40	0.06	0.24
18	0.15	0.36	0.05	0.20

† Fixture description: X, recessed lights which are not vented. The supply and return air registers are below the ceiling or through the ceiling space and grille. Y, vented or free-hanging lights. The supply air registers are below or through the ceiling with the return air registers around the fixtures and through the ceiling space.

Table 4-7 Heat gain from occupants

Activity	Heat gain, W	Sensible heat gain %
Sleeping	70	75
Seated, quiet	100	60
Standing	150	50
Walking, 3 km/h	305	35
Office work	150	55
Teaching	175	50
Retail shop	185	50
Industrial	300–600	35

lights operating 10 and 16 h/d. Additional information covering variations in fixtures, floor mass, and operating periods is available.[4,6]

For heat-producing equipment it is also necessary to estimate the power used along with the period and/or frequency of use in a manner similar to that used for lighting. For equipment having little radiant-energy transmission the CLF can be assumed equal to 1.0.

Table 4-7 shows loads from occupants as a function of their activity. The greatest uncertainty in estimating this load component is the number of occupants. If the number of occupants is unknown, values such as those in Table 4-8 may be used. Since a portion of the heat transferred from occupants is by radiation, the ASHRAE methodology again uses the cooling-load factor for a better representation of actual loads. Table 4-9 gives these values. Thus,

Occupant sensible cooling load in watts
= gain per person from Table 4-7 × number of people × CLF from Table 4-9

For the latent load the CLF is 1.0.

Although there are a number of uncertainties in estimating internal loads, these loads are significant and must be evaluated as carefully as possible.

4-12 Solar loads through transparent surfaces Heat gain due to solar energy incident on a surface will depend upon the physical characteristics of the surface. Surface op-

Table 4-8 Space per occupant[2]

Type of space	Occupancy
Residence	2–6 occupants
Office	10–15 m^2 per occupant
Retail	3–5 m^2 per occupant
School	2.5 m^2 per occupant
Auditorium	1.0 m^2 per occupant

Table 4-9 Sensible-heat cooling-load factors for people[4]

Hours after each entry into space	Total hours in space							
	2	4	6	8	10	12	14	16
1	0.49	0.49	0.50	0.51	0.53	0.55	0.58	0.62
2	0.58	0.59	0.60	0.61	0.62	0.64	0.66	0.70
3	0.17	0.66	0.67	0.67	0.69	0.70	0.72	0.75
4	0.13	0.71	0.72	0.72	0.74	0.75	0.77	0.79
5	0.10	0.27	0.76	0.76	0.77	0.79	0.80	0.82
6	0.08	0.21	0.79	0.80	0.80	0.81	0.83	0.85
7	0.07	0.16	0.34	0.82	0.83	0.84	0.85	0.87
8	0.06	0.14	0.26	0.84	0.85	0.86	0.87	0.88
9	0.05	0.11	0.21	0.38	0.87	0.88	0.89	0.90
10	0.04	0.10	0.18	0.30	0.89	0.89	0.90	0.91
11	0.04	0.08	0.15	0.25	0.42	0.91	0.91	0.92
12	0.03	0.07	0.13	0.21	0.34	0.92	0.92	0.93
13	0.03	0.06	0.11	0.18	0.28	0.45	0.93	0.94
14	0.02	0.06	0.10	0.15	0.23	0.36	0.94	0.95
15	0.02	0.05	0.08	0.13	0.20	0.30	0.47	0.95
16	0.02	0.04	0.07	0.12	0.17	0.25	0.38	0.96
17	0.02	0.04	0.06	0.10	0.15	0.21	0.31	0.49
18	0.01	0.03	0.06	0.09	0.13	0.19	0.26	0.39

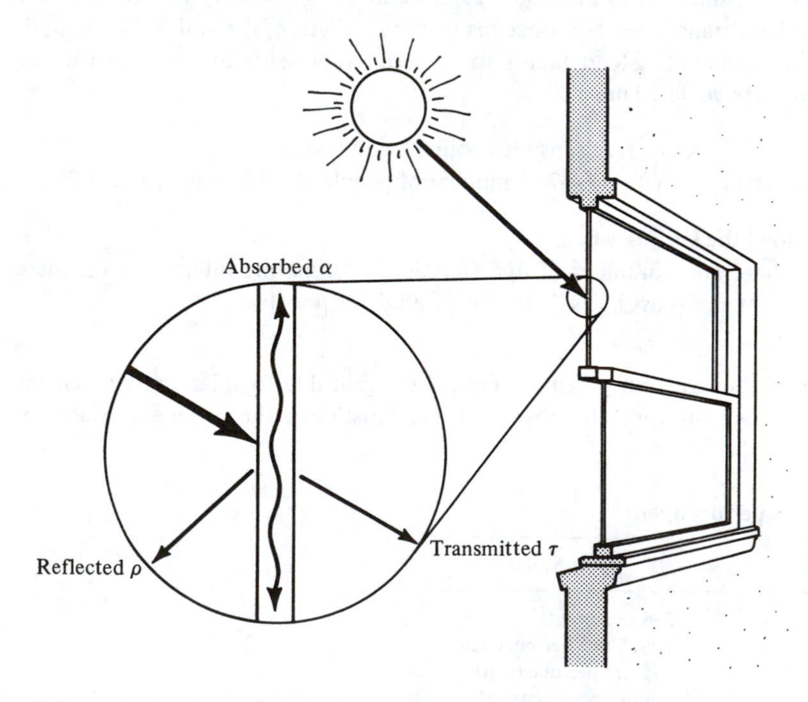

Figure 4-4 Distribution of solar heat striking a transparent surface.

tical properties are described by

$$\tau + \rho + \alpha = 1$$

where τ = transmittance
ρ = reflectance
α = absorptance

The value of each of these terms has a pronounced effect on solar-heat gain.

For transparent surfaces, such as the window shown in Fig. 4-4, the solar energy passing through the surface q_{sg} in watts is

$$q_{sg} = A(\tau I_t + N\alpha I_t) = AI_t(\tau + N\alpha) \tag{4-2}$$

where I_t = irradiation on exterior surface, W/m^2
N = fraction of absorbed radiation transferred by conduction and convection to inside environment
h_o = outside heat-transfer coefficient, W/m^2 · K

Under steady-state conditions N can be shown to be U/h_o. Restating the equation in terms of U and h_o gives

$$q_{sg} = AI_t\left(\tau + \frac{U\alpha}{h_o}\right)$$

The expression $I_t(\tau + U\alpha/h_o)$ for a single sheet of clear window glass is frequently referred to as the solar-heat gain factor (SHGF). Maximum values[4] for the SHGF are given for two latitudes by month and orientation in Table 4-10.

Table 4-10 Maximum solar-heat gain factor for sunlit glass[4] W/m^2

	N/shade	NE/NW	E/W	SE/SW	S	Hor.
			32° north latitude			
Dec	69	69	510	775	795	500
Jan, Nov	75	90	550	785	775	555
Feb, Oct	85	205	645	780	700	685
Mar, Sept	100	330	695	700	545	780
Apr, Aug	115	450	700	580	355	845
May, July	120	530	685	480	230	865
June	140	555	675	440	190	870
			40° north latitude			
Dec	57	57	475	730	800	355
Jan, Nov	63	63	480	755	795	420
Feb, Oct	80	155	575	760	750	565
Mar, Sept	95	285	660	730	640	690
Apr, Aug	110	435	690	630	475	790
May, July	120	515	690	545	350	830
June	150	540	680	510	300	840

Table 4-11 Shading coefficients[4]

		Shading Coefficient				
			Venetian blinds		Roller shades	
Type of glass	Thickness, mm	No indoor shading	Medium	Light	Dark	Light
Single glass						
Regular sheet	3	1.00	0.64	0.55	0.59	0.25
Plate	6–12	0.95	0.64	0.55	0.59	0.25
Heat-absorbing	6	0.70	0.57	0.53	0.40	0.30
	10	0.50	0.54	0.52	0.40	0.28
Double glass						
Regular sheet	3	0.90	0.57	0.51	0.60	0.25
Plate	6	0.83	0.57	0.51	0.60	0.25
Reflective	6	0.2–0.4	0.2–0.33			

A shading coefficient (SC) is used to adjust these SHGF values for other types of glass or to account for inside shading devices. This coefficient is

$$\mathrm{SC} = \frac{\tau + U\alpha/h_o}{(\tau + U\alpha/h_o)_{ss}}$$

where the subscript ss stands for a single sheet of clear glass. Typical values of the shading coefficient for several types of glass with and without internal shading are presented in Table 4-11. If external surfaces shade the window, SHGF values for a north orientation are used for the shaded portion of the window.

The solar energy passing through a window q_{sg} can be expressed as

$$q_{sg} = (\mathrm{SHGF}_{\max})(\mathrm{SC})A$$

One more factor must be considered since the solar energy entering the space does not appear instantaneously as a load on the cooling system. The radiant energy is first absorbed by the surfaces in the space, during which time these surface temperatures increase at a rate dependent on their dynamic thermal characteristics. Thus, the solar energy absorbed is delayed before being transferred to the air in the space by convection. Since this process may involve a significant time lag, it is also usual practice to include a *cooling-load factor* (CLF) in calculating the cooling load attributable to radiation through glass. Values of CLF derived from an extensive computer analysis[6] are presented in Table 4-12.

When estimating solar-heat gain through transparent surfaces, external shading must be considered. Shading from overhangs or other projections, such as shown in Fig. 4-5, can significantly reduce solar-heat gain through a window. The depth of a shadow cast by a horizontal projection above a window can be calculated using the solar altitude angle β and the wall-azimuth angle γ, where β is the angle measured from a horizontal plane on earth up to the sun and γ is the angle between two vertical

Table 4-12 Cooling-load factors for glass with interior shading, north latitudes[4]

Solar Time h	Window facing								
	N	NE	E	SE	S	SW	W	NW	Hor.
6	0.73	0.56	0.47	0.30	0.09	0.07	0.06	0.07	0.12
7	0.66	0.76	0.72	0.57	0.16	0.11	0.09	0.11	0.27
8	0.65	0.74	0.80	0.74	0.23	0.14	0.11	0.14	0.44
9	0.73	0.58	0.76	0.81	0.38	0.16	0.13	0.17	0.59
10	0.80	0.37	0.62	0.79	0.58	0.19	0.15	0.19	0.72
11	0.86	0.29	0.41	0.68	0.75	0.22	0.16	0.20	0.81
12	0.89	0.27	0.27	0.49	0.83	0.38	0.17	0.21	0.85
13	0.89	0.26	0.24	0.33	0.80	0.59	0.31	0.22	0.85
14	0.86	0.24	0.22	0.28	0.68	0.75	0.53	0.30	0.81
15	0.82	0.22	0.20	0.25	0.50	0.83	0.72	0.52	0.71
16	0.75	0.20	0.17	0.22	0.35	0.81	0.82	0.73	0.58
17	0.78	0.16	0.14	0.18	0.27	0.69	0.81	0.82	0.42
18	0.91	0.12	0.11	0.13	0.19	0.45	0.61	0.69	0.25

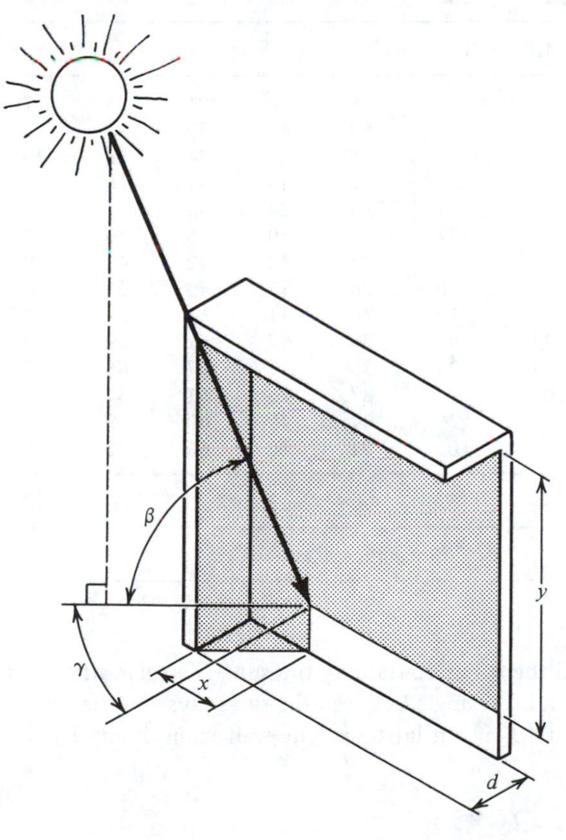

Figure 4-5 Shading angles and dimensions.

Table 4-13 Solar position angles for the twenty-first day of month[4]

		Solar time, A.M.							
Month	Angle	5	6	7	8	9	10	11	12
				32° north latitude					
Dec	β				10	20	28	33	35
	ϕ				54	44	31	16	0
Jan, Nov	β			1	13	22	31	36	38
	ϕ			65	56	46	33	18	0
Feb, Oct	β			7	18	29	38	45	47
	ϕ			73	64	53	39	21	0
Mar, Sep	β			13	25	37	47	55	58
	ϕ			82	73	62	47	27	0
Apr, Aug	β		6	19	31	44	56	65	70
	ϕ		100	92	84	74	60	37	0
May, Jul	β		10	23	35	48	61	72	78
	ϕ		107	100	93	85	73	52	0
Jun	β	1	12	24	37	50	62	74	81
	ϕ	118	110	103	97	89	80	61	0
				40° north latitude					
Dec	β				5	14	21	25	27
	ϕ				53	42	29	15	0
Jan, Nov	β				8	17	24	28	30
	ϕ				55	44	31	16	0
Feb, Oct	β			4	15	24	32	37	39
	ϕ			72	62	50	35	19	0
Mar, Sep	β			11	23	33	42	48	50
	ϕ			80	70	57	42	23	0
Apr, Aug	β		7	19	30	41	51	59	62
	ϕ		99	89	79	67	51	29	0
May, Jul	β	2	13	24	35	47	57	66	70
	ϕ	115	106	97	87	76	61	37	0
Jun	β	4	15	26	37	49	60	69	73
	ϕ	117	108	100	91	80	66	42	0
		7	6	5	4	3	2	1	12
		Solar time, P.M.							

planes, one normal to the wall and the other containing the sun. The solar altitude β and the solar azimuth angle ϕ, which is the angle between the sun's rays and the south, are given in Table 4-13 for 32° and 40° north latitude. The wall-azimuth angle γ can be computed from the equation

$$\gamma = \phi \pm \psi$$

where ψ is the angle a vertical plane normal to the wall makes with the south. The depth of a shadow y below a horizontal projection of width d is given by

$$y = d \frac{\tan \beta}{\cos \gamma}$$

The width of a shadow cast by a vertical projection of depth d is

$$x = d \tan \gamma$$

The SHGF and CLF for a north-facing window (in the northern hemisphere) are used for that portion of the window surface which is shaded.

Example 4-4 A 1.25-m high by 2.5-m wide window is inset from the face of the wall 0.15 m. Calculate the shading provided by the inset at 2 P.M. sun time if the window is facing south at 32° north latitude, August 21.

Solution For a south-facing window, $\psi = 0$ and $\gamma = \phi$. From Table 4-13, $\beta = 56°$ and $\gamma = 60°$; then

$$x = d \tan \gamma = 0.15 \tan 60° = 0.26 \text{ m}$$

$$y = d \frac{\tan \beta}{\cos \gamma} = \frac{0.15 \tan 56°}{\cos 60°} = 0.44 \text{ m}$$

$$\text{Sunlit area} = (2.5 - 0.26)(1.25 - 0.44) = 1.81 \text{ m}^2$$

4-13 Solar loads on opaque surfaces The process of solar-heat gain for an opaque wall is illustrated schematically in Fig. 4-6. A portion of the solar energy is reflected and

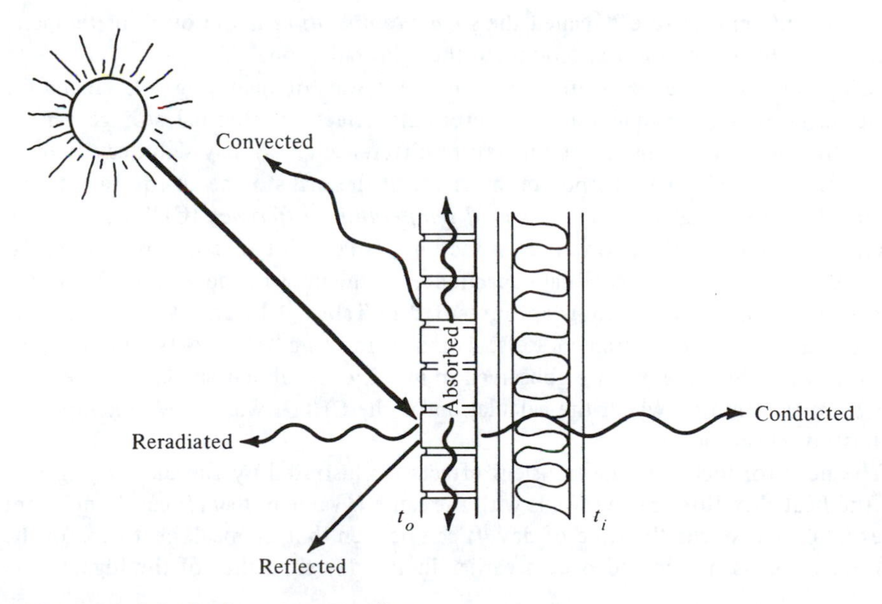

Figure 4-6 Solar loads on opaque surfaces.

the remainder absorbed. Of the energy absorbed some is convected and some reradiated to the outside. The remainder of the absorbed solar energy is transmitted to the inside by conduction or temporarily stored.

The transmissivity τ of an opaque surface is zero, and thus for walls and roofs

$$\rho + \alpha = 1$$

Equation (4-2) therefore reduces to

$$q_w = \frac{U_w \alpha}{h_o} I_t A$$

If the transmission due to the air-temperature difference is included,

$$q_w = \frac{U_w \alpha}{h_o} I_t A + U_w A(t_o - t_i) \tag{4-3}$$

Equation (4-3) can then be rearranged to give

$$q_w = U_w A \left[\left(t_o + \frac{\alpha I_t}{h_o} \right) - t_i \right] \tag{4-4}$$

From Eq. (4-4) it is apparent that if the first term in the brackets is replaced by an equivalent temperature t_e, where

$$t_e = t_o + \alpha I_t / h_o$$

Eq. (4-4) can be rewritten as

$$q_w = U_w A(t_e - t_i)$$

The equivalent temperature t_e called the *sol-air temperature* is the outdoor temperature increased by an amount to account for the solar radiation.

Using the sol-air temperature is a convenient way of including solar loads for opaque surfaces. For opaque walls, however, the effects of thermal storage can be quite pronounced, and using the temperature difference $t_e - t_i$ may significantly overestimate the heat gain. To incorporate the effect of thermal storage an equivalent temperature difference, called the *cooling-load temperature difference* (CLTD), has been developed[4] for commonly used wall cross sections. It takes into account both the solar flux on the surface and the thermal capacitance of the mass of the wall. CLTD values for several wall and roof sections are presented in Tables 4-14 and 4-15. More complete tabulations will be found in Refs. 4 and 6. In using Tables 4-14 and 4-15 the notes following the tables must be observed to modify the tabular values for cases that differ from the ones on which the calculation of the CLTDs was based. Example 4-5 will illustrate this process.

The need for including the transient effects is illustrated by the data in Fig. 4-7. Here the heat flux through two walls with the same U value but significantly different masses is plotted versus the time of day. It can be seen that the peak heat flux for the lower-mass wall is higher and occurs earlier in the day than that of the higher-mass wall.

When the thermal storage is included, the heat gain through the wall is given by

$$q_w = UA(\text{CLTD}) \tag{4-5}$$

Example 4-5 Determine the peak heat gain through a west-facing brick veneer wall (similar in cross section to that shown in Example 4-3), July 21 at 40° north latitude. The inside temperature is 25°C, and the average daily temperature is 30°C.

Table 4-14 Cooling-load temperature difference for flat roofs,[4] K

Roof type†	Mass per unit area, kg/m²	Heat capacity, kJ/m² · K	Solar time 7	8	9	10	11	12	13	14	15	16	17	18	19	20
			Roofs without suspended ceilings													
1	35	45	3	11	19	27	34	40	43	44	43	39	33	25	17	10
2	40	75	-1	2	8	15	22	29	35	39	41	41	39	34	29	21
3	90	90	-2	1	5	11	18	25	31	36	39	40	40	37	32	25
4	150	120	1	0	2	4	8	13	18	24	29	33	35	36	35	32
5	250	230	4	4	6	8	11	15	18	22	25	28	29	30	29	27
6	365	330	9	8	7	8	8	10	12	15	18	20	22	24	25	26
			Roofs with suspended ceilings													
1	45	50	0	5	13	20	28	35	40	43	43	41	37	31	23	15
2	50	85	1	2	4	7	12	17	22	27	31	33	35	34	32	28
3	100	100	0	0	2	6	10	16	21	27	31	34	36	36	34	30
4	150	130	6	4	4	4	6	9	12	16	20	24	27	29	30	30
5	260	240	12	11	11	11	12	13	15	16	18	19	20	21	21	21
6	360	340	13	13	13	12	12	13	13	14	15	16	16	17	18	18

Notes: 1. Directly applicable for the following conditions: inside temperature = 25°C; outside temperature, maximum = 35°C, average = 29°C; daily range = 12°C; and solar radiation typical of July 21 at 40° north latitude. Exact values for the *U* value appropriate for the design conditions being considered should be used for calculations.

2. Adjustments to these CLTD values should be made as follows if the indoor or outdoor design conditions differ from those specified in note 1:

$$\text{CLTD}_{adj} = \text{CLTD} + (25 - t_i) + (t_{av} - 29)$$

where t_i = inside design dry-bulb temperature, °C

t_{av} = average outdoor dry-bulb temperature for design day, °C

3. For roof constructions not listed choose the roof from the table which is of approximately the same density and heat capacity.

4. When the roof has additional insulation, for each $R = 1.2\ \text{m}^2 \cdot \text{K/W}$ in additional insulation use the CLTD for the next thermally heavier roof. For example, for a type 3 roof with an additional $R = 1.2$ insulation, use the type 4 roof CLTD.

† 1 = Sheet steel with 25 to 50 mm insulation, 2 = 25 mm wood with 25 mm insulation, 3 = 100 mm lightweight concrete, 4 = 150 mm lightweight concrete, 5 = 100 mm heavyweight concrete, 6 = roof terrace system.

Table 4-15 Cooling-load temperature difference for sunlit walls[4]

Wall type†	Mass per unit area kg/m²	Heat capacity, kJ/m²·K	Solar time	N	NE	E	SE	S	SW	W	NW
			7	4	15	17	10	1	1	1	1
			8	5	20	26	18	3	3	3	3
			9	5	22	30	24	7	4	5	4
			10	7	20	31	27	12	6	6	6
			11	8	16	28	28	17	9	8	8
			12	10	15	22	27	22	14	10	10
			13	12	14	19	23	25	21	15	12
G	50	15	14	13	15	17	20	26	28	23	15
			15	13	15	17	18	24	33	31	20
			16	14	14	16	16	21	35	37	26
			17	14	14	15	15	17	34	40	31
			18	15	12	13	13	14	29	37	31
			19	12	10	11	11	11	20	27	23
			20	8	8	8	8	8	13	16	14
			$CLTD_{max}$	15	22	31	28	26	35	40	31
			7	1	3	4	2	1	1	2	1
			8	2	8	9	6	1	1	2	1
			9	3	13	16	10	2	2	2	2
			10	4	16	21	15	4	3	3	3
			11	5	17	24	20	7	4	4	4
			12	6	16	25	23	11	6	6	6
			13	8	16	24	24	15	10	8	7
F	200	130	14	9	15	22	23	19	14	11	9
			15	11	15	20	22	21	20	16	12
			16	12	15	19	20	22	24	22	15
			17	12	15	18	19	21	28	27	19
			18	13	14	17	17	19	30	32	24
			19	13	13	15	16	17	29	33	26
			20	13	12	13	14	15	25	30	24
			$CLTD_{max}$	13	17	25	24	22	30	33	26
			7	2	3	3	3	2	4	4	3
			8	2	5	6	4	2	3	3	3
			9	3	8	10	7	2	3	3	3
			10	3	11	15	10	3	3	4	3
			11	4	13	18	14	5	4	4	4
			12	5	14	20	17	7	5	5	5
			13	6	14	21	19	10	7	6	6
E	300	230	14	7	14	21	20	14	10	8	7
			15	8	14	20	20	16	14	11	9
			16	10	15	19	20	18	18	15	11
			17	10	14	18	19	19	21	20	14
			18	11	14	18	18	18	24	24	18
			19	12	14	17	17	17	25	27	21
			20	12	13	15	16	16	24	27	21
			$CLTD_{max}$	12	15	21	20	19	25	27	21

Table 4-15 Cooling-load temperature difference for sunlit walls (cont.)

Wall type†	Mass per unit area, kg/m^2	Heat capacity, $kJ/m^2 \cdot K$	Solar time	Orientation N	NE	E	SE	S	SW	W	NW
			7	3	4	5	5	4	6	7	6
			8	3	4	5	5	4	5	6	5
			9	3	6	7	5	3	5	5	4
			10	3	8	10	7	3	4	5	4
			11	4	10	13	10	4	4	5	4
			12	4	11	15	12	5	5	5	4
			13	5	12	17	14	7	6	6	5
D	390	350	14	6	13	18	16	9	7	6	6
			15	6	13	18	17	11	9	8	7
			16	7	13	18	18	13	12	10	8
			17	8	14	18	18	15	15	13	10
			18	9	14	18	18	16	18	17	12
			19	10	14	17	17	16	20	20	15
			20	11	13	17	17	16	21	22	17
			$CLTD_{max}$	11	14	18	18	16	21	23	18
			7	5	6	7	7	6	9	10	8
			8	4	6	7	6	6	8	9	7
			9	4	6	8	7	5	7	8	6
			10	4	7	9	7	5	7	7	6
			11	4	8	11	9	5	6	7	5
			12	4	10	13	10	5	6	7	5
			13	5	10	14	12	6	6	7	6
C	530	450	14	5	11	15	13	8	7	7	6
			15	6	12	16	14	9	8	8	6
			16	6	12	16	15	11	10	9	7
			17	7	12	17	16	12	12	11	9
			18	8	13	17	16	13	14	13	10
			19	9	13	16	16	14	16	16	12
			20	9	13	16	16	14	18	18	14
			$CLTD_{max}$	9	13	17	16	14	18	20	15

Notes: 1. Reference 4 also shows CLTD values for heavier walls such as 300 mm concrete with interior and exterior finish; also 100 mm face brick with 50 mm insulation and 200 mm concrete.

2. This table is directly applicable for the conditions stated under Note 1 of Table 4-14.

3. The procedure for correcting for indoor and outdoor temperatures differing from standard is given in Note 2 of Table 4-14.

4. Wall constructions not listed can be approximated by using the wall with the nearest values of density and heat capacity.

5. For walls with additional insulation shift to the wall with next higher mass i.e., wall type designated by the preceding letter of the alphabet, for each additional R of 1.2 $m^2 \cdot K/W$. For example, for the addition of $R = 1.2\ m^2 \cdot K/W$ to wall type E, use CLTD values for wall type D.

† G = metal curtain or frame wall with 25 to 75 mm insulation. F = 100-mm concrete block with 25 to 50 mm insulation; or 100 mm face brick with insulation. E = 200-mm concrete block with interior and exterior finish; or 100 mm face brick with 100-mm concrete block and interior finish; or 100-mm concrete wall with interior and exterior finish. D = 100 mm face brick with 200-mm concrete block and interior finish; or 100 mm face brick and 100-mm common brick with interior finish. C = 200-mm concrete wall with interior and exterior finish.

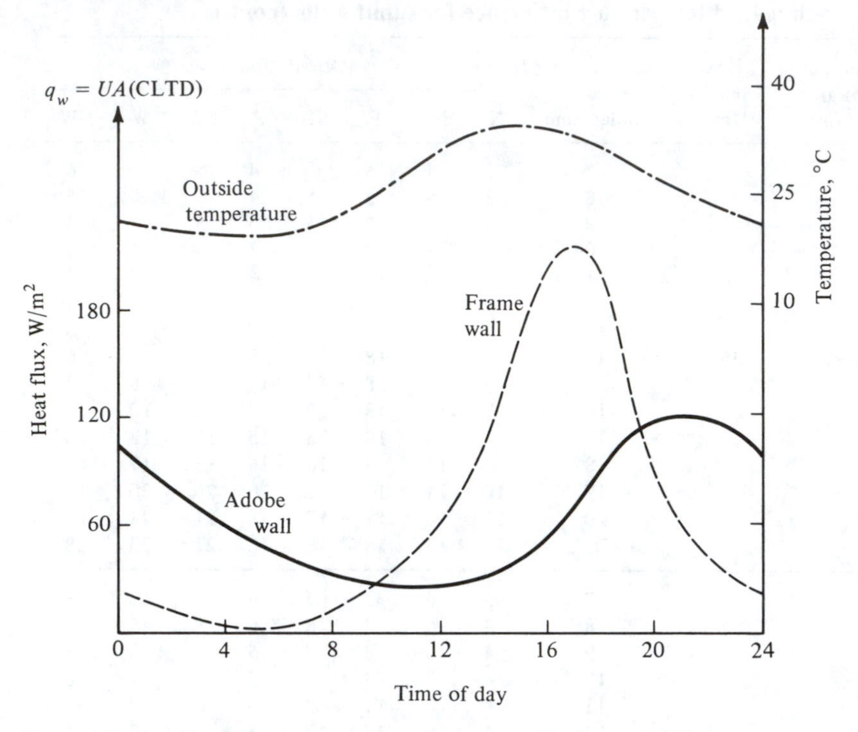

Figure 4-7 Heat flux through two walls with the same U values but different masses.

Solution The wall most nearly matches type F in Table 4-15. The maximum CLTD occurs at 1900 h (7 P.M.) with a value of 33°C. The average outdoor temperature is 30°C rather than 29°C, on which the table is based, so the adjusted CLTD is

$$\text{CLTD} = 33 + (30 - 29) = 34 \text{ K}$$

From Example 4-3 $R = 2.812 \text{ m}^2 \cdot \text{K/W}$, so the $U = 0.356 \text{ W/m}^2 \cdot \text{K}$.

$$\frac{q_{max}}{A} = U(\text{CLTD}) = 0.356(34) = 12.1 \text{ W/m}^2$$

4-14 Summary of procedures for estimating cooling loads The process of estimating cooling loads is similar to that used in determining heating loads. In fact much of the same information is applicable. There is enough difference, however, to make each step worth noting again.

1. Select design values for outdoor summer dry-bulb temperature (2.5 percent value), mean coincident wet-bulb temperature, and the daily average temperature from Table 4-3.
2. Select an indoor design temperature which is appropriate for the activities to be carried out in the space.

3. Determine whether any special conditions exist, such as adjacent unconditioned spaces. Estimate temperatures in the adjacent spaces.
4. On the basis of building plans and specifications, compute heat-transfer coefficients for the building components in each enclosing surface. Any surfaces connecting with spaces to be maintained at the same temperature may be omitted. Note that the only differences between the *U* values calculated here and those for the heating-load estimate are the values used for the surface convection coefficients, which differ in summer and winter and may vary with the direction of heat flow.
5. From the building plans and specifications, system operating schedule, and design values of wind velocity and temperature difference estimate the rate of infiltration and/or ventilation of outside air. For the cooling load the latent load is also included.
6. Determine the additional building characteristics, e.g., location, orientation, external shading, and mass, that will influence solar-heat gain.
7. On the basis of building components and design conditions determine the appropriate cooling-load temperature differences, solar-heat gain factors, and cooling-load factors.
8. On the basis of the heat-transfer coefficients, areas, and temperature differences determined above calculate the rate of heat gain to the space.
9. For spaces with heat gain from internal sources (lights, equipment, or people), apply the cooling-load factor when appropriate.
10. Sum all the pertinent load components to determine the maximum capacity required for heating and cooling. If the building is to be operated intermittently, additional capacity may be required.

The above procedure and the discussion in this chapter has been brief. The reader is directed to the most recent "ASHRAE Handbook, Fundamentals Volume" or similar sources for a more complete discussion of the details and for more extensive tabular data.

PROBLEMS

4-1 The exterior wall of a single-story office building near Chicago is 3 m high and 15 m long. The wall consists of 100-mm face brick, 40-mm polystyrene insulating board. 150-mm lightweight concrete block, and an interior 16-mm gypsum board. The wall contains three single-glass windows 1.5 m high by 2 m long. Calculate the heat loss through the wall at design conditions if the inside temperature is 20°C. *Ans*. 2.91 kW.

4-2 For the wall and conditions stated in Prob. 4-1 determine the percent reduction in heat loss through the wall if (*a*) the 40 mm of polystyrene insulation is replaced with 55 mm of cellular polyurethane, (*b*) the single-glazed windows are replaced with double-glazed windows with a 6-mm air space. (*c*) If you were to choose between modification (*a*) or (*b*) to upgrade the thermal resistance of the wall, which would you choose and why? *Ans*. (*a*) 12.4%

4-3 An office in Houston, Texas, is maintained at 25°C and 55 percent relative humidity. The average occupancy is five people, and there will be some smoking. Calculate the cooling load imposed by ventilation requirements at summer design conditions with supply air conditions set at 15°C and 95 percent relative humidity if (*a*) the recommended rate of outside ventilation air is used and (*b*) if a filtration device of E = 70 percent is used. *Ans.* (*a*) 2.1 kW, (*b*) 1.31 kW.

4-4 A computer room located on the second floor of a five-story office building is 10 by 7 m. The exterior wall is 3.5 m high and 10 m long; it is a metal curtain wall (steel backed with 10 mm of insulating board), 75 mm of glass-fiber insulation, and 16 mm of gypsum board. Single-glazed windows make up 30 percent of the exterior wall. The computer and lights in the room operate 24 h/d and have a combined heat release to the space of 2 kW. The indoor temperature is 20°C.

(*a*) If the building is located in Columbus, Ohio, determine the heating load at winter design conditions. *Ans.* 602 W

(*b*) What would be the load if the windows were double-glazed?

4-5 Compute the heat gain for a window facing southeast at 32° north latitude at 10 A.M. central daylight time on August 21. The window is regular double glass with a 13-mm air space. The glass and inside draperies have a combined shading coefficient of 0.45. The indoor design temperature is 25°C, and the outdoor temperature is 37°C. Window dimensions are 2 m wide and 1.5m high. *Ans.* 750 W

4-6 The window in Prob. 4-5 has an 0.5-m overhang at the top of the window. How far will the shadow extend downward? *Ans.* 0.55 m.

4-7 Compute the instantaneous heat gain for the window in Prob. 4-5 with the external shade in Prob. 4-6. *Ans.* 558 W.

4-8 Compute the total heat gain for the south windows of an office building that has no external shading. The windows are double-glazed with a 6-mm air space and with regular plate glass inside and out. Draperies with a shading coefficient of 0.7 are fully closed. Make the calculation for 12 noon in (*a*) August and (*b*) December at 32° north latitude. The total window area is 40 m^2. Assume that the indoor temperatures are 25 and 20°C and that the outdoor temperatures are 37 and 4°C. *Ans.* (*a*) 9930 W.

4-9 Compute the instantaneous heat gain for the south wall of a building at 32° north latitude on July 21. The time is 4 P.M. sun time. The wall is brick veneer and frame with an overall heat-transfer coefficient of 0.35 $W/m^2 \cdot K$. The dimensions of the wall are 2.5 by 5 m. *Ans.* 87.5 W.

4-10 Compute the peak instantaneous heat gain per square meter of area for a brick west wall similar to that in Example 4-3. Assume that the wall is located at 40° north latitude. The date is July. What time of the day does the peak occur? The outdoor daily average temperature of 30°C and indoor design temperature is 25°C.

REFERENCES

1. Thermal Environmental Conditions for Human Occupancy, Standard 55-81, American Society of Heating, Refrigerating, and Air-Conditioning Engineers, Atlanta, Ga., 1981.
2. Standard for Ventilation Required for Minimum Acceptable Indoor Air Quality, ASHRAE Standard 62-81, American Society of Heating, Refrigerating, and Air-Conditioning Engineers, Atlanta, Ga., 1981.

3. "Handbook and Product Directory, Equipment Volume," American Society of Heating, Refrigerating, and Air-Conditioning Engineers, Atlanta, Ga., 1979.
4. "ASHRAE Handbook, Fundamentals Volume," American Society of Heating, Refrigerating, and Air-Conditioning Engineers, Atlanta, Ga., 1981.
5. C. W. Coblentz and P. R. Achenbach: Field Measurements of Air Infiltration in Ten Electrically-Heated Houses, *ASHRAE Trans.*, vol. 69, pp. 358–365, 1963.
6. W. Rudoy: "Cooling and Heating Load Calculation Manual," American Society of Heating, Refrigerating, and Air-Conditioning Engineers, Atlanta, Ga., 1979.

CHAPTER

FIVE

AIR-CONDITIONING SYSTEMS

5-1 Thermal distribution systems Chapter 4 explained how to compute heating and cooling loads in conditioned spaces. In order to compensate for these loads energy must be transferred to or from the space. In most medium-sized and large buildings the thermal energy is transferred by means of air, water, and occasionally refrigerant. The transfer of energy often requires conveying energy from a space to a central heat sink (refrigeration unit) or conveying heat from a heat source (heater or boiler) to the space. The assembly that transfers heat between the conditioned spaces and the source or sink is called the *thermal distribution system.* Another of its functions is to introduce outdoor ventilation air. The selection of sizes and capacities of the components of the air and water thermal distribution systems is covered in Chaps. 6 and 7, respectively, and the present chapter concentrates on the configuration of the components. We begin with an explanation of the classic single-zone system, which would be used for a large auditorium or a laboratory where precise conditions are to be maintained. Most thermal distribution systems serve multiple zones. The multiple-zone systems that will be studied in this chapter include:

1. Air systems
   - *a.* Terminal reheat
   - *b.* Dual-duct or multizone
   - *c.* Variable-air-volume
2. Water systems
   - *a.* Two-pipe
   - *b.* Four-pipe

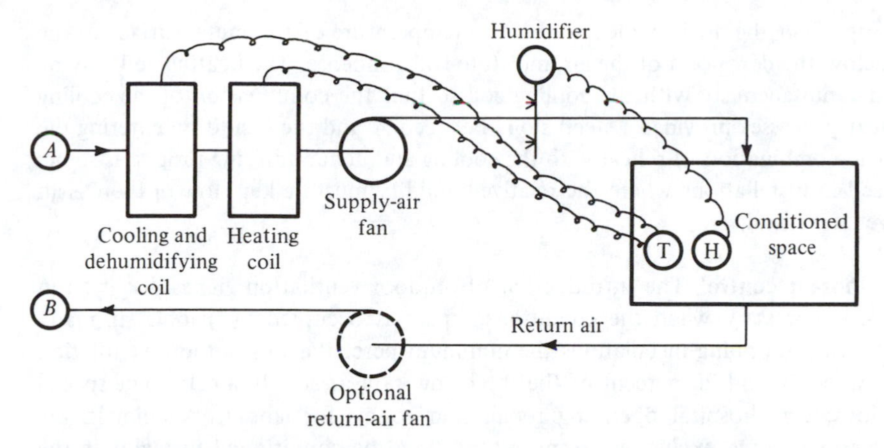

Figure 5-1 A single-zone system.

5-2 Classic single-zone system The elements of the air-conditioning systems that will provide heating (and humidification) or cooling (and dehumidification) are shown in Fig. 5-1. A subsystem of this and most other air-conditioning systems controls the flow rate of outdoor ventilation air. This subsystem interfaces with the facility shown in Fig. 5-1 at points *A* and *B*. Outdoor-air control is discussed in Sec. 5-3. From point *A* the air flows to the cooling coil, heating coil, fan, and humidifier toward the conditioned space. In the return-air line a fan is often installed to avoid excessive air pressure in the conditioned space relative to the outside-air pressure. The temperature control is provided by a thermostat regulating the cooling or heating coil, and the humidity is controlled by a humidistat that regulates the humidifier. Two of the possible operating modes are shown on the psychrometric charts of Fig. 5-2. Figure 5-2*a* shows a heating and humidification process wherein air at point *A* having a low temperature is warmed in the heating coil and humidified by the direct admission of steam. As Prob. 3-9 showed, the process of humidification by direct admission of steam results in little change in dry-bulb temperature.

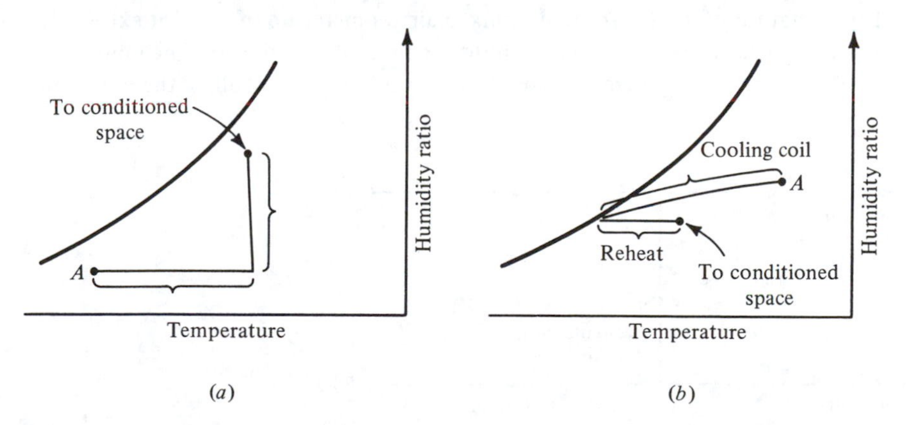

Figure 5-2 (*a*) Heating and humidification and (*b*) cooling and dehumidification with reheat.

In Fig. 5-2*b* the air is cooled, and if the temperature of the metal surface of the coil is below the dew point of the air, moisture will condense. The heating coil may be operated simultaneously with the cooling coil so that the combination of the cooling and reheat processes provides a steep slope between *A* and the condition entering the space. The combination of reheat with the cooling and dehumidifying process is sometimes used in installations where the relative humidity must be kept low or there is an excessive latent load.

5-3 Outdoor-air control The introduction of outdoor ventilation air, as discussed in Sec. 4-4, is necessary when the conditioned space is occupied by people. In many comfort air-conditioning installations the minimum percentage of outdoor ventilation air is between 10 and 20 percent of the total flow rate of supply air. In some special applications, e.g., hospital operating rooms and rooms for laboratory animals, the supply air may come exclusively from outdoors and be conditioned to maintain the specified space conditions. No return air is recirculated in these installations.

The outdoor-air control mechanism that interfaces with the air system of Fig. 5-1 and other air systems presented later in this chapter is shown in Fig. 5-3. The stream of return air at *B* flowing back from the zones divides, some exhausting and some recirculating. The outdoor ventilation air mixes with the recirculated air and flows to the conditioning unit at *A*. Dampers in the outdoor-, exhaust-, and recirculated-air lines regulate the flow rates. The dampers in the outdoor- and exhaust-air lines open and close in unison and in the direction opposite the motion of the recirculated-air dampers.

A standard outdoor-air control plan attempts to maintain the mixed-air temperature at point *A* at approximately 13 to 14°C since the basic function of the air-conditioning system is to provide cooling. Another requirement of the outdoor-air control is to assure that the minimum percentage of outdoor air is maintained. The program to accomplish these several requirements is shown in Fig. 5-4. At a high outdoor temperature the dampers provide the minimum flow rate of outdoor air. At an outdoor air temperature lower than about 24°C (or whatever the return-air temperature is) it is more economical in cooling energy to use 100 percent outdoor air. For outdoor-air temperatures below 13°C the dampers proportion themselves to maintain a mixed temperature of 13°C. To hold a mixed-air temperature of 13°C at extremely low outdoor temperatures, the fraction of outdoor air could drop below the minimum. The controls are therefore designed to hold to that minimum and allow the mixed-air

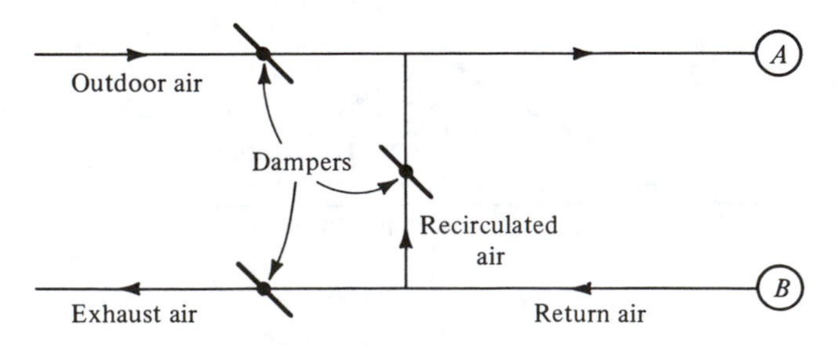

Figure 5-3 Outdoor-air control.

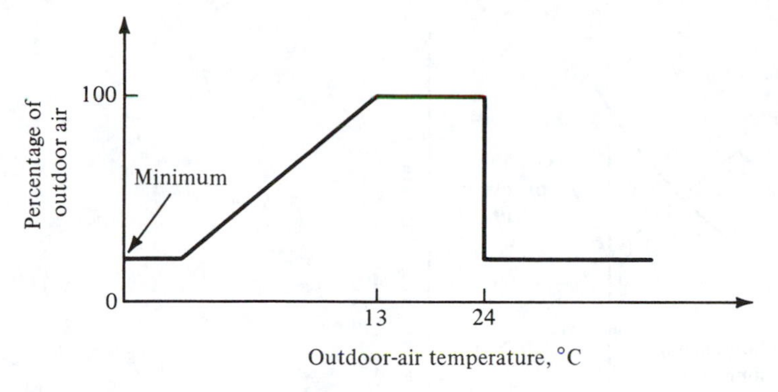

Figure 5-4 Outdoor-air control.

temperature to drop below 13°C. The pneumatic control system to achieve this plan is described in Sec. 9-10.

Example 5-1 If the outdoor-air controller is to maintain a mixed-air temperature of 13°C and a 20 percent minimum percentage of outdoor air when the recirculated-air temperature is 24°C, at what outdoor temperature do the dampers close to the minimum 20 percent position during cold weather?

Solution An energy balance for the mixing process when the dampers have closed to the 20 percent position is

$$0.20t_{od} + 0.80(24) = 1.00(13)$$

The outdoor-air temperature t_{od} is -31°C.

Since this outdoor-air temperature of -31°C is lower than experienced in all but the coldest locations, sometimes the provision to hold the minimum outdoor-air rate at low outdoor temperatures is not even incorporated into the outdoor-air controller.

The shift to minimum outdoor air when the outdoor-air temperature exceeds the recirculated-air temperature is made to conserve energy. At outdoor-air temperatures above this crossover point it is more economical to condition the recirculated air. An air property that is more decisive than temperature in predicting the refrigerating rate at the coil is the enthalpy. The psychrometric chart in Fig. 5-5 shows two triangular regions, *X* and *Y*, where the *enthalpy-control concept* makes judgments differing from the changeover controlled by comparative dry-bulb temperatures. When the outdoor-air conditions lie in region *X*, the temperature changeover would choose outdoor air, even though the outdoor air, because of its high humidity, requires greater cooling capacities than if recirculated air were used. Also, if the outdoor air lies in region *Y*, the temperature changeover uses recirculated air, even though the low-humidity outdoor air would require less cooling energy.

While the traditional practice has been to try to achieve a mixed-air temperature year round of approximately 13°C, current practice is influenced by the desire to

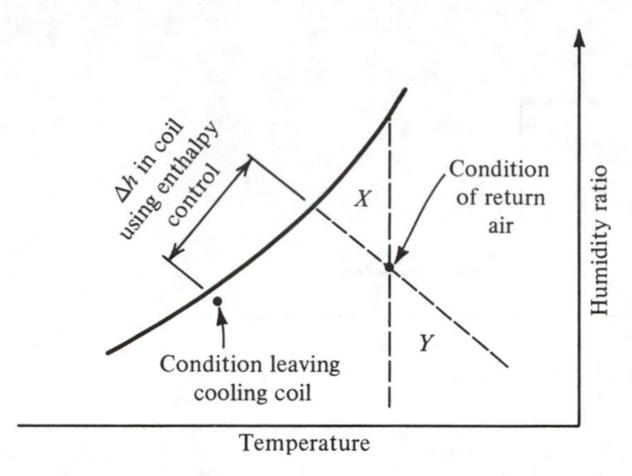

Figure 5-5 Using the comparison of air enthalpies to switch between 100 percent and minimum outdoor air.

conserve energy; so the mixed-air temperature may be reset to a higher value if and when cooling loads can be met by higher-temperature air.

5-4 Single-zone-system design calculations When a conditioned space experiences a net addition of both sensible and latent heat from interior and exterior loads, the supply air must enter the conditioned space with both a temperature and humidity ratio lower than the values to be maintained in the space. If q_s is the sensible cooling load in kilowatts and q_L the latent load, any entering point *i* along the *load-ratio line* shown in Fig. 5-6 will provide the proper proportions of sensible and latent cooling provided that

$$\frac{c_p(t_c - t_i)}{h_c - h_i} = \frac{q_s}{q_s + q_L} \tag{5-1}$$

where h = enthalpy, kJ/kg
t = temperature, °C
c_p = specific heat of air = 1.0 kJ/kg · K

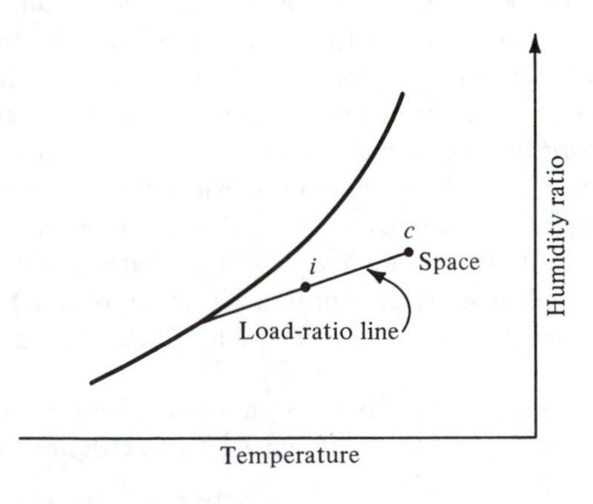

Figure 5-6 Load-ratio line for cooling and dehumidifying assignment.

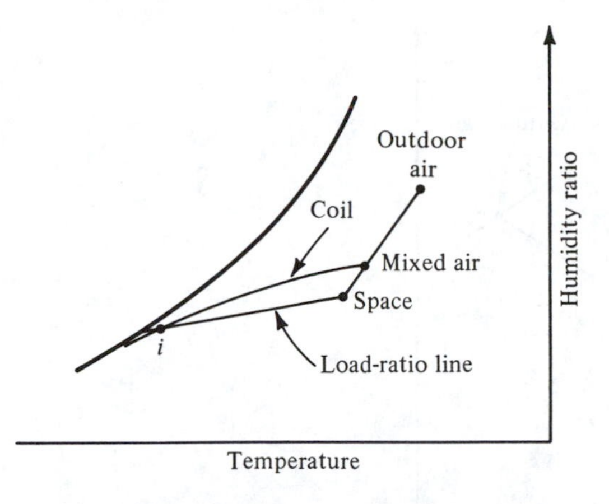

Figure 5-7 State points of air during cooling and dehumidifying with a single-zone air conditioner.

The combination of the supply condition i and the flow rate of the air w must be such that the sensible- and latent-heat loads are satisfied

$$w = \frac{q_s}{c_p(t_c - t_i)} = \frac{q_s + q_L}{h_c - h_i} \quad \text{kg/s} \tag{5-2}$$

The sensible- and latent-heat gains q_s and q_L are only part of the heat the cooling coil must remove, since there is also the cooling load attributable to the ventilation air. This additional load is evident from the psychrometric chart in Fig. 5-7 because the assignment imposed on the coil is to cool and dehumidify the mixed air to a point i that lies on the load-ratio line.

Example 5-2 The sensible- and latent-heat gains in a space served by a single-zone air conditioner are 65 and 8 kW, respectively. The space is to be maintained at 24°C and 50 percent relative humidity. The design conditions of outdoor air are 35°C dry-bulb and 25°C wet-bulb temperatures. For ventilation purposes outdoor air is mixed with recirculated air in a 1:4 proportion. When mixed air at the resulting conditions enters the cooling coil, the outlet air conditions are a function of the temperature of the chilled water supplied to the coil, as indicated in Table 5-1. Determine (*a*) the air conditions entering the coil, (*b*) the air condi-

Table 5-1 Outlet air conditions from cooling coil in Example 5-2

Chilled-water-supply temperature, °C	Air leaving temperatures, °C: Dry bulb	Air leaving temperatures, °C: Wet bulb
4.0	10.7	10.5
5.0	11.6	11.5
6.0	12.5	12.4
7.0	13.3	13.2

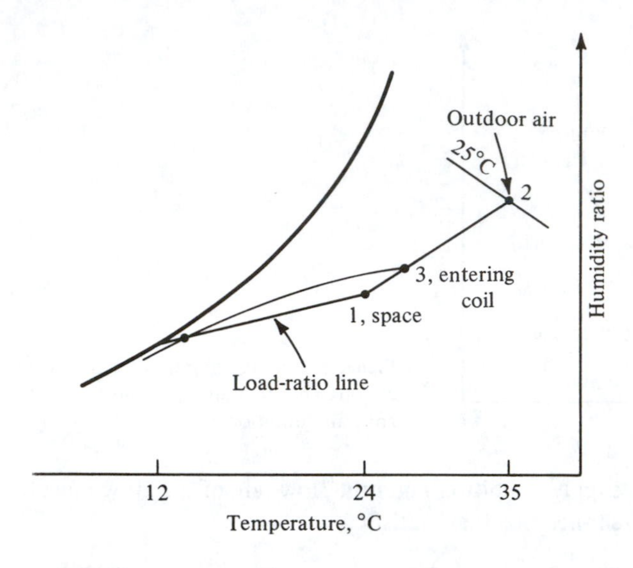

Figure 5-8 Psychrometric chart showing properties of air in Example 5-2.

tions leaving the coil and the required temperature of the supply chilled water, and (*c*) the cooling capacity of the coil.

Solution (*a*) On the psychrometric chart in Fig. 5-8, four parts of return air at point 1 (24°C and 50 percent relative humidity, h = 47.5 kJ/kg, W = 0.0093 kg/kg) mix with one part of outdoor air at point 2 (35°C dry-bulb temperature, 25°C wet-bulb temperature, h = 76.0 kJ/kg, W = 0.016 kg/kg). From energy and mass balances the enthalpy and humidity ratio at point 3 after mixing are

$$h_3 = 0.8(47.5) + 0.2(76.0) = 53.2 \text{ kJ/kg}$$

$$W_3 = 0.8(0.0093) + 0.2(0.016) = 0.00106 \text{ kg/kg}$$

At the point located by these values of h_3 and W_3, other properties can be determined:

Dry-bulb temperature = 26.2°C

Wet-bulb temperature = 18.8°C

(These are the air conditions entering the coil to which the performance data in Table 5-1 apply.)

(*b*) The load-ratio line extends downward and to the left from point 1 with a slope such that the proportions of sensible- and latent-heat removal are satisfied. To fix the load-ratio line choose arbitrarily a temperature t_i (Fig. 5-6) of 14°C and then compute a value of h_i at that temperature. From Eq. (5-1)

$$h_i = 47.5 - 1.0(24 - 14)\frac{65 + 8}{65} = 36.3 \text{ kJ/kg}$$

Connecting the two points 3 and i gives the load-ratio line which intersects the saturation line on Fig. 5-8 at 12°C.

Table 5-1 indicates that a chilled-water temperature of 5.5°C results in outlet-air conditions of 12.1 and 12.0°C dry- and wet-bulb temperatures, respectively, which satisfies the load-ratio line.

(*c*) The enthalpy of air leaving the coil at dry- and wet-bulb temperatures of 12.1 and 12.0°C, respectively, is 34.2 kJ/kg. The mass rate of flow of supply air to the space is, from Eq. (5-2),

$$w = \frac{65 + 8 \text{ kW}}{47.5 - 34.2 \text{ kJ/kg}} = 5.49 \text{ kg/s}$$

The enthalpy of air entering the cooling coil is 53.2 kJ/kg, so the cooling capacity required of this coil is

$$(5.49 \text{ kg/s})(53.2 - 34.2 \text{ kJ/kg}) = 104.3 \text{ kW}$$

The difference between 104.3 and the room load of 73 kW is attributable to the cooling load of the outdoor ventilation air.

5-5 Multiple-zone systems For large buildings it is usually not economically feasible to provide a separate system for each zone. For such cases the basic central-system concept is expanded to meet the cooling and heating requirements of multiple zones. A zone may be a single room, one floor of a building, one side of a building, or the interior space. Essentially a zone is the space controlled by one thermostat.

A wide variety of combinations of duct networks, coil locations, and control strategies are in use, but the most common are:

1. Constant-volume systems
   a. Terminal-reheat
   b. Dual duct or multizone
2. Variable-volume systems
   a. Single-purpose cooling or heating
   b. Cooling with reheat
   c. Dual-duct variable-volume

5-6 Terminal-reheat system The schematic diagram of the terminal-reheat system is shown in Fig. 5-9. All the air is cooled to a temperature of perhaps 13°C to assure dehumidification, and the thermostat in each zone controls the reheat coil associated with that zone so that the temperature of the entering air will be such that the zone temperature is maintained. The reheat coil may be hot water or electric. The advantages of the terminal-reheat system include the small space occupied by ducts and excellent temperature and humidity control over a wide range of zone loads. The primary disadvantage is the relatively high energy requirements for both cooling and heating. The energy penalties may be partially overcome by raising the temperature of cool supply air until one of the reheat coils is completely off. Another means of reducing the energy penalty is to perform the heating with recovered energy[1] from some other part of the system such as the refrigerant condenser or lighting.

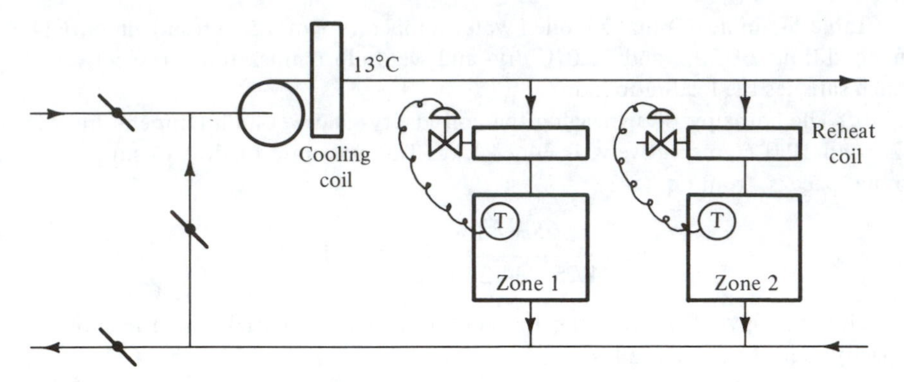

Figure 5-9 A terminal-reheat system.

5-7 Dual-duct or multizone system In the dual-duct system, shown schematically in Fig. 5-10, the air from the supply-air fan divides. Part of the air flows through the heating coil and part through the cooling coil. The thermostat in each zone regulates a mixing box that proportions the flow rate of warm and cool air to maintain the desired temperature in the zone. The dual-duct system is very responsive to changes in load of the zone and can simultaneously accommodate heating in some zones and cooling in others. One disadvantage of the system is the expense of two supply-air ducts, both of which must be large enough to handle all the airflow. As with the terminal-reheat system, there will be periods of simultaneous heating and cooling which reduce the energy efficiency of the dual-duct system. On the other hand, when the outdoor-air temperature is low enough to achieve the 13°C without operating the cooling coil, some energy can be conserved. Good energy effectiveness is also achieved during hot weather if the temperature of the warm duct is set low or—even better—the heating coil is shut down.

The multizone system is thermally identical to the dual-duct system, but the configuration differs in that all the mixing boxes are clustered at the central unit and individual ducts convey the mixed air to each zone.

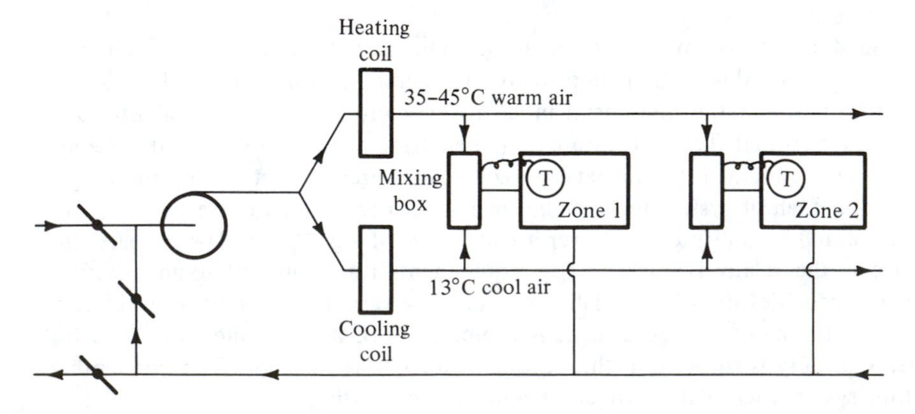

Figure 5-10 Dual-duct system.

Example 5-3 A zone served by the dual-duct system in Fig. 5-10 has a design heating load of 8 kW and a design sensible-cooling load of 6 kW. The zone is to be maintained at 24°C, the temperature of the air in the cool air duct is 13°C, and the temperature in the warm air duct is 40°C. Assume that the temperature of mixed air is 24°C. At a part-load condition when the sensible cooling load is 3kW, what are the heating- and cooling-energy rates attributable to this zone?

Solution It is first necessary to determine the airflow rate to the zone, a value that in the dual-duct system remains constant for all load conditions. To meet the design heating load the required flow rate is (8 kW)/[(40 - 24°C) (1.0 kJ/kg · K)] = 0.5 kg/s, and to meet the design cooling load (6 kW)/[24 - 13) (1)] = 0.55 kg/s. The flow rate needed for the design cooling load controls, so the airflow rate is set at 0.55 kg/s.

When the sensible-cooling load is 3 kW, the air temperature entering the zone is 24°C - (3 kW)/[0.55(1.0)] = 18.55°C. The energy balance of the airstream in and out of the mixing box is

$$w_c(13°\text{C})(1.0) + w_w(40)(1.0) = 0.55(18.55)(1.0)$$

where w_c = flow rate of cool air, kg/s
w_w = flow rate of warm air, kg/s

Since $w_w = 0.55 - w_c$,

$$13w_c + 40(0.55) - 40w_c = 10.20$$

Then w_c = 0.437 kg/s and w_w = 0.113 kg/s. The energy rate needed to bring the 0.113 kg/s up to 40°C in the heating coil is 0.113(40 - 24) = 1.80 kW, and the energy rate in the cooling coil is 0.437(24 - 13) = 4.80 kW.

Because of "thermal bucking" of the two coils, the cooling coil must extract not only the 3-kW cooling load of the zone but also the 1.8 kW introduced by the heating coil.

5-8 Variable-air-volume systems The poor energy characteristics, especially during light heating or cooling loads, of the constant-volume air systems discussed in Secs. 5-6 and 5-7 have shifted preferences in new designs to variable-air-volume (VAV) systems. There are a number of variations of VAV systems and also a number of possible combinations of VAV with other systems. Three important configurations are (1) cooling or heating only, (2) VAV reheat, and (3) VAV dual duct.

In the cooling-only system, as shown in Fig. 5-11, a single stream of cool air serves all the zones, and a thermostat in each zone regulates a damper to control the flow rate of cool air into the zone. The desirable energy characteristic of this system is that at low cooling loads the flow rate of air is reduced so that the required cooling capacity at the coil is correspondingly reduced. The cooling-only VAV system is widely used for interior spaces of buildings with no heating loads and where only cooling loads prevail. The system experiences a problem at very light cooling loads where the airflow rate drops off so much that poor air distribution and/or ventilation results.

The heating-only VAV system has the same structure as that in Fig. 5-11, but instead of a cooling coil a heating coil provides a source of constant-temperature

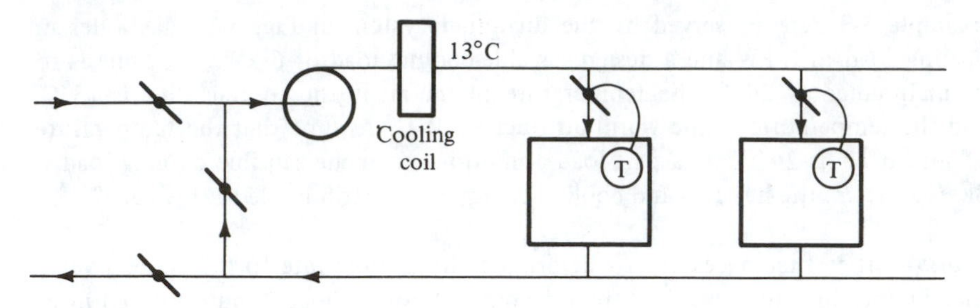

Figure 5-11 A cooling-only VAV system.

warm air. The conditions adaptable to a heating-only VAV system are rather rare in building air-conditioning systems.

The VAV reheat system is identical to the one shown in Fig. 5-11 except that the branch line to each zone contains a reheat coil. The control sequence is that as the cooling load drops off, the damper progressively reduces the flow rate of air until about 25 to 30 percent of full flow rate. At this point the airflow rate remains constant and the reheat coil is activated. There is thus some thermal bucking, just as in the conventional terminal-reheat system, but it occurs at a reduced airflow rate and thus results in only a modest loss of efficiency. The VAV reheat system overcomes a number of deficiencies of the cooling-only VAV system since it provides a means of obtaining adequate air distribution and ventilation without paying the energy penalty incurred in constant-volume reheat applications.

In the VAV dual-duct system, the arrangement is similar to the conventional dual-duct system of Fig. 5-10, except for the flow characteristics of the mixing boxes. Instead of providing a constant flow rate of mixed air, the dampers are arranged so that the warm and cool airflow rates drop appreciably before the other stream begins to supply air. The result, as shown in Fig. 5-12, is that the airflow rate to the zone is variable, but by proper choice of control characteristics the desired minimum airflow

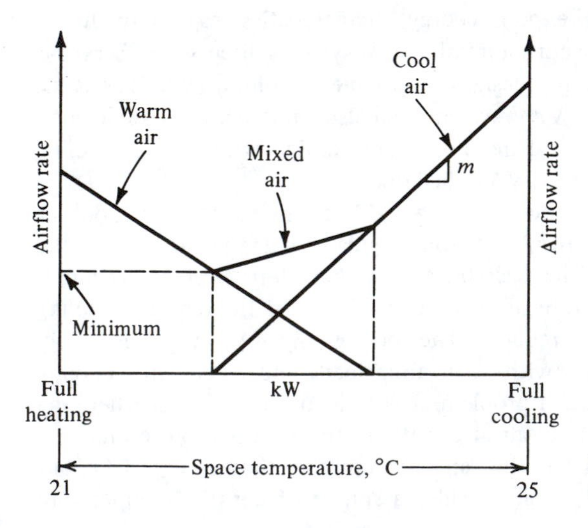

Figure 5-12 Airflow rates in a VAV dual-duct system.

rate is assured. The characteristics of modulating controls, discussed in Chap. 9, lend themselves to achieving the flow rates shown in Fig. 5-12 by permitting a span of space air temperatures. In Fig. 5-12, for example, the VAV mixing box provides the necessary flow rate of warm air for full heating when the space temperature is 21°C. When the space temperature is 25°C, the cool-air damper is open enough to meet full cooling load.

The VAV reheat and dual-duct systems provide all the flexibility of the conventional reheat and dual-duct systems in the sense that a zone can be accommodated as it switches from heating to cooling requirements and that some zones on a system can be provided with heating at the same time that other zones need cooling. There is some thermal bucking in both the VAV reheat and dual-duct systems, but the magnitude of heating- and cooling-energy cancellation is modest because of the low airflow rates at which the thermal bucking occurs.

Example 5-4 In a certain VAV dual-duct system a flow rate of warm air of 0.8 kg/s is required at full heating load when the space temperature is 21°C; the space requires 1.1 kg/s of cool air at full cooling load, which is called for at a space temperature of 25°C. It will be convenient to choose control equipment that has the same slope of the airflow-to-space-temperature lines for both the warm and cool air (one slope is the negative of the other). If the minimum airflow rate is to be 0.3 kg/s, at what temperatures do the warm airflow rate and the cool airflow rate fall to zero?

Solution Let m = slope of the cool-airflow-rate line, as in Fig. 5-12, and $-m$ the slope of the warm-airflow-rate line. Then

$$w_c = c_c + mt_s \quad \text{and} \quad w_h = c_h - mt_s$$

where w_c, w_h = airflow rates of cool and warm air, respectively, kg/s
c_c, c_h = const
t_s = space temperature, °C

The 100 percent cooling and 100 percent heating conditions provide the expressions

$$1.1 = c_c + m(25) \tag{5-3}$$

and

$$0.8 = c_h - m(21) \tag{5-4}$$

The minimum airflow rate will occur when either w_c or w_h has dropped to zero; thus

$$w_c = 0 = c_c + mt_{s0} \tag{5-5}$$

where t_{s0} is the temperature where $w_c = 0$, at which condition

$$w_h = \text{minimum airflow} = 0.3 = c_h - mt_{s0} \tag{5-6}$$

Equations (5-3) to (5-6) are four simultaneous equations that can be solved for the four unknowns c_c, c_h, m, and t_{s0}. The results are

$$c_c = -8.9 \qquad c_h = 9.2 \qquad m = 0.4 \qquad \text{and} \qquad t_{s0} = 22.25°\text{C}$$

The flow rate of warm air drops to zero when t_s rises to 23°C, so between 22.25 and 23°C the mixing of warm and cool air results in the minimum total airflow rate of 0.3 kg/s.

5-9 Water systems Water systems accomplish heating and cooling through the distribution of water alone, although the final heat transfer in the conditioned space must be to or from air. Outdoor air for ventilation must be provided and conditioned in each zone. Fan-coil units, unit ventilators, or convectors are the most common terminal units served by water-piping systems. Water systems occupy relatively little space and are often the lowest first-cost systems available. The systems usually lack humidity control, and ventilation may be uncertain even if outside openings are provided at each terminal unit. Wind pressure, the stack effect in tall buildings, and the possibility of freezing the coils in cold weather all require special precautions when outdoor-air openings are provided. Since condensate drains must be provided at each coil, maintenance is also a more significant factor than in air systems, where the dehumidification can be accomplished at a central location.

Fan-coil units can be served by two- or four-pipe water-distribution systems. A two-pipe system serves units with a single coil, and the system can provide heating or cooling, but it is not possible to heat some zones while simultaneously cooling others.

The four-pipe system (Fig. 5-13) serves fan-coil units with two coils, one for

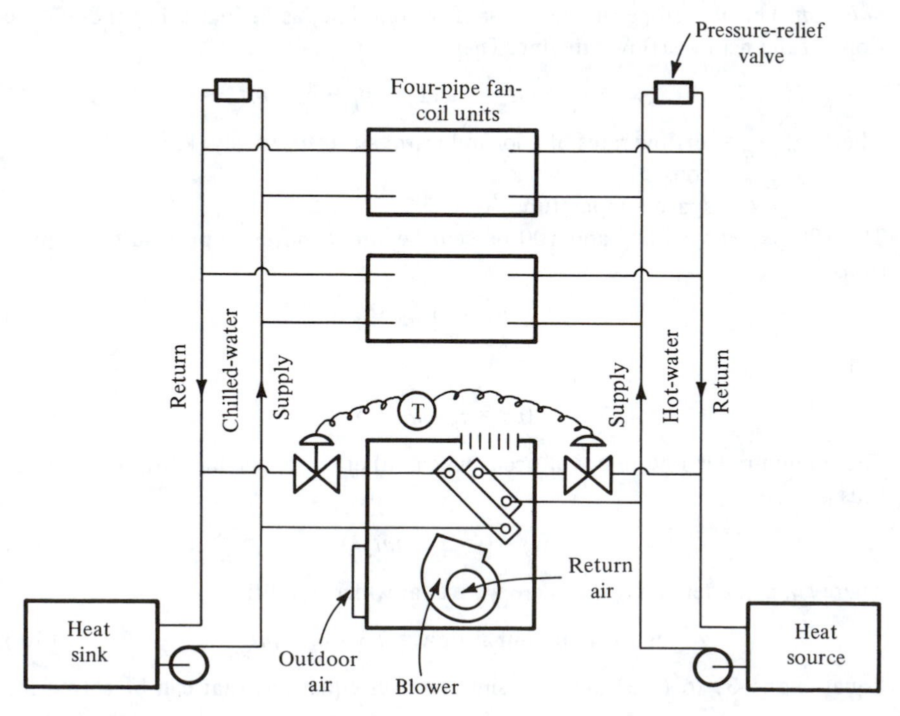

Figure 5-13 The four-pipe water thermal-distribution system.

heating and one for cooling. The hot- and cold-water loops have their own supply and return pipes. The space thermostat regulates the flow of hot and cold water to the coils, but the control is sequenced so that the hot water is off before cold water is admitted to its coil, and vice versa.

Water systems may also serve convectors that have no fans (see Sec. 7-4). Convectors are used almost universally for heating and rarely for cooling, because of the problem of draining the condensate. A popular application of convectors is in perimeter zones of buildings that are also served by a VAV system. In this arrangement the VAV system can be of the cooling-only type with all the heating provided by the convectors.

5-10 Unitary systems The systems described to this point have all incorporated centrally located, field-assembled, heating and cooling equipment. Multiple-unit or unitary systems, on the other hand, are factory-assembled units located in or near the conditioned space. These systems are available as a single package containing direct-expansion evaporating coils, controls, fan, compressor, and condenser, or they may be split units with the compressor and condenser located remotely.

When appropriately applied, multiple-unit systems offer a number of advantages. The fact that they are mass-produced and factory-assembled usually means lower first cost and lower installation costs. With proper selection and control they may also provide relatively low operating costs. By their nature multiple-unit systems are zoned systems. These units are manufactured with matched components and usually with certified ratings and published performance data.

The disadvantages of unitary systems are that there are relatively few options with respect to sizing the evaporator, condenser, fans, compressor, and controls. Since each unit must be capable of meeting the peak load of the space it serves, the installed capacity and connected electrical load are usually larger than with a central system.

Examples of the unitary air conditioner are window units, through-the-wall units, rooftop units, and split systems. Window units are primarily found in residential applications, and they have no ducted air distribution. The appearance and noise of these units limit their application. Through-the-wall units are generally more acceptable than window units from the standpoint of appearance since they are designed into the building. They find application in motels, health-care facilities, schools, and sometimes offices. Rooftop units (as shown in Fig. 1-2) are primarily applied in low-rise buildings with flat roofs, such as stores, shopping centers, and factories. For better air distribution in the conditioned space the conditioned air should be ducted from the rooftop unit to multiple outlets instead of introducing the total airflow rate at one position. The split system, serving such installations as small stores or an office suite, usually conveys the conditioned air through ducts to the spaces served by the conditioner.

PROBLEMS

5-1 A conditioned space that is maintained at 25°C and 50 percent relative humidity experiences a sensible-heat gain of 80 kW and a latent-heat gain of 34 kW. At what temperature does the load-ratio line intersect the saturation line? *Ans.* 9°C

5-2 A conditioned space receives warm, humidified air during winter air conditioning in order to maintain 20°C and 30 percent relative humidity. The space experiences an infiltration rate of 0.3 kg/s of outdoor air and an additional sensible-heat loss of 25 kW. The outdoor air is saturated at a temperature of −20°C (see Table A-2). If conditioned air is supplied at 40°C dry-bulb temperature, what must the wet-bulb temperature of supply air be in order to maintain the space conditions? *Ans.* 18.8°C

5-3 A laboratory space to be maintained at 24°C and 50 percent relative humidity experiences a sensible-cooling load of 42 kW and a latent load of 18 kW. Because the latent load is heavy, the air-conditioning system is equipped for reheating the air leaving the cooling coil. The cooling coil has been selected to provide outlet air at 9.0°C and 95 percent relative humidity. What is (*a*) the temperature of supply air and (*b*) the airflow rate? *Ans.* (*b*) 3.8 kg/s

5-4 In discussing outdoor-air control Sec. 5-3 explained that with outdoor conditions in the *X* and *Y* regions on the psychrometric chart in Fig. 5-5 enthalpy control is more energy-efficient. We now explore some limitations of that statement with respect to the *Y* region. Suppose that the temperature setting of outlet air from the cooling coil is 10°C and that the outlet air is essentially saturated when dehumidification occurs in the coil. If the condition of return air is 24°C and 40 percent relative humidity and the outdoor conditions are 26°C and 30 percent relative humidity, would return air or outside air be the preferred choice? Explain why.

5-5 A terminal reheat system (Fig. 5-9) has a flow rate of supply air of 18 kg/s and currently is operating with 3 kg/s of outside air at 28°C and 30 percent relative humidity. The combined sensible load in the spaces is 140 kW, and the latent load is negligible. The temperature of the supply air is constant at 13°C. An accountant of the firm occupying the building was shocked by the utility bill and ordered all space thermostats set up from 24 to 25°C. What is (*a*) the rate of heat removal in the cooling coil before and after the change and (*b*) the rate of heat supplied at the reheat coils before and after the change? Assume that the space sensible load remains constant at 140 kW. *Ans.* (*a*) 15 kW increase in cooling rate; (*b*) 18 kW increase in heating rate

REFERENCES

1. T. C. Gilles: Energy Reclaiming Modular Self-Contained Multizone Unit, *Proc. Conf. Improv. Effic. HVAC Equip. Components, Purdue Univ., October 7-8, 1974,* pp. 117–125.
2. Variable Air Volume Systems, ASHRAE Symp., *ASHRAE Trans.,* vol. 80, pt. 1, pp. 472–505, 1974.
3. Control Systems to Capitalize on the Inherent Energy-Conserving Features of VAV Systems," ASHRAE Symp., *ASHRAE Trans.,* vol. 83, pt. 1, pp. 581–611, 1977.
4. Control of Variable Air-Volume Terminals, ASHRAE Symp., *ASHRAE Trans.,* vol. 86, pt. 2, pp. 825–858, 1980.
5. "Handbook and Product Directory, Systems Volume," American Society of Heating, Refrigerating, and Air-Conditioning Engineers, Atlanta, Ga., 1980.

CHAPTER

SIX

FAN AND DUCT SYSTEMS

6-1 Conveying air Chapter 5 explained the arrangement of the popular air systems (variable-volume, terminal-reheat, etc.), and this chapter follows up by concentrating on four topics associated with the flow of air in an air system: (1) computing pressure drops of air flowing through ducts and fittings, (2) extending the computation of pressure drops to designing a duct system, (3) understanding the characteristics of a fan, both independently and in conjunction with a duct system, and (4) designing the distribution of air in a conditioned space.

Since the fan motor is a large consumer of energy and the duct system occupies considerable space in the building, the fan-duct system deserves keen attention during design. Unfortunately, because there are dozens (or hundreds) of interrelated decisions in the design of a fan-and-duct system, most designers are satisfied with achieving a workable system and do not proceed further toward a combined optimization of lifetime energy cost, duct-system cost, and building-space cost of the fan and ducts. Computer-aided design (CAD) should bring improvements in the future, but even CAD must apply principles correctly; some of these principles will be explained in this chapter.

6-2 Pressure drop in straight ducts The fundamental equation for computing the pressure drop of a fluid flowing through a straight duct of circular cross section is

$$\Delta p = f \frac{L}{D} \frac{V^2}{2} \rho \qquad (6\text{-}1)$$

where Δp = pressure drop, Pa

f = friction factor, dimensionless

L = length, m

D = inside diameter (ID) of duct, m
V = velocity, m/s
ρ = density of fluid, kg/m^3

The friction factor f is a function of the Reynolds number and the relative roughness of the pipe surface ϵ/D, where ϵ is the absolute roughness in meters. Both graphical and equation representations of the friction factor are available. Equation (6-2) derives from the work of Colebrook,[1]

$$f = \left\{ \frac{1}{1.14 + 2 \log \dfrac{D}{\epsilon} - 2 \log \left[1 + \dfrac{9.3}{\text{Re}\,(\epsilon/D)\sqrt{f}} \right]} \right\}^2 \tag{6-2}$$

This equation is implicit in f, so a trial value of f can be substituted on the right side and an improved value computed from the equation. The definition of the Reynolds number is

$$\text{Re} = \frac{VD\rho}{\mu} \tag{6-3}$$

where μ is the viscosity in pascal-seconds.

The graphical source of the value of f is the traditional Moody chart[2] reproduced in Fig. 6-1. The absolute roughness ϵ of some surfaces is shown in Table 6-1.

Example 6-1 Compute the pressure drop in 15 m of straight circular sheet-metal duct 300 mm in diameter when the flow rate of 20°C air is 0.5 m^3/s.

Solution The velocity V is

$$V = \frac{0.5 \text{ m}^3/\text{s}}{\pi(0.3^2)/4} = 7.07 \text{ m/s}$$

Table 6-1 Absolute roughness of some surfaces

Material	Roughness ϵ, m
Riveted steel	0.0009–0.009
Concrete	0.0003–0.003
Cast iron	0.00026
Sheet metal	0.00015
Commercial steel	0.000046
Drawn tubing	0.0000015

Figure 6-1 Moody chart for determining friction factor.

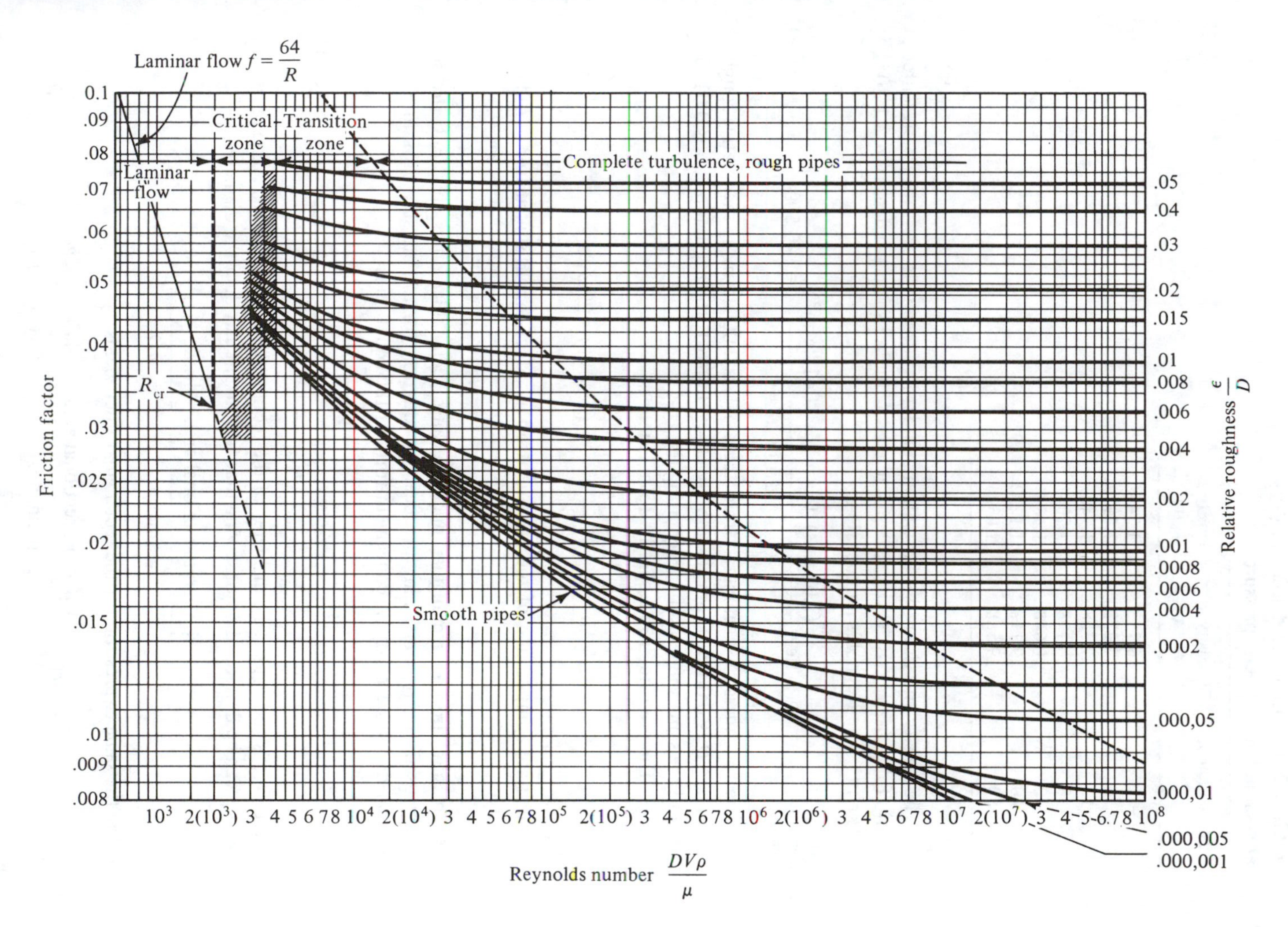
Laminar flow $f = \frac{64}{R}$
Critical zone
Transition zone
Laminar flow
Complete turbulence, rough pipes
R_{cr}
Smooth pipes
Friction factor
0.1
.09
.08
.07
.06
.05
.04
.03
.025
.02
.015
.01
.009
.008
Relative roughness $\frac{\epsilon}{D}$
.05
.04
.03
.02
.015
.01
.008
.006
.004
.002
.001
.0008
.0006
.0004
.0002
.000,05
.000,01
.000,005
.000,001
10^3 $2(10^3)$ 3 4 5 6 7 8 10^4 $2(10^4)$ 3 4 5 6 7 8 10^5 $2(10^5)$ 3 4 5 6 7 8 10^6 $2(10^6)$ 3 4 5 6 7 8 10^7 $2(10^7)$ 3 4 5 6 7 8 10^8
Reynolds number $\frac{DV\rho}{\mu}$

Table 6-2 Viscosity and density of dry air at standard atmospheric pressure

Temperature, °C	Viscosity μ, μPa • s	Density ρ, kg/m^3
−10	16.768	1.3414
0	17.238	1.2922
10	17.708	1.2467
20	18.178	1.2041
30	18.648	1.1644
40	19.118	1.1272
50	19.588	1.0924

The density and viscosity of dry air at standard atmospheric pressure are presented in Table 6-2. At 20°C the viscosity μ is 18.178μPa • s and the density ρ is 1.2041 kg/m^3. The Reynolds number Re is

$$\text{Re} = \frac{(7.07 \text{ m/s})\,(0.3 \text{ m})\,(1.2041 \text{ kg/m}^3)}{18.178\ \mu\text{Pa} \cdot \text{s}} = 140{,}500$$

The roughness of sheet metal, from Table 6-1, is 0.00015 m; so the relative roughness ϵ/D is 0.0005. From the Moody chart, Fig. 6-1, the friction factor is 0.0195, and the same value results when using Eq. (6-2).

The pressure drop Δp in the 15-m length is

$$\Delta p = 0.0195 \frac{15}{0.3} \frac{7.07^2}{2} (1.2041) = 29.3 \text{ Pa}$$

To facilitate computation of the pressure drop in a duct, graphs similar to Fig. 6-2 are available.

6-3 Pressure drop in rectangular ducts Because rectangular ducts (Fig. 6-3) are so widely used in air-conditioning practice, an equation for the pressure drop in a rectangular duct is necessary. A convenient form of the equation is

$$\Delta p = f \frac{L}{D_{eq}} \frac{V^2}{2} \rho \tag{6-4}$$

where D_{eq} is the equivalent diameter of the rectangular duct in meters. An expression for D_{eq} can be developed by observing that for the circular duct the diameter is

$$\frac{4 \times \text{cross-sectional area}}{\text{perimeter}} = \frac{4(\pi D^2/4)}{\pi D} = D$$

Using the same expression for the rectangular duct yields

$$D_{eq} = \frac{4 \times \text{cross-sectional area}}{\text{perimeter}} = \frac{4ab}{2(a+b)} = \frac{2ab}{a+b} \tag{6-5}$$

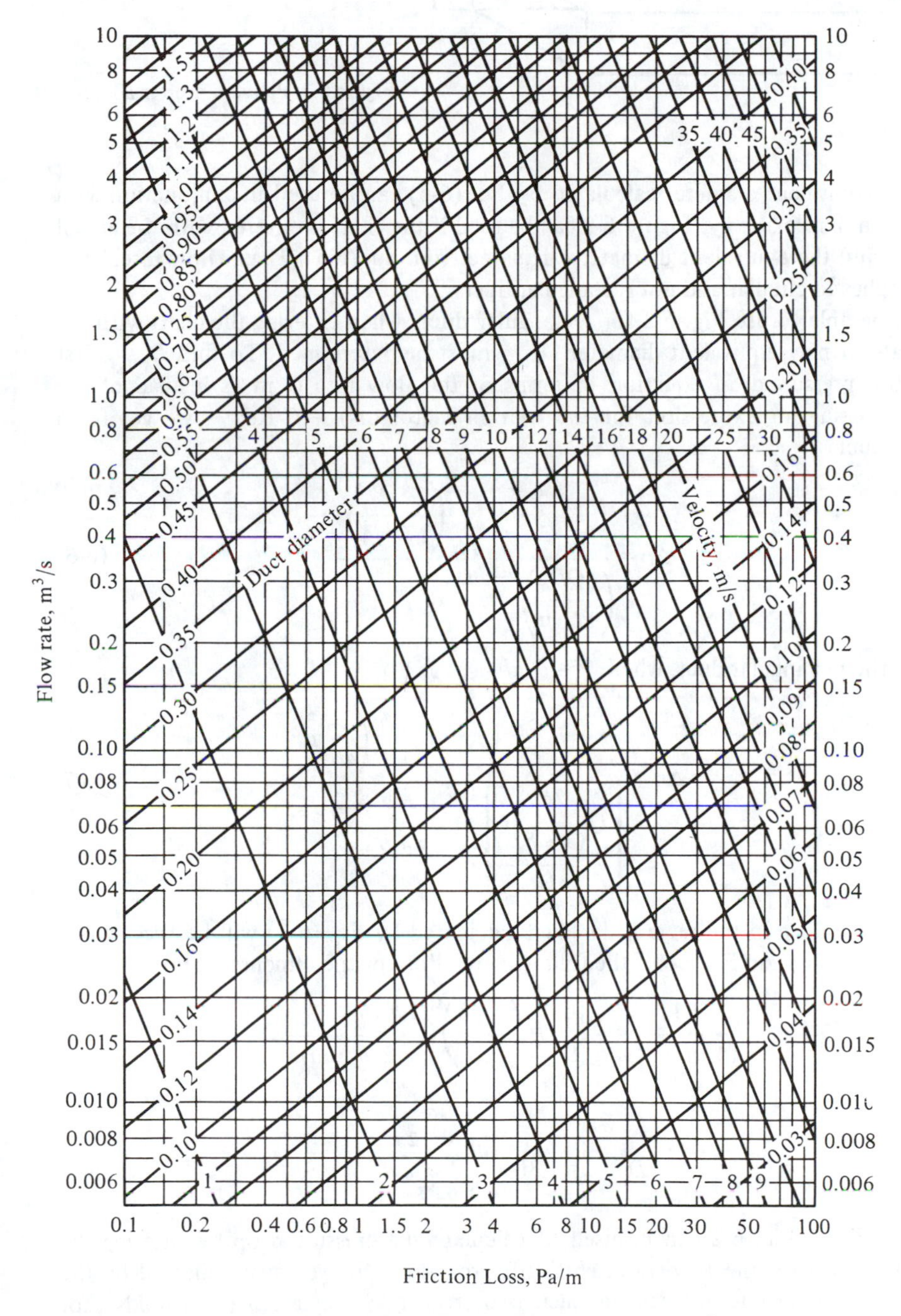

Figure 6-2 Pressure drop in straight, circular, sheet-metal ducts, 20°C air, absolute roughness 0.00015 m.

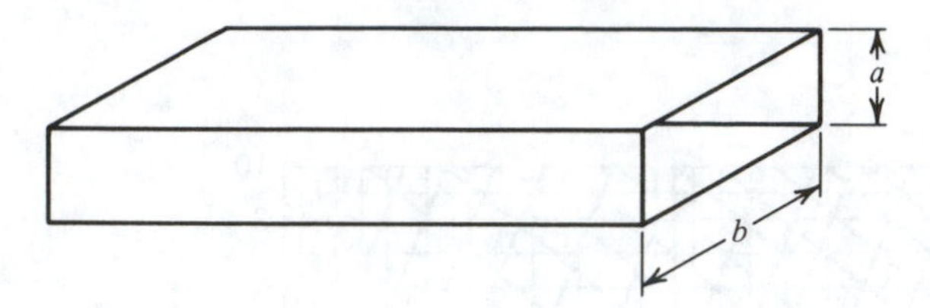

Figure 6-3 Pressure drop in a rectangular duct.

The equivalent diameter calculated by Eq. (6-5) can be used in conjunction with Fig. 6-2 in a special way. Figure 6-2 can be used if the chart is entered with the actual velocity and the equivalent diameter calculated from Eq. (6-5). The airflow rate, however, applies to circular and not rectangular ducts.

To be able to use Fig. 6-2 for rectangular ducts when entering the chart with the flow rate, a new equivalent diameter $D_{eq,f}$ must be determined. To find $D_{eq,f}$ first write the pressure-drop equation in terms of the flow rate Q m^3/s instead of the velocity. Using an expression for the friction factor of $f = C/\text{Re}^{0.2}$, we get for a circular duct

$$\Delta p = \frac{C}{\left(\dfrac{4QD\rho}{\pi D^2 \mu}\right)^{0.2}} \frac{L}{D} \frac{\left(\dfrac{Q}{\pi D^2/4}\right)^2 \rho}{2} \tag{6-6}$$

and for the rectangular duct, where $V = Q/ab$,

$$\Delta p = \frac{C}{\left[\dfrac{Q\left(\dfrac{2ab}{a+b}\right)\rho}{ab\mu}\right]^{0.2}} \frac{\text{L}}{\dfrac{2ab}{a+b}} \frac{\left(\dfrac{Q}{ab}\right)^2 \rho}{2} \tag{6-7}$$

The pressure drop in the rectangular duct calculated by Eq. (6-7) will be the same as that for Eq. (6-6) and Fig. 6-2 if the following relation for the dimensions holds

$$\frac{1}{(4/\pi D)^{0.2} D} \frac{16}{\pi^2 D^4} = \left(\frac{a+b}{2}\right)^{0.2} \frac{a+b}{2ab} \frac{1}{(ab)^2}$$

Thus

$$D_{eq,f} = 1.30 \frac{(ab)^{0.625}}{(a+b)^{0.25}} \tag{6-8}$$

With $D_{eq,f}$ Fig. 6-2 can be used to calculate the pressure drop by entering the chart directly with the flow rate. With this procedure, the velocity indicated by the chart is not applicable but can be calculated from Q/A. Equation (6-8) holds[3] for width-to-height ratios up to about 8:1.

Example 6-2 An airflow rate of 1.5 m^3/s passes through a rectangular duct 0.3 by 0.5 m. Calculate the pressure drop in 40 m of straight duct using (*a*) D_{eq} and (*b*) $D_{eq,f}$.

Solution (*a*)

$$D_{eq} = \frac{2ab}{a+b} = \frac{2(0.3)(0.5)}{0.3+0.5} = 0.375 \text{ m}$$

$$\text{Velocity} = \frac{1.5 \text{ m}^3/\text{s}}{0.3(0.5)} = 10 \text{ m/s}$$

From Fig. 6-2 with a velocity of 10 m/s and an equivalent diameter of 0.375 m the pressure drop is 3.0 Pa/m. Note that the airflow rate of 1.2 m^3/s indicated by the chart is not applicable. In 40 m the pressure drop is 120 Pa.

(*b*)

$$D_{eq,f} = 1.30 \frac{[0.3(0.5)]^{0.625}}{(0.3+0.5)^{0.25}} = 0.42 \text{ m}$$

From Fig. 6-2 with an airflow rate of 1.5 m^3/s and an equivalent diameter of 0.42 m the pressure drop is again 3.0 Pa/m. Note that the velocity indicated on the chart of 10.8 m/s is not applicable. The correct velocity is 10 m/s.

6-4 Pressure drop in fittings An air-handling system consists of straight duct and fittings. In the fittings the air undergoes changes in areas and direction. These fittings include enlargements, contractions, elbows, branches, dampers, filters, and registers. The air-pressure drop in these fittings must be known in order to design a system properly. In actual system design, pressure drop in these fittings may be of more concern than that in the straight duct that connects them. For example, pressure drop in an elbow may be the equivalent of 3 to 12 m of straight duct and could be as high as 20 m. Care in estimating the pressure drop in fittings is therefore justified, although unfortunately the type and quality of construction have a pronounced influence on the pressure drop of a given fitting.

Since fittings occupy only short lengths along the flow network–generally less than 1 m—explanation of the pressure drop cannot be attributed to the drag along the surfaces of the duct. It is due instead to momentum exchanges between portions of the fluid moving at different velocities. More specifically, at some position in the fitting the fluid experiences a sudden expansion, and this process will be used to explain, at least qualitatively, some of the trends in pressure-drop characteristics.

Changes in area and direction that will be explored in this chapter are sudden expansions and contractions, elbows or turns, and branches.

6-5 The $V^2\rho/2$ term A grouping that has already appeared in Eq. (6-1) will recur in the calculation of pressure drops. The pattern that will emerge is that the pressure loss for an incompressible fluid is the product of the $V^2\rho/2$ group and a term that char-

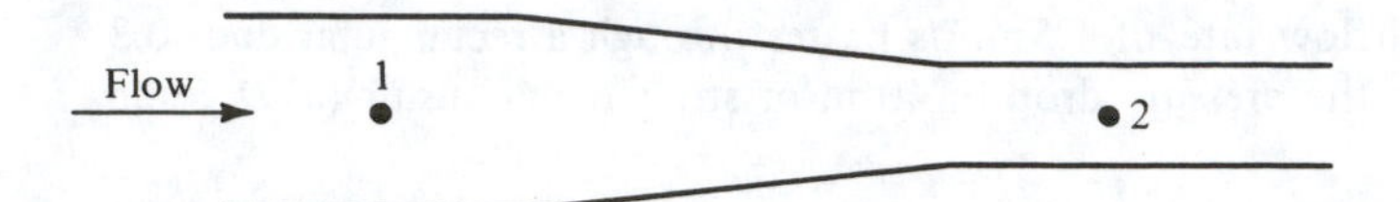

Figure 6-4 Flow through a converging duct section.

acterizes the geometry of the duct or fitting. Thus, from Eq. (6-1)

$$\Delta p = f \underbrace{\frac{L}{D}}_{\text{Geometry}} \frac{V^2 \rho}{2}$$

When air flows frictionlessly through a converging or diverging nozzle, as in Fig. 6-4, the Bernoulli equation (2-8) applies

$$\frac{p_1}{\rho} + \frac{V_1^2}{2} = \frac{p_2}{\rho} + \frac{V_2^2}{2} \tag{2-8}$$

Since

$$\frac{V_2}{V_1} = \frac{A_1}{A_2}$$

$$p_1 - p_2 = \frac{V_1^2 \rho}{2} \underbrace{\left[\left(\frac{A_1}{A_2}\right)^2 - 1\right]}_{\text{Geometry}} \tag{6-9}$$

The next several sections will show that the $V^2\rho/2$ group also appears in expressions for the pressure drops which represent losses, in contrast to Eq. (2-8), where there is no pressure loss, only a pressure conversion.

6-6 Sudden enlargement What is almost a building block for predicting the pressure loss in fittings is the relation describing the pressure loss in a sudden enlargement, as shown in Fig. 6-5. There is a change in area, as in the converging section of Fig. 6-4, but because of losses that prevail the Bernoulli equation no longer applies. It can be amended to the *revised Bernoulli equation*

$$\frac{p_1}{\rho} + \frac{V_1^2}{2} = \frac{p_2}{\rho} + \frac{V_2^2}{2} + \frac{p_{\text{loss}}}{\rho} \tag{6-10}$$

The other principle to be combined with Eq. (6-10) to develop an expression for the

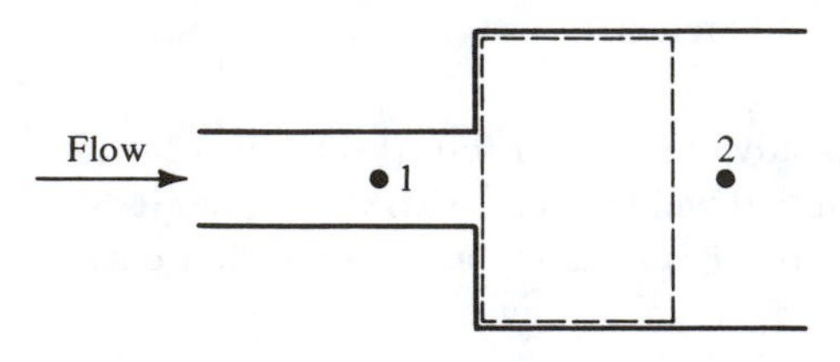

Figure 6-5 A sudden enlargement.

pressure loss is the momentum equation

$$p_1 A_2 - p_2 A_2 = V_2(V_2 A_2 \rho) - V_1(V_1 A_1 \rho) \qquad (6\text{-}11)$$

which states that the difference in force on opposite sides of the control volume shown by the dashed lines in Fig. 6-5 equals the rate of change of momentum. The not so obvious term in Eq. (6-11) is $p_1 A_2$. Due to separation of the flow from the surface at the abrupt enlargement the low pressure p_1 prevails immediately after the expansion and acts over the entire area A_2. It is this separation that causes the pressure loss in the sudden enlargement. Substituting $p_1 - p_2$ from Eq. (6-11) into Eq. (6-10) yields the expression for p_{loss}

$$p_{loss} = \frac{V_1^2 \rho}{2}\left(1 - \frac{A_1}{A_2}\right)^2 \quad \text{Pa} \qquad (6\text{-}12)$$

The pattern appearing in Eq. (6-12) is now familiar in that the loss is the product of the $V^2\rho/2$ group and a term representing the geometry.

Equation (6-12), called the *Borda-Carnot equation,* agrees sufficiently well with experimental results to be used for duct-system design. At high airflow rates the equation gives losses that are several percent high[4] and at low airflow rates, several percent low.

Example 6-3 Air at standard atmospheric pressure and a temperature of 20°C flowing with a velocity of 12 m/s enters a sudden enlargement where the duct area doubles. What is the increase in static pressure of the air as it passes through the enlargement?

Solution From Eq. (6-12) the pressure loss due to the sudden enlargement is

$$p_{loss} = \frac{(12\ \text{m/s})^2\ (1.204\ \text{kg/m}^3)}{2}\left(1 - \tfrac{1}{2}\right)^2 = 21.7\ \text{Pa}$$

Substituting p_{loss} into the revised Bernoulli equation permits computation of the pressure rise

$$p_2 - p_1 = \frac{(V_1^2 - V_2^2)\rho}{2} - p_{loss} = \frac{12^2 - 6^2}{2}(1.204) - 21.7$$

$$= 43.3\ \text{Pa}$$

Instead of the pressure rising 65 Pa, as the Bernoulli equation would indicate for a process with no losses, the actual pressure rise is 43.3 Pa.

6-7 Sudden contraction A sudden contraction occurs in a duct section where the duct size is abruptly reduced in the direction of flow. The flow pattern in a sudden contraction, as shown in Fig. 6-6, consists of a separation of the fluid from the wall upon entering the reduced cross-sectional area, and a vena contracta forms at 1′. The concept in predicting the pressure loss is to propose no loss from positions 1 to 1′ and to treat the flow from positions 1′ to 2 as a sudden enlargement. This logic is quite valid,

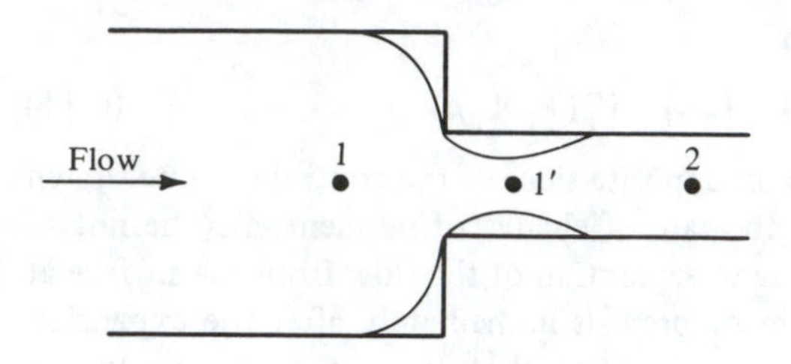

Figure 6-6 A sudden contraction.

since converging, accelerating flow is efficient, while deceleration of fluid is difficult to achieve without losses. Borrowing from Eq. (6-12), then, we find the pressure loss in the contraction to be

$$p_{\text{loss}} = \frac{(V_1')^2 \rho}{2}\left(1 - \frac{A_1'}{A_2}\right)^2 \tag{6-13}$$

The area of the vena contracta can be related to A_2 by defining a contraction co efficient C_c

$$C_c = \frac{A_1'}{A_2} = \frac{V_2}{V_1'} \tag{6-14}$$

Substituting A_1' and V_1' from Eq. (6-14) into Eq. (6-13) yields

$$p_{\text{loss}} = \frac{V_2^2 \rho}{2}\left(\frac{1}{C_c} - 1\right)^2 \tag{6-15}$$

The contraction coefficient is a function of the ratio of the areas, A_2/A_1; it was determined experimentally by Weisbach[5] in 1855 and is shown in Table 6-3.

The form of Eq. (6-15) again shows the pressure drop to be calculated by the $V_2\rho/2$ group multiplied by a geometry factor. The geometry factor reaches a maximum of approximately $\frac{1}{3}$, which may be compared with the maximum value of the

Table 6-3 Contraction coefficients in sudden contractions

$\dfrac{A_2}{A_1}$	C_c	$\left(\dfrac{1}{C_c} - 1\right)^2$
0.1	0.624	0.366
0.2	0.632	0.340
0.3	0.643	0.310
0.4	0.659	0.270
0.5	0.681	0.221
0.6	0.712	0.160
0.7	0.755	0.103
0.8	0.813	0.050
0.9	0.892	0.010
1.0	1.000	0.000

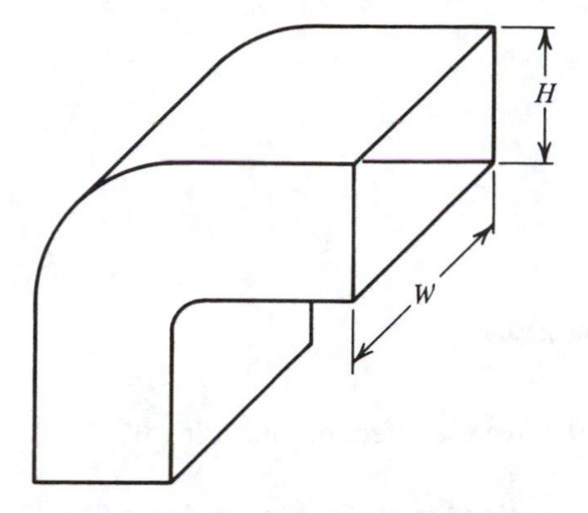

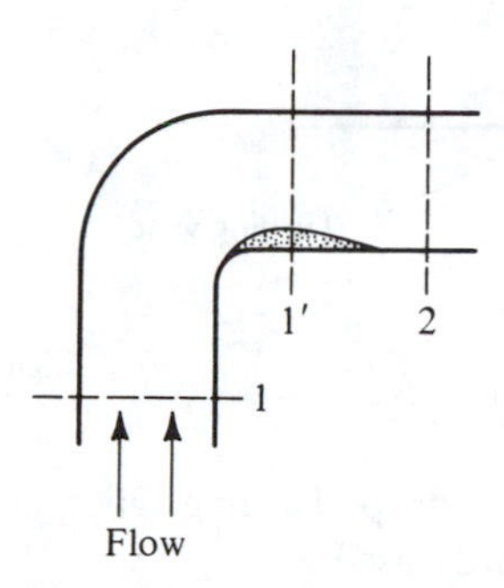

Figure 6-7 Separated flow in an elbow.

geometry factor for a sudden enlargement of 1.0. In any sudden change in area for a given rate of flow there will be a greater pressure loss if the fluid flows from the small to the large area than when it flows in the opposite direction.

6-8 Turns The most common elbows used in duct systems are 90° turns, either circular or rectangular in cross section. Weisbach[5] proposed that the pressure loss in an elbow is due to the sudden expansion from the contracted region in plane 1′ in Fig. 6-7 to the full cross-sectional area of the duct in plane 2. If this proposal is correct, it should be possible to express the pressure loss in terms of the $V^2\rho/2$ group. The magnitude of the Reynolds number has an influence[6] but not a dominant one.

Pressure losses in rectangular elbows determined by Madison and Parker[7] are shown in Fig. 6-8. These data also show that a flat 90° elbow with a large value of W/H suffers less pressure drop than a deep 90° elbow, suggesting that the subdivision of an elbow into multiple elbows of large W/H by installing turning vanes will reduce

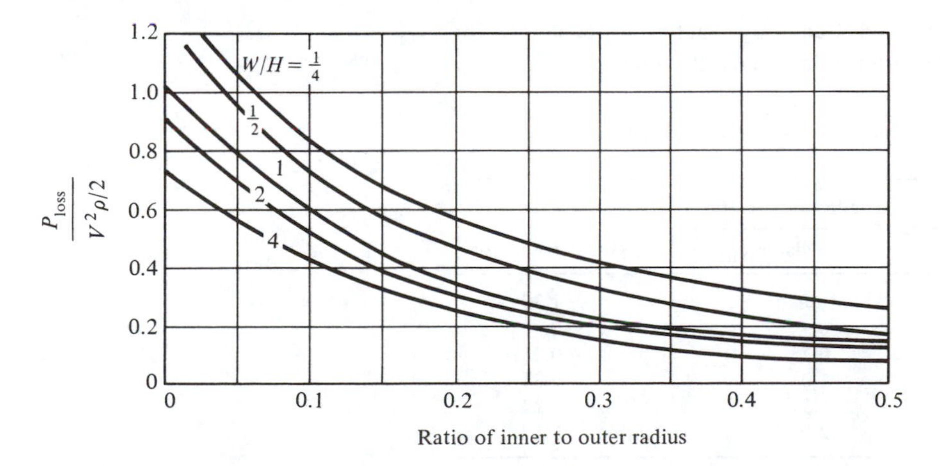

Figure 6-8 Pressure loss in rectangular elbows.

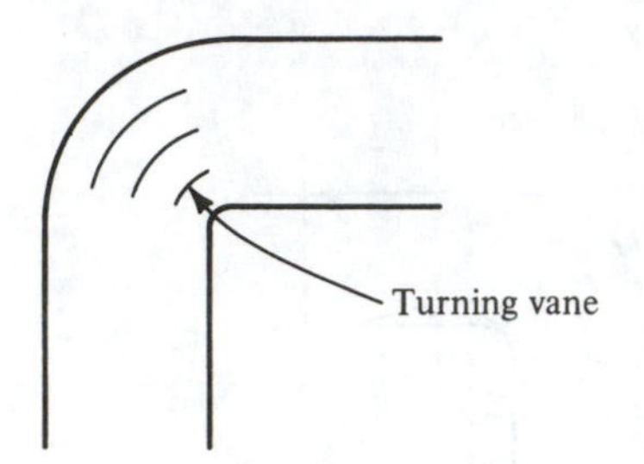

Figure 6-9 Turning vanes.

the pressure drop. Turning vanes, as in Fig. 6-9, have the effect of increasing W/H and are frequently used.

The pressure loss in elbows of circular cross section[8] is summarized in Table 6-4.

6-9 Branch takeoffs When a main duct supplies air to several branch ducts, a takeoff must be provided for each branch, as in Fig. 6-10. From the upstream position u there is a pressure loss both to the downstream position d and into the branch to point b.

Considering first the pressure loss from u to d in the straight section of duct, this loss occurs because the pressure buildup from the higher velocity at u to the lower velocity at d is less than the ideal. The pressure loss in the straight-through section of a branch takeoff is usually small compared with other losses in the system. In many low-velocity designs it is neglected, but an equation[9] that closely approximates tabular data is

$$p_{\text{loss}} = \frac{V_d^2 \rho}{2} (0.4) \left(1 - \frac{V_d}{V_u}\right)^2 \quad \text{Pa} \tag{6-16}$$

The pressure loss[10-12] from u to b expressed in terms of $V_b^2 \rho/2$ for several different angles of takeoff is shown in Fig. 6-11.

Table 6-4 Geometry factor in equation for pressure loss in circular 90° elbows

$$p_{\text{loss}} = \frac{V^2 \rho}{2} \text{(geometry factor)} \quad \text{Pa}$$

Ratio = radius of curvature† / diameter	Geometry factor
Mitered	1.30
0.5	0.90
0.73	0.45
1.0	0.33
1.5	0.24
2.0	0.19

† Measured to the duct centerline.

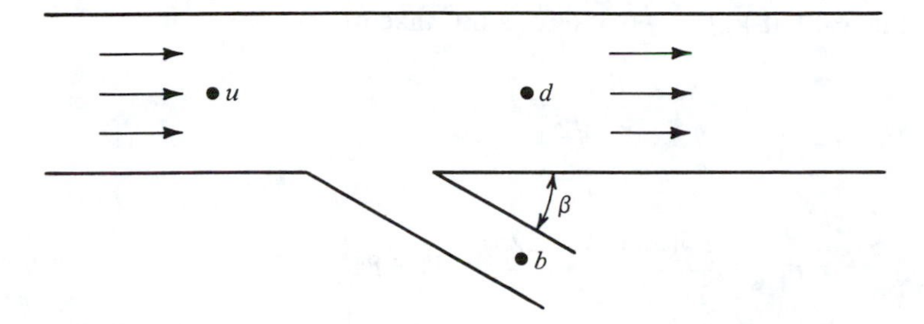

Figure 6-10 Branch takeoff.

Example 6-4 A 60°, 30- by 30-cm branch takeoff leaves a 30- by 50-cm trunk duct. The size of the downstream section is also 30 by 50 cm. The upstream flow rate is 1.5 m^3/s, and the branch flow rate is 0.5 m^3/s. The upstream pressure is 500 Pa and the air temperature is 15°C. (*a*) What is the pressure following the straight-through section, and (*b*) what is the pressure in the branch line?

Solution Velocities

$$V_u = 10 \text{ m/s} \qquad V_d = 6.67 \text{ m/s} \qquad V_b = 5.56 \text{ m/s} \qquad \rho = 1.225 \text{ kg/m}^3$$

(*a*) From Eq. (6-16)

$$p_{\text{loss}} = \frac{6.67^2(1.225)(0.4)}{2}\left(1 - \frac{6.67}{10}\right)^2 = 1.2 \text{ Pa}$$

Substituting into the revised Bernoulli equation (6-10) gives

$$p_2 = \rho\left(\frac{p_1}{\rho} + \frac{V_1^2}{2} - \frac{V_2^2}{2} - \frac{p_{\text{loss}}}{\rho}\right)$$

$$p_d = 500 + \frac{10^2(1.225)}{2} - \frac{6.67^2(1.225)}{2} - 1.2 = 533 \text{ Pa}$$

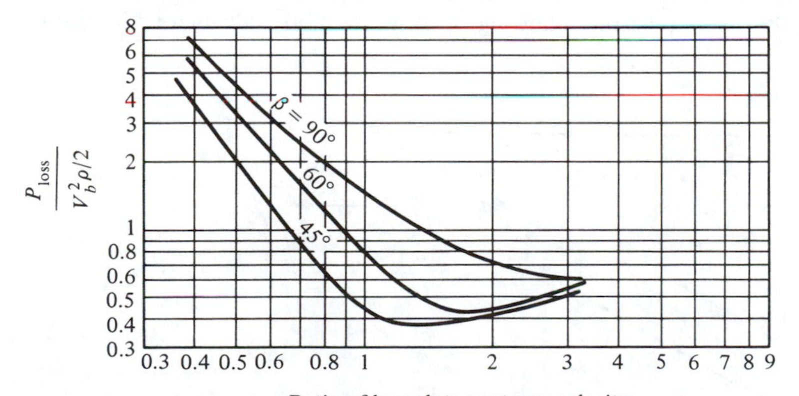

Figure 6-11 Pressure loss from the upstream position to the branch duct.

(*b*) From Fig. 6-11 at $V_b/V_u = 0.556$ for a 60° takeoff

$$\frac{p_{\text{loss}}}{V_b^2\rho/2} = 2.5$$

therefore

$$p_{\text{loss}} = \frac{2.5(5.56^2)\,(1.225)}{2} = 47.3 \text{ Pa}$$

Substituting into the revised Bernoulli equation (6-10) gives

$$p_b = 1.225\left(\frac{500}{1.225} + \frac{10^2}{2} - \frac{5.56^2}{2} - \frac{47.3}{1.225}\right) = 495 \text{ Pa}$$

In addition to circular and rectangular cross sections of ducts and fittings, oval ducts are sometimes used to meet space limitations. Data on the pressure loss in elbows and branch fittings of the oval cross sections are available.[13]

6-10 Branch entries The branch takeoffs examined in the previous section appear in supply-air systems. In return-air systems, branch entries (Fig. 6-12) bring air from the various branches to the main duct. Just as for the branch takeoff, there is a pressure loss that must be incorporated into the revised Bernoulli equation in the straight-through section of the main as well as from the branch. The straight-through pressure loss can be estimated by using the momentum equation

$$V_d^2 A_d \rho - V_u^2 A_d \rho - V_b^2 A_b \rho \cos\beta = (p_u - p_d)A_d \tag{6-17}$$

If $\beta = 90°$, the combination of Eq. (6-17) and the revised Bernoulli equation (6-10) gives

$$p_{\text{loss}} = \frac{V_d^2\rho}{2}\left[1 - \left(\frac{V_u}{V_d}\right)^2\right] \tag{6-18}$$

To calculate the p_{loss} from the branch duct at position b to the main duct at position d, Healy[14] proposed the following equation for a 90° side inlet with main-to-

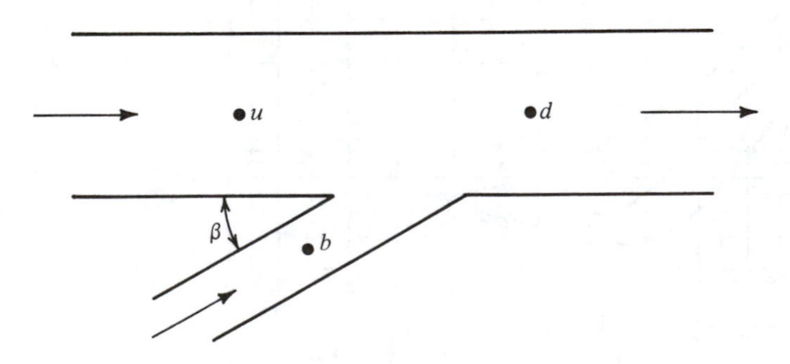

Figure 6-12 A branch entry.

branch area ratios greater than 4:

$$p_{\text{loss}} = \frac{V_d^2 \rho}{2}\left[1.5\left(\frac{A_d}{A_b}\right)^2 - 1\right] \tag{6-19}$$

6-11 Design of duct systems A duct system consists of necessary sections of straight duct, elbows, branch outlets and inlets, dampers, and such terminal units as registers and diffusers. In designing the duct system the pressure drop contributed by heat-transfer coils and filters must also be included. The chief requirements of a duct system are (1) that it convey specified rates of airflow to prescribed locations, (2) that it be economical in combined first cost, fan-operating cost, and cost of building space occupied, and (3) that it not transmit or generate objectionable noise. The noise characteristics are explored further in Chap. 21.

Since there are so many decisions to make in selecting the size of ducts and fittings, and since each decision affects the remainder of the system, the design of a duct system is (or at least could be) a sophisticated operation. The design procedures outlined next are simply methodical procedures for arriving at duct sizes that are reasonable with respect to space requirements and velocities. There are three major design techniques in use: (1) velocity method, (2) equal-friction method, and (3) static-regain method. The first two will be presented in this chapter, followed by an introduction to duct-system optimization.

6-12 Velocity method In this method of duct design the velocities in the mains and branches are selected and the pressure drops in all runs calculated. The fan is selected to provide a pressure sufficient to meet the requirements of the run with the highest pressure drop. Standard practice calls for installation of a balancing damper in each branch line, and the damper in the run calling for the highest pressure difference is left completely open while the other dampers are throttled to reduce the flow rate to the design value.

No fixed recommendations can be given for the velocity to be selected because the choice is a function of the economics, space limitations, and type of acoustic treatment. High velocities result in high pressure drops that are costly in fan power if they occur in the critical run. High velocities also result in increased noise generation. On the other hand, high velocities permit smaller ducts, which are lower in first cost and require less space in the building. For air-duct systems in public buildings with no extensive acoustic treatment, typical velocities in the main duct are of the order of 5 to 8 m/s and in the branch ducts 4 to 6 m/s. Typical velocities for residential systems are lower and velocities for industrial buildings are higher than those quoted for public buildings.

In the duct system shown schematically in Fig. 6-13, the airflow-rate requirements at outlets 1 to 5 will all be known from the cooling- and heating-load calculation, so the flow rate in each of the sections A to I can be computed. In applying the velocity method of duct design, velocities will be selected for all the sections and the pressure drop in each run calculated using applicable relations for straight ducts, elbows, and branch takeoffs and manufacturer's data for other components, e.g., coils and filters.

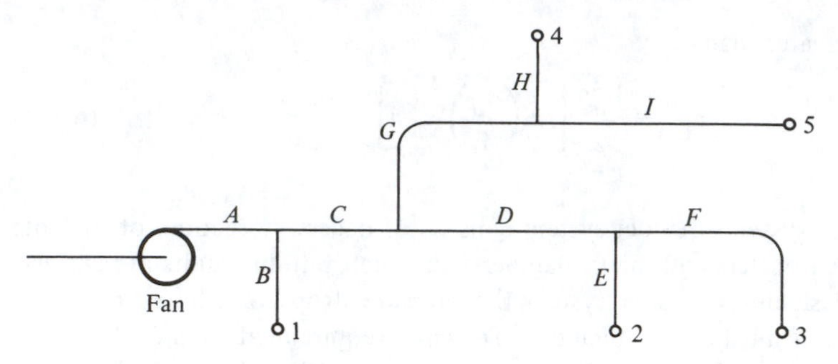

Figure 6-13 Multibranch duct system.

Suppose that the calculation results in the pressure drops shown in Table 6-5. A fan could be selected to develop 92 Pa at the total flow rate, the balancing damper left open in run *A-C-G-H* and the dampers in the other runs partially closed to provide 92 Pa pressure drop in all these runs at the desired flow rates.

An improved design results if one or more sections of run *A-C-G-H* are enlarged to reduce the critical pressure drop. It is also possible to reduce the size of sections in the other runs while staying within velocity constraints, since excess pressure drop is available.

6-13 Equal-friction method There are several versions of the equal-friction method, but one that often results in a superior design is to select the pressure drop to be available in the duct system and size the ducts to dissipate this pressure. The steps in this design method are as follows:

1. Decide what pressure drop will be available.
2. Compute the equivalent length of all runs (the sum of the length of straight duct plus the equivalent length of fittings).
3. Divide the available pressure drop by the equivalent length of the run having the longest equivalent length.
4. With the pressure gradient from step 3 and the flow rate in each section of the longest run, select the duct size of all those sections, using Fig. 6-2.
5. For the remaining sections, select the size to use the available pressure drop but stay within velocities appropriate for noise restrictions.

Table 6-5 Pressure drops in the system in Fig. 6-13

Run	Pressure drop, Pa
A-B	28
A-C-D-E	58
A-C-D-F	43
A-C-G-H	92
A-C-G-I	80

In step 2 the equivalent length can be computed by dividing the coefficient of the $V^2\rho/2$ term in the equation for the pressure loss of the fitting by f/D for duct of the same size. Elbows might be equivalent to 3 to 12 m and branch takeoffs 20 m of straight duct. In step 5 some sections of the main duct might have been selected in sizing the sections in the critical run. If the critical run in the system shown in Fig. 6-13 is *A-C-G-H*, for example, the size of section *A* will be specified in step 4, so the pressure available to section *B* should be computed and the size of *B* chosen to dissipate the available pressure.

The equal-friction method usually results in better design than the velocity method because most of the available pressure is dissipated in friction in the ducts and fittings rather than in balancing dampers. The size and cost of the duct are consequently reduced.

6-14 Optimization of duct systems The principal contributors to the owning cost of a duct system are the costs of the duct and installation, insulation, sound attenuation, energy to drive the fan, and space requirements. The objective of an optimization procedure is to minimize total owning cost. A detailed optimization study may be difficult to perform, and in the design of small duct systems it may require more in engineering cost than it saves in owning cost. On the other hand, in large systems an optimization study may be a good investment, particularly if the value of the building space is considered.

A simple example of an optimization procedure would be to select the duct diameter that minimizes the initial plus operating cost of a duct system consisting of a fan and a length of straight duct. Total cost of the system is the sum of the first cost and the operating cost:

$$\text{Total cost} = C = \text{first cost} + \text{lifetime operating cost}$$

Estimators often use the mass of metal in the duct system as a guide to the cost of the duct system, taking into consideration that the cost of installation might be about 6 times the actual cost of the metal. For circular duct, initial cost could be represented by the expression

$$\text{Initial cost} = (\text{thickness})\,(\pi D)\,(L)\,(\text{density of metal})\,(\text{installed cost/kg}) \qquad (6\text{-}20)$$

where D = diameter of the duct, m
L = length of the duct, m

For a given thickness of material, the expression for the initial cost can be simplified to

$$\text{Initial cost} = C_1 DL \qquad (6\text{-}21)$$

where C_1 is the constant combining the constant terms of Eq. (6-20).

Operating cost of the duct system is the energy cost per hour multiplied by the number of hours of operation expected during the amortization period. The electric power required can be calculated by dividing the air power by the efficiencies of the fan and motor. Since air power is the product of the volume rate of flow and the pressure difference, as will be shown in Sec. 6-16, the operating cost can be expressed as

$$\text{Operating cost} = C_2 H \,\Delta p\, Q \tag{6-22}$$

where H = number of hours of operation during amortization period
C_2 = constant including motor and fan efficiencies and cost of electric energy

Pressure drop Δp, Eq. (6-1), can be replaced by

$$\Delta p = f \frac{L}{D} \frac{V^2}{2} \rho = f \frac{L}{D} \frac{Q^2 \rho}{(\pi^2 D^4/16)2} \tag{6-23}$$

Assuming that the friction factor and the air density are constant, Eq. (6-23) can be substituted into Eq. (6-22) and the constants regrouped into C_3

$$\text{Operating cost} = C_3 \text{LH} \frac{Q^3}{D^5} \tag{6-24}$$

The total cost, which is the sum of the first and operating costs, is

$$C = C_1 DL + \frac{C_3 HLQ^3}{D^5} \tag{6-25}$$

When Eq. (6-25) is differentiated and equated to zero, the optimum diameter is found to be

$$D_{\text{opt}} = \left(\frac{5C_3 HQ^3}{C_1} \right)^{1/6} \tag{6-26}$$

In the above differentiation, the cost of the fan and motor was assumed to be constant. With small duct sizes a further reduction in duct size might require a larger fan and motor, so that a reduction in duct size would increase the first cost of the fan and duct system, rather than decreasing it as Eq. (6-21) suggests.

In actual duct systems with elbows, branches, and other fittings, optimization might not be feasible by purely analytical techniques. Several different duct designs might have to be investigated in order to select an optimum.

6-15 System balancing After a commercial air-handling system has been installed, measurements of actual rates of airflow at supply- and return-air registers should be taken. The dampers should then be adjusted so that the rates of airflow correspond to those specified in the plans. Since balancing an air-handling system in a large building may cost thousands of dollars, a duct system designed so that it is nearly balanced when installed will reduce this cost. The designer is at the mercy of the installers, however, because the quality of construction can appreciably affect the pressure drop, especially in fittings. Nevertheless, an investment in engineering time in the design of the duct system will usually result in a better-operating system.

6-16 Centrifugal fans and their characteristics Air enters a centrifugal fan, as shown in Fig. 6-14, axially, then turns and moves in an approximate radial direction into the blades. The air leaving the blades enters the scroll, which channels the air around the

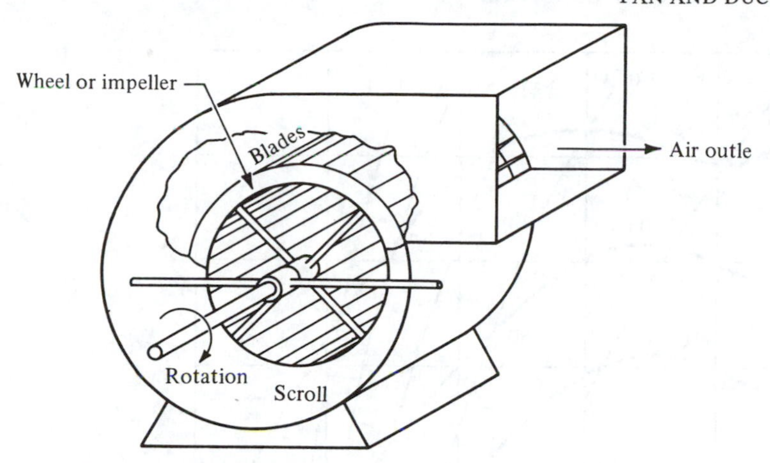

Figure 6-14 A centrifugal fan.

periphery of the wheel and directs it toward the outlet. A fan may have a single or double inlet, depending upon whether the air enters the impeller from one or from both sides. The usual direction of discharge is horizontal, but certain applications call for discharge to be nonhorizontal.

Four types of blading are common in centrifugal fans, radial, forward-curved, backward-curved, and airfoil. The forward-curved blade fan is commonly used in low-pressure air-conditioning systems and is the only type considered in this chapter. Airfoil- or backward-curved-blade fans are used in high-volume or high-pressure systems where increased efficiency is important.

The typical shape of the pressure-flow characteristics of a forward-curved-blade fan operating at various speeds is shown in Fig. 6-15. The characteristic dip in static pressure at low flow rates results[15] because the channels between the blades partially fill with eddies rather than directed flow.

The power required by the fan is also shown on Fig. 6-15. The power ideally required has two components: that needed to raise the pressure and that needed to provide the kinetic energy setting the air in motion. The power required to raise the pressure derives from the expression for an isentropic compression

$$w \int v \, dp$$

where w = mass rate of flow, kg/s
v = specific volume, m^3/kg

The specific volume changes little because the changes in absolute pressure are small in a fan, so v can be removed from within the integral, giving the ideal power required to raise the pressure as

$$\text{Power to raise pressure} = Q(p_2 - p_1) \quad \text{W} \tag{6-27}$$

where Q = volume rate of flow, m^3/s
$p_2 - p_1$ = pressure rise, Pa

Since the power required to provide kinetic energy to the air is $wV^2/2$, the combina-

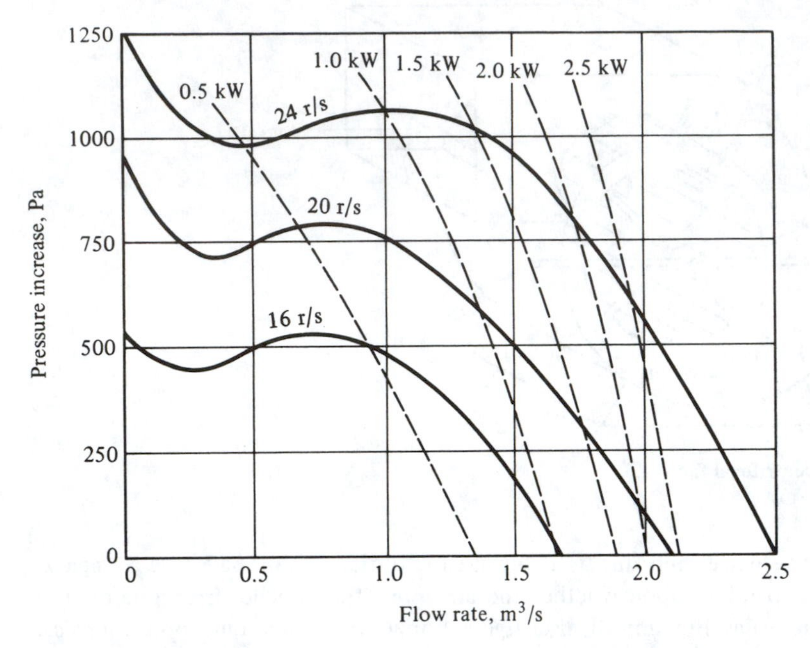

Figure 6-15 Performance characteristics of a forward-curved-blade centrifugal fan. The wheel diameter and the width are both 270 mm. The dimensions of the outlet are 0.517 by 0.289 m.

tion of the two power quantities is

$$\text{Power}_{\text{ideal}} = Q(p_2 - p_1) + \frac{wV^2}{2} \quad \text{W} \tag{6-28}$$

The fan efficiency is defined as the ratio of the ideal to actual power

$$\text{Efficiency } \eta = \frac{\text{power}_{\text{ideal}}}{\text{power}_{\text{actual}}}$$

Example 6-5 Compute the efficiency of the fan whose characteristics are shown in Fig. 6-15 when it operates at 20 r/s and delivers 1.5 m^3/s.

Solution When the rotative speed is 20 r/s and the flow rate is 1.5 m^3/s, the fan can elevate the pressure of the air 500 Pa. The ideal power required to raise the pressure is (1.5 m^3/s) (500 Pa) = 750 W.

Assuming an air density of 1.2 kg/m^3, the mass rate of flow is (1.5 m^3/s) (1.2 kg/m^3) = 1.8 kg/s. Since the area of the fan outlet is 0.517(0.289) = 0.149 m^2, the velocity is (1.5 m^3/s)/(0.149 m^2) = 10.1 m/s. The power required to provide the kinetic energy to the air is (1.8 kg/s) ($10.1^2/2$) = 91 W.

Figure 6-15 shows the power required by the fan at the operating point to be 1.2 kW; the fan efficiency is therefore

$$\eta = \frac{750 + 91}{1200} = 70\%$$

6-17 Fan laws The fan laws are a group of relationships that predict the effect on fan performance of changing such quantities as the conditions of the air, operating speed, and size of the fan. Because the laws applying to fan size are of particular interest to a fan designer and not a user, we do not consider them. Laws applying to a user are important and are shown below with the following notation:

Q = volume rate of flow, m^3/s
ω = rotative speed, r/s
ρ = air density, kg/m^3
SP = static pressure increase through fan, Pa
P = power required by the fan, W

These laws apply to what is called a *constant system,* i.e., one with no changes in the duct and fittings; ~ is read "varies as."

Law 1 Variation in rotative speed, constant air density

$$Q \sim \omega \qquad SP \sim \omega^2 \qquad P \sim \omega^3$$

Law 2 Variation in air density, constant volume flow rate

$$Q = \text{const} \qquad SP \sim \rho \qquad P \sim \rho$$

Law 3 Variation in air density, constant static pressure

$$Q \sim \frac{1}{\sqrt{\rho}}$$

$$SP = \text{const}$$

$$\omega \sim \frac{1}{\sqrt{\rho}}$$

$$P \sim \frac{1}{\sqrt{\rho}}$$

The usefulness of these laws lies in their ability to predict changes from a base condition. Law 1 indicates what happens when the fan speed is changed. Law 2 permits computation of the change in static pressure and power at constant speed and law 3 the change in speed necessary to maintain a constant pressure rise when the density of air is different from that in the base condition.

These rules are called fan laws, but the explanation for them lies in the characteristics of duct systems as well as those of fans. The three laws can be derived from a basic performance characteristic of a fan (α is read "is proportional to").

Fan: $$Q \propto \omega \tag{6-29}$$

along with a characteristic of

Duct and fittings: $$SP \propto \frac{V^2 \rho}{2} \tag{6-30}$$

(from Secs. 6-2, 6-5, and 6-7 to 6-9) and

Power: $$P = Q(\text{SP}) + \frac{QV^2\rho}{2} \qquad (6\text{-}31)$$

from Eq. (6-28). For a constant system the velocity V is proportional to Q. Law 1 follows directly from Eqs. (6-29) to (6-31). Law 2 applies to a constant value of Q, so with a constant system V is also constant. With V constant in Eq. (6-30) SP is proportional to ρ. In Eq. (6-31) SP can be replaced by a constant multiplied by ρ, in the first term and with Q constant P is proportional to ρ. Law 3 specifies that the SP remains constant, and so Eq. (6-30) requires that $V^2\rho/2$ remain constant; thus V and Q vary as $1/\sqrt{\rho}$. Equation (6-29) states the proportionality of ω and Q, and ω therefore also varies as $1/\sqrt{\rho}$. Finally since the Q term is the only one varying on the right side of Eq. (6-31), P also varies as $1/\sqrt{\rho}$.

Example 6-6 The motor driving a fan is rated at 15 A and is currently drawing 11 A while providing a rotative speed of the fan of 15 r/s. The airflow rate delivered by the fan is to be increased as much as possible. What is the permissible rotative speed of the fan while staying within the rating of the motor, and what percentage increase in airflow rate is possible?

Solution Fan law 1 states that the power varies as ω^3; so

$$\omega_2 = \omega_1\left(\frac{p_2}{p_1}\right)^{1/3} = (15\text{ r/s})\left(\frac{15}{11}\right)^{1/3} = 16.6\text{ r/s}$$

Since flow rate is proportional to the rotative speed, the percentage increase in flow rate is

$$\frac{Q_{\text{new}} - Q_{\text{orig}}}{Q_{\text{orig}}} = \frac{16.6 - 15}{15} = 10.6\%$$

6-18 Air distribution in rooms All the topics studied so far in this chapter have dealt with equipment that conveys air from the conditioned space back through the conditioning equipment and delivers the treated air to the space again. Another crucial design requirement is handling the air properly within the conditioned space. This activity, called *air distribution,* must satisfy several overriding requirements:

1. The flow rate in combination with the temperature difference between the supply air and return air must compensate for the net heat loss or gain in the space.
2. The velocity must not be higher than approximately 0.25 m/s in the occupied regions of the room (below head level), particularly when the supply air is cool.
3. There should be some motion of air in the room to break up temperature gradients in the room such as warm air at the ceiling and cold air at the floors during a heating situation.

To accomplish these objectives the designer must properly select the location and the design of the supply-air diffuser and the location of the return-air grille. The ac-

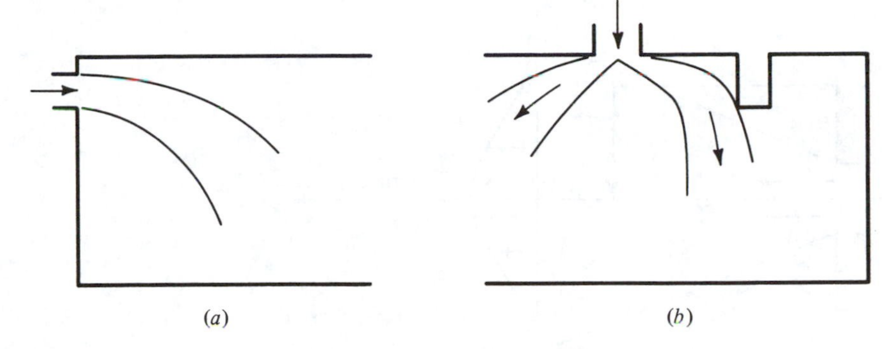

Figure 6-16 (*a*) Drop of a cool air jet and (*b*) deflection when striking a surface.

curate prediction of velocities and temperatures in a conditioned space would require a monstrous computer program that applies the fundamental laws of continuity and momentum transfer. The designer can usually achieve satisfactory air distribution by understanding the influence of the following principles: (1) behavior of a free-stream jet, (2) velocity distribution at an air inlet such as the return-air grille, (3) buoyancy, and (4) deflection. The last three principles will be described briefly, and Sec. 6-19 will present some relations for free-stream jets, which are the building blocks for understanding supply-air inlets.

The highest air velocities in the neighborhood of a return-air grille are at the face of the grille; the velocities fall off very rapidly when moving away from the grille. The effect of velocities controlled by the return-air grilles is in such a confined region that the location and type of return-air grille is not crucial in achieving good air distribution, but the face velocities should be low enough to prevent excessive air noise. The location of the grille (floor-mounted, high-sidewall, etc.) has a slight influence on the overall motion of air in the room, but again it is minor.

Thanks to buoyancy, an airstream cooler than the air in the room will drop, as shown in Fig. 6-16*a*, and a warm stream will rise. The situation to be avoided is the discharge of cool air into a room where buoyancy will cause it to drop and strike at the occupant level.

When a stream of air strikes a solid surface, as in Fig. 6-16*b*, it deflects at the same angle as light would. Thus a stream of air directed against a wall or concrete beam at the ceiling may deflect onto occupants before the stream has been properly diffused.

6-19 Circular and plane jets An understanding of the behavior of circular and plane jets explains many of the characteristics of commercial diffusers. From the combined solution of the momentum and continuity equations the expression[16] for velocities in a circular jet (Fig. 6-17) is

$$u = \frac{7.41 u_o \sqrt{A_o}}{x[1 + 57.5(r^2/x^2)]^2} \tag{6-32}$$

where u = velocity in jet at x and r, m/s

u_o = velocity at outlet, m/s

A_o = area of outlet, m^2

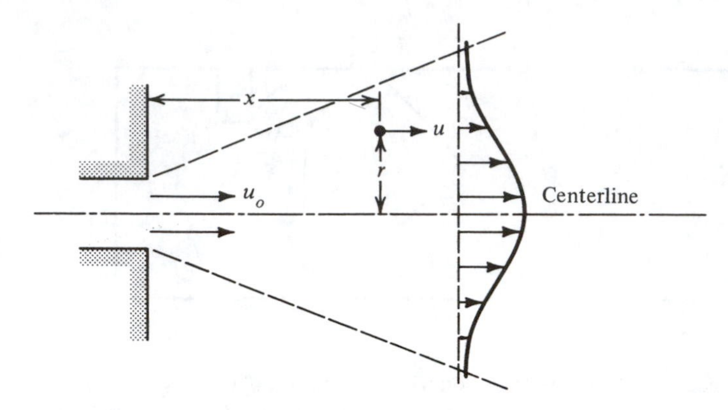

Figure 6-17 A circular jet.

x = distance along centerline from opening, m
r = radial distance from centerline, m

Equation (6-32) permits computation of specific values of velocities and can also be used to predict several trends: (1) the centerline velocity decays as the distance from the opening increases, as illustrated in Example 6-7, (2) the jet spreads as it moves from the outlet, (3) air is entrained as the jet moves from the opening (see Prob. 6-13), and (4) a jet of large diameter (large A_o) sustains its velocity better than a small-diameter jet.

Example 6-7 An air jet issues from a 100-mm-diameter opening with a velocity of 2.1 m/s. What is the centerline velocity 1 and 2 m from the opening?

Solution Along the centerline $r = 0$; so Eq. (6-32) reduces to

$$u_{\text{cl}} = \frac{7.41 u_o \sqrt{A_o}}{x} = \frac{7.41(2.1)\sqrt{0.00785}}{x} = \frac{1.379}{x}$$

$$u_{\text{cl}} = \begin{cases} 1.38 \text{ m/s} \\ 0.69 \text{ m/s} \end{cases} \quad \text{for} \quad \begin{matrix} x = 1 \text{ m} \\ x = 2 \text{ m} \end{matrix}$$

The equation[15] for velocity in a plane jet typical of the jet that prevails when air issues from a long, narrow slot is

$$u = \frac{2.40 u_o \sqrt{b}}{\sqrt{x}} \left[1 - \tanh^2\left(7.67 \frac{y}{x}\right)\right] \tag{6-33}$$

where b = width of opening, m
y = normal distance from centerplane to point where velocity u is being computed

A comparison of Eq. (6-33) for the plane jet with Eq. (6-32) for the circular jet shows that the centerline velocity decreases more rapidly in the circular than in the plane jet as the jet moves away from the outlet. The plane jet entrains less air than the circular jet and is therefore not decelerated as rapidly.

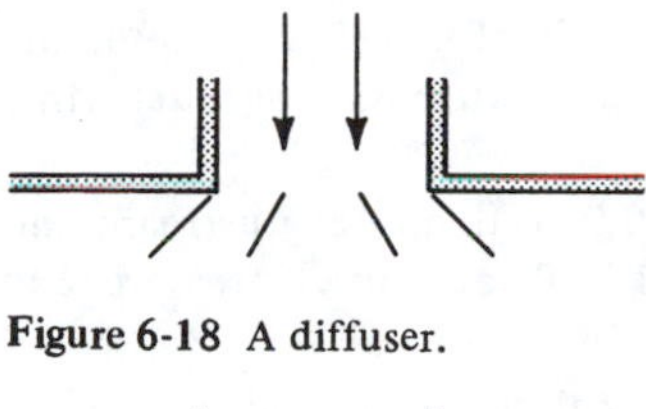

Figure 6-18 A diffuser.

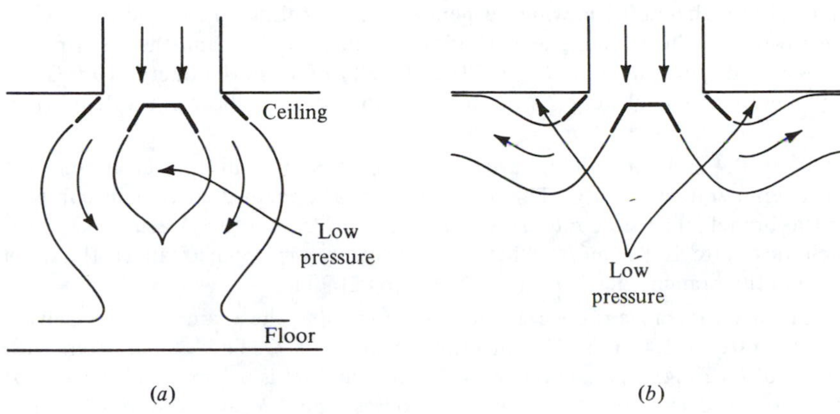

Figure 6-19 Some possible consequences of entrainment.

6-20 Diffusers and induction While long narrow slots are used in actual installations, large circular openings as air outlets are rare because of the long distance a circular jet would penetrate into the room. Instead, diffusers (see the cross section in Fig. 6-18) are common. They provide air patterns in which the velocity decays before the air reaches occupied regions of the room. The diffuser thus expands the influence of the supply outlet over a large region and breaks up temperature gradients.

The phenomenon of entrainment has its desirable features, but entrainment also produces undesirable results. In Fig. 6-19*a* the air from a circular diffuser at the inner cones attempts to entrain air and forms a pocket of low pressure which causes the cone to converge into one large jet. This situation can result in cool supply air dumping down on the region immediately below the diffuser. In Fig. 6-19*b* the cones flip up to the ceiling and the pocket at the ceiling is maintained at low pressure because of entrainment of the upper surface of the cone.

PROBLEMS

6-1 Compute the pressure drop of 30°C air flowing with a mean velocity of 8 m/s in a circular sheet-metal duct 300 mm in diameter and 15 m long, using (*a*) Eqs. (6-1) and (6-2) and (*b*) Fig. 6-2. *Ans.* (*a*) 36 Pa

6-2 A pressure difference of 350 Pa is available to force 20°C air through a circular sheet-metal duct 450 mm in diameter and 25 m long. Using Eq. (6-1) and Fig. 6-1, determine the velocity. *Ans.* 25.6 m/s

6-3 A rectangular duct has dimensions of 0.25 by 1 m. Using Fig. 6-2, determine the pressure drop per meter length when 1.2 m^3/s of air flows through the duct. *Ans.* 0.65 Pa/m

6-4 A sudden enlargement in a circular duct measures 0.2 m diameter upstream and 0.4 m diameter downstream. The upstream pressure is 150 Pa and downstream is 200 Pa. What is the flow rate of 20°C air through the fitting? *Ans.* 0.467 m^3/s

6-5 A duct 0.4 m high and 0.8 m wide, suspended from a ceiling in a corridor, makes a right-angle turn in the horizontal plane. The inner radius is 0.2 m, and the outer radius is 1.0 m, measured from the same center. The velocity of air in the duct is 10 m/s. To how many meters of straight duct is the pressure loss in this elbow equivalent? *Ans.* 15 m.

6-6 An 0.3- by 0.4-m branch duct leaves an 0.3- by 0.6-m main duct at an angle of 60°. The air temperature is 20°C. The dimensions of the main duct remain constant following the branch. The flow rate upstream is 2.7 m^3/s, and the pressure is 250 Pa. The branch flow rate is 1.3 m^3/s. What is the pressure (*a*) downstream in the main duct and (*b*) in the branch duct? *Ans.* (*a*) 346 Pa, (*b*) 209 Pa

6-7 In a branch entry, an airflow rate of 0.8 m^3/s joins the main stream to give a combined flow rate of 2.4 m^3/s. The air temperature is 25°C. The branch enters with an angle $\beta = 30°$ (see Fig. 6-12). The area of the branch duct is 0.1 m^2, and the area of the main duct is 0.2 m^2 both upstream and downstream. What is the reduction in pressure between points u and d in the main duct? *Ans.* 95 Pa

6-8 A two-branch duct system of circular duct is shown in Fig. 6-20. The fittings have the following equivalent length of straight duct: upstream to branch, 4 m; elbow, 2 m. There is negligible pressure loss in the straight-through section of the branch. The designer selects 4 Pa/m as the pressure gradient in the 12- and 15-m straight sections. What diameter should be selected in the branch section to use the available pressure without dampering? *Ans.* 0.35 m

6-9 A duct system consists of a fan and a 25-m length of circular duct that delivers 0.8 m^3/s of air. The installed cost is estimated to be $115 per square meter of sheet metal, the power cost is 6 cents per kilowatthour, the fan efficiency is 55 percent, and the motor efficiency 85 percent. There are 10,000 h of operation during the amortization period. Assume $f = 0.02$. What is the optimum diameter of the duct? *Ans.* 0.24 m

6-10 Measurements made on a newly installed air-handling system were: 20 r/s fan speed, 2.4 m^3/s airflow rate, 340 Pa fan-discharge pressure, and 1.8 kW supplied to the motor. These measurements were made with an air temperature of 20°C, and the system is eventually to operate with air at a temperature of 40°C. If the fan speed remains at 20 r/s, what will be the operating values of (*a*) airflow rate, (*b*) static pressure, and (*c*) power? *Ans.* (*c*) 1.685 kW

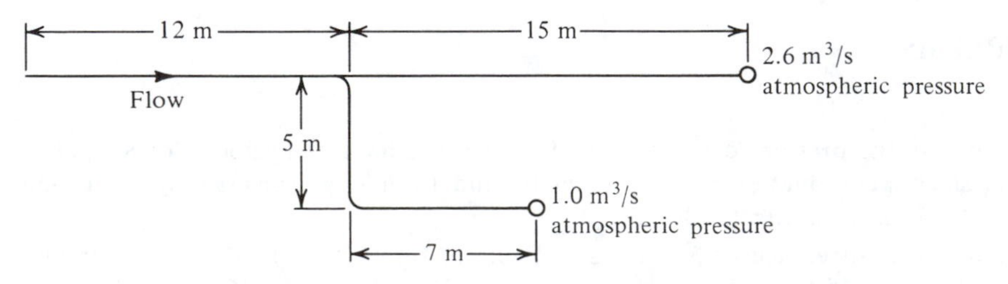

Figure 6-20 Duct system in Prob. 6-8.

6-11 A fan-duct system is designed so that when the air temperature is 20°C, the mass flow rate is 5.2 kg/s when the fan speed is 18 r/s and the fan motor requires 4.1 kW. A new set of requirements is imposed on the system: the operating air temperature is changed to 50°C, and the fan speed is increased so that the same mass flow of air prevails. What are the revised fan speed and power requirement? *Ans.* 19.8 r/s, 4.95 kW

6-12 An airflow rate of 0.05 m^3/s issues from a circular opening in a wall. The centerline velocity of the jet is to be reduced to 0.75 m/s at a point 3 m from the wall. What should be the outlet velocity u_o of this jet? *Ans.* 1.84 m/s

6-13 Section 6-19 points out that jets entrain air as they move away from their inlet into the room. The *entrainment ratio* is defined as the ratio of the air in motion at a given distance x from the inlet to the airflow rate at the inlet Q_x/Q_o. Use the expression for the velocity in a circular jet, Eq. (6-32), multiplied by the area of an annular ring $2\pi r\ dr$ and integrate r from 0 to ∞ to find the expression for Q_x/Q_o. *Ans.* $0.405x/\sqrt{A_o}$

6-14 From the equation for velocities in a plane jet, determine the total included angle between the planes where the velocities are one-half the centerline velocities at that x position. *Ans.* 13.2°

REFERENCES

1. C. F. Colebrook: Turbulent Flow in Pipes with Particular Reference to the Transition Region between Smooth and Rough Pipe Flows, *J. Inst. Civ. Eng.*, vol. 12, no. 4, pp. 133–156, February 1939.
2. L. B. Moody: Friction Factors for Pipe Flow, *ASME Trans.*, vol. 66, p. 671, 1944.
3. R. G. Huebscher: Friction Equivalents for Round, Square, and Rectangular Ducts, *ASHVE Trans.*, vol. 54, p. 101, 1948.
4. A. P. Kratz and J. R. Fellows: Pressure Losses Resulting from Changes in Cross-Sectional Area in Air Ducts, *Univ. Ill. Eng. Exp. Stn. Bull. 300,* 1938.
5. J. Weisbach: "Die experimental Hydraulik," Engelhardt, Freiberg, 1855.
6. A. Hoffman: Der Verlust in 90-Degree-Rohrkruemmern mit gleichbleibendem Kreisquerschnitt, *Mitt. Hydraul. Inst. Tech. Hochsch. Muenchen,* no. 3, 1929.
7. R. D. Madison and J. R. Parker: Pressure Losses in Rectangular Elbows, *ASME Trans.*, vol. 58, pp. 167–176, 1936.
8. D. W. Locklin: Energy Losses in 90-degree Duct Elbows, *ASHVE Trans.*, vol. 56, p. 479, 1950.
9. "ASHRAE Handbook, Fundamentals Volume" chap. 33, American Society of Heating, Refrigerating, and Air-Conditioning Engineers, Atlanta, Ga., 1981.
10. H. H. Korst, N. A. Buckley, S. Konzo, and R. W. Roose: Fitting Losses for Extended-Plenum Forced Air Systems, *ASHVE Trans.* vol. 56, p. 295, 1950.
11. E. Kinne: Beitrage zur Kenntnis der hydraulischen Verluste in Abzweigstucken, *Mitt. Hydraul. Inst. Tech. Hochsch. Muenchen,* vol. 4, pp. 70–93, 1931.
12. F. Petermann: Der Verlust in schiefwinkligen Rohrverzweigungen, *Mitt. Hydraul. Inst. Tech. Hochsch. Muenchen,* vol. 3, pp. 98–117, 1929.
13. J. R. Smith and J. W. Jones: Pressure Loss in High Velocity Flat Oval Duct Fittings, *ASHRAE Trans.*, vol. 82, pt. I, pp. 244–255, 1976.
14. J. H. Healy, M. N. Patterson, and E. J. Brown: Pressure Losses through Fittings Used in Return Air Duct Systems, *ASHRAE Trans.*, vol. 68, p. 281, 1962.
15. B. Eck: "Ventilatoren," 4th ed., Springer-Verlag, Berlin, 1962.
16. H. Schlichting: "Boundary Layer Theory," 7th ed., McGraw-Hill, New York, 1979.

CHAPTER

SEVEN

PUMPS AND PIPING

7-1 Water and refrigerant piping The most common heat-conveying media in air-conditioning and refrigeration systems are air, water, and refrigerants. Airflow systems were studied in Chap. 6, and this chapter concentrates on piping systems for water and pumps for motivating the flow of water in what are often called *hydronic systems.* This chapter also covers guidelines for selecting the size of refrigerant pipes. Special procedures are needed for the design of steam systems and the return of condensate, but this subject will not be treated. Although steam systems are quite common in industrial facilities, in air-conditioning systems hot water has almost completely supplanted steam as a heat-conveying medium.

The requirements of a water-distribution system are that it provide the necessary flow rate to all the heat exchangers, that it be safe, and that its life-cycle cost (including both first and operating costs) be low. In selecting sizes of refrigerant pipes there are some standard recommendations which are heavily influenced by the refrigerant pressure drop. Some pressure drop is expected, but the pipe size should ensure that it will not be excessive, which would result in high operating cost.

This chapter first compares air and water as media for conveying heat, citing reasons why air would be used in one situation and water in another. Next, water heaters (popularly and incorrectly called boilers) are described and thereafter the other elements of a water-distribution system, including the heat exchangers, pipes, and pumps. Finally, the joint working of these elements is explored in the design of water-distribution systems.

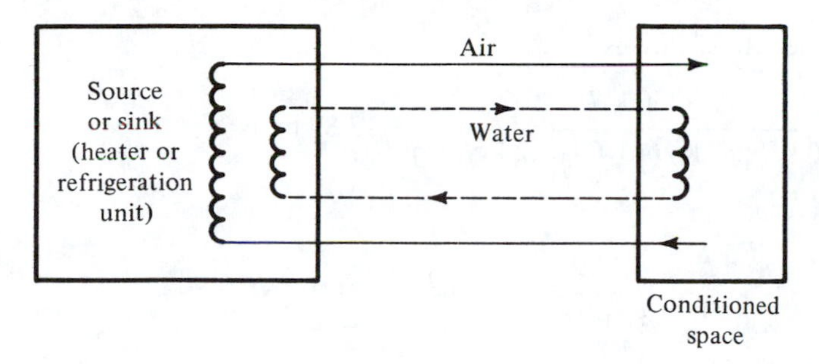

Figure 7-1 Concept of heat-conveying process in an air-conditioning system.

7-2 Comparison of water and air as heat-conveying media The final transfer of heat is almost always either from or to the air in the conditioned space. The device for providing heat (heat source) is usually an electric or fuel-fired furnace, and the device for extracting heat (heat sink) is a refrigeration unit. When a heat pump is used, the same equipment is both the source and sink of energy. While the source or sink of heat may sometimes be located in the conditioned space, the typical situation is that the source and sink be remotely located from the conditioned space, so that either air or water is heated or cooled at the source or sink and cooled or heated, respectively, at the conditioned space, as shown schematically in Fig. 7-1. Air could be heated or cooled at the source or sink and delivered directly to the conditioned space, or water could be heated or cooled and subsequently heat or cool the air in the conditioned space.

The advantages of the water- over the air-distribution arrangement are (1) that the size of the heat source is smaller, (2) that less space is required by water pipes than by air ducts, and (3) that a higher temperature of water than of air is practical in heating, since the pipes, which are small, are easier to insulate than ducts.

Example 7-1 A heat-transfer rate of 250 kW is to be effected through a change in medium temperature of 15°C. What must the cross-sectional area be to convey this energy flow if (*a*) a water pipe is used and the water velocity is 1 m/s and (*b*) an air duct is used and the air velocity is 10 m/s?

Solution (*a*) For water the volume rate of flow is

$$\frac{250 \text{ kW}}{(4.19 \text{ kJ/kg} \cdot \text{K})(15°\text{C})(1000 \text{ kg/m}^3)} = 0.00398 \text{ m}^3/\text{s}$$

The cross-sectional area is

$$\frac{0.00398}{1 \text{ m/s}} = 0.00398 \text{ m}^2$$

(*b*) For air the volume rate of flow is

$$\frac{250\ \text{kW}}{(1.0\ \text{kJ/kg}\cdot\text{K})\,(15^\circ\text{C})\,(1.2\ \text{kg/m}^3)} = 13.89\ \text{m}^3/\text{s}$$

The cross-sectional area is

$$\frac{13.89}{10\ \text{m/s}} = 1.38\ \text{m}^2$$

which is 347 times the area of the water pipe.

The strengths and weaknesses of the two conveying media usually resolve into the following system choices. Small plants, such as residential and small commercial, use air throughout the system and no water whatsoever. The distances the heat must be transferred are short and the order of magnitude of the capacities does not result in excessive sizes of heat sources and ducts. Large air-conditioning systems, on the other hand, use hot- and chilled-water distribution. Fuel-fired or electric heat sources to heat air directly are not generally available in large sizes. The hot or chilled water may be piped directly to coils in the conditioned space; another popular arrangement is to pipe water to air heating and cooling coils, each serving one floor or one section of the building.

7-3 Water heaters The combustion of fuel (natural gas, coal, oil, etc.) and electric resistance heating are the major sources of energy for heating water in hydronic systems. Fuel-fired water heaters are usually constructed of steel under strict safety codes. One way of classifying heaters is according to the operating pressure (and thus the permissible water temperature). The lowest-pressure group provides water at a temperature of approximately 100°C, and so the pressure is at atmospheric or slightly above. Higher-pressure heaters serve the systems discussed in Sec. 7-5.

The efficiency of fuel-fired water heaters is defined as the rate of energy supplied to the water divided by the rate of energy input based on the lower heating value of the fuel. The lower heating value is the heat of combustion with the assumption that the water in the flue gas leaves as vapor. One of the losses of water heaters is a *standby loss* that occurs when the burner is shut off but air from the equipment room flows by natural convection up past the hot surfaces of the heater and carries heat out the stack. Many fuel-fired water heaters are equipped with dampers that close when the burner turns off, and some larger heaters are equipped with modulating air control that regulates the flow of combustion air so that there is enough air for complete combustion but not enough to carry heat away in the excess air. Typical efficiencies of commercial waters heaters are approximately 80 percent.

The size of the heater is sometimes chosen larger than the maximum design heating capacity in order to have excess capacity to bring the temperature of the building up after a night or weekend setback. When this extra heater capacity is required, it is important that the piping distribution system also be sized larger than design;[1] otherwise the distribution system becomes a bottleneck in transferring the extra capacity to the conditioned space.

Figure 7-2 A fire-tube water heater. *(Cleaver-Brooks Division of Aqua-Chem, Inc.)*

Figure 7-2 shows a firetube heater in which the combustion gases flow through the tubes, and the water being heated is in a shell that surrounds the tubes.

7-4 Heat distribution from hot-water systems The principal types of heat exchangers used to transfer energy from the hot water are coils in warm air ducts, fan-coil units, or natural-convection convectors located in the conditioned space. The convectors may be of the cabinet type or of the baseboard type (Fig. 7-3). The baseboard convector consists of a tube with fins attached. Hot water flows through the tube, and the air

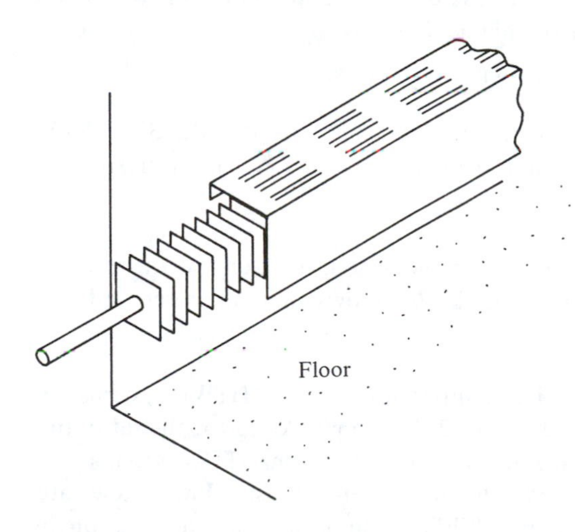

Figure 7-3 A baseboard convector.

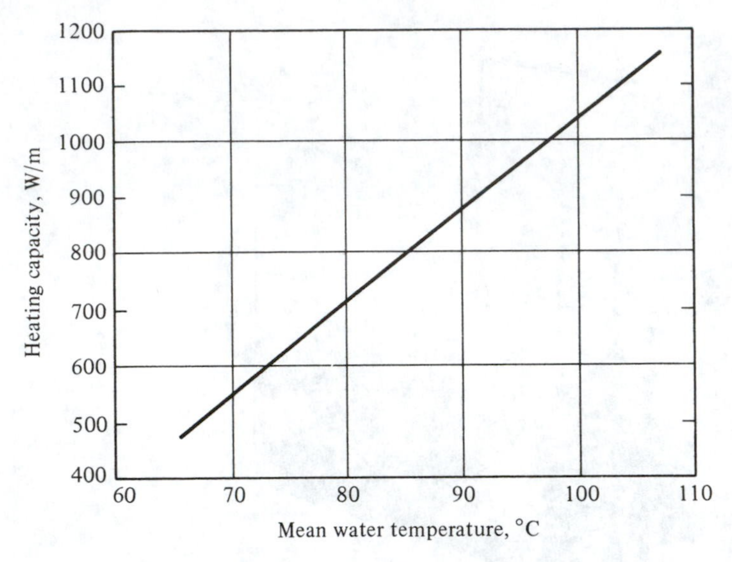

Figure 7-4 Heating capacity of a certain baseboard convector, based on room air entering at 18°C.

being heated flows by natural convection up over the tube and fins and out the louvers at the top of the enclosure. The performance of a certain baseboard convector is shown in Fig. 7-4.

Example 7-2 What mean water temperature is needed in the baseboard convector of Fig. 7-4 to compensate for the heat loss from a single-pane glass wall when the indoor and outdoor design air temperatures are 21 and -23°C, respectively? The height of the glass is 2.4 m, and the convectors are placed along the entire length of wall.

Solution The U value of single-pane glass is 6.2 $W/m^2 \cdot K$ from Table 4-4, so the heat-transfer rate through each square meter of glass is

$$6.2[21 - (-23)] = 273 \text{ W}$$

Each 1-m length of glass wall experiences a design heat loss of 2.4(273) = 655 W. Figure 7-4 shows that the convector can provide the 655 W with a mean water temperature of 77°C.

The heating load through an expanse of single-pane glass is one of the worst in building design, but the calculation in Example 7-2 shows that the baseboard convectors can handle this extreme load.

7-5 High-temperature water systems High-temperature water (HTW) systems are those operating with supply water in the 180 to 230°C range. At 230°C the saturation pressure of water is 2800 kPa. A fundamental reason for using HTW systems is to make it possible to transfer the required rate of heat energy using lower flow rates than if the supply temperature were lower. While steam generators are occasionally

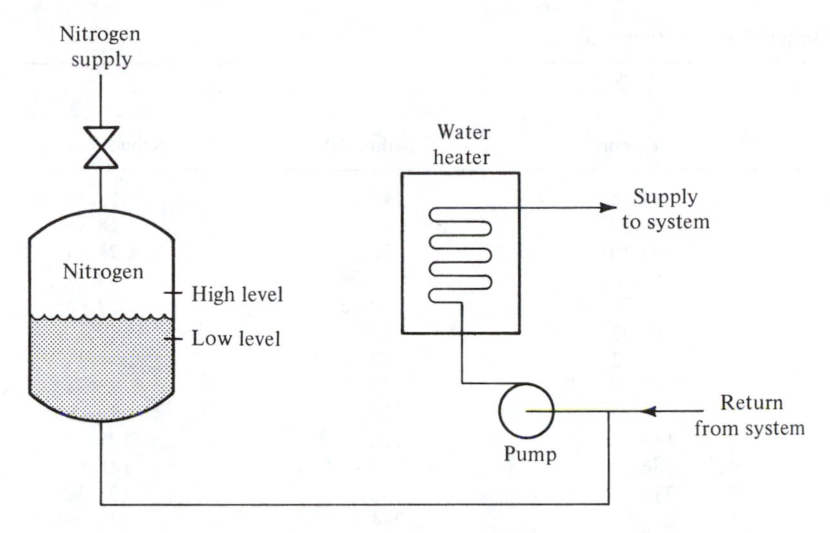

Figure 7-5 Nitrogen pressurization of a high-temperature water system.

used as the heater in a HTW plant, systems are usually pressurized so that no vaporization of water takes place in the heater, and in no case should flashing into vapor be permitted in the distribution system. One common method of pressurization is through the use of high-pressure nitrogen,[2] as shown in Fig. 7-5. A pressurization chamber is connected to the system, and the level of liquid is maintained in this vessel between the high and low limits. Should the system pressure fall too low, nitrogen is added to the system. If there should be a leak of water somewhere in the system, both the pressure and water level are likely to drop and water should be replaced.

7-6 Available pipe and tubing Standard sizes of copper tubes and of steel pipe used in refrigeration and air-conditioning systems are shown in Tables 7-1 and 7-2. Of the thicknesses shown, type L copper tubing and Schedule 40 pipe thicknesses are the most common.

Table 7-1 Dimensions of copper tubes

	ID, mm			ID, mm	
OD, mm	Type K	Type L	OD, mm	Type K	Type L
9.53	7.75	8.00	53.98	49.76	50.42
12.70	10.21	10.92	66.68	61.85	62.61
15.88	13.39	13.84	79.38	73.84	74.80
19.05	16.56	16.92	92.08	85.98	87.00
22.23	18.92	19.94	104.8	97.97	99.19
28.58	25.27	26.04	130.2	122.1	123.8
34.93	31.62	32.13	155.6	145.8	148.5
41.28	37.62	38.23	206.4	192.6	196.2
			257.2	240.0	244.5
			308.0	287.4	293.8

Table 7-2 Dimensions of steel pipe

Nominal size, mm	OD, mm	ID, mm	
		Schedule 40	Schedule 80
15	21.34	15.80	13.88
20	26.67	20.93	18.85
25	33.40	26.64	24.30
35	42.16	35.04	32.46
40	48.26	40.90	38.10
50	60.33	52.51	49.25
60	73.03	62.65	59.01
75	88.90	77.92	73.66
100	114.3	102.3	97.18
125	141.3	128.2	122.2
150	168.3	154.1	146.4
200	219.1	202.7	193.7
250	273.0	254.5	242.9
300	323.9	303.3	289.0

7-7 Pressure drop of water flowing in pipes The equation for drop in pressure of a fluid flowing in a straight pipe is the same as for airflow, Eq. (6-1) repeated here

$$\Delta p = f \frac{L}{D} \frac{V^2}{2} \rho \qquad (7\text{-}1)$$

Many engineers calculating the pressure drop of water in pipes express the diameter D in millimeters, which makes the units of Δp kilopascals, a convenient unit for water pressure.

Example 7-3 Compute the pressure drop when 3.0 L/s of 80°C water flows through a steel pipe with a nominal diameter of 50 mm (ID = 52.5 mm) that is 40 m long.

Solution From Table 7-3 at 80°C the pertinent properties of water are

$$\rho = 971.64 \text{ kg/m}^3 \qquad \mu = 0.358 \text{ mPa} \cdot \text{s}$$

Table 7-3 Density and viscosity of water at various temperatures

t, °C	Viscosity, mPa · s	Density, kg/m^3	t, °C	Viscosity, mPa · s	Density, kg/m^3
0	1.790	999.84	60	0.476	983.19
10	1.310	999.70	70	0.406	977.71
20	1.008	998.21	80	0.358	971.63
30	0.803	995.64	90	0.319	965.16
40	0.656	992.22	100	0.282	958.13
50	0.552	988.04			

For steel $\epsilon = 0.000046$ m from Table 6-1. The velocity is

$$V = \frac{0.003 \text{ m}^3/\text{s}}{\pi(0.0525^2)/4} = 1.386 \text{ m/s}$$

$$\frac{\epsilon}{D} = \frac{0.000046}{0.0525} = 0.00088$$

$$\text{Re} = \frac{(1.386 \text{ m/s})(0.0525 \text{ m})(971.63 \text{ kg/m}^3)}{0.358 \text{ mPa} \cdot \text{s}} = 197{,}500$$

From the Moody chart (Fig. 6-1) at Re = 197,500 and $\epsilon/D = 0.00088$, $f = 0.0208$

$$\Delta p = 0.0208 \frac{40}{52.5 \text{ mm}} \frac{1.386^2}{2} (971.6) = 14.8 \text{ kPa}$$

Many designers find a pressure-drop chart like that in Fig. 7-6 convenient when many calculations must be performed. Such a chart can apply only to one temperature of water, because the water temperature affects both the density and viscosity. The density appears directly in Eq. (7-1), and both the density and viscosity influence the Reynolds number and thus the friction factor. The best single parameter besides the temperature that correlates the pressure drop at other temperatures is the velocity. Figure 7-7 shows a graph of the correction factors to be applied to Fig. 7-6 for temperatures other than the 20°C for which Fig. 7-6 applies.

Example 7-4 Use Figs. 7-6 and 7-7 to solve Example 7-3.

Solution For pipe with a nominal diameter of 50 mm and a flow rate of 3 L/s, Fig. 7-6 shows a pressure drop of 425 Pa/m and a velocity of 1.4 m/s. The correction factor to be applied to 425 Pa/m at a temperature of 80°C and a velocity of 1.4 m/s is found from Fig. 7-7 to be 0.885

$$\Delta p = (425 \text{ Pa/m})(0.885)(40 \text{ m}) = 15.1 \text{ kPa}$$

compared with 14.8 kPa from Example 7-3.

7-8 Pressure drop in fittings One approach to computing the pressure drop caused by fittings (elbows, tees, open valves, etc.) in a piping system is to express their pressure drop in terms of the equivalent length of straight pipe that would cause the same pressure drop. The usefulness of this method is that the section of the piping system in which the diameter and flow are constant can be considered as one length of straight pipe. Table 7-4 shows some equivalent lengths of straight pipe.[3]

7-9 Refrigerant piping Refrigeration systems and components will be explained in Chap. 10 and thereafter, but it is appropriate to discuss here the three major pipe sections in a basic refrigeration system. As shown in Fig. 7-8, they are the piping for the discharge gas, for liquid, and for suction gas. Somewhat different considerations apply to the selection of sizes of these three different pipe sections.

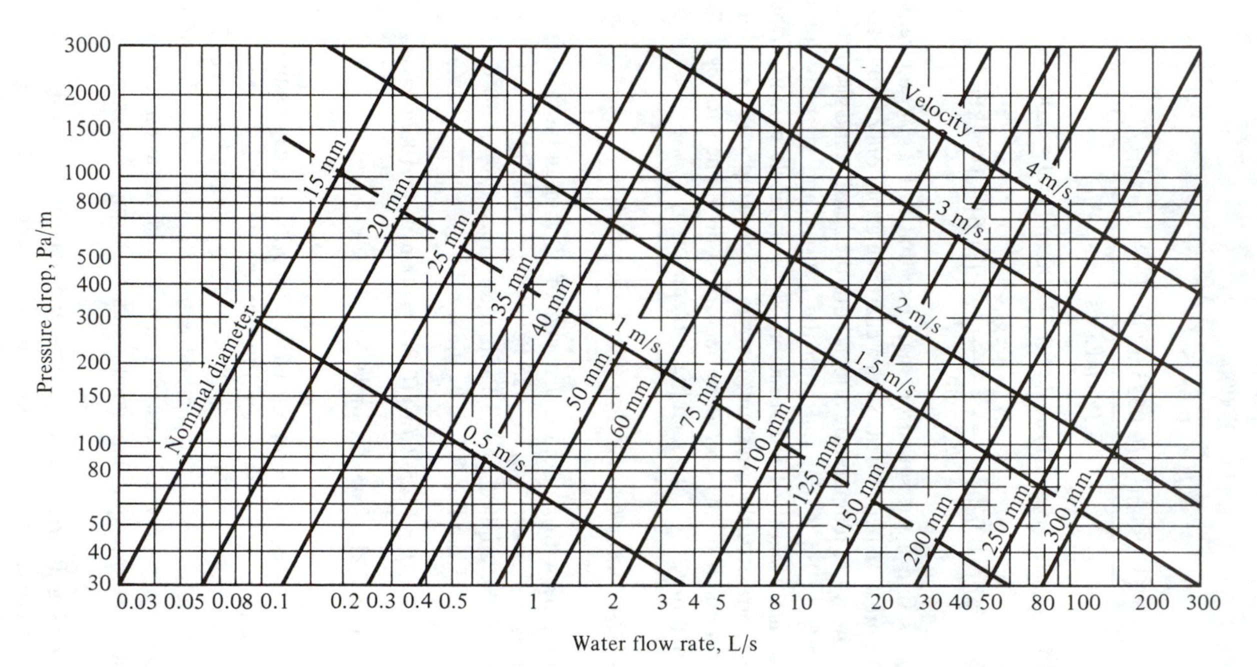

Figure 7-6 Pressure drop for water flowing in Schedule 40 steel pipes, using Eq. (7-1), for a water temperature of 20°C.

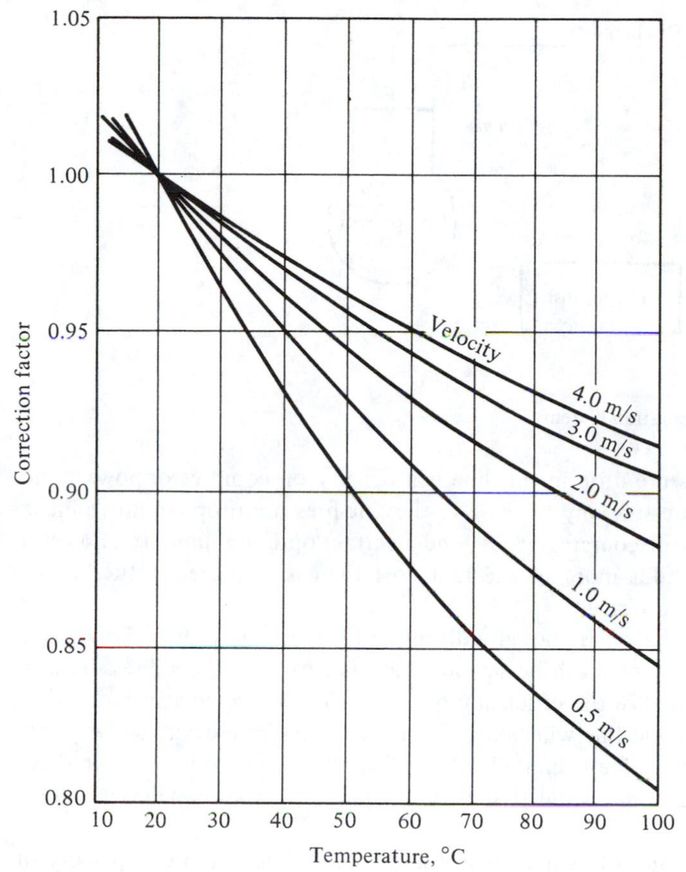

Figure 7-7 Multiplying factor for the pressure drops read from Fig. 7-6 in order to correct for temperature.

Table 7-4 Equivalent lengths in straight pipe of several fittings, meters†

Pipe diameter, mm	90° elbow	45° elbow	Tee: Side branch	Tee: Straight branch	Open globe valve
15	0.6	0.4	0.9	0.2	5
20	0.8	0.5	1.2	0.2	6
25	0.9	0.6	1.5	0.3	8
35	1.2	0.7	1.8	0.4	11
40	1.5	0.9	2.1	0.5	14
50	2.1	1.2	3.0	0.6	17
60	2.4	1.5	3.7	0.8	20
75	3.0	1.8	4.6	0.9	24
100	4.3	2.4	6.4	1.2	38
125	5.2	3.0	7.6	1.5	43
150	6.1	3.7	9.1	1.8	50

† Data from Plumbing Manual, *U.S. Natl. Bur. Std. Rep.* BM566.

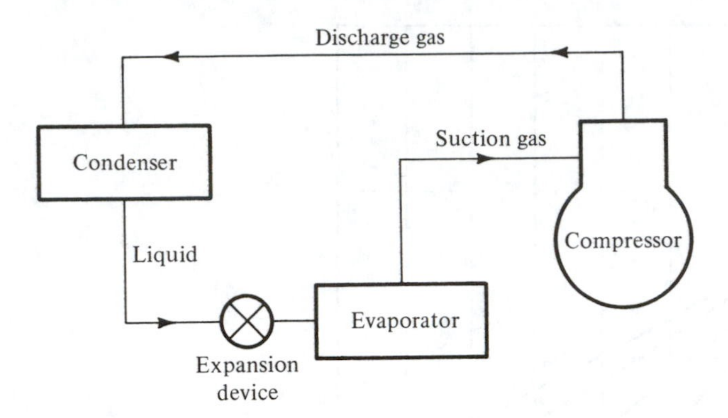

Figure 7-8 Piping in a refrigeration system.

Discharge line The pressure drop in this line is a penalty on compressor power, since for a given condenser pressure any increase in the line-pressure drop requires a higher discharge pressure from the compressor. Beyond a certain optimum pipe size, however, additional enlargement adds more to the first cost than is recovered in the lifetime pumping cost of the compressor.

Liquid line Since this pipe carries liquid with much higher density than the vapor in the other sections, its diameter will be smaller than that of the others. Pressure drop in this line does not penalize the efficiency of the cycle because what pressure drop does not occur in the liquid line will take place in the expansion device anyway. The pressure drop in the liquid line is limited for a different reason, however: if the pressure drops so much that some liquid flashes into vapor, the expansion device will not work properly.

Suction line Pressure drop in this line, as in the discharge line, imposes a penalty on efficiency because this pressure drop reduces the entering pressure to the compressor. There is a limitation on how large a suction line can be chosen, however, imposed by the need in many refrigeration systems to carry lubricating oil from the evaporator back to the compressor. Velocities in vertical suction lines are often maintained at 6 m/s and higher in order to facilitate oil return.

The pressure drop in refrigerant lines can be computed using Eq. (7-1) with refrigerant properties in Table 15-5 (for viscosities) and the densities from tables in the appendix. The pressure drop corresponding to various refrigerating capacities can also be obtained from Ref. 4.

7-10 Pump characteristics and selection The most useful performance data of a pump are the pressure differences it is capable of developing at various flow rates. Of equal importance is the knowledge of the power requirement at the design condition and at other possible operating points. Typical performance characteristics of a centrifugal pump are shown in Fig. 7-9. Pump manufacturers often show in their catalogs the Δp curve and the power (or required motor size) at various positions along the curve. The isoefficiency curves (curves of constant efficiency) shown in Fig. 7-9 are not normally given in catalogs but are presented here to increase understanding of pump performance.

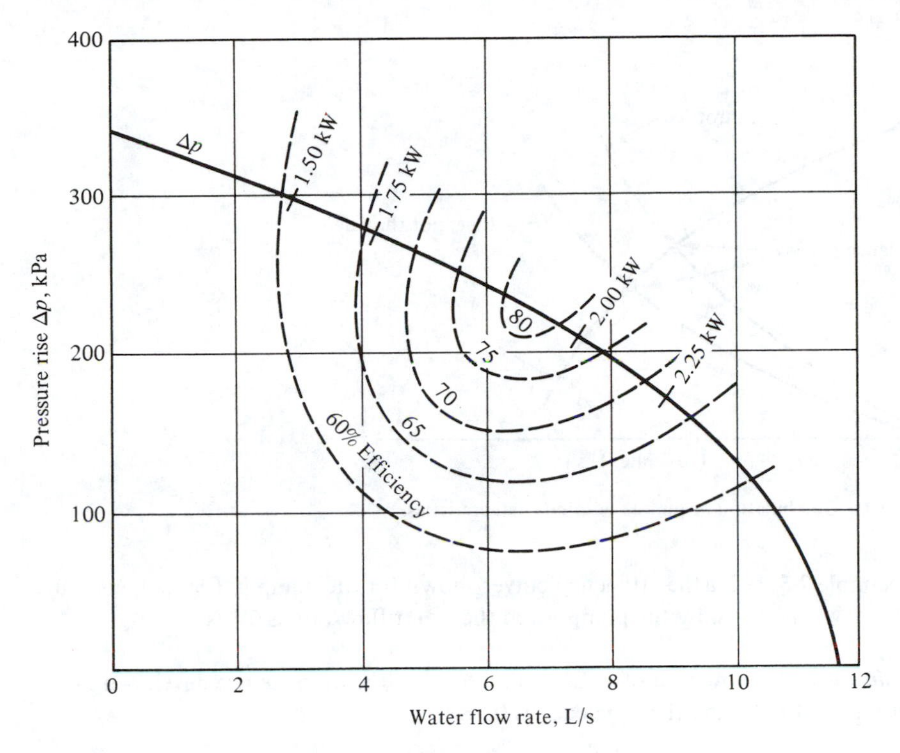

Figure 7-9 Performance characteristics of a centrifugal pump.

The power required in a perfect pumping or compression process P_i is the integral of $v\,dp$

$$P_i = w \int_{p_1}^{p_2} v\,dp \tag{7-2}$$

where P_i = ideal power, W
p_1 = entering pressure, Pa
p_2 = leaving pressure, Pa
w = mass rate of flow, kg/s
v = specific volume, m^3/kg

Since the specific volume of a liquid experiences a negligible change as it passes through the pump, v can be moved outside the integral sign and combined with w to give Q, the volume rate of flow in cubic meters per second. The expression then becomes

$$P_i = Q(p_2 - p_1) \tag{7-3}$$

The power P required in the actual pumping process where there are losses is

$$P = \frac{Q(p_2 - p_1)}{\eta/100} \tag{7-4}$$

where η is the efficiency, percent.

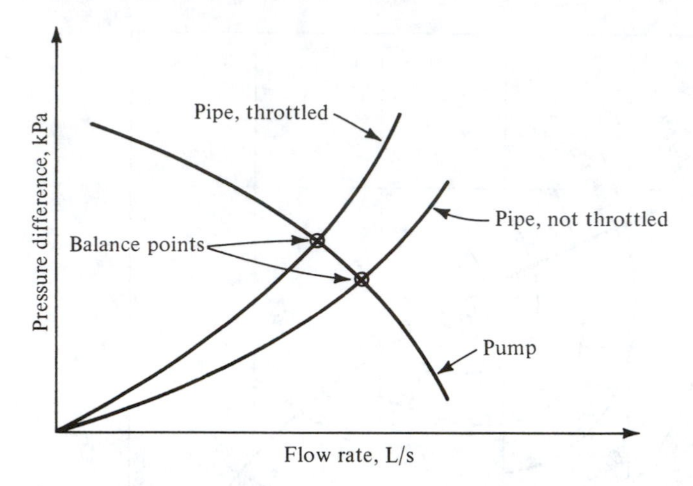

Figure 7-10 Combination of pump and pipe characteristics.

Example 7-5 Using the efficiency curves shown for the pump in Fig. 7-9, compute the power required by the pump when the water flow rate is 6 L/s.

Solution At a flow rate of 6 L/s = 0.006 m^3/s, the pressure rise developed in the pump is 240 kPa and the efficiency is 0.78

$$\text{Power} = \frac{0.006(240{,}000)}{0.78} = 1846 \text{ W}$$

The performance of the pump usually must be considered in combination with the characteristics of the pipe network it serves. A visualization of the combined pump and pipe characteristics is available from a graph of pressure difference versus flow rate, as in Fig. 7-10. The pump curve has the shape already shown in Fig. 7-9. The pressure difference experienced by the pipe increases as the square of the flow rate. This relationship is predictable from Eq. (7-1) for straight pipe, and Sec. 7-8 suggests that the pressure drop in fittings also increases as the square of the flow rate. The pressure drop of a fitting is thus translatable into a length of straight pipe.

The intersection of the pipe characteristics with the pump characteristic in Fig. 7-10 is called the *balance point* because here the flow rate and pressure difference of the two components are satisfied. Figure 7-10 shows a balance point for a pipe with no throttling (valves fully open). If the flow rate is greater than desired, a valve can be partially closed to introduce additional resistance in the pipeline. The result is a lower flow rate and higher pressure difference. It is of interest to refer back to Fig. 7-9 to see what happens to the required power as the pipe network is throttled. Instead of the increased power requirement intuition leads one to expect, the power drops.

7-11 Design of a water-distribution system In closed water systems the major components are straight pipe, valves, fittings, pump(s), heat exchangers, and an expansion tank. The design process includes both the determination of sizes and their arrange-

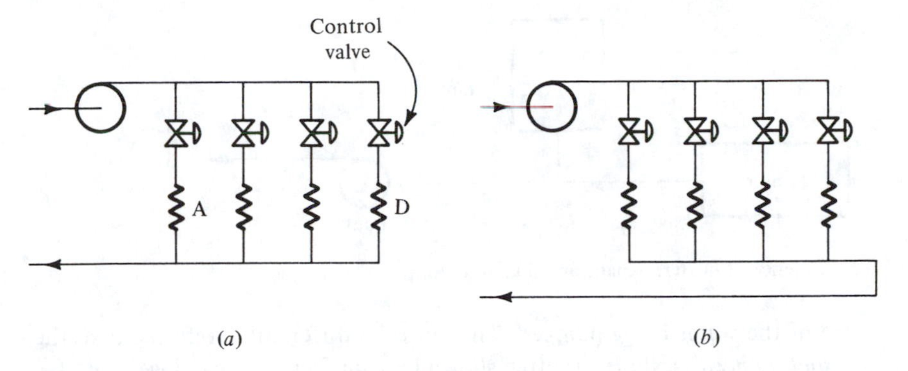

Figure 7-11 Piping arrangements: (*a*) direct return and (*b*) reversed return.

ment. Some major tasks in the design process are to decide on the location of the components, select the pipe size, select the pump, and choose the size of the expansion tank.

Two basic piping arrangements are the *direct-return* and the *reversed-return plans* (Fig. 7-11). A drawback of the direct-return system is that the pressure difference available to the various heat exchangers is nonuniform. Heat exchanger *A* in Fig. 7-11*a* has available a greater pressure difference than heat exchanger *D*. The control valve of heat exchanger *A* might have to be nearly closed; in that condition its operation is not stable, and heat exchanger *D* may have insufficient pressure difference available to provide the required flow rate. The reversed-return system provides essentially uniform pressure difference to all heat exchangers. The disadvantage of the reversed-return arrangement is the additional pipe required compared with the direct-return arrangement.

Another basic question of component location is the relative location of the heater, expansion tank, and pump.[5] Several principles and operating characteristics should be observed. When the pump is set in operation, the relative pressures change throughout the system, a high pressure developing at the pump outlet and a low pressure at the pump inlet. The one location where the absolute pressure stays constant is at the expansion tank. When these two facts are combined, the preferred location of the expansion tank can be chosen. If the expansion tank is placed after the pump, the outlet pressure remains constant and the inlet pressure drops, which may cause cavitation, described in the next paragraph. A rule is to "pump away from the expansion tank." Next try the sequence of tank, pump, and water heater. The heater then receives a high pressure during operation, and this pressure may be high enough to open the relief valve. The usual sequence is shown in Fig. 7-12 as heater, expansion tank, and pump.

In pumping hot water care should be taken to prevent cavitation at the pump. Cavitation is caused by the liquid water flashing into vapor at localized regions of low pressure. Cavitation results in poor performance of the pump and accelerated wear. The critical location for cavitation is at the pump inlet, where the pressure is low and localized regions of high velocity further reduce the pressure. To prevent cavitation the pressure at the pump inlet must be kept a certain magnitude higher than the satura-

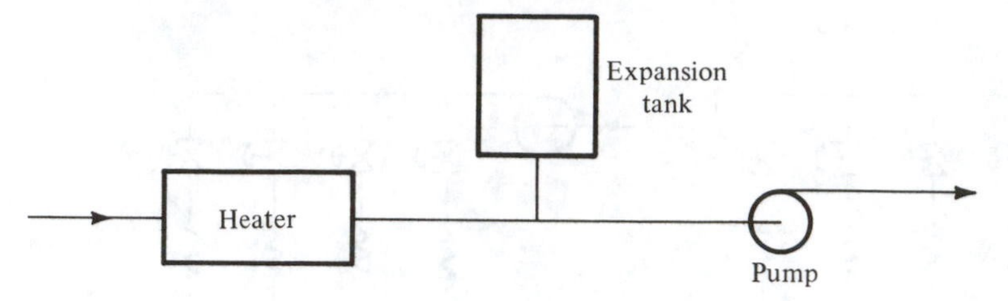

Figure 7-12 Sequence of heater, expansion tank, and pump.

tion pressure of the water being pumped. This pressure differential, referred to as the *net positive suction head*[6] (NPSH), is often shown in manufacturers' catalogs.

7-12 Sizing the expansion tank The purpose of the expansion tank is to provide a cushion of air that accommodates the change in volume of the water as it changes temperature. An equation for the expansion-tank size adapted from Ref. 7 is

$$V_t = \frac{\Delta v}{v_c} \frac{V_s}{p_i/p_c - p_i/p_h} \tag{7-5}$$

where Δv = difference in specific volume of liquid water between operating and filling temperatures, m^3/kg

v_c = specific volume of liquid water at filling temperature, m^3/kg

V_s = volume of system, m^3

p_i = pressure in expansion tank when water first enters, kPa abs

p_c = pressure in expansion tank before raising water temperature, kPa abs

p_h = pressure in expansion tank when water in system is hot, kPa abs

The logic behind Eq. (7-5) can be established by tracing the sequence of events when the tank is first filled and the water temperature of the system is first raised. The system diagram in Fig. 7-13 shows three stages, represented by three levels of water in the expansion tank; *A*, *B*, and *C*. At stage *A* the level of water has just sealed off the air in the tank and the pressure in the tank is p_i. As filling of the system continues and air is vented from the piping, a static head imposes a higher pressure p_c at the lower-level expansion tank as the air in the tank is compressed. After filling the

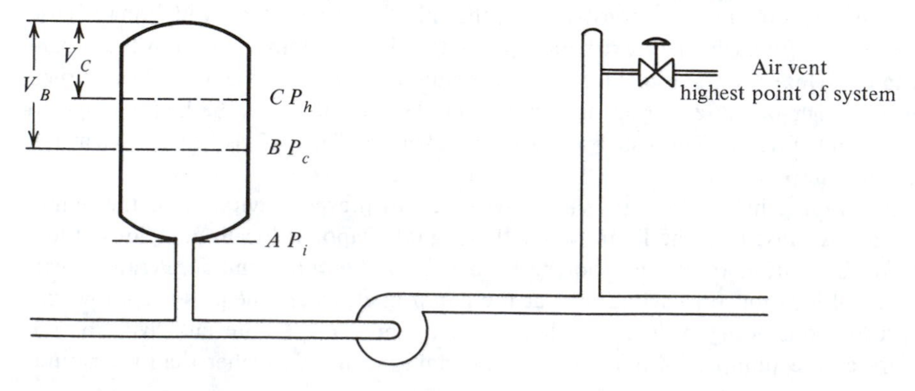

Figure 7-13 Water levels and pressures in an expansion tank.

system the water is brought up to temperature and in the process expands, compressing the air in the tank to p_h and raising the level in the tank to C. During operation the level in the tank should vary between B and C with the pressure varying from p_c to p_h.

In Eq. (7-5) the term $(\Delta v/v_c)V_s$ is the volume change in the system as the water is heated from its filling temperature to operating temperature; thus

$$\frac{\Delta v}{v_c} V_s = V_B - V_C \tag{7-6}$$

The difference in the pressure ratios, assuming a constant temperature of the air in the tank, is

$$\frac{1}{p_i/p_c - p_i/p_h} = \frac{1}{V_B/V_t - V_C/V_t} = \frac{V_t}{V_B - V_C} \tag{7-7}$$

The product of the terms in Eqs. (7-6) and (7-7) yields the tank volume V_t.

Example 7-6 What is the size of an expansion tank for a hot-water system with a volume of 7.6 m^3 if the highest point in the system is 12 m above the expansion tank, the system is filled with 20°C water, the operating temperature is to be 90°C, and the maximum pressure in the system is to be 250 kPa gauge?

Solution The specific volumes of liquid water are found in Table A-1, 0.0010017 m^3/kg at 20°C, and 0.0010361 m^3/kg at 90°C. The change in water volume to be accommodated in the expansion tank is

$$\frac{\Delta v}{v_c} V_s = \frac{0.0010361 - 0.0010017}{0.0010017} (7.6 \text{ m}^3) = 0.261 \text{ m}^3$$

Assume that both atmospheric pressure and p_i are 101 kPa absolute. After filling with cold water the additional pressure due to the 12-m column of water is

$$\frac{(12 \text{ m})(9.807 \text{ m/s}^2)}{0.0010017} = 117.5 \text{ kPa}$$

and so

$$p_c = 117.5 + 101 = 218.5 \text{ kPa abs}$$

$$p_h = 250 + 101 = 351 \text{ kPa abs}$$

$$V_t = \frac{0.261 \text{ m}^3}{(101/218.5) - 101/351} = 1.496 \text{ m}^3$$

PROBLEMS

7-1 A convector whose performance characteristics are shown in Fig. 7-4 is supplied with a flow rate of 0.04 kg/s of water at 90°C. The length of the convector is 4 m, and the room-air temperature is 18°C. What is the rate of heat transfer from the convector to the room air? *Ans.* 2.92 kW

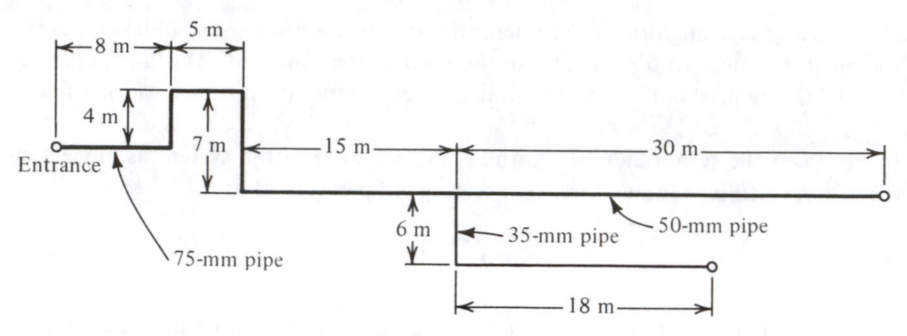

Figure 7-14 Piping system in Prob. 7-3.

7-2 Compute the pressure drop in pascals per meter length when a flow rate of 8 L/s of 60°C water flows through a Schedule 40 steel pipe of nominal diameter 75 mm (*a*) using Eq. (7-1) and (*b*) using Figs. 7-6 and 7-7. *Ans.* (*a*) 334 Pa/m

7-3 In the piping system shown schematically in Fig. 7-14 the common pipe has a nominal 75 mm diameter, the lower branch 35 mm, and the upper branch 50 mm. The pressure of water at the entrance is 50 kPa above atmospheric pressure, and both branches discharge to atmospheric pressure. The water temperature is 20°C. What is the water flow rate in liters per second in each branch? *Ans.* total flow = 6.9 L/s

7-4 A centrifugal pump with the characteristics shown in Fig. 7-9 serves a piping network and delivers 10 L/s. An identical pump is placed in parallel with the original one to increase the flow rate. What is (*a*) the new flow rate in liters per second and (*b*) the total power required by the two pumps? *Ans.* (*b*) 3.9 kW

7-5 An expansion tank is to be sized so that the change in air volume between the cold-water condition (25°C) and the operating water temperature (85°C) is to be one-fourth the tank volume. If p_i = 101 kPa abs and p_c = 180 kPa abs, what will p_h be? *Ans.* 325 kPa abs

REFERENCES

1. W. S. Harris, C. O. Pedersen, and W. F. Stoecker: Hot Water and Steam Heating Selection Factors, I: Theoretical Development; II: Experimental Verification and Application, *ASHRAE Trans.*, vol. 78, pt. 2, pp. 67–91, 1972.
2. "Handbook and Product Directory, Systems Volume," chap. 17, American Society of Heating, Refrigerating, and Air-Conditioning Engineers, Atlanta, Ga., 1980.
3. "ASHRAE Handbook, Fundamentals Volume," chap. 32, American Society of Heating, Refrigerating, and Air-Conditioning Engineers, Atlanta, Ga, 1981.
4. D. D. Wile, Refrigerant Line Sizing, *ASHRAE Spec. Publ.* 185, 1977.
5. G. A. Israel, Jr.: Centrifugal Pump Basics, *Energy Eng.*, vol. 77, no. 6, pp. 19–52, October-November, 1980.
6. G. F. Carlson: NPSH: It Shouldn't Mean Not Pumping So Hot, *Heat., Piping Air Cond.*, vol. 51, no. 4, pp. 65–72, April 1979.
7. "Handbook and Product Directory, Systems Volume," chap. 15, American Society of Heating, Refrigerating, and Air-Conditioning Engineers, Atlanta, Ga., 1980.

CHAPTER

EIGHT

COOLING AND DEHUMIDIFYING COILS

8-1 Types of cooling and dehumidifying coils One of the frequent assignments of a refrigeration or air-conditioning system is to reduce the temperature of an airstream. A natural concomittant of dropping the temperature of the air is removing moisture from it. In cooling the air in a low-temperature refrigerated warehouse the dehumidification process forms frost on the coil, which is usually an undesirable by-product of the temperature-reduction process. In a comfort or industrial air-conditioning application the dehumidification is usually a desirable objective. This chapter concentrates on the cooling and dehumidification of air in the 5 to 35°C temperature range.

The focus of this chapter is on the air side of a heat exchanger on the other side of which cold water or cold refrigerant flows. How the cold water or refrigerant is produced is treated in later chapters. Most air-cooling coils consist of tubes with fins attached to the outside of the tubes to increase the area on the air side where the convection coefficient is generally much lower than on the refrigerant or water side. Refrigerant or water flows inside the tubes, and air flows over the outside of the tubes and the fins. When a refrigerant evaporates in the tubes, the coil is called a *direct-expansion coil.* When, on the other hand, a secondary refrigerant, such as chilled water, carries away the heat, this water is chilled by an evaporator in the machine room. A chilled-water coil is shown in Fig. 8-1. The air-conditioning systems in many large buildings use central water chillers and distribute chilled water throughout the building.

8-2 Terminology Several terms and features of coil construction should be explained:

Face area of the coil. The cross-sectional area of the airstream at the entrance of the coil

Face velocity of the air. The volume rate of airflow divided by the face area

Surface area of the coil. The heat-transfer area in contact with the air

Number of rows of tubes. The number of rows in the direction of airflow

Figure 8-1 A chilled-water coil for cooling and dehumidifying air. *(Bohn Heat Transfer Division of Gulf & Western.)*

8-3 Condition of air passing through the coil (ideal) A condition curve is a series of points on the psychrometric chart representing the condition of the air as it passes through the coil. Some thermodynamics texts show the line 1-2-3 on Fig. 8-2 as the process. This condition curve shows a drop in temperature at constant humidity ratio until the air becomes saturated. From point 2 to point 3 the state of the air follows

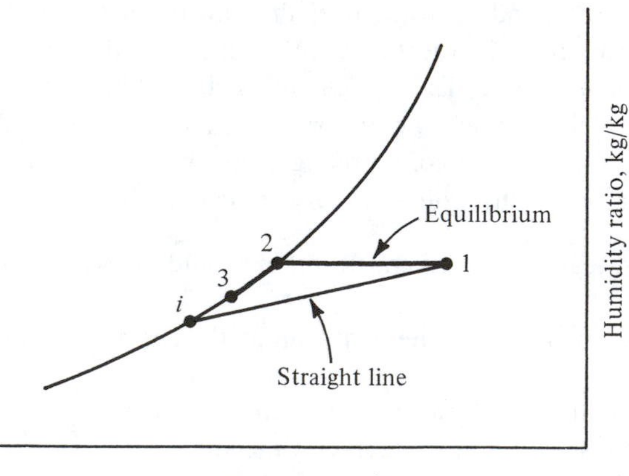

Figure 8-2 Idealized coil condition curves.

along the saturation line until the air leaves the coil. Path 1-2-3 would occur only if the entire air mass were at a uniform temperature and vapor pressure as it is cooled. Such is not the case, however, because gradients of both temperature and vapor pressure exist in the air passing through the coil.

Another idealization is the straight line from point 1 to point *i*, first introduced in Sec. 3-8. This curve is an idealization because this process occurs only when the wetted surface exists at a constant temperature throughout the coil. The actual coil condition curve will be found to lie somewhere between the two idealizations of Fig. 8-2.

8-4 Heat and mass transfer An elementary cooling and dehumidifying coil is shown in Fig. 8-3, where the successive transport processes are (1) combined heat- and mass-transfer process from the air to the wetted surface, (2) conduction through the water film and metal, and (3) convection to the refrigerant or chilled water. For a differential area of coil, two equations are available for the rate of heat transfer, dq W, for an element of area dA m^2

$$dq = \frac{h_c\, dA}{c_{pm}} (h_a - h_i) \tag{8-1}$$

where h_c = convection coefficient, W/m^2 · K
c_{pm} = specific heat of air mixture, kJ/kg · K
h_a = enthalpy of air, kJ/kg
h_i = enthalpy of saturated air at wetted-surface temperature, kJ/kg

The second equation expresses the rate of heat transfer to the refrigerant or chilled water

$$dq = h_r\, dA_i (t_i - t_r) \tag{8-2}$$

where t_r = temperature of refrigerant or chilled water, °C
t_i = temperature of wetted surface, °C
dA_i = refrigerant- or chilled-water-side area, m^2
h_r = combined conductance through wetted surface, metal of fins and tube, and refrigerant or water boundary layer inside tubes, W/m^2 · K

The term $1/h_r A_i$ is the thermal resistance between the wetted surface and the refrigerant or water inside the tube. The value of h_r is less than the refrigerant-side coeffi-

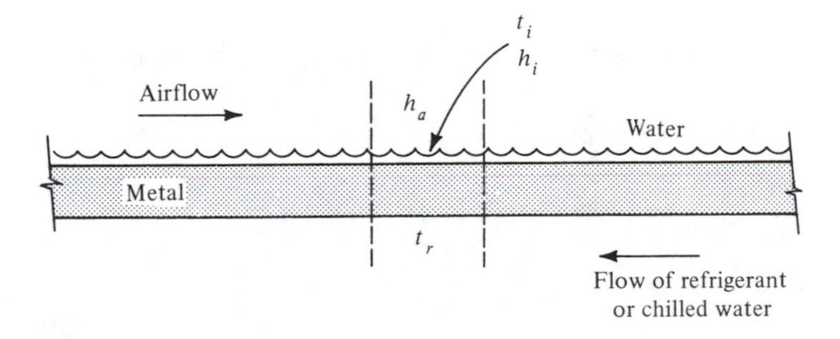

Figure 8-3 Transport processes in a coil.

cient because of the additional resistance to heat transfer provided by the metal and water film on the air side.

Equating Eqs. (8-1) and (8-2) gives

$$\frac{t_i - t_r}{h_a - h_i} = \frac{h_c}{c_{pm} h_r} \frac{A}{A_i} = R \tag{8-3}$$

When the enthalpy of the air, the temperature of the refrigerant, and the ratio R are known for a certain position in the coil, the temperature of the wetted surface can be calculated. Determining the temperature of the wetted surface is a key step in analyzing the performance of the coil. In Eq. (8-3) when t_r, h_a, and R are known, two variables are still unknown, t_i and h_i. The enthalpy of saturated air h_i is a function of the temperature of the wetted surface t_i, and values are available from Table A-2, but h_i can also be related to t_i by the cubic equation

$$h_i = 9.3625 + 1.7861t_i + 0.01135t_i^2 + 0.00098855t_i^3 \tag{8-4}$$

which is applicable between 2 and 30°C.

Substitution of Eq. (8-4) into Eq. (8-3) yields a nonlinear equation

$$\frac{t_i}{R} - \frac{t_r}{R} - h_a + 9.3625 + 1.7861t_i + 0.01135t_i^2 + 0.00098855t_i^3 = 0 \tag{8-5}$$

which can be solved for t_i.

Example 8-1 At one position in a cooling and dehumidifying coil which has an R value of 0.22, h_a = 85.5 kJ/kg and t_r = 9.0°C. What are the values of t_i and h_i?

Solution One technique for solving the nonlinear equation (8-5) uses the Newton-Raphson technique. If a function of x is written in such a form that $f(x) = 0$, the procedure is to choose a trial value of x and compute f and df/dx. The improved value of x is then

$$x_{\text{new}} = x_{\text{old}} - \frac{f}{df/dx}$$

The iterations continue until they are sufficiently close to convergence.

Equation (8-5) for this problem can be written

$$f = \frac{t_i}{0.22} - \frac{9.0}{0.22} - 85.5 + 9.3625 + 1.7861t_i + 0.01135t_i^2 + 0.00098855t_i^3$$

where $f = 0$ for the correct value of t_i

$$\frac{df}{dt_i} = \frac{1}{0.22} + 1.7861 + 0.0227t_i + 0.002966t_i^2$$

Try t_i = 20°C; then

$$f = 22.0329 \quad \text{and} \quad \frac{df}{dt_i} = 17.236$$

and so

$$t_{i,\text{new}} = 20 - \frac{22.0329}{7.9718} = 17.236$$

Another iteration yields

$$f = 0.5188, \qquad \frac{df}{dt_i} = 7.604$$

$$t_{i,\text{new}} = 17.236 - \frac{0.5188}{7.604} = 17.17°\text{C}$$

From Eq. (8-4)

$$h_i = 48.37 \text{ kJ/kg}$$

8-5 Calculating the surface area of a coil The foregoing relations can now be applied to compute the surface area of a coil when the entering conditions and flow rate of the air, the temperatures of the chilled water or refrigerant, and the heat-transfer conditions are known.[1]

Example 8-2 A counterflow chilled-water coil is to cool 2.5 kg/s of air from an entering condition of 30°C dry-bulb and 21°C wet-bulb temperature to a final wet-bulb temperature of 13°C. Chilled water enters the coil at 7°C and leaves at 12°C. The ratio of outside to inside surface area is 16:1, h_c = 55 W/m^2 · K, h_r = 3 kW/m^2 · K, and c_{pm} = 1.02 kJ/kg · K. Calculate (*a*) the required surface area and (*b*) the dry-bulb temperature of the leaving air.

Solution Consider the coil to be as shown in Fig. 8-4, having countercurrent flow of air and chilled water on opposite sides of the metal. The coil will be divided into two increments of area, and each part will be calculated separately. Arithmetic-mean temperature differences will be used to express the rate of heat transfer to the chilled water, and arithmetic-mean enthalpy differences will be used to express the heat and mass transfer from the air to the wetted surface.

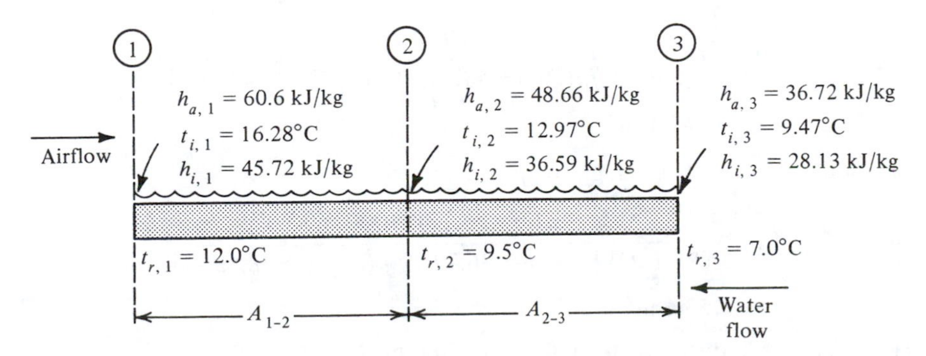

Figure 8-4 State points of air and chilled water for coil in Example 8-2.

Label the air inlet (and water outlet) of the coil as position 1 and the air outlet (and water inlet) as 3. An arbitrary position 2 is taken where one-half the heat has been transferred. At 2 the temperature of the chilled water $t_{r,2}$ is midway between 7.0 and 12°C, and the enthalpy of the air $h_{a,2}$ is midway between $h_{a,1}$ and $h_{a,3}$. The enthalpy of entering air at 30 and 21°C dry- and wet-bulb temperatures, respectively, is, from the psychrometric chart in Fig. 3-1, equal to 60.6 kJ/kg. The wet-bulb temperature of the leaving air is 13°C, and since this air will be close to saturation, from Table A-2 or Fig. 3-1, $h_{a,3}$ = 36.72 kJ/kg. The air enthalpy at point 2 is (60.6 + 36.72)/2 = 48.66 kJ/kg.

The value of R in Eq. (8-3) is

$$R = \frac{h_c}{c_{pm} h_r} \frac{A}{A_i} = \frac{55(16)}{1.02(3000)} = 0.2876 \text{ kg} \cdot \text{K/kJ}$$

When the value of R, the air enthalpy $h_{a,1}$, and the chilled-water temperature $t_{r,1}$ are known, the interface conditions $t_{i,1}$ and $h_{i,1}$ can be computed by the method illustrated in Example 8-1. The interface values for the three points are:

Position	h_a	t_r	t_i	h_i
1	60.6	12.0	16.28	45.72
2	48.66	9.5	12.97	36.59
3	36.72	7.0	9.47	28.13

(*a*) The coil area could be calculated using either the air-side or the chilled-water-side heat-transfer relations. Choosing the air-side calculation, designate $A_{1\text{-}2}$ as the area between positions 1 and 2 in which the rate of heat transferred $q_{1\text{-}2}$ is

$$q_{1\text{-}2} = (2.5 \text{ kg/s})(60.6 - 48.66) = 29.85 \text{ kW} = 29{,}850 \text{ W}$$

Another expression for this rate of heat transfer comes from the rate equation,

$$q_{1\text{-}2} = \frac{h_c A_{1\text{-}2}}{c_{pm}} \times \text{mean-enthalpy difference}$$

Equating the two expressions for $q_{1\text{-}2}$ and solving for $A_{1\text{-}2}$ gives

$$A_{1\text{-}2} = \frac{29{,}850}{(55/1.02)\,[(60.6 + 48.66)/2 - (45.72 + 36.59)/2]} = 41.1 \text{ m}^2$$

In a similar manner $A_{2\text{-}3}$ can be computed

$$A_{2\text{-}3} = \frac{2.5(48.66 - 36.72)\,(1000 \text{ W/kW})}{(55/1.02)\,[(48.66 + 36.72)/2 - (36.59 + 28.13)/2]} = 53.6 \text{ m}^2$$

The surface area of the coil is 41.1 + 53.6 = 94.7 m^2.

(*b*) *After the areas* $A_{1\text{-}2}$ and $A_{2\text{-}3}$ have been calculated, sensible-heat relationships can be used to predict the dry-bulb temperatures at positions 2 and 3. For area $A_{1\text{-}2}$, the sensible-heat transfer in kilowatts is

$$Q_s = (2.5 \text{ kg/s})c_{pm}(1000 \text{ W/kW})(t_1 - t_2, \text{K})$$

where c_{pm} is in kJ/kg · K,

$$= A_{1\text{-}2}h_c\left(\frac{t_1 + t_2}{2} - \frac{t_{i,1} + t_{i,2}}{2}\right)$$

and also

$$2.5(1.02)(1000)(30.0 - t_2) = 41.1(55)\left(\frac{30.0 + t_2}{2} - \frac{16.28 + 12.97}{2}\right)$$

$$t_2 = 20.56°\text{C}$$

For area $A_{2\text{-}3}$

$$2.5(1.02)(1000)(20.56 - t_3) = 53.6(55)\left(\frac{20.56 + t_3}{2} - \frac{12.97 + 9.47}{2}\right)$$

$$t_3 = 13.72°\text{C}$$

Improved accuracy will be achieved if the coil is divided into a large number of small increments. If just one increment of area had been used in Example 8-2, the calculated area would have been 94.35 m^2, and if four increments had been used, the total area would amount to 91.14 m^2. The calculation of the outlet dry-bulb temperature is also affected by the number of increments. For one, two, and four increments of area, the calculated outlet dry-bulb temperatures are 12.73, 13.72, and 14.12°C, respectively.

In Example 8-2 water is the cooling medium, and its temperature changes continuously as it flows through the coil. In a direct-expansion coil a constant refrigerant temperature is assumed. The same calculation method is used for the direct-expansion coil, and even though the refrigerant temperature remains constant, the wetted-surface temperature progressively drops in the direction of airflow.

8-6 Moisture removal The rate of moisture removal from the air in an increment of area can be determined after the area and the wetted-surface temperatures have been calculated. The mass balance specifies that

$$\text{Rate of water removal} = G(W_1 - W_2) \qquad \text{kg/s}$$

where G is the airflow rate in kilograms per second.

Also the equation for the rate of mass transfer first proposed in Sec. 3-14 can be integrated for the increment of area $A_{1\text{-}2}$; when the arithmetic-mean difference in humidity ratios is used it becomes

$$\text{Rate of water removal} = \frac{h_c A_{1\text{-}2}}{c_{pm}}\left(\frac{W_1 + W_2}{2} - \frac{W_{i,1} + W_{i,2}}{2}\right) \qquad \text{kg/s}$$

where $W_{i,1}$ is the humidity ratio (kg/kg) of saturated air at the wetted-surface temperature at point 1. Equating the two expressions for the rate of water removal gives

$$G(W_1 - W_2) = \frac{h_c A_{1\text{-}2}}{c_{pm}} \left(\frac{W_1 + W_2}{2} - \frac{W_{i,1} + W_{i,2}}{2} \right) \tag{8-6}$$

8-7 Actual coil condition curves Figure 8-2 illustrated two idealized condition curves showing the state of air passing through the coil. Path 1-2-3 would occur if the entire mass of air were in equilibrium while being cooled. Path 1-*i* would occur if a wetted surface of constant temperature cooled the air. The actual curve takes a shape shown in Fig. 8-5; the curve becomes steeper as the air progresses through the coil. Typical outlet conditions after the air passes through two, four, six, and eight rows of tubes are shown by points *b, c, d,* and *e,* respectively.

The reason for the progressive increase in slope of the coil condition curve as the air flows through the coil is that the temperature of the wetted surface diminishes from the air inlet to the air outlet. In Example 8-2, $t_{i,3}$ is less than $t_{i,2}$ which in turn is less than $t_{i,1}$. The actual coil condition curve could be imagined as a composite of a series of straight lines. Line *ab* in Fig. 8-5 is in contact with a wetted surface at temperature *x*. The next segment of the condition curve is *bc,* where the wetted surface is at temperature *y*. In addition to observing the change in slope of the condition curve, one also sees that the points representing the outlet conditions from successive rows of tubes become closer. Usually more cooling is performed, for example, in the first row of tubes in a coil than in the last because the enthalpy difference between the air and wetted surface is less at the air outlet.

If the state points of the air in Example 8-2 at points 1, 2, and 3 are plotted on a psychrometric chart, they outline a curve similar to the one in Fig. 8-5. The example used chilled water as a refrigerant with its lowest temperature at the air outlet. Does a direct-expansion coil give a curved line for the coil condition curve? Yes, because even in a direct-expansion coil the temperature of the wetted surface drops in the direction of the airflow. The truth of this statement can be shown by reference to Eq. (8-3). The values of R and t_r are constant throughout the direct-expansion coil, and changes in

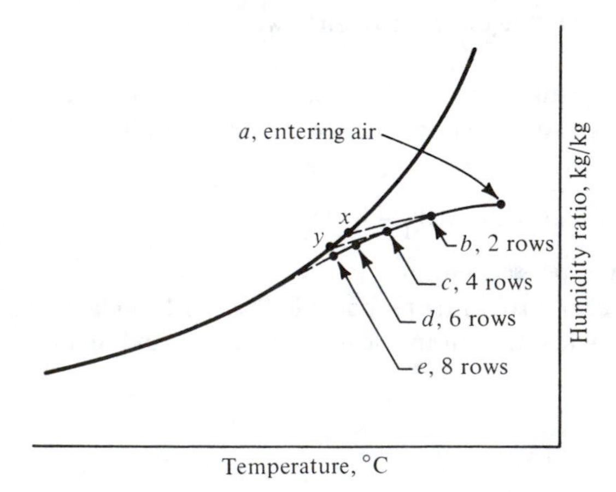

Figure 8-5 Actual coil condition curve.

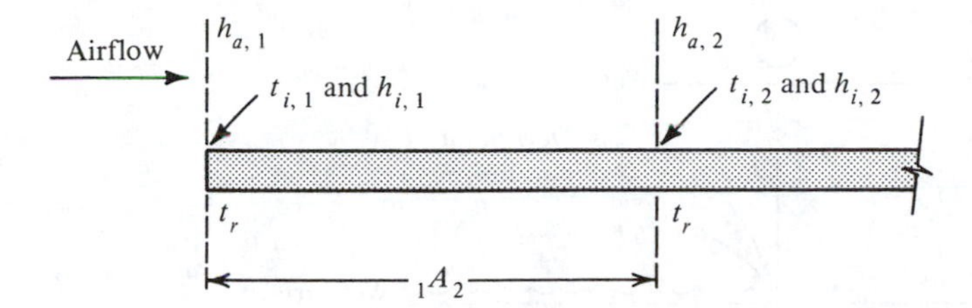

Figure 8-6 Determining outlet conditions by a stepwise calculation.

t_i and h_i must be in the same direction. When the enthalpy of air h_a decreases as the air flows through the coil, t_i and h_i must also decrease. Data from manufacturers' catalogs of the outlet air conditions from successive rows of tubes in a direct-expansion coil confirm a curved line, provided that the surfaces of the coil are wet.

8-8 Solving for outlet conditions Example 8-2 showed how the required area of a coil could be calculated to give a desired rate of heat transfer. Such a calculation might be made in the selection of the coil. Another important calculation, however, is to predict the outlet conditions with a given coil, refrigerant temperature, and inlet-air condition.

A stepwise solution of a direct-expansion coil is possible by dividing the known coil area into several sections. The first such section is $A_{1\text{-}2}$ in Fig. 8-6. Four equations can be written for the first increment of areas

$$G(h_{a,1} - h_{a,2}) = q \quad \text{kW} \tag{8-7}$$

$$\frac{h_c A_{1\text{-}2}}{c_{pm}}\left(\frac{h_{a,1} + h_{a,2}}{2} - \frac{h_{i,1} + h_{i,2}}{2}\right) = q \quad \text{W} \tag{8-8}$$

$$\frac{h_r A_{1\text{-}2}}{A/A_i}\left(\frac{t_{i,1} + t_{i,2}}{2} - t_r\right) = q \quad \text{W} \tag{8-9}$$

$$h_{i,2} = f(t_{i,2}) \quad \text{from Eq. (8-4)} \tag{8-10}$$

where A/A_i is the ratio of air-side to refrigerant-side areas. The wetted-surface conditions $t_{i,1}$ and $h_{i,1}$ can be determined as in Example 8-1. The four remaining unknowns, q, h_2, $h_{i,2}$, and $t_{i,2}$, can be solved from the four simultaneous equations (8-7) to (8-10). Kusuda[2] has developed a graphical method for making this calculation.

8-9 Partially dry coil So far in this chapter we have considered only coils whose entire surface is wet. All the coil surface will be wet if the entire surface is below the dew-point temperature of the entering air. Sometimes, however, the coil surfaces at the air inlet are dry, and condensation does not begin until farther along in the coil. Condensation begins when the surface temperature of the coil drops to the dew-point temperature of the entering air, as illustrated in Fig. 8-7. On the psychrometric chart in Fig. 8-8, sensible cooling with no dehumidification occurs along the dry portion of the coil, 1-2. In heat-transfer calculations the coil must be treated as two separate surfaces, using applicable equations for the dry and the wet sections.

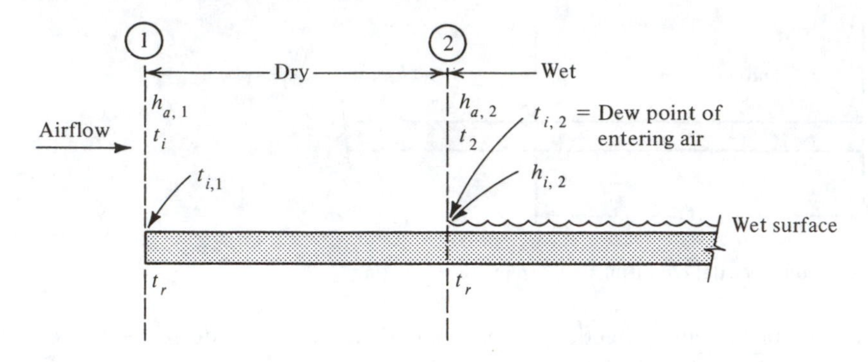

Figure 8-7 A partially dry coil.

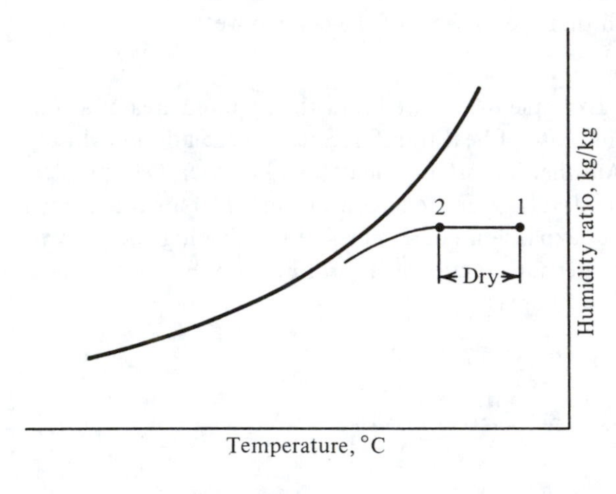

Figure 8-8 Coil condition curve of a partially dry coil.

Example 8-3 The airflow rate through a direct-expansion coil is 0.32 kg/s, and the entering conditions of the air are 30°C dry-bulb temperature and 20°C wet-bulb temperature. The refrigerant temperature is 10°C, h_r = 2400 W/m^2 · K, h_c = 100 W/m^2 · K, and the ratio of external to internal surface area is 18.0. (*a*) What is the dry-bulb temperature of the air when condensation begins? (*b*) How much surface area of the coil is dry?

Solution (*a*) From the psychrometric chart (Fig. 3-1) $h_{a,1}$ = 57.2 kJ/kg. The dew-point temperature of entering air is $t_{i,2}$ = 15.0°C. At the point where condensation begins, both the following equations apply

$$(t_2 - t_{i,2})h_c\, dA = (t_{i,2} - t_r)\frac{h_r\, dA\, A_i}{A} \tag{8-11}$$

and

$$(h_{a,2} - h_{i,2})\frac{h_c\, dA}{c_{pm}} = (t_{i,2} - t_r)\frac{h_r\, dA\, A_i}{A} \tag{8-12}$$

Equations (8-11) and (8-12) show, incidentally, that for the special case where $W_2 = W_{i,2}$

$$t_2 - t_{i,2} = \frac{h_{a,2} - h_{i,2}}{c_{pm}}$$

Using Eq. (8-11) gives

$$(t_2 - 15.0)(100) = (15.0 - 10.0)\frac{2400}{18.0}$$

$$t_2 = 21.7°\text{C}$$

(*b*) The same value of h_r is assumed to apply to the dry area even though h_r includes the resistance of the condensed water film. Using the arithmetic-mean temperature difference, we have

$$G(c_{pm})(t_1 - t_2) = \frac{1}{(1/h_c) + A/A_i h_r} A_{1\text{-}2}\left(\frac{t_1 + t_2}{2} - t_r\right)$$

or

$$(0.32 \text{ kg/s})(1020 \text{ J/kg} \cdot \text{K})(30.0 - 21.7)$$

$$= \frac{1}{1/100 + 18/2400} A_{1\text{-}2}\left(\frac{30.0 + 21.7}{2} - 10.0\right)$$

$$A_{1\text{-}2} = 2.99 \text{ m}^2$$

8-10 Coil performance from manufacturers' catalogs The methods of analyzing the performance of cooling and dehumidifying coils outlined in Secs. 8-4 to 8-9 are not the routine procedures used by designers in selecting coils. The purpose of the previous sections is to explain the trends in the performance of coils and to equip the designer to calculate the performance of coils for unusual applications when catalog data are not available. For conventional applications the manufacturers of coils present performance data to simplify coil selection in both tabular and graphical form. Table 8-1 is an excerpt from a coil catalog. The complete catalog gives the coil performance at other dry- and wet-bulb temperatures of entering air and at additional face velocities. Several characteristics of coil performance can be demonstrated by the data from Table 8-1: (1) a plot on the psychrometric chart of the points representing the condition of the air at the outlet from successive rows of tubes shows curvature similar to Fig. 8-5, (2) each successive row of tubes removes less heat than its predecessor (this fact can be shown by determining the reduction in enthalpy of air through each successive row of tubes), (3) a lower refrigerant temperature with a given face velocity causes a greater ratio of latent- to sensible-heat removal, and (4) an increase in the face velocity increases the capacity but also increases the dry- and wet-bulb temperatures of the outlet air.

Table 8-2 shows the performance of the same coil as Table 8-1 with the same entering wet-bulb temperature but a different dry-bulb temperature. Comparison of Table 8-2 with the section of Table 8-1 at the same refrigerant temperature and face

Table 8-1 Performance of a Trane Company direct-expansion refrigerant 22 cooling coil

Air enters at 30°C dry-bulb temperature and 21.7°C wet-bulb temperature

2.0 m/s face velocity			3.0 m/s face velocity		
Rows of tubes	Final DBT, °C	Final WBT, °C	Rows of tubes	Final DBT, °C	Final WBT, °C
1.7°C refrigerant temperature					
2	17.0	16.2	2	18.6	17.3
3	14.7	14.1	3	16.3	15.6
4	12.6	12.3	4	14.6	14.0
6	9.8	9.6	6	11.7	11.4
8	7.9	7.8	8	9.7	9.5
4.4°C refrigerant temperature					
2	18.2	17.1	2	19.7	18.0
3	16.1	15.3	3	17.5	16.5
4	14.3	13.8	4	15.9	15.2
6	11.8	11.5	6	13.5	13.1
8	10.2	9.9	8	11.7	11.4
7.2°C refrigerant temperature					
2	19.6	17.9	2	21.1	18.7
3	17.5	16.5	3	18.9	17.5
4	16.1	15.3	4	17.4	16.4
6	13.9	13.4	6	15.4	14.7
8	12.4	12.1	8	13.9	13.4

velocity shows that the final wet-bulb temperatures are approximately the same for a given number of rows of tubes.

Figure 8-9 shows a comparison of the condition curves when air enters a given coil at different dry-bulb temperatures but at the same wet-bulb temperature. The essentially identical final wet-bulb temperatures in Table 8-2 and the corresponding section of Table 8-1 can be explained by recalling the following facts. The enthalpy difference between the air and the wetted surface controls the rate of heat transfer; so when the refrigerant temperature and airflow rate are specified and the enthalpy of the entering air is fixed by specifying the entering wet-bulb temperature, the outlet enthalpy and wet-bulb temperature will be fixed, regardless of the dry-bulb temperatures.

PROBLEMS

8-1 A cooling and dehumidifying coil is supplied with 2.4 m^3/s of air at 29°C dry-bulb and 24°C wet-bulb temperatures, and its cooling capacity is 52 kW. The face velocity is 2.5 m/s, and the coil is of the direct-expansion type provided with re-

Table 8-2 Coil performance from the same catalog as Table 8-1

Air enters at 35.6°C dry-bulb temperature, 21.7°C wet-bulb temperature, and 2.0 m/s face velocity; 1.7°C refrigerant temperature

Rows of tubes	Final DBT, °C	Final WBT, °C
2	18.7	16.1
3	15.1	14.0
4	12.9	12.2
6	9.9	9.6
8	7.9	7.7

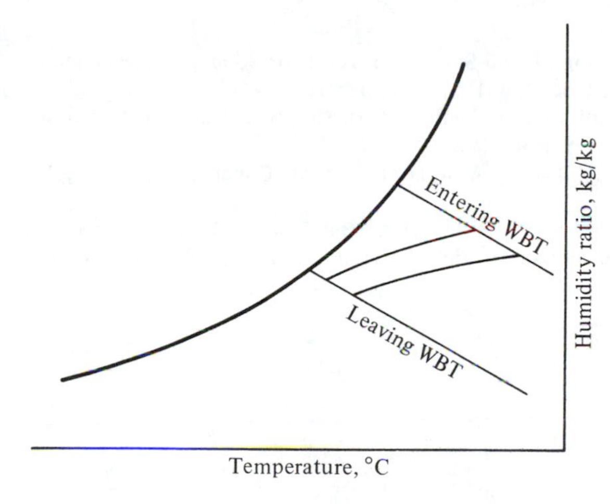

Figure 8-9 Coil condition curves with same entering wet-bulb temperatures, face velocities, and refrigerant temperatures.

frigerant evaporating at 7°C. The coil has an air-side heat-transfer area of 15 m^2 per square meter of face area per row of tubes. The ratio of air-side to refrigerant-side area is 14. The values of h_r and h_c are 2050 and 65 W/m^2 • K, respectively. Calculate (*a*) the face area, (*b*) the enthalpy of the outlet air, (*c*) the wetted-surface temperatures at the air inlet, air outlet, and at the point where the enthalpy of air is midway between its entering and leaving conditions, (*d*) the total surface area, (*e*) the number of rows of tubes, and (*f*) the outlet dry-bulb temperature of the air. *Ans.* (*a*) 0.96 m^2; (*b*) 53.2 kJ/kg; (*c*) 17.25, 15.44, 13.56°C; (*d*) 42.7 m^2; (*e*) 3; (*f*) 20.1°C

8-2 For the area $A_{1\text{-}2}$ in Example 8-2 using the entering conditions of the air and the wetted-surface temperatures at points 1 and 2, (*a*) calculate the humidity ratio of the air at point 2 using Eq. (8-6), and (*b*) check the answer with the humidity ratio determined from the dry-bulb temperature and enthalpy at point 2 calculated in Example 8-2. *Ans.* 0.0111 kg/kg

8-3 A direct-expansion coil cools 0.53 kg/s of air from an entering condition of 32°C dry-bulb and 20°C wet-bulb temperature. The refrigerant temperature is 9°C, h_r = 2 kW/m^2 • K, h_c = 54 W/m^2 • K, and the ratio of air-side to refrigerant-side areas is 15. Calculate (*a*) the dry-bulb temperature of the air at which condensation begins and (*b*) the surface area in square meters of the portion of the coil that is dry. *Ans.* (*a*) 25.7°C, (*b*) 4.51 m^2

8-4 For a coil whose performance and conditions of entering air are shown in Table 8-1, when the face velocity is 2 m/s and the refrigerant temperature is 4.4°C, calculate (*a*) the ratio of moisture removal to reduction in dry-bulb temperature in the first two rows of tubes in the direction of airflow and in the last two rows and (*b*) the average cooling capacity of the first two and the last two rows in kilowatts per square meter of face area. *Ans.* (*a*) 0.0000932 and 0.00056 kg/kg · K (*b*) 33.1 and 8.91 kW

8-5 An airflow rate of 0.4 kg/s enters a cooling and dehumidifying coil, which for purposes of analysis is divided into two equal areas, $A_{1\text{-}2}$ and $A_{2\text{-}3}$. The temperatures of the wetted coil surfaces are $t_{i,1}$ = 12.8°C, $t_{i,2}$ = 10.8°C, and $t_{i,3}$ = 9.2°C. The enthalpy of entering air $h_{a,1}$ = 81.0 and $h_{a,2}$ = 64.5 kJ/kg. Determine $h_{a,3}$. *Ans.* 52.25 kJ/kg

REFERENCES

1. J. McElgin and D. C. Wiley: Calculation of Coil Surface Areas for Air Cooling and Dehumidification, *Heat. Piping Air Cond.*, vol. 12, no. 3, p. 195, March 1940.
2. T. Kusuda: Graphical Method Simplifies Determination of Air Coil Wet Heat Transfer Surface Temperature, *Refrig. Eng.*, vol. 65, no. 5, p. 41, May 1957.
3. F. C. McQuiston and J. D. Parker: "Heating, Ventilating and Air Conditioning," chap. 14, Wiley, New York, 1977.
4. Air Cooling and Dehumidifying Coils, chap. 6 in "ASHRAE Handbook and Product Directory, Equipment Volume," American Society of Heating, Refrigerating and Air-Conditioning Engineers, Atlanta, Ga., 1979.

CHAPTER
NINE
AIR-CONDITIONING CONTROLS

9-1 What controls do The three major assignments of the control system of an air-conditioning plant are

1. To regulate the system so that comfortable conditions are maintained in the occupied space
2. To operate the equipment efficiently
3. To protect the equipment and the building from damage and the occupants from injury

From a functional standpoint a control and a control system can only reduce capacity and never increase it. For example, a valve in a water line can only reduce the flow rate by closing, and its presence can never increase the flow rate. So in the usual situation the air-conditioning system has its maximum cooling and heating capacity when operating with no control; the action of the control system only reduces that capacity.

The major elements in a basic control system are shown in Fig. 9-1, where the controlled condition, e.g., a temperature, is perceived by a sensor and converted into a pneumatic pressure or voltage, which is compared with a pressure or voltage representing the desired condition. If the two signals do not match, the actuator repositions a valve, damper, or similar element that has the potential for changing the temperature.

There are various approaches to the study of control systems. One is highly mathematical, concentrating on the dynamic behavior of control systems and stressing changes with respect to time. The emphasis in this chapter is on a *logical* rather than *mathematical* treatment and *steady-state* rather than *dynamic* behavior. Dynamic considerations, such as instability of the control loop, seem to occur only in a few air-conditioning applications. When dynamic problems do occur, they may be particularly

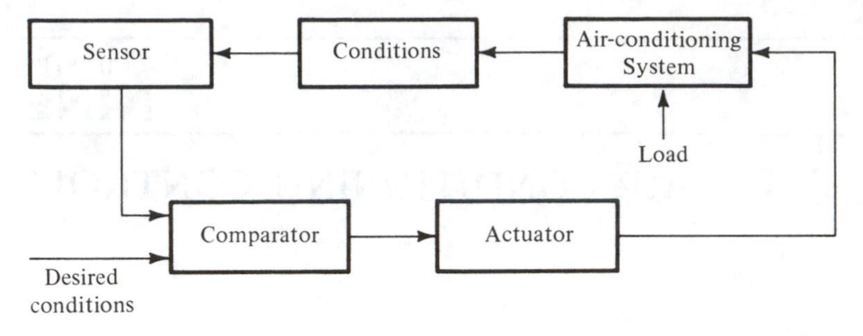

Figure 9-1 Elements of a basic control system.

troublesome and the mathematical background then becomes crucial. Nevertheless, the steady-state, logic approach is the starting point for understanding and designing air-conditioning controls.

9-2 Pneumatic, electric, and electronic control There are several types of sensors, actuators and other hardware: pneumatic, electric, and electronic. The standard type used in large-building air-conditioning installations has been and still is pneumatic, where the physical variables are transduced into an air pressure, signals are transmitted by means of the magnitude of air pressure, and dampers, valves, and other actuators are powered by air pressure. The distinction between electric and electronic systems is somewhat arbitrary, since all electric systems are truly electronic. But "electronic" usually refers to the incorporation of solid-state devices. The electric systems have always competed with the pneumatic ones and have predominated in small-building air conditioning. There are several reasons why pneumatic systems have prevailed so stubbornly in large systems: (1) the pneumatic system provides modulating control quite naturally, (2) they are easier for most service personnel to understand, maintain, and repair, and (3) air pressure operating through a piston and cylinder still remains the most expedient means of providing the magnitude of needed power to operate valves and dampers.

There is no requirement that a system be all pneumatic or all electric; it can be hybrid, and in such a combination the sensors and transmission of the control signals may be electric or electronic while the final driving force at the actuator is pneumatic. The interface between the electric and pneumatic portions of the system might be an electric-to-pneumatic transducer that converts a voltage into a pressure. Central supervisory and computer control systems in large buildings dovetail with the electric portion of hybrid control systems, but the local control loop may be pneumatic.

Since a thorough treatment of control devices and systems would fill an entire book, no one chapter can provide an exhaustive coverage of controls. This chapter seeks to provide an introduction to the strategy of control systems. Pneumatic control will be emphasized, because the functions provided by the pneumatic elements are basically the same as provided by electric or electronic ones.

9-3 Pneumatic control hardware The designer of a control system has available several dozen control components, many of which exist in several versions. The designer's challenge is to select and arrange these components to perform the required function(s) of the control system. A list of the most commonly used pneumatic control components is given below. These components are described later in the chapter, and techniques of combining them will be explained. Common to all pneumatic systems is the air-supply system, consisting of the compressor, storage tank, and air filter. In some installations where the air lines are subject to low temperatures, a refrigerated aftercooler condenses and removes much of the water in the air so that it will not freeze in the distribution lines. In some cases an oil separator is advisable, particularly if the compressor discharges oil into the air.

The major control components are:

1. Valves for liquids
   a. Two-way
   b. Three-way mixing
   c. Three-way bypass
2. Valves for control air
   a. Two-way solenoid
   b. Three-way solenoid
3. Dampers to restrict the flow of air
4. Manual pressure regulator for control air
5. Pressure regulators for working fluids, e.g., steam
6. Differential-pressure regulators
7. Velocity sensors
8. Thermostats
9. Temperature transmitters
10. Receiver-controllers
11. Humidistats
12. Master and submaster controllers
13. Reversing relays for control pressure
14. Pressure selectors
15. Pneumatic electric switches
16. Freezestats

9-4 Direct- and reverse-acting thermostats One type of thermostat, typical of room thermostats, is shown schematically in Fig. 9-2 and pictorially in Fig. 9-3. Air pressure at approximately 135 kPa gauge is supplied to the thermostat, a small quantity of which bleeds through the orifice into the low-pressure chamber. The bimetallic strip, composed of two materials with different coefficients of thermal expansion, opens and closes the port, venting or trapping air in the low-pressure chamber. When the bimetallic strip closes the port, pressure builds up in the low-pressure chamber, closing the valve to the control-air line.

A *direct-acting* thermostat is one which provides an increase in control pressure upon an increase in temperature. A *reverse-acting* thermostat is one which decreases the control pressure upon an increase in temperature. In the thermostat of Fig. 9-2, if the bimetallic strip closes the port upon an increase in room temperature, the thermostat is reverse-acting.

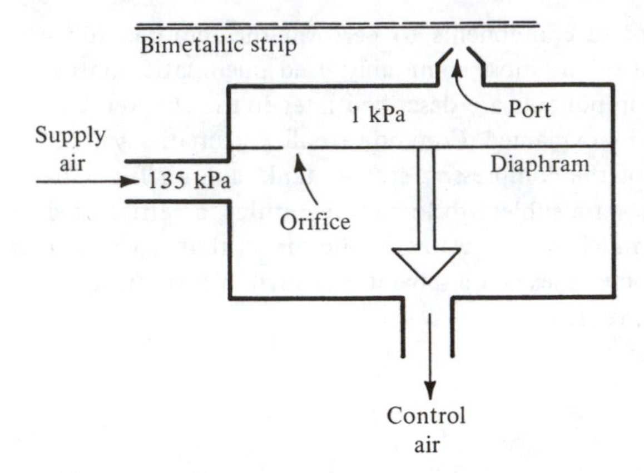

Figure 9-2 Room thermostat using a bimetallic strip.

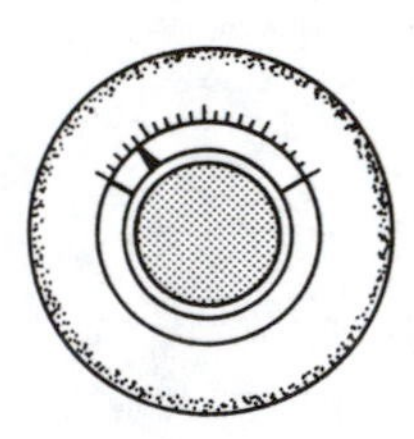

Figure 9-3 Room thermostat.

9-5 Temperature transmitter with receiver-controller An alternate set of hardware to function as a thermostat is the combination of a *temperature transmitter* and *receiver-controller*, shown in Fig. 9-4. The duty of the temperature transmitter is to convert a temperature into a pneumatic pressure, while the function of the receiver-controller is to amplify this pneumatic pressure and provide for selection of a set point.

Temperature transmitters are available in different temperature ranges, for example, 15 to 30°C, 10 to 35°C, 0 to 100°C, and the transmitter sends a proportional

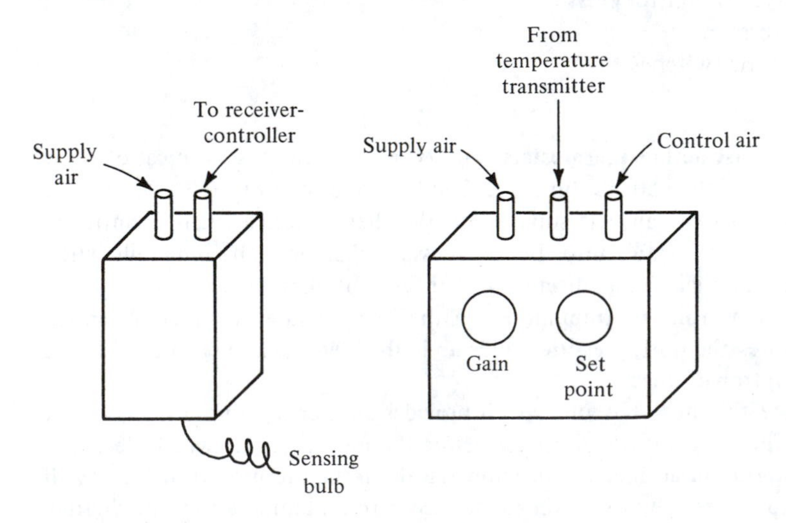

Figure 9-4 Temperature transmitter and receiver-controller.

signal of 20 to 100 kPa for the particular temperature range of the transmitter. The transmitter with the 15 to 30°C range, for example, will deliver a pressure of 20 kPa when it senses 15°C and 100 kPa when it senses 30°C and provide a linear variation of pressure to temperature between those limits.

The receiver-controller expects to receive a 20 to 100 kPa signal from the temperature transmitter and then performs the following functions: (1) passes on a control signal that is either direct- or reverse-acting with reference to that from the temperature transmitter and (2) selects a small range of the transmitter output and amplifies it. For example, the receiver-controller can translate the 40- to 50-kPa range from the transmitter into a 30- to 70-kPa output pressure.

It may seem that the combination of the temperature transmitter and receiver-controller is a complex replacement for the simple thermostat of Sec. 9-4, but the combination has several advantages. Control using the combination is normally more accurate than with a single thermostat because the transmitter need provide only a very small flow rate of air to the receiver-controller, in contrast to the single thermostat, where the one instrument must supply all the air needed to operate a valve or damper. When the control settings are to be grouped, the transmitter can be located close to the temperature being sensed, while the receiver-controllers for a number of temperature controls can be grouped in one location. A further advantage of the separate components is that a pressure gauge, calibrated in temperature, can monitor the output of the transmitter to provide an indication of the temperature being controlled.

9-6 Liquid valves Two-way throttling and three-way liquid valves are available; the latter are of two types, three-way mixing and three-way bypass valves.

Throttling valves are shown functionally and symbolically in Fig. 9-5; Fig. 9-6 shows two-way and three-way valves. Control pressure feeds to the top of the diaphragm, pushing the valve stem downward. Withdrawal of the control pressure permits the compressed spring to return the stem to its upward extremity. With no control pressure, the valve in Fig. 9-5*a* will be in its open position, so the valve is called a normally open (n.o.) valve.

In Fig. 9-5*a* and *b* the valves hold against the fluid pressure; i.e., the liquid flow tends to push the valve stem into the open position. With this arrangement the valve will open more easily and will not slam shut from a nearly closed position.

Typical spring ranges for valves are 28 to 55 and 62 to 90 kPa. The meaning of a

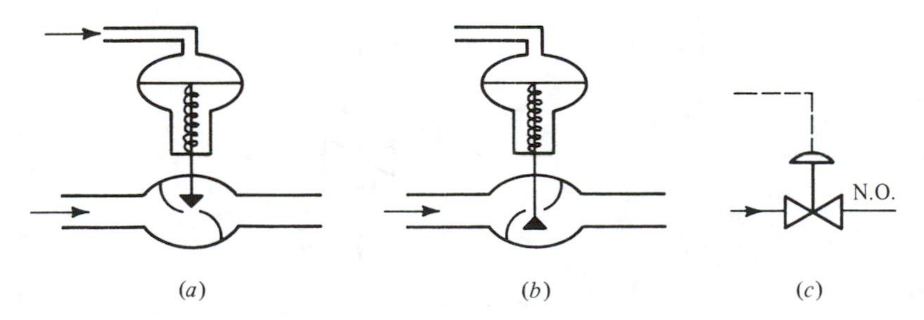

Figure 9-5 Throttling valves: (*a*) normally open, (*b*) normally closed, (*c*) symbol.

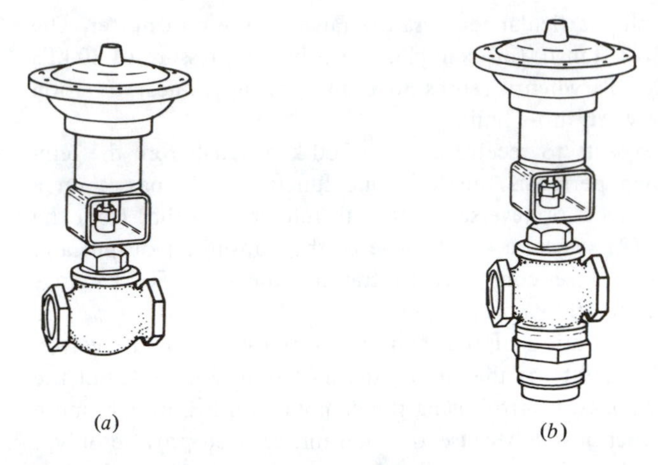

Figure 9-6 (*a*) Two-way and (*b*) three-way valves.

28- to 55-kPa spring range on a normally open valve, for example, is that the valve is fully open if the control pressure is 28 kPa or below. As the control pressure increases from 28 to 55 kPa, the valve moves from an open to a closed position. Control pressures above 55 kPa keep the valve in a closed position.

The symbols for two types of three-way valves are shown in Fig. 9-7. The mixing valve in Fig. 9-7*a* has two inlet streams and one outlet or common stream, while the bypass valve has one inlet and two outlet streams. For the mixing valve in Fig. 9-7*a*, a control pressure lower than the spring range opens the horizontal inlet completely and closes off the vertical inlet. As the control pressure increases through the spring range, the horizontal inlet gradually closes while the vertical inlet opens.

Applications of these two types of valves are shown in Fig. 9-8, where the mixing valve serves a coil and is placed in the return line, the bypass valve being placed in the supply line.

When should two- and three-way valves be used? The two-way valve usually costs least, with the three-way mixing next, and three-way bypass the most expensive. If two-way valves are used to control a number of coils, the system flow rate drops as the load decreases, because the only flow passing between the supply and return-water lines must pass through the valves and coils. When either of the three-way valves is

Figure 9-7 Three-way valves: (*a*) mixing and (*b*) bypass.

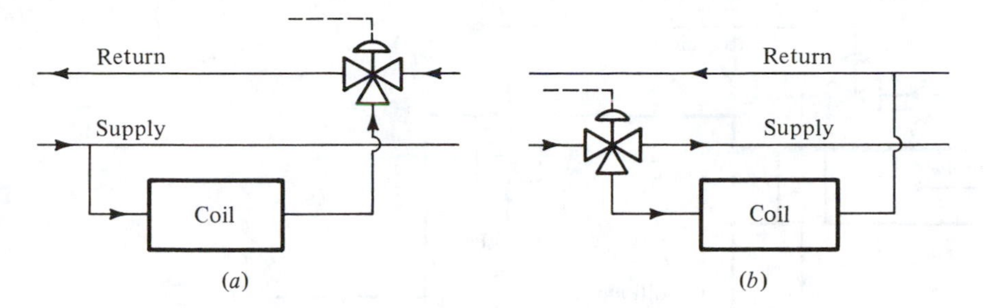

Figure 9-8 Application of (*a*) mixing and (*b*) bypass valves.

used, the system flow remains approximately constant and the temperature of the return water moves closer to that of the supply water at part load.

Whether it is an advantage or disadvantage to maintain a nearly constant flow rate even at low loads depends upon the system. If the flow rate of water drops too low, the pump may overheat, because the pump continues to deliver mechanical energy into a low flow rate of water. This situation can be prevented by using several pumps in parallel and then shutting pumps down as the flow rate drops. Another disadvantage of low flow rates with chilled water is that the poor distribution of velocities in the evaporator at low flow may result in localized freezing of water, which might burst a tube.

The constant system flow rate that occurs when three-way valves are used is a disadvantage in some large installations, such as central chilled-water plants, because all pumps must continue to operate, sending flow through all the chillers, in order to maintain the high flow rate.

9-7 Fail-safe design The loss of supply-air pressure is a situation that must be anticipated in the design of the control system. In moderate and cold climates the status to which various elements should revert upon loss of supply-air pressure is as follows:

Heating coils. Normally open valves
Cooling coils. Not crucial, either normally open or normally closed valves
Humidification. Normally closed valves
Outdoor-air inlet and exhaust air. Normally closed dampers

Example 9-1 Specify for the terminal-reheat control shown in Fig. 9-9 whether the thermostat should be reverse- or direct-acting and whether the valve should be normally open or normally closed.

Solution To meet the fail-safe condition in moderate and cold climates, a *normally open valve* should be chosen. An increase in pneumatic pressure closes the valve; closing the valve provides less heat to the space; and less heat to the space is called for when the temperature in the space rises. Since an increase in space temperature therefore calls for an increase in control pressure, a direct-acting thermostat is needed.

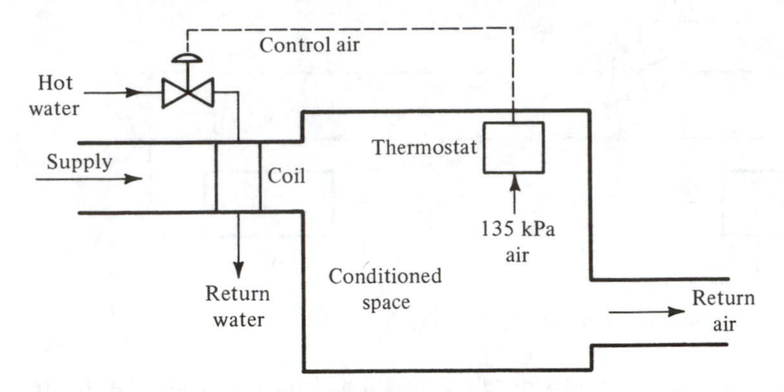

Figure 9-9 Terminal-reheat control system.

Another workable valve-thermostat combination would be a normally closed valve with a reverse-acting thermostat, but this combination would not provide the desired fail-safe position.

9-8 Throttling range Air-conditioning control systems do not regulate the variable they are controlling to a precise value. A fundamental characteristic of control systems is that there must be a deviation from the set point in order to instigate an action. The normal operation of a modulating pneumatic temperature control, for example, means that the temperature will be different at light load and at heavy load.

A measure of change in the controlled variable from zero load to full load is called the *throttling range*. In an air-temperature controller regulating the outlet temperature from a cooling coil, for example, the temperature of the leaving air must drop in order to close the valve regulating chilled water to the coil. If the air temperature is 13°C when the water valve is fully open and 10°C in order to close the valve, the throttling range is 3 K.

Example 9-2 The air-temperature control of a heating coil served by hot water, as shown in Fig. 9-10, consists of a temperature transmitter, receiver-controller, and

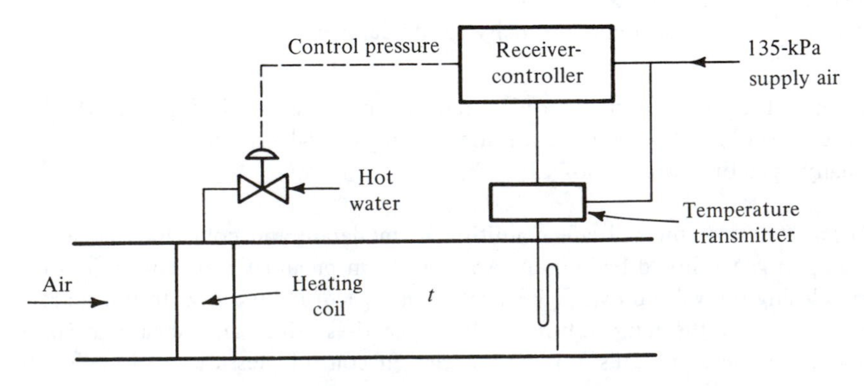

Figure 9-10 Air-temperature control of a heating coil.

water valve. The temperature transmitter has a range of 10 to 65°C during which its pressure output changes from 20 to 100 kPa. The receiver-controller is set with a gain of 10 to 1, and the spring range of the hot-water valve is 60 to 90 kPa. What is the throttling range of this assembly?

Solution If the hot-water valve is of the normally open type, it will be fully open when zero control pressure is applied and remain fully open until the control pressure increases to 60 kPa. Thereupon the valve begins to close and will be completely closed when the control pressure is 90 kPa and above. Between fully open and fully closed positions, then, the control pressure changes 30 kPa. Since the gain of the receiver-controller is 10 to 1, to achieve the 30-kPa change in control pressure, the pressure range from the temperature transmitter must be 30/10 = 3 kPa. This pressure change in the temperature transmitter corresponds to a temperature change of

$$(3\text{ kPa})\,\frac{65 - 10^\circ\text{C}}{100 - 20\text{ kPa}} = 2\text{ K}$$

Thus the temperature of the air leaving the coil must rise 2 K from the condition where the valve is fully open until the valve is fully closed, and the throttling range is 2 K.

9-9 Dampers Dampers are throttling devices for air and consist of pivoted metal plates, as shown in Fig. 9-11*a* and schematically in Fig. 9-11*b*. Dampers are installed in such locations as the supply line of a variable-air-volume system or the outdoor, recirculated, and exhaust-air ducts, as shown in Fig. 9-12. The damper operator (Fig. 9-13), is a piston-cylinder combination in which the piston is spring-loaded so that it moves to the far-left position when no air pressure is applied. A typical spring range for a damper operator is 55 to 90 kPa.

In the control subsystem shown in Fig. 9-12, the dampers in the recirculated-air duct move to their closed position while the other dampers move to their open position and vice versa. It might seem desirable to link all the dampers mechanically and

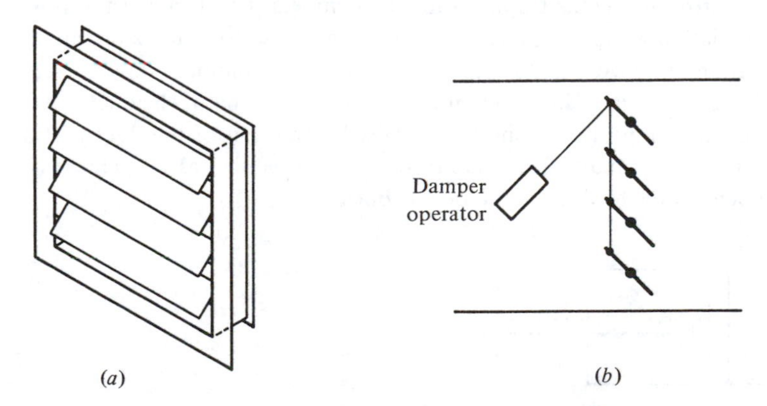

Figure 9-11 (*a*) A damper assembly and (*b*) schematic representation.

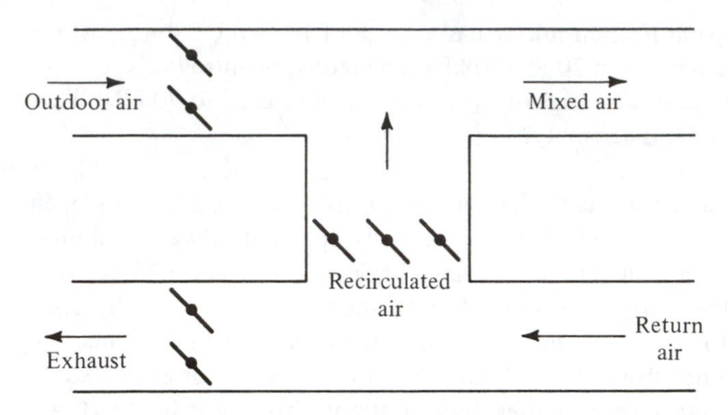

Figure 9-12 Outdoor-, exhaust-, and recirculated-air control.

use just one damper operator. Only on very small units, however, is such a practice followed because of the complexity in setting up the linkage. Normally all the exhaust dampers are linked, all the recirculated-air dampers are linked, and all the outdoor-air dampers are linked, each set having its own damper operator, but there are exceptions to this practice. Dampers covering large areas (larger than about 4 m^2, for example) are often divided into several sections, each regulated by its own damper operator.

Typical flow characteristics[1] of a damper are shown in Fig. 9-14, giving the percent of maximum flow as a function of the damper position in degrees of opening. The curves are applicable when the pressure difference across the damper is constant. Curve *A* represents the condition where the open-damper resistance is in the range of 0.25 to 0.50 percent of the total system resistance; for curve *B* the resistance is in the range of 1.5 to 2.5 percent. Clearly characteristics *B* are preferable to those of *A*, since in *A* a 10° opening (approximately 11 percent) results in 28 percent of the full flow. The damper provides very little control in the 45 to 90° range, during which span the flow changes only from 85 to 100 percent. The dampers can be selected to approximate curve *B* rather than curve *A* chiefly by selecting a smaller cross-sectional area for the dampers.

9-10 Outdoor-air control A frequent application of dampers is to regulate the flow rate of outdoor ventilation air (as discussed in Chap. 5) and also to assure a minimum flow rate of ventilation air. Two possibilities for providing the minimum flow rate are shown in Fig. 9-15. Consider the plan shown in Fig. 9-15*a*. A portion of the duct is left open so that when the dampers are completely closed a fraction of the duct is still open. Presumably if the minimum flow of outdoor air is specified as 20 percent of the maximum, 20 percent of the duct would be left open.

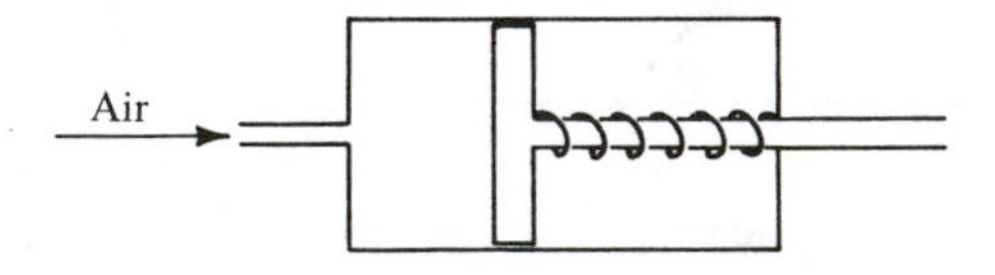

Figure 9-13 A damper operator.

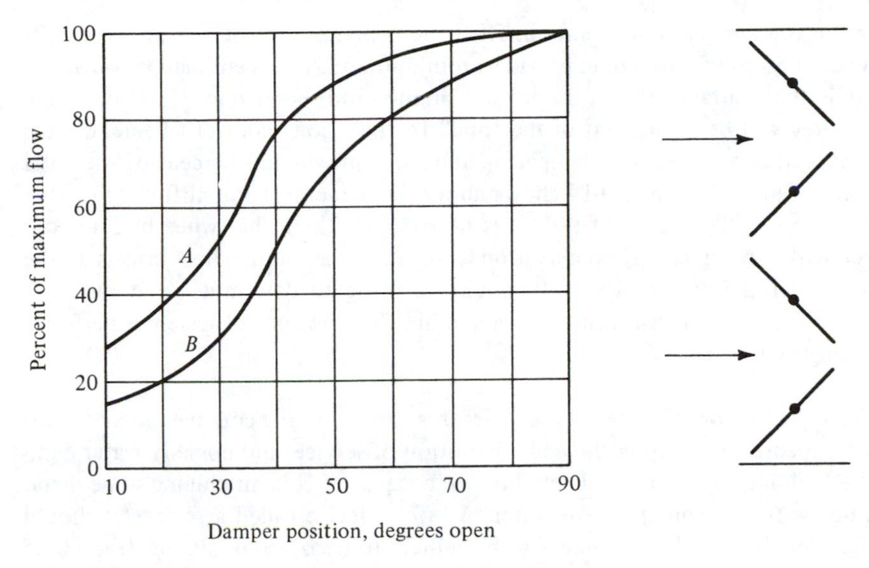

Figure 9-14 Flow characteristics of an opposed-leaf damper.

The disadvantage of the scheme in Fig. 9-15*a* is that there are times when the ducts to the outside (both the outdoor-air duct and the exhaust-air duct) should be completely closed. If the system is shut down and the fans are idle, it is conceivable in cold weather that low-temperature air could blow in through the ducts and freeze the water in a coil. To guard against this danger, a damper should be installed in the minimum-flow passage, as in Fig. 9-15*b*; it should be completely open when the fan is in operation and completely closed when the fan is shut down. Furthermore this damper should be a normally closed damper so that its fail-safe position is to close off the access of outdoor air in the event of a failure of control-air pressure.

There is yet another frequently used approach to supplying the minimum flow rate of outdoor air. Suppose that the dampers in the outdoor-air and exhaust ducts are positioned by damper operators with a 55- to 90-kPa spring range. The damper will be closed when the control pressure is 55 kPa and completely open when the pressure is 90 kPa. If a minimum of 20 percent outdoor air is specified, the control system would be designed so that a pressure no less than 62 kPa would be supplied to the operator,

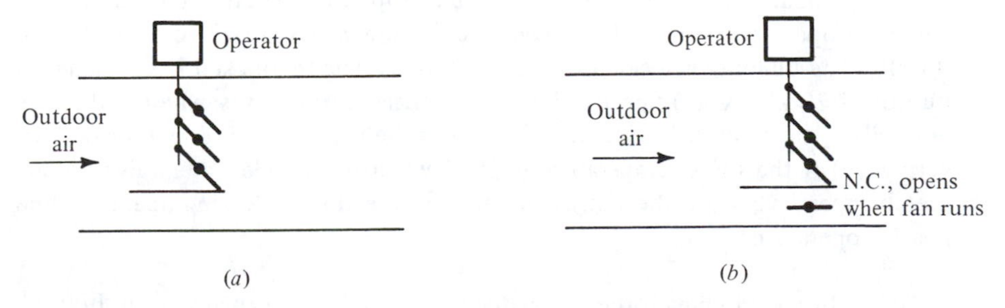

Figure 9-15 Providing minimum ventilation air: (*a*) not recommended; (*b*) preferable.

except when the fan is not in operation. The logic is that a control pressure of 62 kPa in the range of 55 to 90 kPa would provide a minimum of 20 percent damper opening.

What is the assurance that if 20 percent of the outdoor-air duct is left open the minimum flow will be 20 percent of the total? In the second control scheme, what is the assurance that a 20 percent damper opening will provide 20 percent of the total airflow rate? Curve *A* in Fig. 9-14 shows about 37 percent of full airflow when the dampers are open 20 percent of 90°. It is no surprise, then, that when building operators embark on an energy-conservation program, one of the first targets is the flow rate of ventilation air. A survey should be made to determine what outdoor-air flow rates actually prevail, instead of assuming that the control system is performing according to design.

9-11 Freeze protection If water in a coil freezes, a tube in the coil may burst, resulting in the expense of repairing the coil, disruption of service, and possibly water damage to the building. Water in a coil could freeze because of cold air chilling some metal surfaces below the freezing point of water. A safety device called a *freezestat* should be installed on all coils that are likely to be subject to freezing conditions. It is a long flexible tube that is clamped to the outlet face of the coil. If a short length (about 15 cm) of this tube drops to a temperature of 0°C, the freezestat opens an electric switch incorporated into the system which takes at least two actions: (1) shuts off fans and (2) closes outdoor-air and exhaust dampers.

In most cases preheat coils (if they are used in the system) are equipped with freezestats, and even some cooling coils that are not drained of their water in the winter should be so equipped. Many designers provide a circulating-water pump on hot-water preheat coils in order to prevent low water velocities. It is during periods of low water velocity that freezeups are most likely to occur. Locating the preheat coil in the mixed-air stream has the advantage of warming the outdoor air before reaching the coil and thus helps to prevent freezeups.

9-12 Sequencing of operations The appropriate selection of spring ranges for valves or dampers makes a sequencing between heating and cooling possible.

Example 9-3 A four-pipe system serves a heating and cooling coil, as shown in Fig. 9-16, packaged in a single cabinet in a conditioned space. One thermostat is to regulate the two valves so that as the temperature in the space drops, the cooling coil gradually closes off and on a further drop in temperature the heating coil begins to open. The spring ranges available for the valves are 28 to 55 and 62 to 90 kPa. The building is located in a cold climate. The thermostat has a change in output of 20 kPa/K. (*a*) Specify the spring ranges of the valves, whether they are normally open or normally closed, and whether the thermostat is direct- or reverse-acting. (*b*) If the space temperature is 25°C when the chilled-water valve is completely open, what are the ranges of temperature during cooling operation and heating operation?

Solution In a cold climate the hot-water valve should be normally open. In order for an increase in temperature to close the valve the control pressure must increase

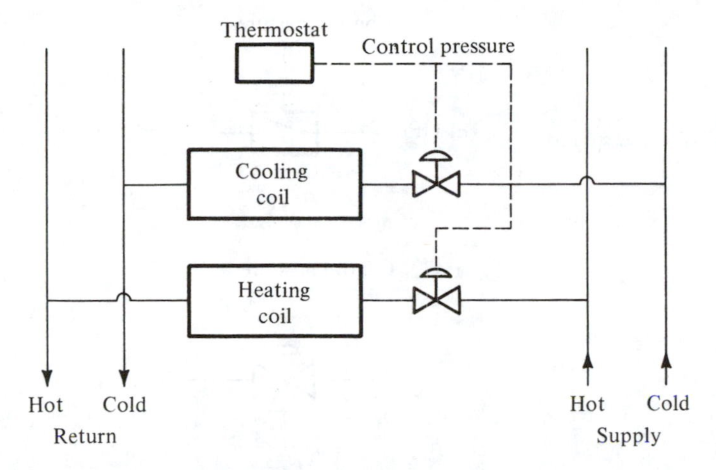

Figure 9-16 Sequenced heating and cooling control.

with an increase in temperature, and so the thermostat must be direct-acting. Assign the lowest spring range (28 to 55 kPa) to the heating coil and the highest to the cooling coil. The chilled-water valve should be closed at a control pressure of 62 kPa and fully open at 90 kPa, so it is a normally closed valve.

The space temperature is 25°C when the chilled-water valve is completely open (90 kPa); the temperature ranges of this sequence are as shown in Table 9-1.

9-13 Other valves, switches, and controls A number of additional valves, switches, and other devices are available to the control-system designer. Some are shown schematically in Fig. 9-17 and described below.

Manual pressure-setting switch (Fig. 9-17*a*). Provides a constant output pressure that is manually adjustable.

Pressure-electric switch (Fig. 9-17*b*). An electric switch that changes its position from open to closed (normally open) or from closed to open (normally closed) upon an increase in pressure above the set point.

Electric air switch (Fig. 9-17*c*). A two-way solenoid valve for interruption of control-air pressure.

Table 9-1 Temperature ranges in sequenced operation of Example 9-3

Control pressure, kPa	Temperature, °C	
90	25	Cooling
62	$25 - (90 - 62)\dfrac{1}{20\text{ kPa/K}} = 23.6$	
55	$25 - (90 - 55)\left(\frac{1}{20}\right) = 23.25$	Heating
28	$25 - (90 - 28)\left(\frac{1}{20}\right) = 21.9$	

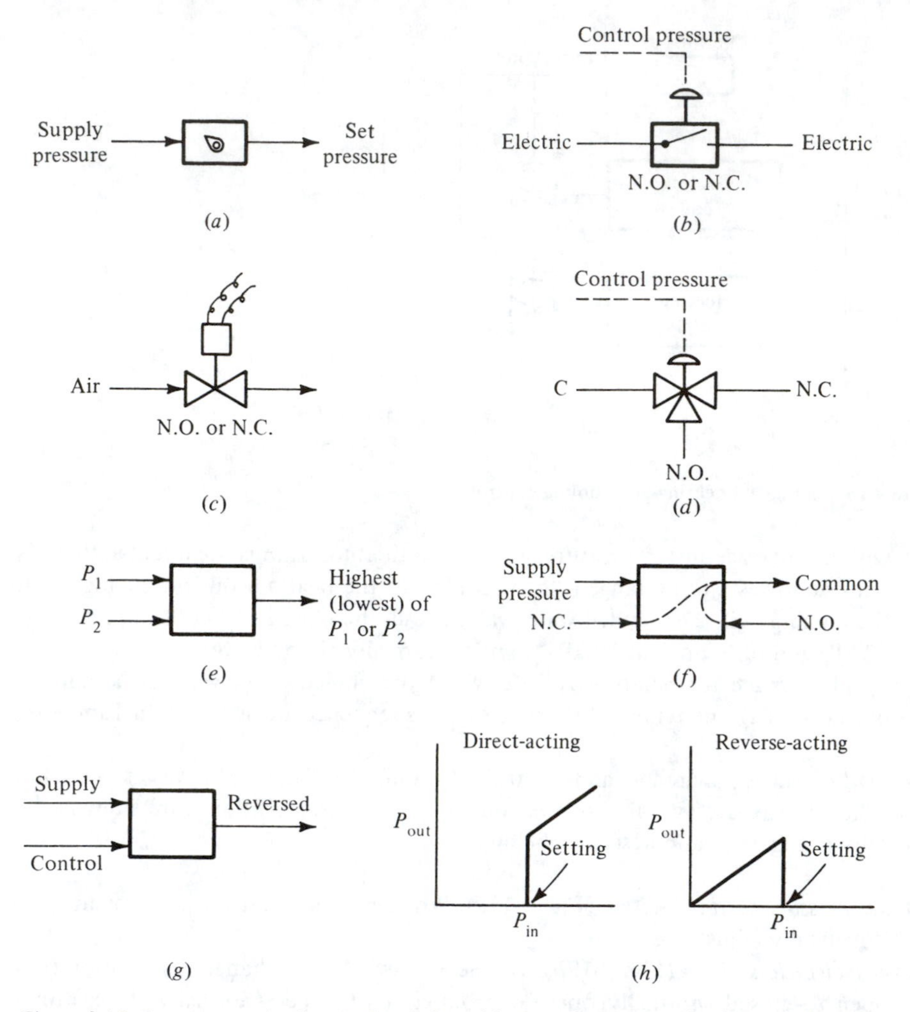

Figure 9-17 Some control valves and switches: (*a*) manual pressure-selecting switch, (*b*) pressure-electric switch, (*c*) electric air switch, (*d*) pneumatic air switch, (*e*) pressure selector, (*f*) diverting relay, (*g*) reversing relay, (*h*) switching valve.

Pneumatic air switch (Fig. 9-17*d*). Three-way, two-position valve that selects one of two different air pressures dependent upon the control pressure.

Pressure selector (Fig. 9-17*e*). A high-pressure selector chooses the highest of two incoming pressures and passes it on. A low-pressure selector performs the opposite function.

Diverting relay (Fig. 9-17*f*). At a low supply pressure (100 kPa, for example) the common line is connected with the normally open port, and with a high supply pressure (160 kPa, for example) the common is connected with the normally closed port.

Reversing relay (Fig. 9-17*g*). Receives an input pressure (20 to 100 kPa, for example) and converts the signal into an output pressure of 100 to 20 kPa.

Switching valve (Fig. 9-17*h*). A direct-acting switching valve transmits the input pressure beyond the set value; a reverse-acting switching valve transmits the input pressure up to the set value.

9-14 Building up a control system Two additional control devices (humidistats and master-submaster thermostats) will be introduced in the next several sections, but enough control elements have now been provided to construct realistic control systems. Complicated functions can often be developed by a logical (and sometimes clever) combination of basic elements.

Example 9-4 The outdoor-air control shown in Fig. 9-18 is to be regulated so that when the outdoor-air temperature is above 13°C, 100 percent outdoor air is used. Below outdoor-air temperatures of 13°C the dampers mix outdoor and recirculated air so that a t_{mix} of 13°C is maintained. The outdoor-air dampers must be open 20 percent or more to provide minimum ventilation air.

Draw the control diagram, specifying the spring ranges on the dampers, whether they are normally open or normally closed, and whether thermostats are direct- or reverse-acting. Design for fail-safe conditions in cold climates.

Solution A suitable control scheme is shown in Fig. 9-19. The fail-safe position for the dampers is normally closed for the outdoor and exhaust dampers and normally open for the recirculated air. The thermostat is direct-acting so that on a rise in temperature the control pressure increases, opening the outdoor and exhaust dampers and closing the recirculated-air dampers. At 13°C outdoor temperature the outdoor-air dampers are completely open. Outdoor temperatures above 13°C increase the control pressure, which keeps the outdoor dampers open. As the outdoor temperature drops below 13°C, the dampers modulate to maintain a mix temperature of 13°C until the control pressure drops to 68 kPa (20 percent of the span between the spring range of 62 to 90 kPa). At control pressures below 68 kPa the high-pressure selector chooses the 68 kPa from the manual pressure regulator.

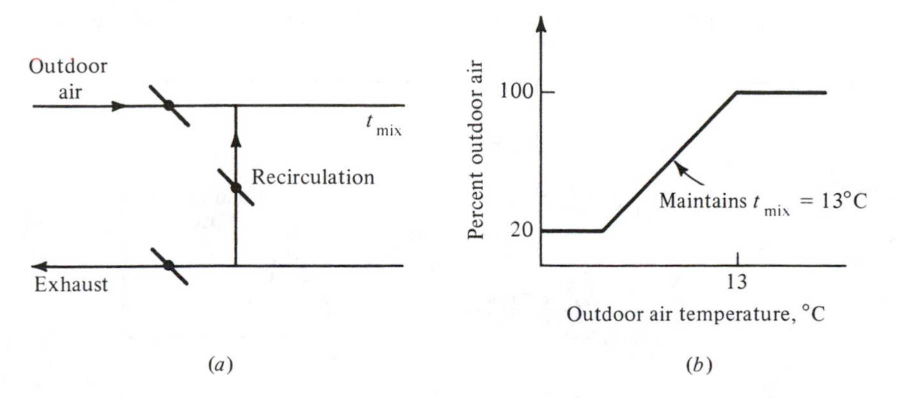

Figure 9-18 (*a*) Outdoor-air control and (*b*) desired damper position in Example 9-4.

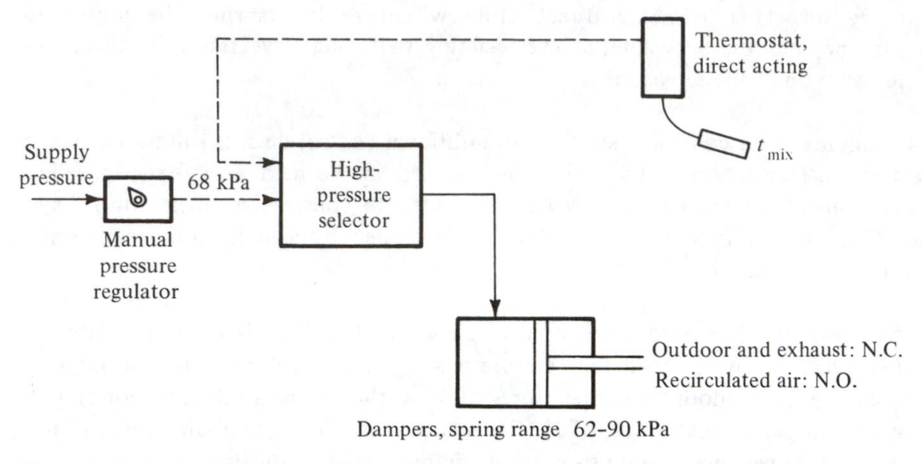

Figure 9-19 Control diagram for Example 9-4.

9-15 Humidistats and humidifiers Without humidification many buildings would experience low humidities during the winter. The outdoor air brought in for ventilation has a low humidity ratio, and relative humidities in unhumidified buildings of 10 percent are not uncommon.

Humidistats are usually located in the conditioned space or in the return line, as shown in Fig. 9-20. Two commonly used types of humidistats are mechanical and electric. In the mechanical type a change in relative humidity changes a mechanical property (such as the length of nylon sensing elements) which actuates a control. In the electric type the change in relative humidity changes an electrical property of a substance, such as its resistance or capacitance.

A popular type of humidifier is one which admits steam directly into the supply airstream, as shown in Fig. 9-20. The valve regulating the steam to the humidifier is of the normally closed type to avoid excessive moisture upon a failure of control pressure. In addition, a normally closed shutoff valve is placed in the steam line and opened only when the fan is operating. One of the typical problems in the control of humidifiers is overshooting because of the lag experienced between the time the humidifier increases the rate of humidification and the increase is sensed by the

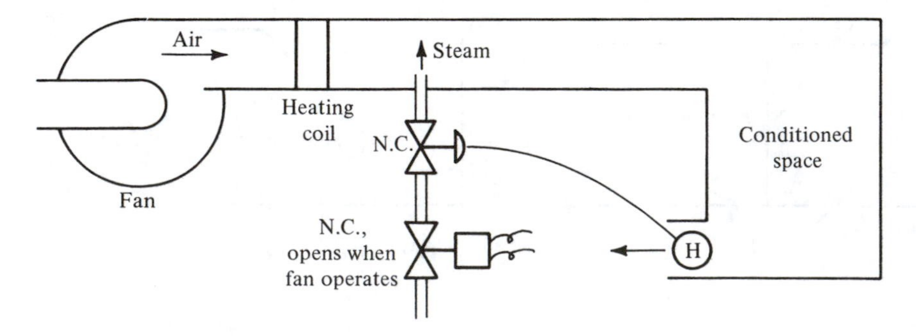

Figure 9-20 Humidistat and humidifier.

humidistat. One technique is to use two humidistats, one as shown in Fig. 9-20 and another in the supply-air duct following the humidifier. The space humidistat resets the duct humidistat, which directly controls the humidifier. If a nominal 40 percent relative humidity in the space is desired, for example, the space humidity sets the duct humidistat as follows:

Space humidity, %	35	40	45
Setting of duct humidistat, %	60	40	20

This scheme affords direct control on the humidity of the supply air.

There is an important fact to consider with respect to humidification, though not particularly related to its control. If the windows in the building are single-pane and the outdoor temperatures are cold, moisture added by the humidifier will condense on the inside of the windows; the process becomes one of pouring moisture in at the humidifier and condensing it out of the windows without ever appreciably increasing the relative humidity.

9-16 Master and submaster thermostats It is often advantageous to change the *setting* of one control automatically in response to another variable. For example, the air temperature in the warm duct of a dual-duct system can be programmed so that the air temperature is lowered as the outdoor temperature increases (and the heating load decreases). One version of the receiver-controller, as in Fig. 9-21, has settings for gain, set point, and reset.

The master transmitter is usually direct-acting, so that an increase in the tempera-

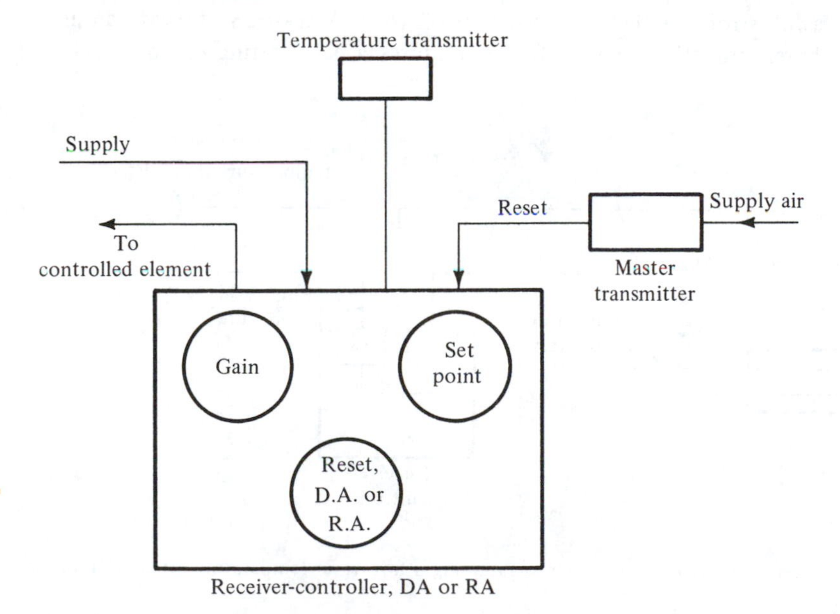

Figure 9-21 A receiver-controller with master-submaster capabilities.

ture it senses increases the pressure fed to the receiver-controller. The reset capability of the receiver-controller may be chosen as either direct- or reverse-acting and the dial can be set for a ratio of change of reset pressure to a change in set-point pressure. An example will serve to illustrate the application of the master-submaster function.

Example 9-5 The valve in a steam line serving a hot-water heater is to be regulated so that the water temperature leaving the heater is programmed on the basis of outdoor-air temperature as follows with a linear variation between:

Outdoor temperature, °C	20	-20
Hot-water supply temperature, °C	40	90

The range of the temperature transmitter in the hot-water supply line is 30 to 100°C and that of the outdoor-air-temperature transmitter is -30 to 30°C, both providing air pressure outputs in this range of 20 to 100 kPa. Specify the gain, set-point, and reset ratios and whether the receiver-controller and the reset are direct- or reverse-acting.

Solution The control system is shown schematically in Fig. 9-22. The steam valve is chosen as normally open so that hot water will still be provided if there is a failure of control pressure. One of the available spring ranges, 62 to 90 kPa, is arbitrarily chosen. Because the valve is normally open, the response of the receiver-controller to the hot-water-temperature sensor must be direct-acting. The output of the temperature transmitter is (100 - 20 kPa)/(100 - 30°C) = 1.14 kPa/K. To achieve a full stroke of the steam valve of 62 to 90 kPa or 28 kPa with an arbitrarily chosen throttling range of 5 K requires a gain setting on the receiver-

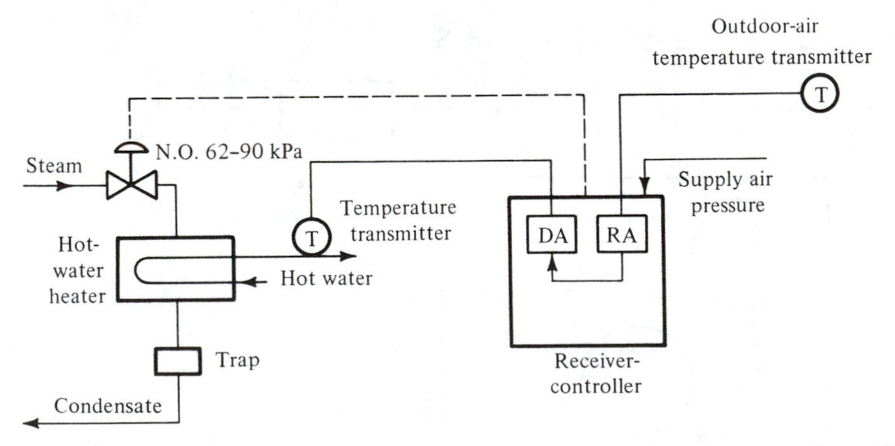

Figure 9-22 Master-submaster control for programming a hot-water temperature using outdoor-air temperature.

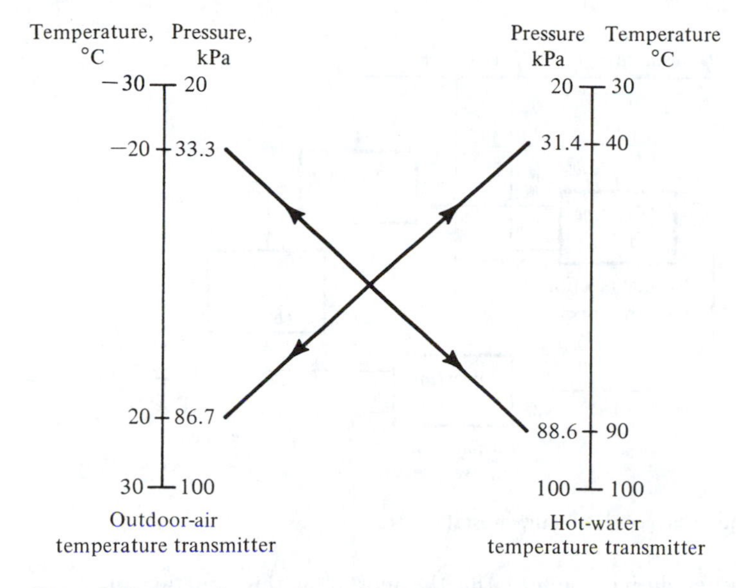

Figure 9-23 Reset through the outdoor-air temperature.

controller of

$$\text{Gain} = \frac{28 \text{ kPa}}{(5 \text{ K})(1.14 \text{ kPa/K})} = 4.91 \approx 5$$

Next turn to the reset capabilities. The outdoor-air temperature transmitter (Fig. 9-23) emits a pressure of 33.3 kPa when the outdoor temperature is -20°C and 86.7 when the outdoor temperature is +20°C. Since this change in pressure must be used to reset the hot-water-temperature controller, the 90°C point must be selected from the span of the hot-water-temperature transmitter when the outdoor-air-temperature transmitter is 33.3 kPa. Similarly the 40°C control (31.4 kPa) must be selected when the pressure output of the outdoor-air-temperature transmitter is 86.7 kPa. The reset ratio is the ratio of the pressure differences

$$\text{Reset ratio} = \frac{88.6 - 31.4 \text{ kPa}}{86.7 - 33.3 \text{ kPa}} = 1.07 \approx 1$$

Since the low pressure from the outdoor-air-temperature transmitter must select a high pressure from the hot-water-temperature transmitter, the reset must be *reverse-acting.*

9-17 Summer-winter changeover Many controls must function in opposite directions for cooling and heating. For example, suppose that a fan-coil unit in a conditioned space is served by a normally open valve and hot water is supplied to the coil during

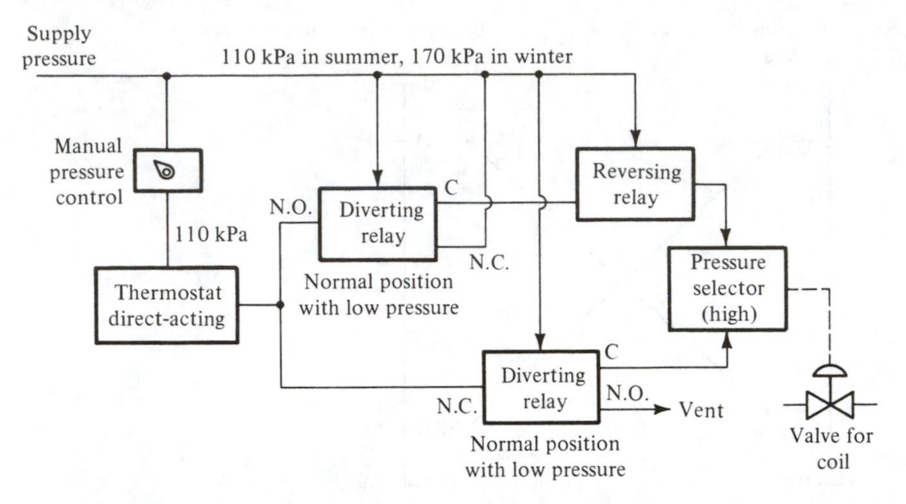

Figure 9-24 Summer-winter changeover of a thermostat control.

heating and chilled water during cooling. The thermostat in this situation must be direct-acting for heating and reverse-acting for cooling. A capability that can be built into a pneumatic control system is to use the level of supply pressure as an indicator of whether summer or winter operation is desired. Typically a supply pressure of 110 kPa is provided for summer operation and 170 kPa for winter. Special thermostats are available to convert from reverse- to direct-acting as the supply pressure is switched from 110 to 170 kPa and thus satisfy the requirement described above. It is also possible to build up a system with the components previously described in this chapter. Such a scheme is shown in Fig. 9-24, where the thermostat operates direct-acting when the supply pressure is 170 kPa but delivers control pressure through a reversing relay when the supply pressure is 110 kPa. The diverting relays connect the common port to the normally open port when the supply pressure is 110 kPa and the common to the normally closed when the supply pressure is 170 kPa.

9-18 Valve characteristics and selection When a control valve is selected to regulate the hot- or chilled-water flow through a coil, as shown schematically in Fig. 9-25, there are essentially three items to be specified: pipe size, C_v value, and characteristics (quick-opening, linear, or equal percentage).

1. The pipe size is often specified to match that of the pipe to and from the valve, although sometimes the pipe size of the valve is smaller in order to reduce the cost of the valve.
2. The C_v value is defined by the equation

$$\text{Flow rate} = C_v \sqrt{\Delta p} \qquad \text{L/s} \tag{9-1}$$

   where Δp is the pressure drop in kilopascals across the valve when the valve is in its wide-open position.
3. The three different valve characteristics commonly available are shown in Fig. 9-26.

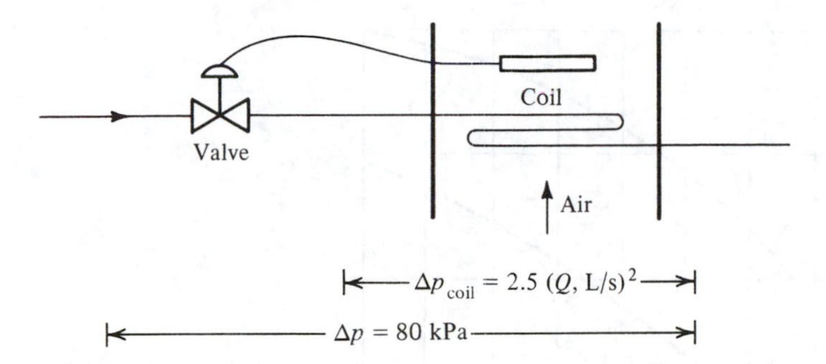

Figure 9-25 Valve and coil combination.

The valve manufacturer provides a certain characteristic by means of the design chosen for the seat and the plug. The influences of the choice of valve characteristic and C_v value are demonstrated by two different choices of C_v, designated case A and case B. Suppose that the coil has a pressure drop

$$\Delta p_{coil} = 2.5Q^2 \quad \text{kPa}$$

where Q = flow rate, L/s and the available pressure difference across both the coil and valve is constant at 80 kPa, as shown in Fig. 9-25. A valve with linear characteristics is used in both cases, but the C_v in case A is 0.6 and in case B C_v = 1.2.

For the valve with linear characteristics,

$$Q = \frac{\text{percent stem stroke}}{100} C_v \sqrt{\Delta p}$$

or

$$\text{Percent stem stroke} = \frac{100Q}{C_v \sqrt{\Delta p}} \tag{9-2}$$

For a given flow rate, say, 2 L/s, the percent stem stroke can be computed for the valve. The pressure drop through the coil would be $2.5(2.0^2) = 10$ kPa, requiring 80 - 10 =

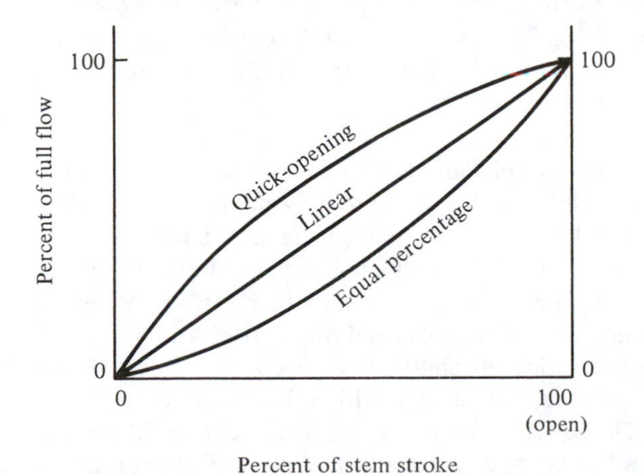

Figure 9-26 Three valve characteristics.

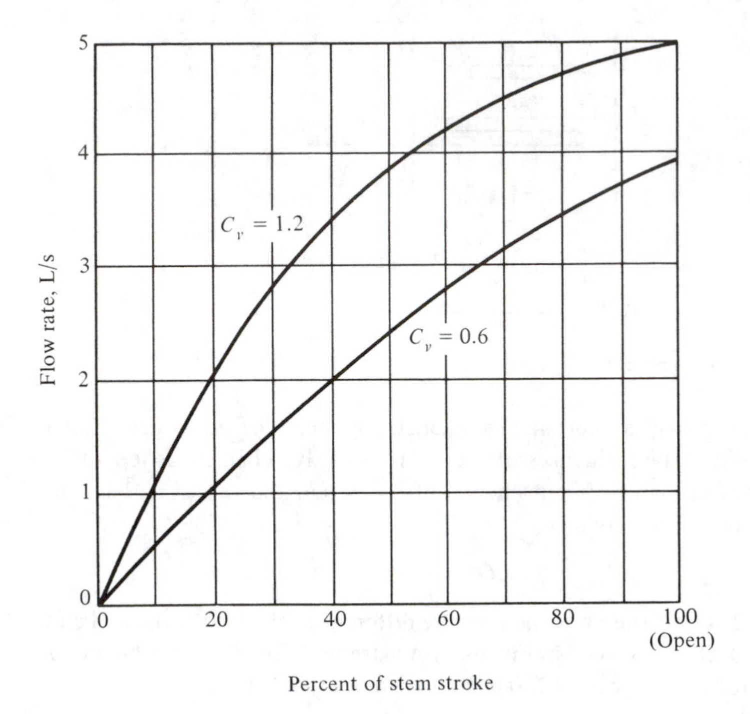

Figure 9-27 Flow-stem-stroke relationship of a control valve and coil combination.

70 kPa to be dissipated in the valve. In case A with C_v = 0.6, Eq. (9-2) indicates the percent stem stroke to be

$$\text{Percent stem stroke} = \frac{100(2)}{0.6\sqrt{70}} = 39.8\%$$

The relationship of the percent stem stroke to the flow rate through the coil-valve combination for the two cases is shown in Fig. 9-27, from which two observations can be made: (1) even though the flow-stem stroke of the valves is linear, the flow-stem stroke characteristic of the coil-valve combination is not; (2) the combination with the valve of the high C_v yields the characteristic with less linearity. A linear relation is desired, as explained in the next section.

9-19 Stability of an air-temperature control loop Most commercial air-temperature control loops are in danger of becoming unstable if improperly designed or adjusted. During unstable behavior the control "hunts," i.e., the leaving air temperature from the coil and the flow rate of the fluid the valve regulates oscillate. Instability is undesirable because it may result in uncomfortable conditions in the space, excessive energy requirements, and accelerated wear of the valve and other components.

The fundamental means of preventing instability is to keep the combined gain around the loop low. This can be done, but usually with other penalties, so that a compromise must be reached. The selection of the valve is one choice influencing stability. The slope of the curves in Fig. 9-27 represents a portion of the gain in the entire control loop. A large change in the flow rate for a given change in percent stem stroke represents a high gain. The choice of the low value of C_v, namely 0.6, is thereby preferred to the higher C_v in Fig. 9-27. Furthermore, the loss of linearity with the high C_v results in a steep slope (high gain) near the closed position of the valve. It is not uncommon for a loop to be stable at moderate and heavy loads and unstable at light loads.

Restoring the combined coil-valve performance to a linear relationship can often be achieved by using a valve with *equal-percentage characteristics* (Fig. 9-26), and manufacturers provide valves with these characteristics precisely to make the combined characteristics nearly linear (see Prob. 9-6). Choosing a low C_v for the valve is not without its penalty, because the maximum flow rate through the valve-coil combination with a wide-open valve is reduced.

The other principal means of adjusting the gain of the loop is to adjust the gain of the controller.[3,4] A compromise must be reached in setting the gain of the controller also, because although it helps form a stable loop, a low setting of the gain increases the throttling range (Sec. 9-8), which in some cases may result in extra energy requirements.

9-20 Temperature reset based on zone load In the discussion in Chap. 5 of energy characteristics of some systems that experience simultaneous heating and cooling, such as dual-duct and terminal reheat, it was pointed out that resetting the supply temperatures at light load will conserve energy. In the dual-duct system shown schematically in Fig. 9-28, for example, the hot duct should supply air at the lowest temperature possible and the cold duct air at the highest temperature possible. A pneumatic control system can be developed that decreases the warm-air supply temperature so that one warm air damper in a mixing box is completely open. Suppose that the spring range on a damper operator at the mixing box is from 35 to 60 kPa and that the space thermostat provides those control pressures with space temperatures of 23 and 22°C, respectively. The schedule would then be:

Zone temperature, °C	Control pressure, kPa	Warm-air damper
24	10	Closed
23	35	Closed
22	60	Open
21	85	Open

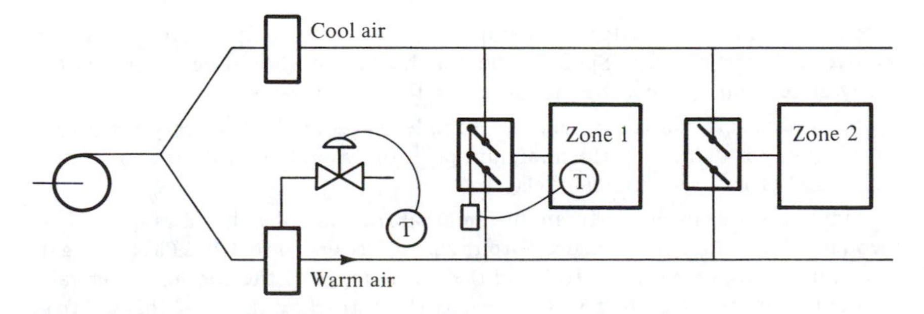

Figure 9-28 Reset of warm-air supply temperature based on zone load.

The control pressures from all zones are connected to a pressure selector that selects the highest pressure and uses it to reset the hot-duct temperature in a schedule typically as follows:

Highest control pressure, kPa	50	55	60
Hot-duct setting, °C	30	35	40†

† Design setting.

When at least one zone is calling for full heating, the hot-duct setting moves to its design value of 40°C, but when all warm-air dampers are partially closed, the setting of the warm-air temperature drops. A comparable control scheme can be applied to raise the cool-air supply temperature so that one zone has a fully open cool-air damper.

9-21 Electric, electronic, and computer control This chapter concentrated on pneumatic controls because of their widespread application in large-building air conditioning. Residential and small commercial air-conditioning systems are regulated mostly by on-off electric controls. Pneumatic systems begin appearing when the system consists of more than several dozen components. Pneumatic control systems are inherently modulating in character, which is a desirable feature. In very large buildings or in multibuilding complexes computer control becomes competitive. In computer-controlled systems the transmission of information is in digital form or characterized by the frequency of pulses. The computer is programmed to perform the standard sequences described in this chapter but is also capable of much more sophisticated decisions and calculations. Even in computer control systems the final exercise of power at a damper or valve might be pneumatic.

When an air-conditioning system is functioning improperly, the control system is often the first segment blamed—rightly or wrongly. Many engineers yearn for simple control systems that are easy to understand and maintain. Unfortunately, to achieve good control of the space conditions and do so in an energy-effective manner the control system must be sophisticated. Educating the building operators in the functioning of the control system is imperative.

PROBLEMS

9-1 A space thermostat regulates the damper in the cool-air supply duct and thus provides a variable airflow rate. Specify whether the damper should be normally open or normally closed and whether the thermostat is direct- or reverse-acting.

9-2 On the outdoor-air control system of Example 9-4, add the necessary features to close the outdoor-air damper to the minimum position when the outdoor temperature rises above 24°C. *Ans.* Use a diverting relay

9-3 The temperature transmitter in an air-temperature controller has a range of 8 to 30°C through which range the pressure output changes from 20 to 100 kPa. If the gain of the receiver-controller is set at 2 to 1 and the spring range of the cooling-water valve the controller regulates is 28 to 55 kPa, what is the throttling range of this control? *Ans.* 3.7 K

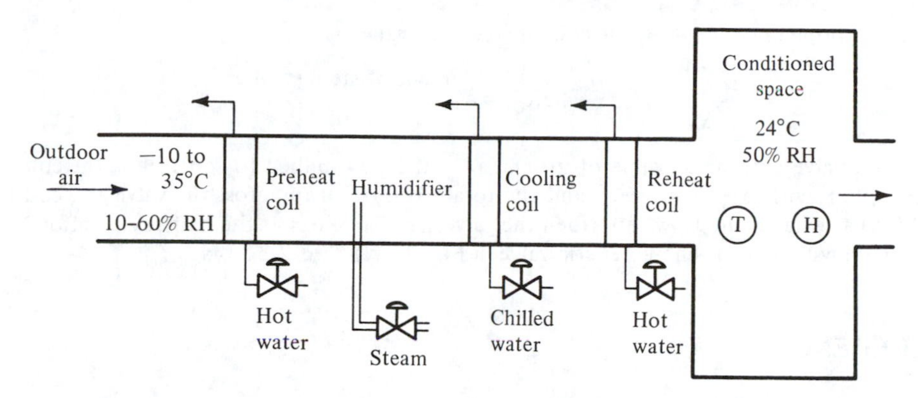

Figure 9-29 Controlled-environment space in Prob. 9-4.

9-4 The air supply for a laboratory (Fig. 9-29) consists of a preheat coil, humidifier, cooling coil, and heating coil. The space is to be maintained at 24°C, 50 percent relative humidity the year round, while the outdoor supply air may vary in relative humidity between 10 and 60 percent and the temperature from −10 to 35°C. The spring ranges available for the valves are 28 to 55 and 62 to 90 kPa. Draw the control diagram, adding any additional components needed, specify the action of the thermostat(s) and humidistat, the spring ranges of the valves, and whether they are normally open or normally closed.

9-5 A face-and-bypass damper assembly at a cooling coil is sometimes used in humid climates to achieve greater dehumidification for a given amount of sensible cooling, instead of permitting all the air to pass over the cooling coil. Given the hardware in Fig. 9-30, arrange the control system to regulate the temperature at 24°C and the relative humidity at 50 percent. If both the temperature and humidity cannot be maintained simultaneously, the temperature control should override the humidity control. The spring ranges available for the valve and damper are 28 to 55 and 48 to 76 kPa. Draw the control diagram and specify the action of the thermostat and humidistat, whether the valve is normally open or normally closed, and which damper is normally closed.

9-6 Section 9-18 described the flow characteristics of a coil regulated by a valve with linear characteristics. The equation of the flow-stem position for another type of

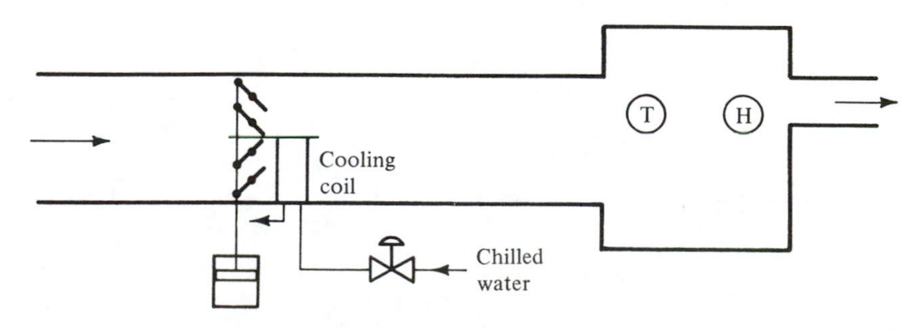

Figure 9-30 Face and bypass damper control.

valve mentioned in Sec. 9-18, the equal-percentage valve, is

$$\frac{Q}{C_v \sqrt{\Delta p}} = A^x \qquad \text{where } x = \frac{\text{percent of stem stroke}}{100} - 1$$

If such a valve with an A value of 20 and a C_v of 1.2 is applied to controlling the coil in Fig. 9-25 with $\Delta p_{\text{coil}} = 2.5Q^2$ and the total pressure drop across the valve and coil of 80 kPa, what is the flow rate when the valve stem stroke is at the halfway position? (Compare with a linear-characteristic valve in Fig. 9-27.) *Ans.* 2.21 L/s

REFERENCES

1. E. J. Brown and J. R. Fellows: Pressure Losses and Flow Characteristics of Multiple-Leaf Dampers, *ASHAE Trans.*, vol. 64, pp. 299–318, 1958.
2. D. H. Spethmann: Humidity Control Comes of Age, *Heat., Piping, Air Cond.*, vol. 45, no. 3, pp. 103–109, March 1973.
3. W. F. Stoecker: Stability of an Air-Temperature Control Loop, *ASHRAE Trans.*, vol. 84, Pt. 1, pp. 35–53, 1978.
4. D. C. Hamilton, R. G. Leonard, and J. T. Pearson: Dynamic Response Characteristics of Discharge Air Temperature Control System at Near Full and Part Heating Load, *ASHRAE Trans.*, vol. 80, Pt. 1, pp. 180–194, 1974.
5. J. P. Kettler: System Control, *Build. Syst. Des.*, vol. 69, pp. 19–21, August 1972.
6. N. J. Janisse: How to Control Air Systems, *Heat., Piping, Air Cond.*, vol. 41, no. 4, pp. 129–136, April 1969.
7. G. Shavit: Enthalpy Control Systems: Increased Energy Conservation, *Heat., Piping, Air Cond.*, vol. 46, no. 1, pp. 117–122, January 1974.
8. R. W. Haines: "Control Systems for Heating, Ventilating and Air Conditioning," 2d ed., Van Nostrand Reinhold, New York, 1977.
9. Most of the major control companies have developed booklets on the fundamentals of control of thermal systems in buildings.

CHAPTER

TEN

THE VAPOR-COMPRESSION CYCLE

10-1 Most important refrigeration cycle The vapor-compression cycle is the most widely used refrigeration cycle in practice. In this cycle a vapor is compressed, then condensed to a liquid, following which the pressure is dropped so that fluid can evaporate at a low pressure. In this chapter the study progresses from the classical Carnot cycle to the actual vapor cycle. The modifications of the Carnot cycle are dictated by practical considerations.

10-2 Carnot refrigeration cycle The Carnot cycle is one whose efficiency cannot be exceeded when operating between two given temperatures. The Carnot cycle operating as a heat engine is familiar from the study of thermodynamics. The Carnot heat engine is shown schematically in Fig. 10-1*a*, with the corresponding temperature-entropy diagram in Fig. 10-1*b*. The Carnot heat engine receives energy at a high level of temperature, converts a portion of the energy into work, and discharges the remainder to a heat sink at a low level of temperature.

The Carnot refrigeration cycle performs the reverse effect of the heat engine, because it transfers energy from a low level of temperature to a high level of temperature. The refrigeration cycle requires the addition of external work for its operation. The diagram of the equipment and the temperature-entropy diagram of the refrigeration cycle are shown in Fig. 10-2*a* and *b*.

The processes which constitute the cycle are:

1-2. Adiabatic compression
2-3. Isothermal rejection of heat
3-4. Adiabatic expansion
4-1. Isothermal addition of heat

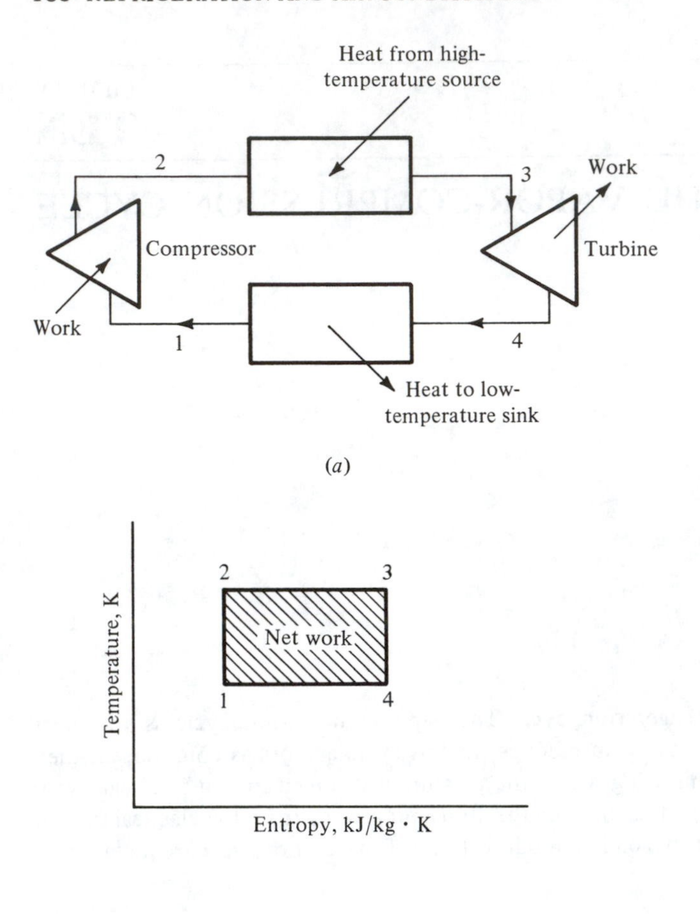

Figure 10-1 *(a)* Carnot heat engine; *(b)* temperature-entropy diagram of the Carnot heat engine.

All the processes in the Carnot cycle are thermodynamically reversible. Processes 1-2 and 3-4 are consequently isentropic.

The withdrawal of heat from the low-temperature source in process 4-1 is the refrigeration step and is the entire purpose of the cycle. All the other processes in the cycle function so that the low-temperature energy can be discharged to some convenient high-temperature heat sink.

The Carnot cycle consists of reversible processes which make its efficiency higher than could be achieved in an actual cycle. A reasonable question is: Why discuss the Carnot cycle if it is an unattainable ideal? There are two reasons: (1) it serves as a standard of comparison, and (2) it provides a convenient guide to the temperatures that should be maintained to achieve maximum effectiveness.

10-3 Coefficient of performance Before any evaluation of the performance of a refrigeration system can be made, an effectiveness term must be defined. The index of

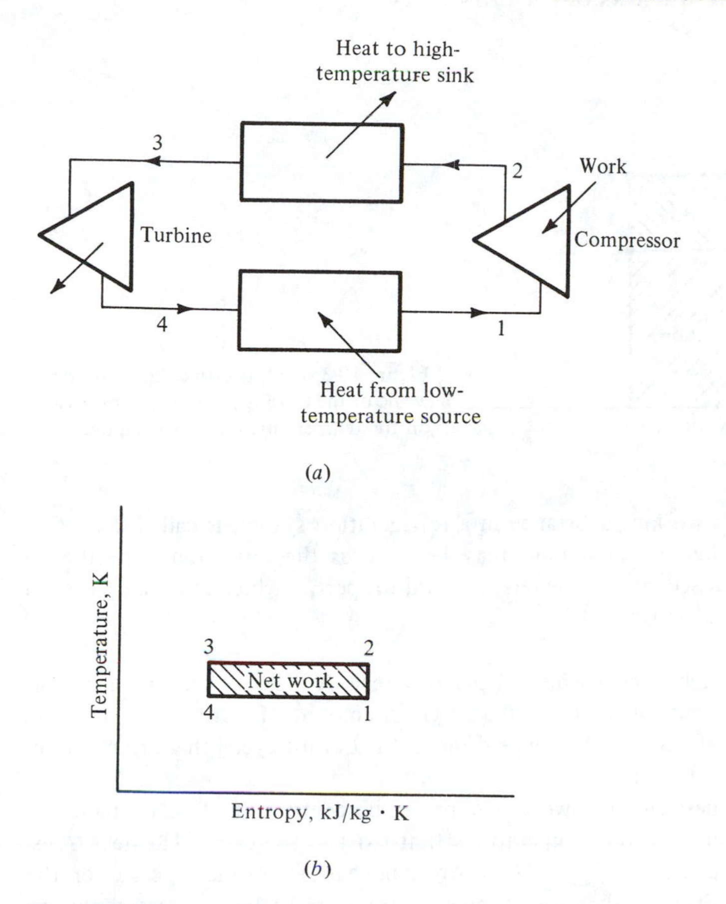

Figure 10-2 (*a*) Carnot refrigeration cycle; (*b*) temperature-entropy diagram of the Carnot refrigeration cycle.

performance is not called efficiency, however, because that term is usually reserved for the ratio of output to input. The ratio of output to input would be misleading applied to a refrigeration system because the output in process 2-3 is usually wasted. The concept of the performance index of the refrigeration cycle is the same as efficiency, however, in that it represents the ratio

$$\frac{\text{Magnitude of desired commodity}}{\text{Magnitude of expenditure}}$$

The performance term in the refrigeration cycle is called the *coefficient of performance,* defined as

$$\text{Coefficient of performance} = \frac{\text{useful refrigeration}}{\text{net work}}$$

The two terms which make up the coefficient of performance must be in the same units, so that the coefficient of performance is dimensionless.

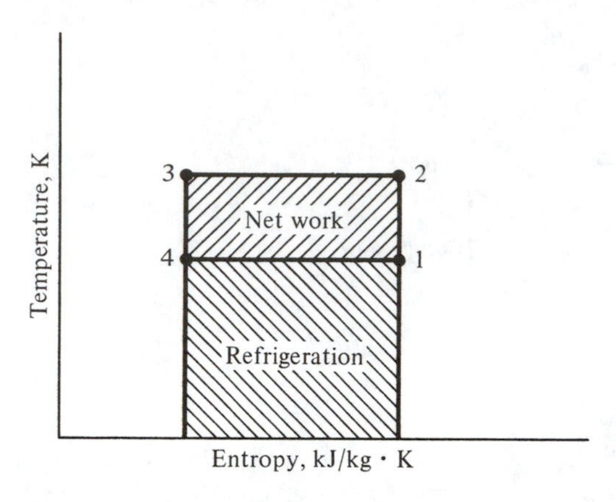

Figure 10-3 Useful refrigeration and net work of the Carnot cycle shown by areas on the temperature-entropy diagram.

10-4 Refrigerant The working substance in a refrigeration system is called the *refrigerant.* One of a number of compounds may be used as the refrigerant. Specific refrigerants will be discussed later in the chapter, and properties which make a successful refrigerant are examined in Chap. 15.

10-5 Conditions for highest coefficient of performance A high coefficient of performance is desirable because it indicates that a given amount of refrigeration requires only a small amount of work. What can be done in the Carnot cycle, then, to maintain a high coefficient of performance?

To answer the question, first we can express the coefficient of performance of the Carnot cycle in terms of the temperatures that exist in the cycle. The heat transferred in a reversible process is $q_{rev} = \int T\, ds$. Areas beneath reversible processes on the temperature-entropy diagram therefore represent transfers of heat. Areas shown in Fig. 10-3 can represent the amount of useful refrigeration and the net work. The useful refrigeration is the heat transferred in process 4–1, or the area beneath line 4-1. The area under line 2-3 represents the heat rejected from the cycle. The difference between the heat rejected from the cycle and heat added to the cycle is the net heat which for a cyclic process equals the net work. The area enclosed in rectangle 1-2-3-4 represents the net work. An expression for the coefficient of performance of the Carnot refrigeration cycle is therefore

$$\text{Coefficient of performance} = \frac{T_1(s_1 - s_4)}{(T_2 - T_1)(s_1 - s_4)} = \frac{T_1}{T_2 - T_1}$$

The coefficient of performance of the Carnot cycle is entirely a function of the temperature limits and can vary from zero to infinity.

A low value of T_2 will make the coefficient of performance high. A high value of T_1 increases the numerator and decreases the denominator, both of which increase the coefficient of performance. The value of T_1, therefore, has a more pronounced effect upon the coefficient of performance than T_2.

To summarize, for a high coefficient of performance (1) operate with T_1 high and (2) operate with T_2 low.

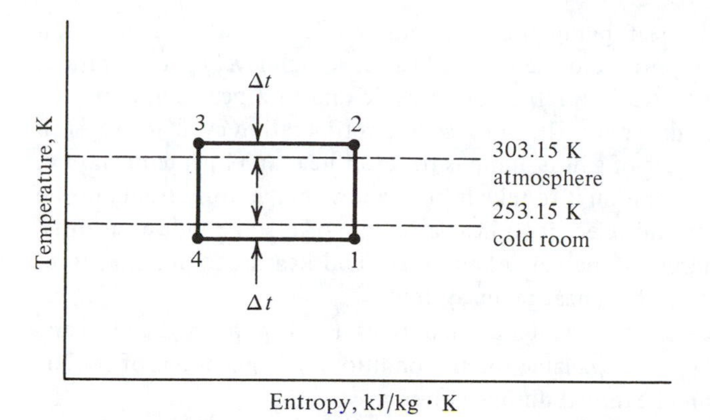

Figure 10-4 Temperature requirements imposed upon a refrigeration cycle.

10-6 Temperature limitations If we left the analysis here, we would leave the false impression that we have complete control over T_1 and T_2. If this were true, T_1 could simply be set equal to T_2, which would make the coefficient of performance equal to infinity.

Closer study shows that certain temperature requirements are always imposed upon the refrigeration system. For example, if the refrigeration system must maintain a cold room at -20°C and can reject heat to the atmosphere at 30°C, these two temperatures are limitations within which the cycle must abide. The two temperatures are shown as dashed lines in Fig. 10-4, expressed in kelvin. During the heat-rejection process, the refrigerant temperature must be higher than 303.15 K. During the refrigeration process, the refrigerant temperature must be lower than 253.15 K in order to transfer heat from the cold room to the refrigerant. The cycle that results is the one shown in Fig. 10-4. It should not be called a Carnot cycle because all processes in the Carnot cycle are reversible and transfers of heat with a difference in temperature are irreversible processes. The cycle is now merely a rectangular cycle on the temperature-entropy plane.

Temperature T_2 should be kept low, but it cannot be reduced below 303.15 K. Temperature T_1 should be kept high, but it can be increased no higher than 253.15 K. What control do we have, then, over the temperature? The answer is that we can concentrate on keeping the Δt as small as possible. Reduction of Δt can be accomplished by increasing A or U in the equation

$$q = UA\ \Delta t$$

where q = heat, W
U = overall heat-transfer coefficient, $W/m^2 \cdot K$
A = heat-transfer area, m^2
Δt = temperature change, K

In order to decrease Δt to zero, either U or A would have to be infinite. Since infinite values of U and A would also require an infinite cost, the actual selection of equipment always stops short of reducing Δt to zero.

10-7 Carnot heat pump A heat pump uses the same equipment as a refrigeration system but operates for the purpose of delivering heat at a high level of temperature. Even though the equipment used in a refrigeration cycle and in a heat pump may be identical, the objectives are different. The purpose of a refrigeration cycle is to absorb heat at a low temperature; that of a heat pump is to reject heat at a high temperature. An example of heat-pump operation is to take heat at a low temperature from outside air, the earth, or well water and reject it to heat a building. In some industrial situations, cooling may be required in one part of the plant and heating in another. Both these functions might be served by a heat-pump system.

A plant can be constructed to operate alternately as a heat pump and a refrigeration system. Units of this type are available for air-conditioning applications of cooling a building during summer and heating it during winter.

The performance of a heat pump is expressed by the *performance factor*. In keeping with the practice of defining the performance index as the amount of the desired commodity divided by the amount of expenditure, the performance factor is

$$\text{Performance factor} = \frac{\text{heat rejected from cycle}}{\text{work required}}$$

The quantities of energy which make up the performance factor can be represented by areas on the temperature-entropy diagram of the Carnot cycle, as shown in Fig. 10-5. The area under line 2-3 represents the heat rejected from the cycle, and the area enclosed in rectangle 1-2-3-4 represents the net work. The performance factor is therefore

$$\text{Performance factor} = \frac{T_2(s_1 - s_4)}{(T_2 - T_1)(s_1 - s_4)} = \frac{T_2}{T_2 - T_1}$$

The refrigeration cycle with the same temperatures as in Fig. 10-5 would have a coefficient of performance of $T_1/(T_2 - T_1)$. Therefore

$$\text{Performance factor} = \frac{T_2}{T_2 - T_1} = \frac{T_2}{T_2 - T_1} - \frac{T_2 - T_1}{T_2 - T_1} + 1$$

$$= \frac{T_1}{T_2 - T_1} + 1 = \text{coefficient of performance} + 1$$

The performance factor can therefore vary from 1 to ∞.

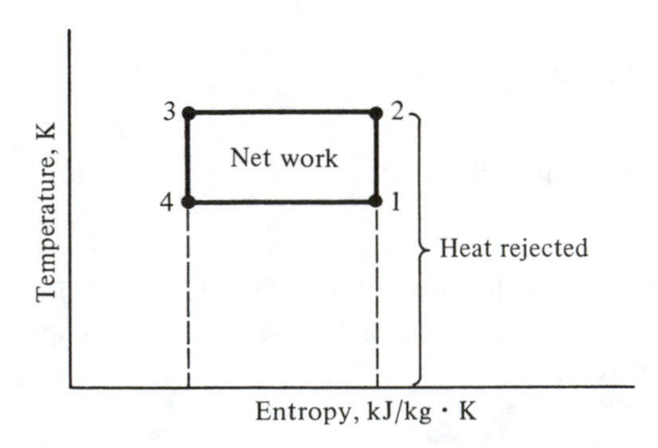

Figure 10-5 Carnot heat-pump cycle.

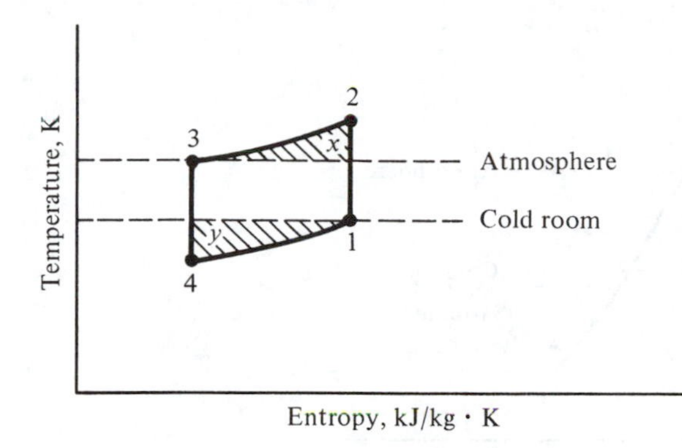

Figure 10-6 Refrigeration cycle when a gas is the refrigerant.

10-8 Using vapor as a refrigerant Because the Carnot refrigeration cycle is the most efficient cycle, every attempt should be made to reproduce it with actual equipment. Certainly the reversible processes cannot be duplicated, but at least the rectangular shape of the cycle on the temperature-entropy diagram should be maintained. Doing so means that all the heat can be received at one temperature level and rejected at another. If a gas, such as air, is used as the refrigerant, the cycle would appear as in Fig. 10-6 rather than as the rectangle of the Carnot cycle. The isentropic compression and expansion are processes 1-2 and 3-4, respectively. Processes 2-3 and 4-1 are constant-pressure cooling and heating processes, respectively. This cycle differs from the Carnot cycle operating between the same two temperatures by the addition of areas x and y. At point 4 the temperature must be lower than the cold-room temperature so that as the gas receives heat in the constant-pressure process it rises to a temperature no higher than that of the cold room. For similar reasons T_2 must be above the atmospheric temperature. The effect of area x is to increase the work required, which decreases the coefficient of performance. The effect of area y is to increase the work required and to reduce the amount of refrigeration. Both these effects of area y reduce the coefficient of performance.

Instead of a gas, a refrigerant may be used that condenses during the heat-rejection process and boils during the heat-addition, or refrigeration, process. Such a refrigerant could therefore operate between liquid and vapor states. With this refrigerant, the Carnot cycle can fit between the saturated-liquid and saturated-vapor lines, as shown in Fig. 10-7. Processes 2-3 and 4-1 take place at constant temperature since constant-pressure processes in the mixture region proceed at constant temperature. Process 2-3 is a condensation process, and the vessel in which it occurs is a *condenser.* Process 4-1 is a boiling process, and it takes place in the *evaporator.*

10-9 Revisions of the Carnot cycle While the cycle shown in Fig. 10-7 offers a high coefficient of performance, practical considerations require certain revisions, described in the next two selections; Sec. 10-10 discusses the changes in the compression process, 1-2, and Sec. 10-11 considers the expansion process, 3-4.

10-10 Wet compression versus dry compression The compression process, 1-2, in Fig. 10-7, is called *wet compression* because the entire process occurs in the mixture region with droplets of liquid present. When a reciprocating compressor is used, several

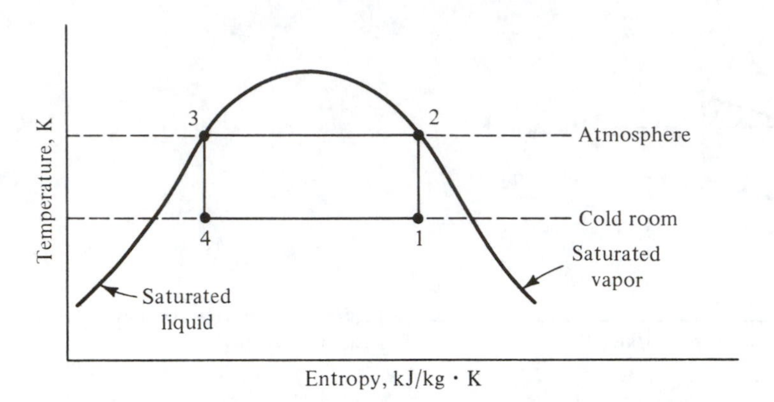

Figure 10-7 Carnot refrigeration cycle when a condensing and evaporating fluid is the refrigerant.

factors discourage the practice of wet compression. One is that liquid refrigerant may be trapped in the head of the cylinder by the rising piston, possibly damaging the valves or the cylinder head. Even though the point at the end of the compression shown as point 2 in Fig. 10-7 is saturated vapor and thus should be free from liquid, such is not the actual case. During compression the droplets of liquid are vaporized by an internal heat-transfer process which requires a finite amount of time. High-speed compressors are susceptible to damage by liquid because of the short time available for heat transfer. In a compressor that has a rotative speed of 30 r/s, for example, the compression takes place in $\frac{1}{60}$ s. At the end of compression, point 2 on the saturated-vapor line represents only average conditions of a mixture of superheated vapor and liquid. Another possible danger of wet compression is that the droplets of liquid may wash the lubricating oil from the walls of the cylinder, accelerating wear. Because of these disadvantages *dry compression*, which takes place with no droplets of liquid present, is preferable to wet compression. If the refrigerant entering the compressor is saturated vapor, as in Fig. 10-8, the compression from point 1 to 2 is called dry compression.

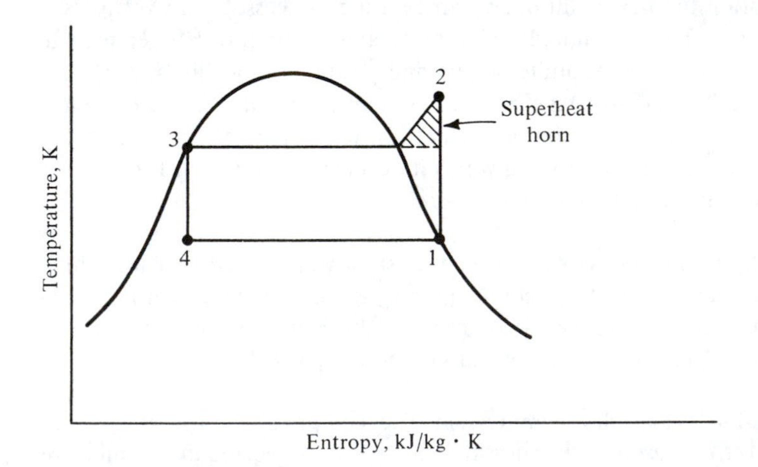

Figure 10-8 Revision of the Carnot refrigeration cycle by using dry compression.

With dry compression the cycle loses the rectangular shape of the Carnot cycle. Compression of a dry vapor results in a temperature at point 2 which is higher than the condensing temperature. The refrigerant therefore leaves the compressor superheated. The area of the cycle which is above the condensing temperature is sometimes called the *superheat horn*. On the temperature-entropy diagram it represents the additional work required for dry compression.

10-11 Expansion process Another revision is made on the Carnot cycle to alter the expansion process. The Carnot cycle demands that the expansion take place isentropically and that the resulting work be used to help drive the compressor. Practical difficulties, however, militate against the expansion engine: (1) the possible work that can be derived from the engine is a small fraction of that which must be supplied to the compressor, (2) practical difficulties such as lubrication intrude when a fluid of two phases drives the engine, and (3) the economics of the power recovery have in the past not justified the cost of the expansion engine. The possibility of using an expansion engine should continue to be studied, however, as the cost of energy increases.

The necessity still remains of reducing the pressure of the liquid in process 3-4. A throttling device, such as a valve or other restriction, is almost universally used for this purpose. Barring changes in potential and kinetic energy and with no transfer of heat, $h_3 = h_4$; that is, the process is isenthalpic. The constant-enthalpy throttling process is irreversible, and during the process the entropy increases. The throttling process takes place from 3 to 4 in Fig. 10-9.

10-12 Standard vapor-compression cycle The standard vapor-compression cycle is shown on the temperature-entropy diagram in Fig. 10-9. The processes constituting the standard vapor-compression cycle are:

1-2. Reversible and adiabatic compression from saturated vapor to the condenser pressure

2-3. Reversible rejection of heat at constant pressure, causing desuperheating and condensation of the refrigerant

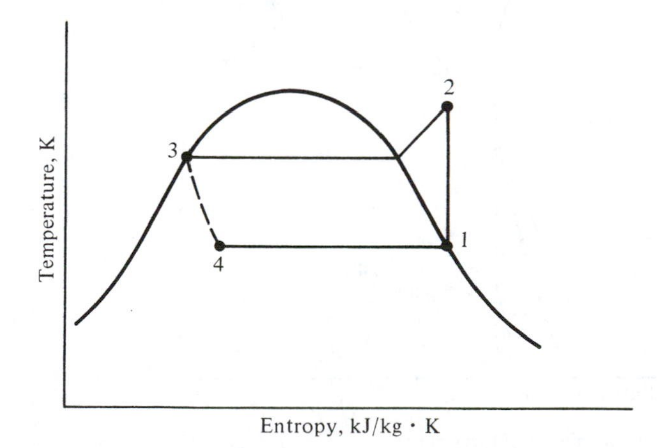

Figure 10-9 The standard vapor-compression cycle.

3-4. Irreversible expansion at constant enthalpy from saturated liquid to the evaporator pressure
4-1. Reversible addition of heat at constant pressure causing evaporation to saturated vapor

10-13 Properties of refrigerants The only properties of refrigerants that have been discussed so far are the characteristic temperature-entropy relationships of saturated liquid and vapor. Other thermodynamic properties are necessary for refrigeration work. All the common refrigerants for vapor-compression systems exhibit similar characteristics although the numerical values of the properties vary from one refrigerant to another.

The pressure-enthalpy diagram is the usual graphic means of presenting refrigerant properties. In other thermodynamic work the temperature-entropy, pressure-volume, or enthalpy-entropy diagrams may be more popular. In refrigeration practice, the enthalpy is one of the most important properties sought, and the pressure can usually be determined most easily. A skeleton pressure-enthalpy diagram is shown in Fig. 10-10. The pressure is the ordinate and the enthalpy the abscissa.

With the saturated-vapor and saturated-liquid lines as the reference, lines of constant temperature, entropy, and specific volume appear on the diagram. The constant-temperature line is horizontal in the mixture region because here the temperature must correspond with the saturation pressure. The subcooled-liquid or compressed-liquid region is to the left of the saturated-liquid line. In this region the constant-temperature line is practically vertical. The temperature of a compressed liquid therefore determines the enthalpy and not the pressure. This statement conforms to the standard practice in using steam tables at moderate pressures. To find the enthalpy of liquid water that is subcooled, the enthalpy is read as the enthalpy of saturated liquid at the existing

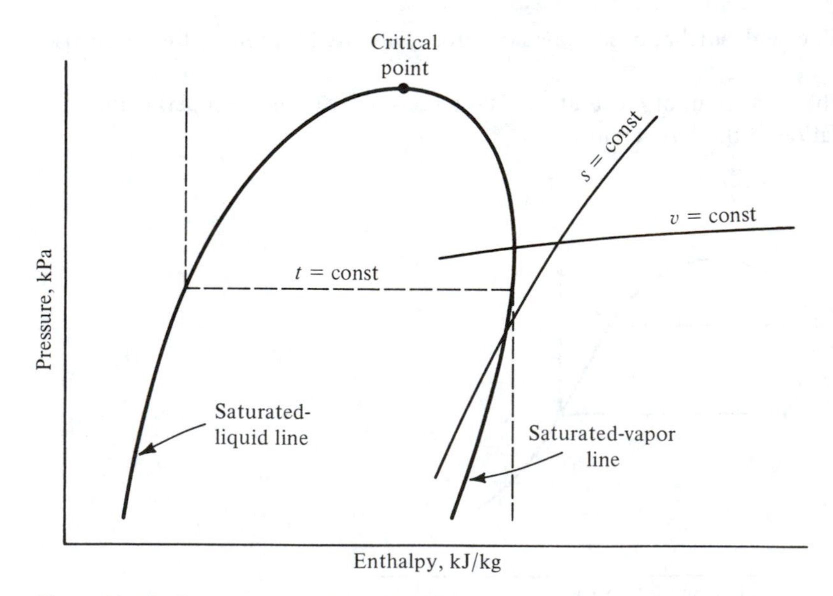

Figure 10-10 The pressure-enthalpy diagram of a refrigerant.

temperature, even though the actual pressure is higher than the saturation pressure. The superheat region is to the right of the saturated-vapor line. In the superheat region the line of constant temperature drops first slightly to the right and then vertically. When the line of constant temperature becomes vertical, $\Delta h = (\text{const})\,(\Delta t)$, the typical relationship of enthalpy and temperature of a perfect gas.

The line of constant specific volume slopes upward to the right. Lines of higher specific volumes are found at progressively lower pressures.

The line of constant entropy runs upward to the right. A reversible and adiabatic compression, which is isentropic, shows the expected increase in enthalpy as the pressure increases during a compression.

Pressure-enthalpy charts for the superheated region of ammonia, refrigerant 11, refrigerant 12, refrigerant 22, and refrigerant 502 are shown in the appendix Figs. A-1 to A-5. Tabular property data are shown for these refrigerants in Tables A-3 to A-8. All the tables pertain to liquid and saturated vapor except for Table A-7 which applies to superheated refrigerant 22 vapor. Refrigerant 22, for example, is the generic name for a refrigerant that is marketed under such names as Freon 22† and Genetron 22.‡ Ammonia is the refrigerant used in many industrial refrigeration systems. A more thorough comparison of the various refrigerants will be found in Chap. 15.

What would be the appearance of the standard vapor-compression cycle on the pressure-enthalpy diagram? Figure 10-11*a* shows the processes which constitute the cycle, and Fig. 10-11*b* is a schematic diagram of the equipment. Process 1-2 is the isentropic compression along the constant-entropy line from saturated vapor to the condenser pressure. Process 2-3 is the constant-pressure desuperheating and condensation, which is a straight horizontal line on the pressure-enthalpy diagram. The throttling process, 3-4, is one of constant enthalpy and therefore is vertical on the chart. Finally, the evaporation process 4-1 is a straight horizontal line because the flow of refrigerant through the evaporator is assumed to be at a constant pressure.

10-14 Performance of the standard vapor-compression cycle With the help of the pressure-enthalpy diagram, the significant quantities of the standard vapor-compression cycle will be determined. These quantities are the work of compression, the heat-rejection rate, the refrigerating effect, the coefficient of performance, the volume rate of flow per kilowatt of refrigeration, and the power per kilowatt of refrigeration.

The work of compression in kilojoules per kilogram is the change in enthalpy in process 1-2 of Fig. 10-11*a* or $h_1 - h_2$. This relation derives from the steady-flow energy equation

$$h_1 + q = h_2 + w$$

where changes in kinetic and potential energy are negligible. Because in the adiabatic compression the heat transfer q is zero, the work w equals $h_1 - h_2$. The difference in enthalpy is a negative quantity, indicating that work is done on the system. Even though the compressor may be of the reciprocating type, where flow is intermittent

† Freon is a registered trademark of the Freon Division, E. I. du Pont de Nemours & Company.

‡ Genetron is a registered trademark of the Speciality Chemicals Division, Allied Chemical Corporation.

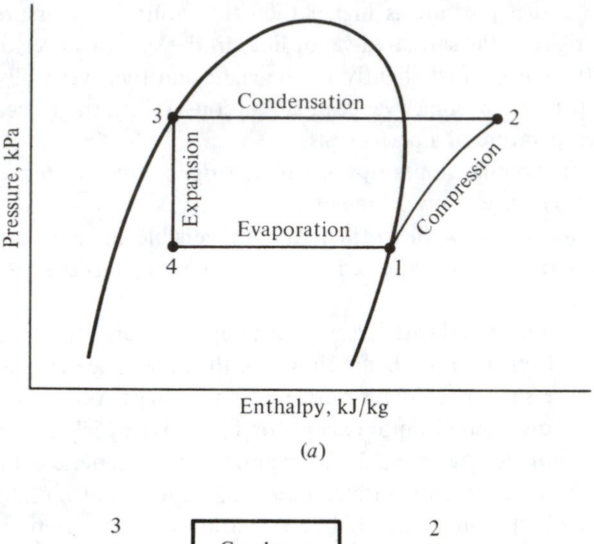

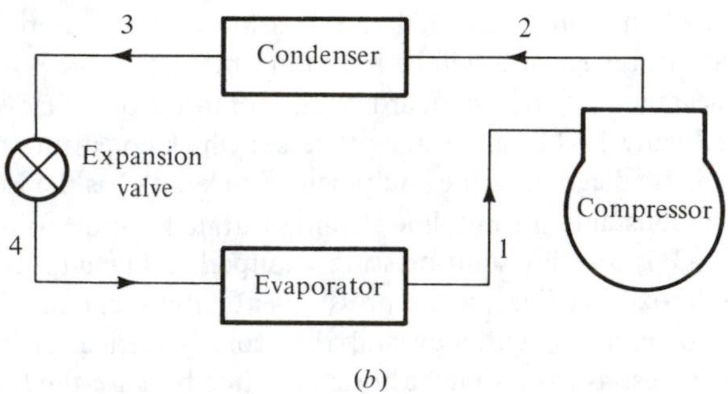

Figure 10-11 (*a*) The standard vapor-compression cycle on the pressure-enthalpy diagram; (*b*) flow diagram.

rather than steady, process 1-2 still represents the action of the compressor. At a short distance in the pipe away from the compressor, the flow has smoothed out and approaches steady flow. Knowledge of the work of compression is important because it may be one of the largest operating costs of the system.

The heat rejection in kilojoules per kilogram is the heat transferred from the refrigerant in process 2-3, which is $h_3 - h_2$. This knowledge also comes from the steady-flow energy equation, in which the kinetic energy, potential energy, and work terms drop out. The value of $h_3 - h_2$ is negative, indicating that heat is transferred from the refrigerant. The value of the heat rejection is used in sizing the condenser and calculating the required flow quantities of the condenser cooling fluid.

The refrigerating effect in kilojoules per kilogram is the heat transferred in process 4-1, or $h_1 - h_4$. Knowledge of the magnitude of the term is necessary because performing this process is the ultimate purpose of the entire system.

The coefficient of performance of the standard vapor-compression cycle is the refrigerating effect divided by the work of compression:

$$\text{Coefficient of performance} = \frac{h_1 - h_4}{h_2 - h_1}$$

Sometimes the volume flow rate is computed at the compressor inlet or state point 1. The volume flow rate is a rough indication of the physical size of the compressor. The greater the magnitude of the term, the greater the displacement of the compressor in cubic meters per second must be.

The power per kilowatt of refrigeration is the inverse of the coefficient of performance, and an efficient refrigeration system has a low value of power per kilowatt of refrigeration but a high coefficient of performance.

An example will illustrate the calculations for determining the performance of a standard vapor-compression cycle.

Example 10-1 A standard vapor-compression cycle developing 50 kW of refrigeration using refrigerant 22 operates with a condensing temperature of 35°C and an evaporating temperature of -10°C. Calculate (*a*) the refrigerating effect in kilojoules per kilogram, (*b*) the circulation rate of refrigerant in kilograms per second, (*c*) the power required by the compressor in kilowatts, (*d*) the coefficient of performance, (*e*) the volume flow rate measured at the compressor suction, (*f*) the power per kilowatt of refrigeration, and (*g*) the compressor discharge temperature.

Solution As the first step in the solution, sketch the pressure-enthalpy diagram (Fig. 10-12) and determine from Tables A-6 and A-7 and Fig. A-4 the enthalpies at key points. The value of h_1 is the enthalpy of saturated vapor at -10°C, which is 401.6 kJ/kg.

To find h_2 move at a constant entropy from point 1 until reaching the saturation pressure corresponding to 35°C. This condensing pressure is 1354 kPa, and the value of h_2 is 435.2 kJ/kg.

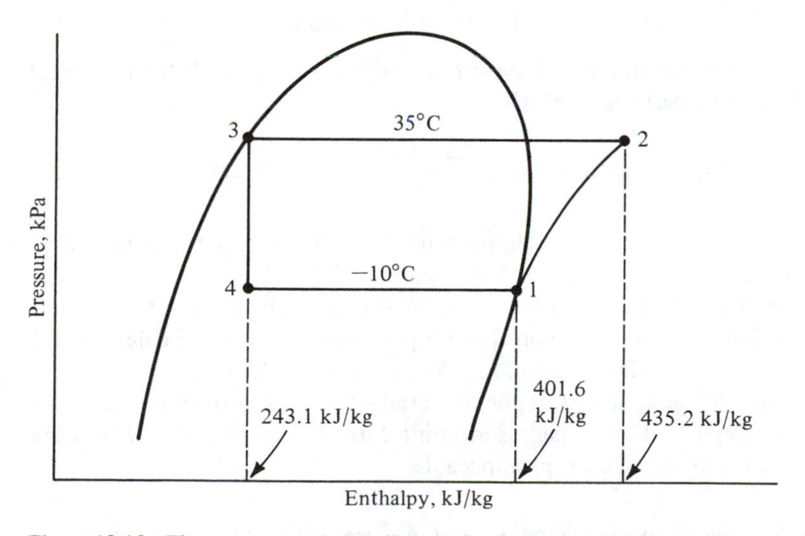

Figure 10-12. The pressure-enthalpy diagram for the system in Example 10-1.

The values of h_3 and h_4 are identical and are equal to the enthalpy of saturated liquid at 35°C, which is 243.1 kJ/kg. Therefore

$$h_1 = 401.6 \text{ kJ/kg} \qquad h_2 = 435.2 \text{ kJ/kg}$$

$$h_3 = h_4 = 243.1 \text{ kJ/kg}$$

(*a*) The refrigerating effect is

$$h_1 - h_4 = 401.6 - 243.1 = 158.5 \text{ kJ/kg}$$

(*b*) The circulating rate of refrigerant can be calculated by dividing the refrigerating capacity by the refrigerating effect

$$\text{Flow rate} = \frac{50 \text{ kW}}{158.5 \text{ kJ/kg}} = 0.315 \text{ kg/s}$$

(*c*) The power required by the compressor is the work of compression per kilogram multiplied by the refrigerant flow rate

$$\text{Compressor power} = (0.315 \text{ kg/s})\,(435.2 - 401.6 \text{ kJ/kg}) = 10.6 \text{ kW}$$

(*d*) The coefficient of performance is the refrigerating rate divided by the compressor power

$$\text{Coefficient of performance} = \frac{50 \text{ kW}}{10.6 \text{ kW}} = 4.72$$

(*e*) The volume rate of flow at the compressor inlet requires knowledge of the specific volume of the refrigerant at point 1. From Table A-6 or Fig. A-4 this value is 0.0654 m^3/kg, and so

$$\text{Volume flow rate} = (0.315 \text{ kg/s})\,(0.0654 \text{ m}^3\text{/kg})$$

$$= 0.0206 \text{ m}^3\text{/s} = 20.6 \text{ L/s}$$

(*f*) The compressor power per kilowatt of refrigeration (which is the reciprocal of the coefficient of performance) is

$$\text{Power of refrigeration} = \frac{10.6 \text{ kW}}{50 \text{ kW}} = 0.212 \text{ kW/kW}$$

(*g*) The compressor discharge temperature is the temperature of superheated vapor at point 2 which from Fig. A-4 is found to be 57°C.

All the properties in Example 10-1 could be extracted from Table A-6 except h_2 and t_2, which are in the superheat region. The properties at point 2 can be determined either from the pressure-enthalpy diagram, Fig. A-4, or from Table A-7. More complete tables of properties of superheated vapor are available,[1] and corresponding tables exist for other refrigerants. The properties at point 2 are determined by interpolating in Table A-7 at the pressure and entropy applicable.

10-15 Heat exchangers Some refrigeration systems use a liquid-to-suction heat exchanger, which subcools the liquid from the condenser with suction vapor coming

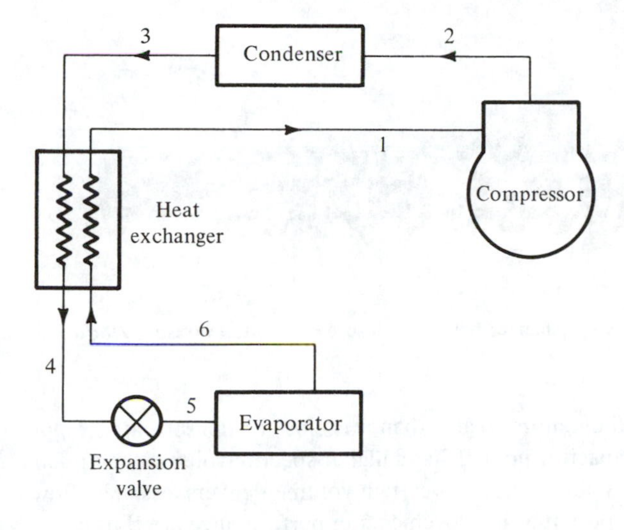

(*a*)

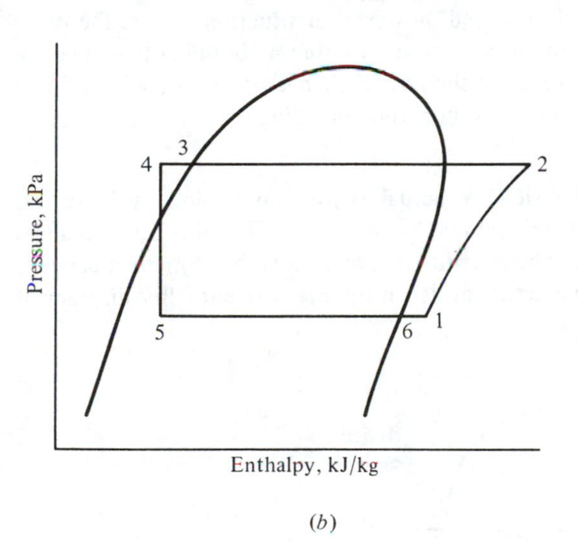

(*b*)

Figure 10-13 (*a*) Refrigeration system with a heat exchanger to subcool the liquid from the condenser. (*b*) Pressure-enthalpy diagram of the system using a heat exchanger shown in (*a*).

from the evaporator. The arrangement is shown in Fig. 10-13*a* and the corresponding pressure-enthalpy diagram in Fig. 10-13*b*.

Saturated liquid at point 3 coming from the condenser is cooled to point 4 by means of vapor at point 6 being heated to point 1. From a heat balance, $h_3 - h_4 = h_1 - h_6$. The refrigerating effect is either $h_6 - h_5$ or $h_1 - h_3$. Figure 10-14 shows a cutaway view of a liquid-to-suction heat exchanger.

Compared with the standard vapor-compression cycle, the system using the heat exchanger may seem to have obvious advantages because of the increased refrigerating effect. Both the capacity and the coefficient of performance may seem to be improved. This is not necessarily true, however. Even though the refrigerating effect is increased, the compression is pushed farther out into the superheat region, where the work of

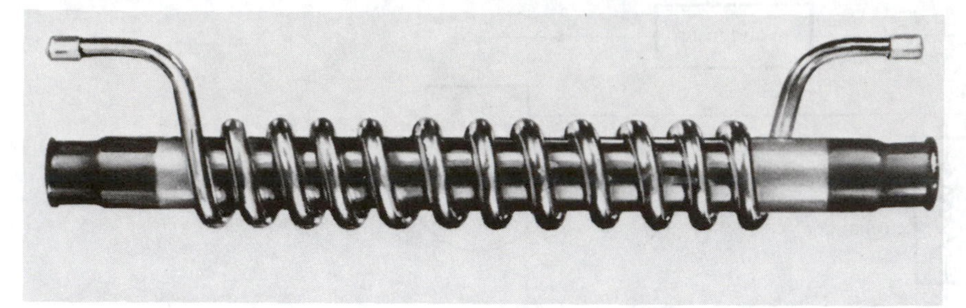

Figure 10-14. A liquid-to-suction heat exchanger before enclosure with outer housing. *(Refrigeration Research, Inc.)*

compression in kilojoules per kilogram is greater than it is close to the saturated-vapor line. From the standpoint of capacity, point 1 has a higher specific volume than point 6, so that a compressor which is able to pump a certain volume delivers less mass flow if the intake is at point 1. The potential improvements in performance are thus counterbalanced, and the heat exchanger probably has negligible thermodynamic advantages.

The heat exchanger is definitely justified, however, in situations where the vapor entering the compressor must be superheated to ensure that no liquid enters the compressor. Another practical reason for using the heat exchanger is to subcool the liquid from the condenser to prevent bubbles of vapor from impeding the flow of refrigerant through the expansion valve.

10-16 Actual vapor-compression cycle The actual vapor-compression cycle suffers from inefficiencies compared with the standard cycle. There are also other changes from the standard cycle, which may be intentional or unavoidable. Some comparisons can be drawn by superimposing the actual cycle on the pressure-enthalpy diagram of the standard cycle, as in Fig. 10-15.

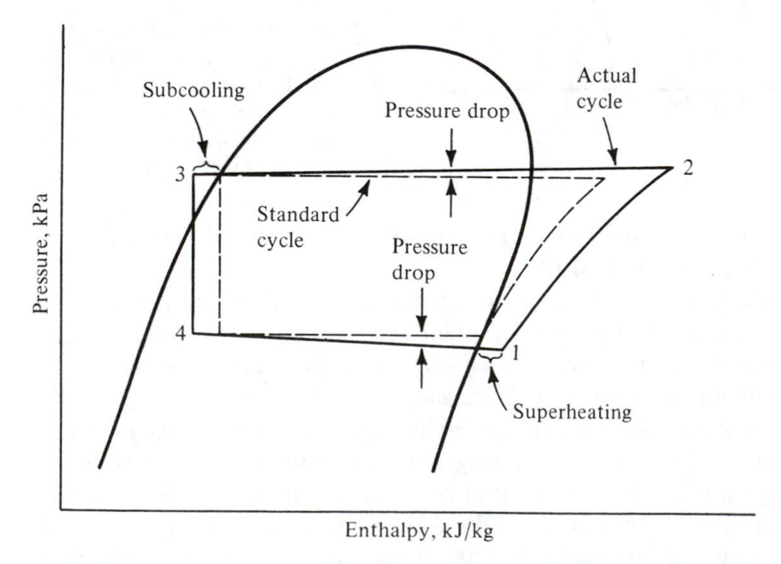

Figure 10-15 Actual vapor-compression cycle compared with standard cycle.

The essential differences between the actual and the standard cycle appear in the pressure drops in the condenser and evaporator, in the subcooling of the liquid leaving the condenser, and in the superheating of the vapor leaving the evaporator. The standard cycle assumes no drop in pressure in the condenser and evaporator. Because of friction, however, the pressure of the refrigerant drops in the actual cycle. The result of these drops in pressure is that the compression process between 1 and 2 requires more work than in the standard cycle. Subcooling of the liquid in the condenser is a normal occurrence and serves the desirable function of ensuring that 100 percent liquid will enter the expansion device. Superheating of the vapor usually occurs in the evaporator and is recommended as a precaution against droplets of liquid being carried over into the compressor. The final difference in the actual cycle is that the compression is no longer isentropic and there are inefficiencies due to friction and other losses.

PROBLEMS

10-1 A Carnot refrigeration cycle absorbs heat at -12°C and rejects it at 40°C.

(*a*) Calculate the coefficient of performance of this refrigeration cycle.

(*b*) If the cycle is absorbing 15 kW at the -12°C temperature, how much power is required?

(*c*) If a Carnot heat pump operates between the same temperatures as the above refrigeration cycle, what is the performance factor?

(*d*) What is the rate of heat rejection at the 40°C temperature if the heat pump absorbs 15 kW at the -12°C temperature? *Ans.* 18 kW

10-2 If in a standard vapor-compression cycle using refrigerant 22 the evaporating temperature is -5°C and the condensing temperature is 30°C, sketch the cycle on pressure-enthalpy coordinates and calculate (*a*) the work of compression, (*b*) the refrigerating effect, and (*c*) the heat rejected in the condenser, all in kilojoules per kilogram, and (*d*) the coefficient of performance. *Ans.* (*d*) 6.47

10-3 A refrigeration system using refrigerant 22 is to have a refrigerating capacity of 80 kW. The cycle is a standard vapor-compression cycle in which the evaporating temperature is -8°C and the condensing temperature 42°C.

(*a*) Determine the volume flow of refrigerant measured in cubic meters per second at the inlet to the compressor.

(*b*) Calculate the power required by the compressor.

(*c*) At the entrance to the evaporator what is the fraction of vapor in the mixture expressed both on a mass basis and a volume basis? *Ans.* (*c*) 0.292, 0.971

10-4 Compare the coefficient of performance of a refrigeration cycle which uses wet compression with that of one which uses dry compression. In both cases use ammonia as the refrigerant, a condensing temperature of 30°C, and an evaporating temperature of -20°C; assume that the compressions are isentropic and that the liquid leaving the condenser is saturated. In the wet-compression cycle the refrigerant enters the compressor in such a condition that it is saturated vapor upon leaving the compressor. *Ans.* 4.42 versus 4.02

10-5 In the vapor-compression cycle a throttling device is used almost universally to reduce the pressure of the liquid refrigerant.

(*a*) Determine the percent saving in net work of the cycle per kilogram of refrigerant if an expansion engine could be used to expand saturated liquid refrigerant

22 isentropically from 35°C to the evaporator temperature of 0°C. Assume that compression is isentropic from saturated vapor at 0°C to a condenser pressure corresponding to 35°C. *Ans.* 12.9%

(*b*) Calculate the increase in refrigerating effect in kilojoules per kilogram resulting from use of the expansion engine.

10-6 Since a refrigeration system operates more efficiently when the condensing temperature is low, evaluate the possibility of cooling the condenser cooling water of the refrigeration system in question with another refrigeration system. Will the combined performance of the two systems be better, the same, or worse than one individual system? Explain why.

10-7 A refrigerant 22 vapor-compression system includes a liquid-to-suction heat exchanger in the system. The heat exchanger warms saturated vapor coming from the evaporator from -10 to 5°C with liquid which comes from the condenser at 30°C. The compressions are isentropic in both cases listed below.

(*a*) Calculate the coefficient of performance of the system without the heat exchanger but with the condensing temperature at 30°C and an evaporating temperature of -10°C. *Ans.* 5.46

(*b*) Calculate the coefficient of performance of the system with the heat exchanger. *Ans.* 5.37

(*c*) If the compressor is capable of pumping 12.0 L/s measured at the compressor suction, what is the refrigeration capacity of the system without the heat exchanger? *Ans.* 30.3 kW

(*d*) With the same compressor capacity as in (*c*), what is the refrigeration capacity of the system with the heat exchanger? *Ans.* 29.9 kW

REFERENCE

1. Thermodynamic Properties of "Freon" 22 Refrigerant, *Tech. Bull.* T-22-SI, Du Pont de Nemours International S.A., Geneva.

CHAPTER

ELEVEN

COMPRESSORS

11-1 Types of compressors Each of the four components of a vapor-compression system—the compressor, the condenser, the expansion device, and the evaporator—has its own peculiar behavior. At the same time, each component is influenced by conditions imposed by the other members of the quartet. A change in condenser-water temperature, for example, may change the rate of refrigerant the compressor pumps, which in turn may require the expansion valve to readjust and the refrigerant in the evaporator to change pressure. We shall first study the components of the vapor-compression cycle singly, analyzing their performance as individuals, and then observe how they interact with each other as a system. The compressor is the first component to be analyzed.

The heart of the vapor-compression system is the compressor. The four most common types of refrigeration compressors are the *reciprocating, screw, centrifugal,* and *vane.* The reciprocating compressor consists of a piston moving back and forth in a cylinder with suction and discharge valves arranged to allow pumping to take place. The screw, centrifugal, and vane compressors all use rotating elements, the screw and vane compressors are positive-displacement machines, and the centrifugal compressor operates by virtue of centrifugal force. The four parts of this chapter examine each of these types of compressors.

PART I: RECIPROCATING COMPRESSORS

The workhorse of the refrigeration industry is the reciprocating compressor, built in sizes ranging from fractional-kilowatt to hundreds of kilowatts refrigeration capacity. Modern compressors are single-acting and may be single-cylinder or multicylinder. In multicylinder compressors the cylinders are in V, W, radial, or in-line arrangements.

Figure 11-1 A 16-cylinder reciprocating compressor for ammonia. *(Vilter Manufacturing Corporation.)*

The compressor in Fig. 11-1 has 16 cylinders, 2 in each of the heads. During the suction stroke of the piston, low-pressure refrigerant gas is drawn in through the suction valve, which may be located in the piston or in the head. During the discharge stroke the piston compresses the refrigerant and then pushes it out through the discharge valve, which is usually located in the cylinder head.

Following the trend of most rotative machinery, the operating speed of compressors has generally increased in the past 20 years. From the slow speeds of early compressors of about 2 or 3 r/s, the speeds have increased until compressors today operate at speeds as high as 60 r/s.

11-2 Hermetically sealed compressors A compressor whose crankshaft extends through the compressor housing so that a motor can be externally coupled to the shaft is called an *open-type compressor*. A seal must be used where the shaft comes through the compressor housing to prevent refrigerant gas from leaking out or air from leaking in if the crankcase pressure is lower than atmospheric. Even though designers have continually developed better seals, piercing of the housing always represents a source of leakage. To avoid leakage at the seal, the motor and compressor are often enclosed in the same housing, as shown in the cutaway view in Fig. 11-2.

Improved techniques for insulating the motor electrically have allowed motors to operate even though they are in contact with the refrigerant. In many designs the cold suction gas is drawn across the motor to keep the motor cool. Almost all small motor-compressor combinations used in refrigerators, freezers, and residential air conditioners are of the hermetic type. The only connections to the compressor housing are the

Figure 11-2 Cutaway view of a hermetically sealed compressor. *(Carlyle Compressor Company, Carrier Corporation.)*

suction and discharge fittings and electric terminals. Moisture in the system can be damaging to the motor; therefore dehydration of hermetic units before charging is essential. On larger hermetically sealed units the cylinder heads are usually removable so that the valves and pistons can be serviced. This type of unit is called *semihermetic.*

11-3 Condensing units The compressor and the condenser of a system are conveniently combined into a *condensing unit* (Fig. 11-3). The motor, compressor, and condenser may be compactly mounted on the same frame and located remotely from the expansion valve and evaporator.

11-4 Performance Two of the most important performance characteristics of a compressor are its refrigeration capacity and its power requirement. These two characteristics of a compressor operating at constant speed are controlled largely by the suction and discharge pressures. An analysis will be made first of an ideal reciprocating compressor because it affords a clearer understanding of the effects of these two pressures. Trends established from a study of the ideal compressor hold true for the actual compressor, although adjustments must be made in the numerical quantities. These adjustments will be examined in the discussion of the actual compressor.

11-5 Volumetric efficiency Volumetric efficiencies are the bases for predicting performance of reciprocating compressors. Two types of volumetric efficiencies will be

Figure 11-3 Compressor and condenser combined into a condensing unit. *(The Trane Company.)*

considered in this chapter, actual and clearance. The *actual volumetric efficiency* η_{va} is defined by

$$\eta_{va} = \frac{\text{volume flow rate entering compressor, m}^3/\text{s}}{\text{displacement rate of compressor, m}^3/\text{s}} \times 100 \qquad (11\text{-}1)$$

where the displacement rate is the volume swept through by the pistons in their suction strokes per unit time.

Clearance volumetric efficiency depends on the reexpansion of gas trapped in the clearance volume and can be best explained by showing a pressure-volume diagram of a compressor, as in Fig. 11-4. The maximum volume in the cylinder, which occurs when the piston is at one end of its stroke, is V_3. The minimum volume, or *clearance volume,* is V_c, which occurs at the other end of the piston stroke. The discharge pressure is p_d.

In the first instance, assume that the suction pressure is p_1. Gas trapped in the clearance volume must first expand to volume V_1 before the pressure in the cylinder is low enough for the suction valves to open and draw in more gas. The volume of gas drawn into the cylinder will be $V_3 - V_1$, and the clearance volumetric efficiency η_{vc} for this case is $(V_3 - V_1)(100)/(V_3 - V_c)$. When the suction pressure is p_2, the intake portion of the stroke is reduced to $V_3 - V_2$. In the extreme case where the suction pressure has dropped to p_3, the piston uses its entire stroke to reexpand the gas in the clearance volume and the clearance volumetric efficiency is 0 percent.

The clearance volumetric efficiency can be expressed in another way, illustrated in Fig. 11-4 using p_1 as the suction pressure. The percent clearance m, which is constant for a given compressor, is defined as

$$m = \frac{V_c}{V_3 - V_c} 100 \qquad (11\text{-}2)$$

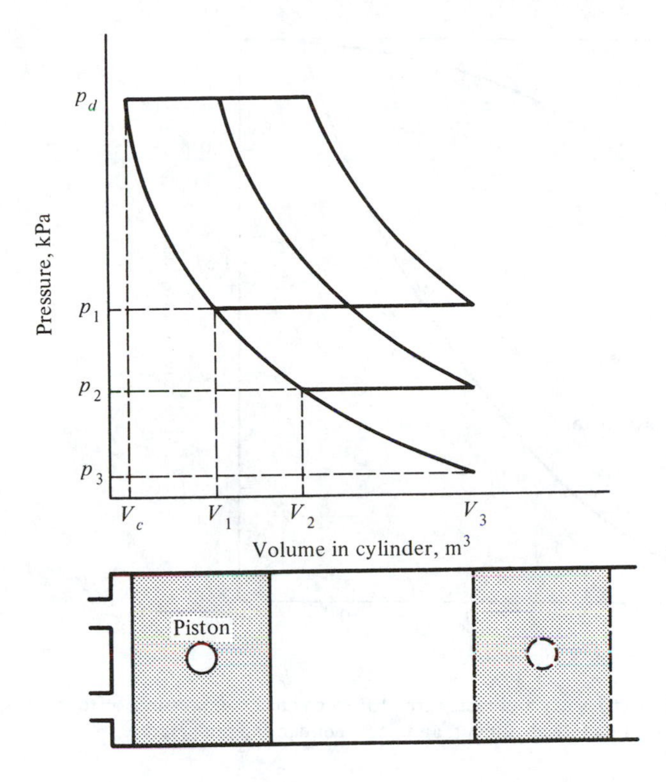

Figure 11-4 Pressure-volume diagram of an ideal compressor.

Adding $V_c - V_c$ to the numerator of the expression for η_{vc} gives

$$\eta_{vc} = \frac{V_3 - V_c + V_c - V_1}{V_3 - V_c} 100 = 100 + \frac{V_c - V_1}{V_3 - V_c} 100 \tag{11-3}$$

and

$$\eta_{vc} = 100 - \frac{V_1 - V_c}{V_3 - V_c} 100 = 100 - \frac{V_c}{V_3 - V_c}\left(\frac{V_1}{V_c} - 1\right) 100$$

Therefore

$$\eta_{vc} = 100 - m\left(\frac{V_1}{V_c} - 1\right) \tag{11-4}$$

If an isentropic expansion is assumed between V_c and V_1,

$$\frac{V_1}{V_c} = \frac{v_{suc}}{v_{dis}} \tag{11-5}$$

where v_{suc} = specific volume of vapor entering compressor
v_{dis} = specific volume of vapor after isentropic compression to p_d

Values of the specific volumes are available from the pressure-enthalpy diagram of the refrigerant or from tables of properties of superheated vapor.

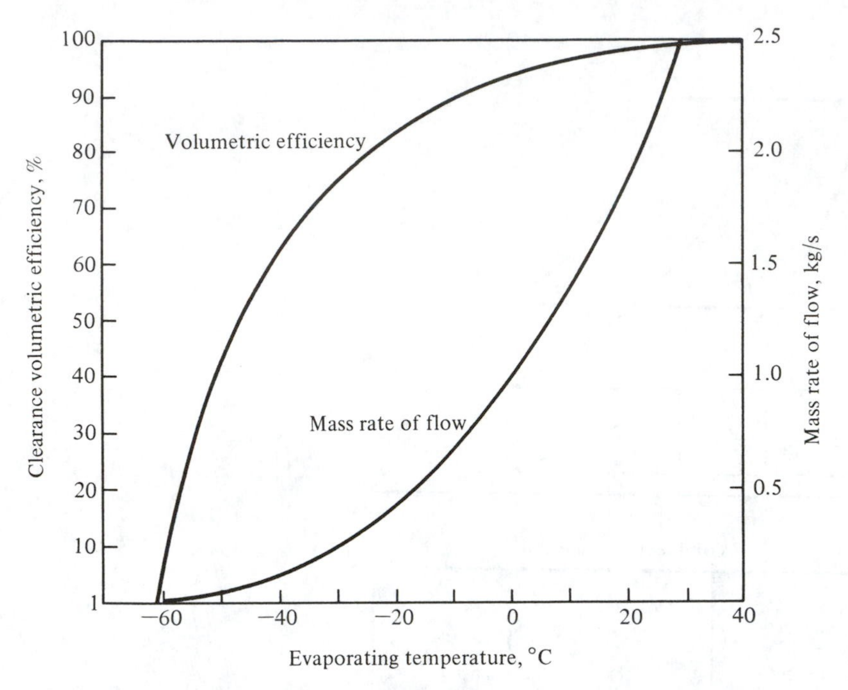

Figure 11-5 Clearance volumetric efficiency and mass rate of flow of ideal compressor, Refrigerant 22, 4.5 percent clearance, 50 L/s rate of displacement, and 35°C condensing temperature.

Substituting Eq. (11-5) into Eq. (11-4) gives

$$\eta_{vc} = 100 - m\left(\frac{v_{\text{suc}}}{v_{\text{dis}}} - 1\right) \tag{11-6}$$

11-6 Performance of the ideal compressor In the ideal compressor, the compression of the gas and the reexpansion of gas trapped in the clearance volume are both isentropic. The reexpansion of the trapped gas is the only factor which influences volumetric effiency in the ideal compressor.

In the next few pages the effect of suction pressure on the performance of an ideal compressor will be studied. Figure 11-5 shows the effect of evaporating temperature on clearance volumetric efficiency. The volumetric efficiencies are calculated from Eq. (11-6) and apply to a refrigerant 22 compressor with a clearance of 4.5 percent operating with a condensing temperature of 35°C. The clearance volumetric efficiency is zero when the evaporating temperature is -61°C, at which temperature the saturation pressure corresponds to p_3 in Fig. 11-4. When the suction pressure and discharge pressure are the same (same evaporating and condensing pressure), the volumetric efficiency is 100 percent.

The mass rate of flow controls the capacity and power requirement more directly than the volume rate of flow. The mass rate of flow, w kg/s, through a compressor is proportional to the displacement rate in liters per second and the volumetric efficiency and inversely proportional to the specific volume of gas entering the compressor. In equation form

$$w = \text{displacement rate} \times \frac{\eta_{vc}/100}{v_{suc}} \tag{11-7}$$

Using Eq. (11-7) and an assumed rate of displacement of 50 L/s, the mass rate of flow can be calculated and plotted as done in Fig. 11-5. As the suction pressure drops, the specific volume entering the compressor increases, which, together with the volumetric efficiency, reduces the mass rate of flow at low evaporating temperatures.

11-7 Power requirement The power required by the ideal compressor is the product of the mass rate of flow and the increase in enthalpy during the isentropic compression,

$$P = w\,\Delta h_i \tag{11-8}$$

where P = power, kW
w = mass rate of flow, kg/s
Δh_i = isentropic work of compression, kJ/kg

Figure 11-6 shows the variation in Δh_i as the evaporating temperature changes. The

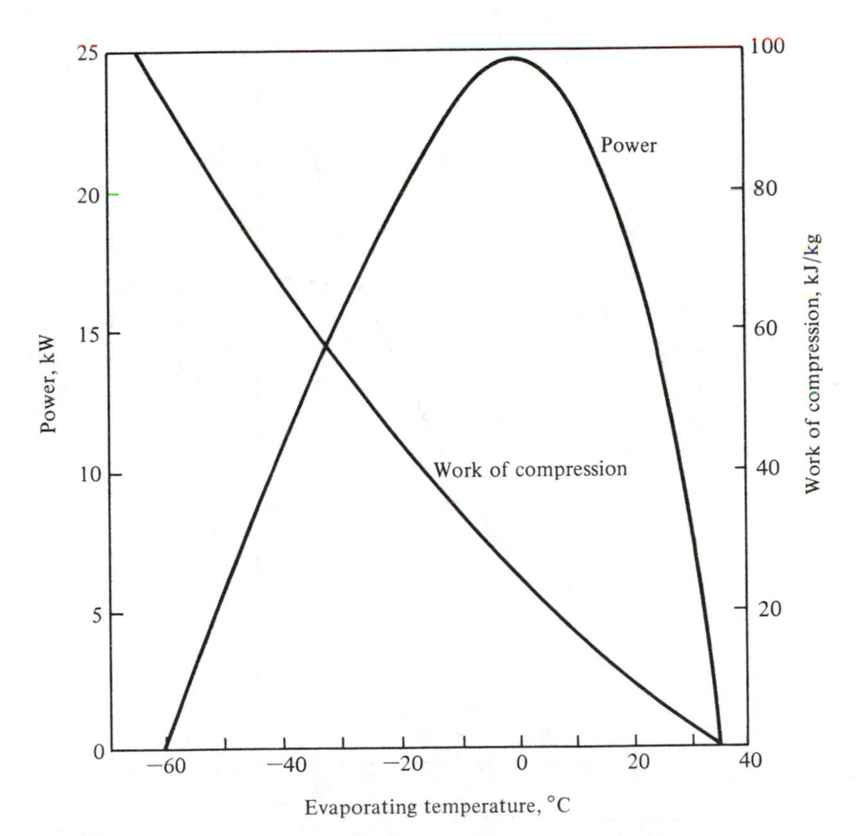

Figure 11-6 Work of compression and power required by an ideal compressor, Refrigerant 22, 4.5 percent clearance, 50 L/s displacement rate, and 35°C condensing temperature.

value of Δh_i is large at low evaporating temperatures and drops to zero when the suction pressure equals the discharge pressure (when the evaporating temperature equals the condensing pressure). The curve of the power requirement in Fig. 11-6 therefore shows a zero value at two points, where the evaporating temperature equals the condensing temperature and where the mass rate of flow is zero. Between the two extremes the power requirement reaches a peak.

The power curve merits close attention because it has important implications. Most refrigeration systems operate on the left side of the peak of the power curve. During the period of pulldown of temperature following start-up with a warm evaporator, however, the power requirement passes through its peak and may demand more power than the motor, which is selected for design conditions, is capable of supplying steadily. Sometimes motors have to be oversized just to take the system down through the peak in the power curve. To avoid oversizing the motor, the suction pressure is sometimes reduced artificially by throttling the suction gas until the evaporator pressure drops below the peak in the power curve.

During regular operation heavy refrigeration loads raise the evaporating temperature, which increases the power requirement of the compressor and may overload the motor.

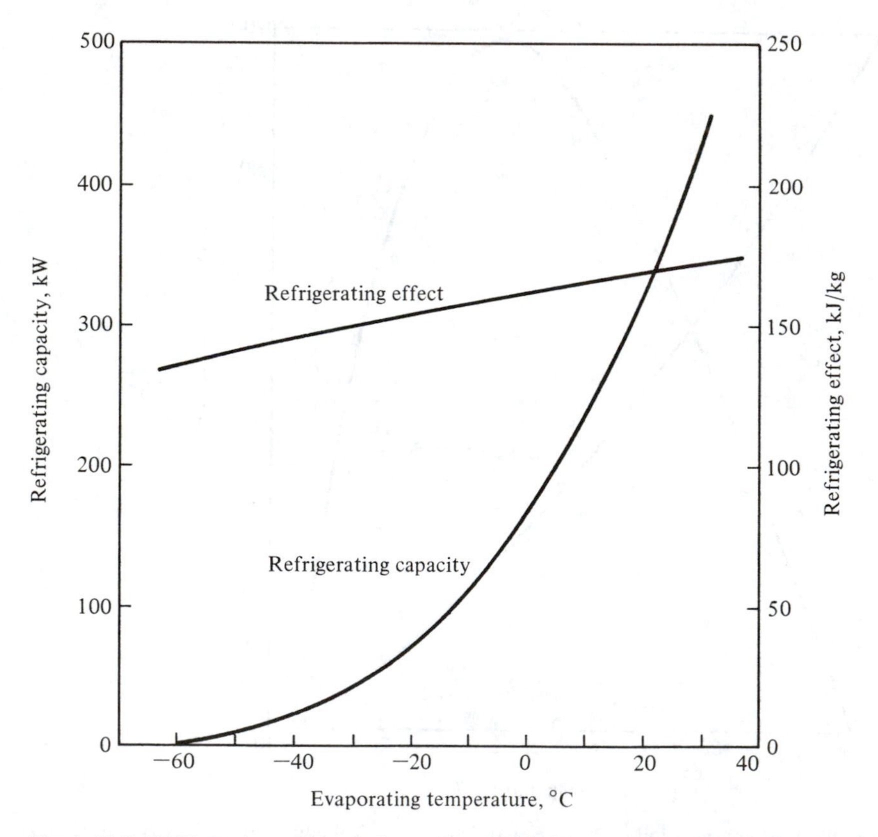

Figure 11-7 Refrigerating effect and capacity of ideal compressor, Refrigerant 22, 4.5 percent clearance, 50 L/s rate of displacement, and 35°C condensing temperature.

11-8 Refrigeration capacity The refrigeration capacity q is

$$q = w(h_1 - h_4) \qquad \text{kW} \tag{11-9}$$

where h_1 and h_4 are the enthalpies in kilojoules per kilogram of the refrigerant leaving and entering the evaporator, respectively. The refrigerating effect; $h_1 - h_4$, increases slightly with an increase in suction pressure, as Fig. 11-7 shows, provided that the enthalpy entering the expansion valve remains constant. The increase is due to the slightly higher enthalpy of saturated vapor at higher evaporating temperatures.

Figure 11-7 also shows the refrigeration capacity calculated with Eq. (11-9). The capacity is zero at the point where the mass rate of flow is zero. The refrigerating capacity can be doubled, for example, by increasing the evaporating temperature from 0 to 20°C.

11-9 Coefficient of performance and volume flow rate per kilowatt of refrigeration The coefficient of performance can be derived from the refrigerating capacity of Fig. 11-7 and the power from Fig. 11-6. The result, displayed in Fig. 11-8, shows a progressive increase as the evaporating temperature increases. The volume flow rate per unit refrigeration capacity is an indication of the physical size or speed of the compressor

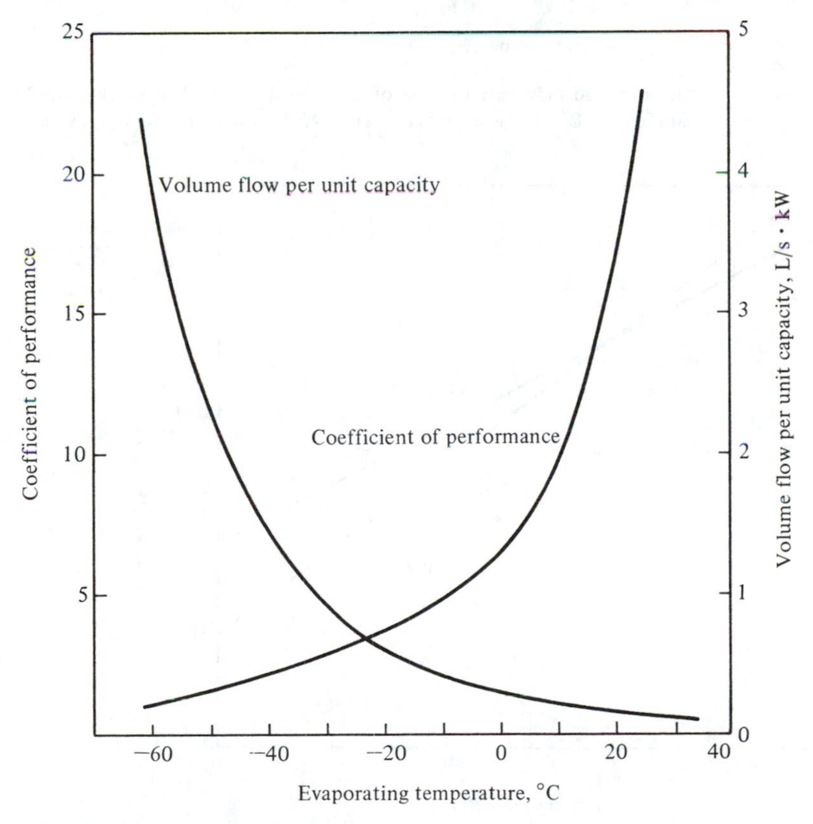

Figure 11-8 Coefficient of performance and volume flow per kilowatt of refrigeration for ideal compressor, Refrigerant 22, 4.5 percent clearance, 50 L/s displacement rate, and 35°C condensing temperature.

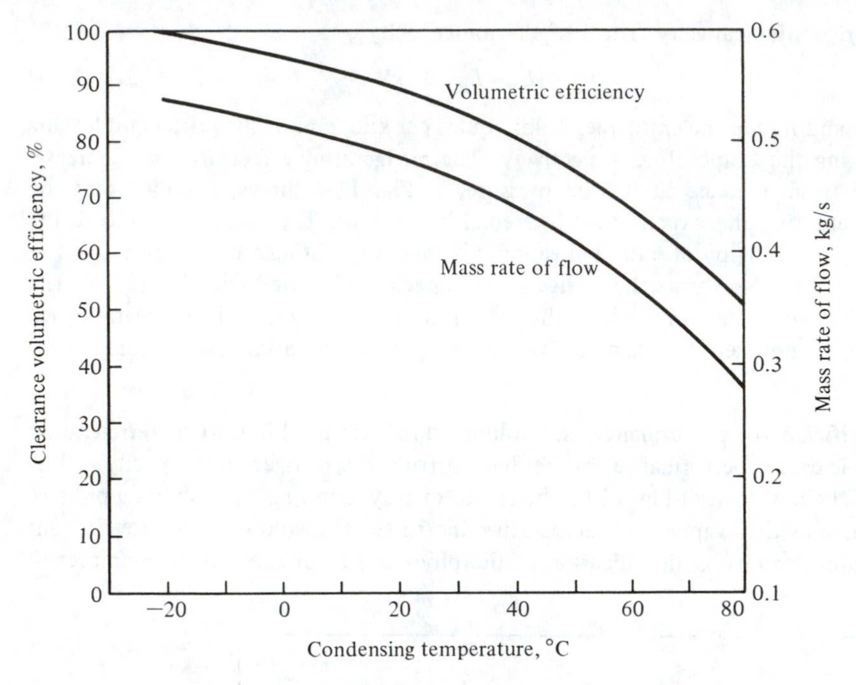

Figure 11-9 Volumetric efficiency and mass rate of flow of refrigerant for an ideal Refrigerant 22 compressor, 4.5 percent clearance, 50 L/s displacement rate, and -20°C evaporating temperature.

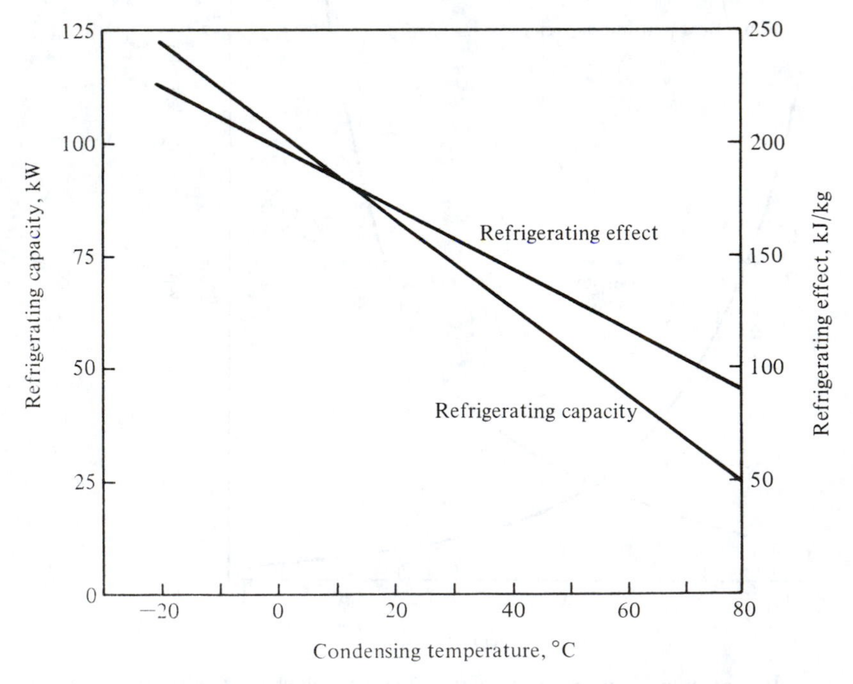

Figure 11-10 Refrigeration effect and refrigerating capacity for an ideal Refrigerant 22 compressor, 4.5 percent clearance, 50 L/s displacement rate, and -20°C evaporating temperature.

necessary to develop 1 kW of refrigeration. A large volume flow must be pumped for a given capacity at low evaporating temperatures because of the high specific volume.

11-10 Effect of condensing temperature Most refrigerating systems reject heat to the atmosphere, and the ambient conditions change throughout the year. Process refrigeration plants that operate year round are particularly subject to a wide range of condensing temperatures. The response of a reciprocating compressor to changes in condensing temperature can be analyzed similarly to the evaporating temperature. Figure 11-9 shows the clearance volumetric efficiency as calculated from Eq. (11-6) for a compressor with an evaporating temperature of -20°C. As the condensing temperature increases, the volumetric efficiency drops off. Because the specific volume of the refrigerant at the compressor suction remains constant, only the volumetric efficiency affects the mass rate of flow, which shows a corresponding decrease as the condensing temperature increases. Figure 11-10 shows such a progressive decrease. The refrigerating capacity is the product of the refrigerating effect and the mass rate of flow, both of which decrease with increasing condensing temperature. The result is that the refrigerating capacity drops rather rapidly on an increase in condensing temperature.

The remaining important characteristic is the power, shown on Fig. 11-11. The

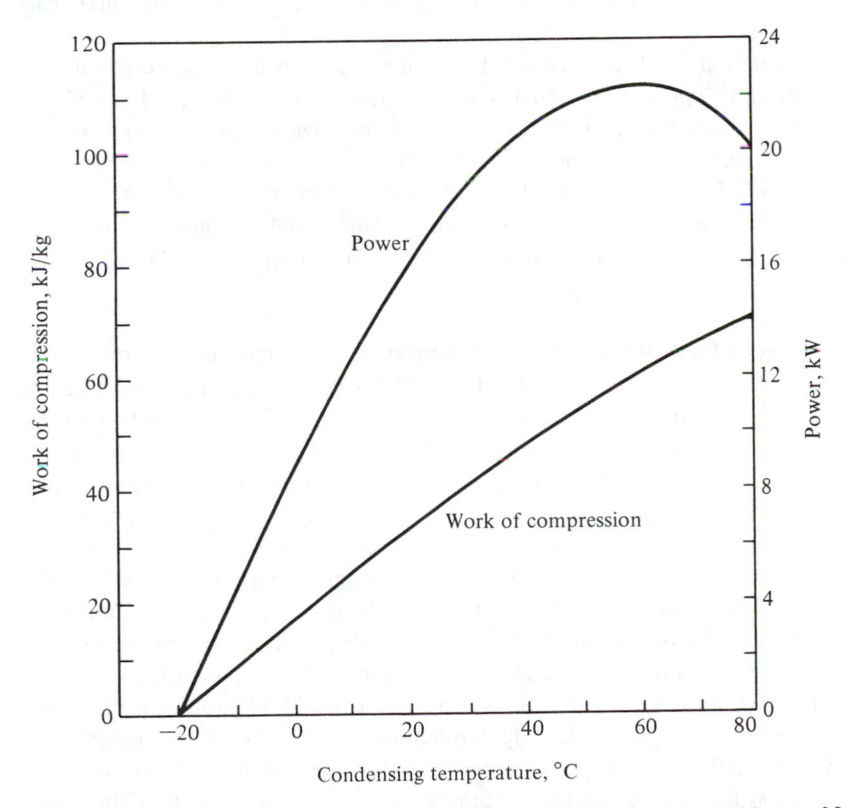

Figure 11-11 Work of compression and compressor power for an ideal Refrigerant 22 compressor, 4.5 percent clearance, 50 L/s displacement rate, and -20°C evaporating temperature.

compressor power is the product of the work of compression in kilojoules per kilogram and the mass rate of flow. The work of compression in kilojoules per kilogram increases and the mass rate of flow decreases as the condensing temperature increases, so that the power increases to a peak and then begins to drop off, a trend similar to the power as a function of the evaporating temperature shown in Fig. 11-6.

A few comments on the significance of the trends in Figs. 11-9 to 11-11 follow. The peaking of the power can occur in real compressors as well as the ideal ones, but only when pumping from low evaporating temperatures. Single-stage compression from -20°C evaporating temperature to a 60°C condensing temperature, which resulted in the peak in Fig. 11-11, is not common. With more moderate differences between the condensing and evaporating temperatures the expectation is that the power required by the compressor will increase with an increase in condensing temperature, although the increase may be slight. The refrigerating capacity always decreases with an increase in condensing temperature. Another important characteristic, not shown on the graphs, is the coefficient of performance, which decreases monotonically as the condensing temperature increases.

From the standpoint of power and efficiency, a low condensing temperature is desirable; thus the condenser should use the coldest air or water available, should operate with the maximum airflow or water flow that is economical, and should have its surfaces kept clean. Air or noncondensable gases in the condenser also cause high condenser pressures.

All the calculations in Figs. 11-9 to 11-11 are based on a clearance volume of 4.5 percent. McGrath[1] pointed out that increasing the percent clearance from about 4 percent, which is customary, to about 15 percent will give a nearly constant power requirement regardless of the discharge pressure for air-conditioning applications. This design feature could be used to prevent overloading the compressor motor during hot weather, when the condensing temperature rises. This benefit would be achieved, however, at the expense of reduced capacity for a given compressor displacement due to the reduced volumetric efficiency.

11-11 Performance of actual reciprocating compressors The trends in performance of reciprocating compressors developed analytically and shown in Figs. 11-5 to 11-11 are recognizable in the performance of real compressors. Actual performance data are used later in the analysis of the complete vapor compression system. Figure 14-1 is a graph from catalog data and shows refrigerating capacity and power requirements as a function of evaporating and condensing temperatures. The refrigerating capacity increases with an increase in evaporating temperature and decreases with an increase in condensing temperature. The power required by the compressor in general increases with an increase in evaporating temperature, except at lower condensing temperatures, where the peak that was first shown in Fig. 11-6 is evident. Increasing the evaporating temperature above 0°C with a condensing temperature of 25°C, as in Fig. 14-1, results in a continued decrease in power requirement. Figure 11-11 shows a peak power requirement reached at high condensing temperatures. The corresponding range in Fig. 14-1 is to the left of the graph, where at very low evaporating temperatures the curves for power at the various condensing temperatures are pulling together and could even cross over each other.

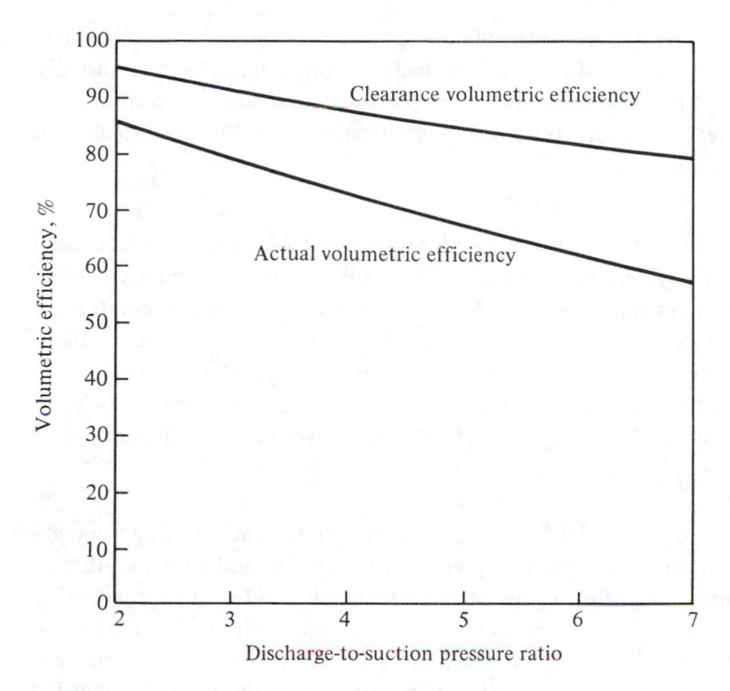

Figure 11-12 Clearance and actual volumetric efficiency of a Refrigerant 22 high-speed compressor. The clearance volume assumed for calculating the clearance volumetric efficiency is 4.5 percent.

11-12 Actual volumetric efficiency The prediction of volumetric efficiency on the basis of reexpansion of vapor in the clearance volume is a good start toward predicting the actual volumetric efficiency. Other factors that influence the volumetric efficiency are the pressure drop across the suction and discharge valves, leakage past the rings of the piston, and leakage back through the discharge and suction valves. Also cylinder heating of the suction gas reduces the volumetric efficiency, since immediately upon entering the cylinder the gas is warmed and expanded. The specific volume of the gas inside the cylinder is consequently higher than when entering the compressor, which is the position on which the volumetric efficiency is based. All the above-mentioned factors result in a lower actual volumetric efficiency than that predicted by the re-expansion of clearance gas alone. Figure 11-12 shows the actual volumetric efficiency[2] compared with the clearance volumetric efficiency.

The abscissa in Fig. 11-12 is the discharge-to-suction pressure ratio, a convenient parameter on which to base the volumetric performance of the compressor. The curve for the actual volumetric efficiency as a function of the pressure ratio applies to a wide variety of evaporating and condensing temperatures. When this curve is available, along with the knowledge of the displacement rate of the compressor, the refrigerating capacity of the compressor can be calculated over a wide variety of conditions.

11-13 Compression efficiency The compression efficiency η_c in percent is

$$\eta_c = \frac{\text{isentropic work of compression, kJ/kg}}{\text{actual work of compression, kJ/kg}} \times 100$$

where the works of compression are referred to the same suction and discharge pressures. The compression efficiencies for open-type reciprocating compressors are usually in the range of 65 to 70 percent. Some of the processes that reduce the compression efficiency from its ideal value of 100 percent are friction of rubbing surfaces and pressure drop through valves.

Example 11-1 Catalog data for a six-cylinder refrigerant 22 compressor operating at 29 r/s indicate a refrigerating capacity of 96.4 kW and a power requirement of 28.9 kW at an evaporating temperature of 5°C and condensing temperature of 50°C. The performance data are based on 3°C liquid subcooling and 8°C superheating of the suction gas entering the compressor. The cylinder bore is 67 mm and the piston stroke is 57 mm. Compute (*a*) the clearance volumetric efficiency if the clearance volume is 4.8 percent, (*b*) the actual volumetric efficiency, and (*c*) the compression efficiency.

Solution The state of the refrigerant leaving the evaporator and entering the compressor is 5°C saturation temperature (pressure of 584 kPa) and a temperature of 13°C. At this state the following properties prevail: h = 413.1 kJ/kg; v = 43.2 L/kg; and s = 1.7656 kJ/kg · K. Following an isentropic compression to a saturation temperature of 50°C (pressure = 1942 kPa) the properties of the refrigerant are h = 444.5 kJ/kg and v = 14.13 L/kg. The enthalpy of the liquid leaving the condenser and entering the evaporator is the enthalpy of liquid at 47°C = 259.1 kJ/kg.

(*a*) The clearance volumetric efficiency is

$$100 - 4.8\left(\frac{43.2}{14.13} - 1\right) = 90.1\%$$

(*b*) The compressor displacement rate is

$$(6 \text{ cyl})\,(29 \text{ r/s})\left(\frac{0.067^2\pi}{4}\,\text{m}^3/\text{cyl} \cdot \text{r}\right)(0.057) = 0.03497 \text{ m}^3/\text{s} = 34.97 \text{ L/s}$$

The actual rate of refrigerant flow is

$$\frac{96.4 \text{ kW}}{413.1 - 259.1 \text{ kJ/kg}} = 0.6260 \text{ kg/s}$$

The actual volumetric flow rate of the refrigerant measured at the compressor suction is

$$(0.6260 \text{ kg/s})\,(43.2 \text{ L/kg}) = 27.04 \text{ L/s}$$

The actual volumetric efficiency is, then,

$$\frac{27.04 \text{ L/s}}{34.97 \text{ L/s}} \times 100 = 77.3\%$$

(*c*) The compression efficiency is the isentropic work of compression divided by the actual work of compression. The latter is

$$\frac{28.9 \text{ kW}}{0.6260 \text{ kg/s}} = 46.2 \text{ kJ/kg}$$

so that

$$\eta_c = \frac{444.5 - 413.1 \text{ kJ/kg}}{46.2 \text{ kJ/kg}} \times 100 = 68\%$$

11-14 Compressor discharge temperatures If the discharge temperature of the refrigerant from the compressor becomes too high, it may result in breakdown of the oil, causing excessive wear or reduced life of the valves, particularly the discharge valves. In general the higher the pressure ratio, the higher the discharge temperature, but the properties of the refrigerant are also crucial. Figure 11-13 shows the discharge temperatures for four refrigerants following isentropic compression from saturated vapor at 0°C to various condensing temperatures. Refrigerants 12 and 502 have low discharge temperatures while refrigerant 22 experiences higher temperatures. Since the highest temperatures of the four refrigerants shown is ammonia, ammonia compressors are equipped with water-cooled heads. The water lines are visible on the compressor shown in Fig. 11-1.

11-15 Capacity control If a refrigeration system is operating in a steady-state mode and the refrigeration load decreases, the inherent response of the system is to decrease

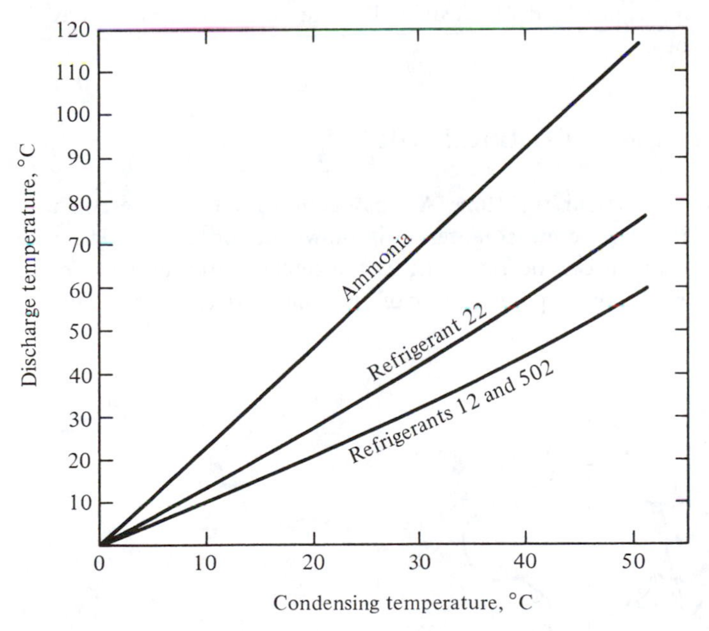

Figure 11-13 Discharge temperatures following isentropic compression from saturated vapor at 0°C.

the evaporating temperature and pressure. This change in the evaporator condition results in reduced compressor capacity, which ultimately matches the reduced refrigeration load. The reduction in evaporator temperature may be undesirable for several reasons. In air conditioning, the coil may collect frost and block the airflow, further reducing the evaporator pressure. Stored fresh food and many other products may be damaged by low temperatures. If the evaporator chills a liquid, the liquid may freeze and burst a tube in the evaporator.

Several methods are commonly used to reduce the compressor capacity:

1. In *cycling* the compressor stops and starts as needed. The method works well in small systems.
2. Back-pressure regulation throttles the suction gas between the evaporator and the compressor to keep the evaporator pressure constant. This method gives good control of the evaporator temperature but is inefficient.
3. Bypassing the discharge gas back to the suction line usually affords precise capacity reduction, but the method is inefficient and the compressor often runs hot. A preferred bypass circuit delivers the discharge gas that is bypassed back to the entrance of the evaporator.
4. Another method is cylinder unloading on a multicylinder compressor by automatically holding the suction valve open or diverting the discharge gas from a cylinder back to the suction line before compression. In the compressor of Fig. 11-1 there are two horizontal lines carrying high-pressure oil from the oil pump at the right end of the compressor to hold the suction valves open when the unloaders are activated. The loss in efficiency with cylinder unloading is moderate. A step control may be provided which unloads more and more cylinders as the suction pressure drops.

PART II: ROTARY SCREW COMPRESSORS

11-16 How the screw compressor functions A cross-sectional view of the two principal rotating elements of the screw compressor is shown in Fig. 11-14. The male rotor with four lobes, shown on the right, drives the female rotor in a stationary housing. Figure 11-15 shows an exploded view of the major parts of the complete

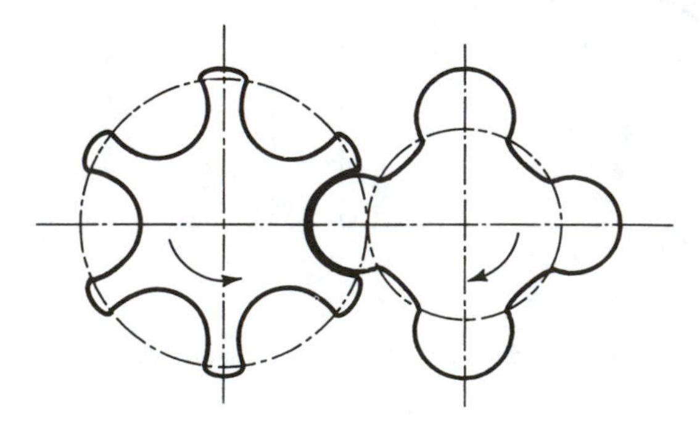

Figure 11-14 Cross section of the two rotors of a screw compressor.

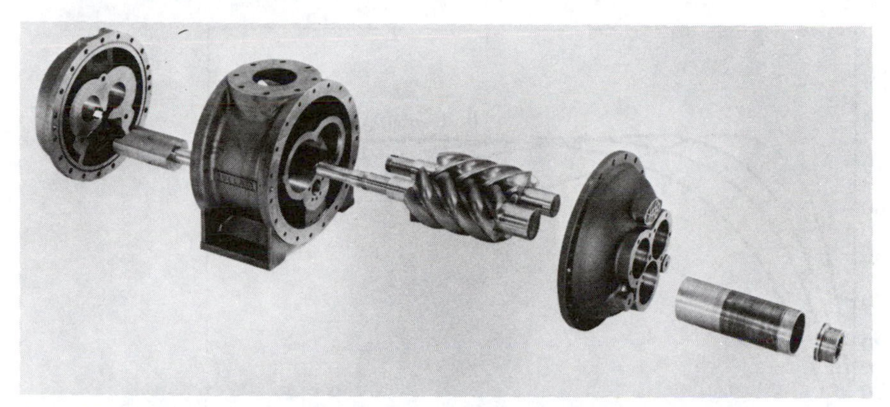

Figure 11-15 Exploded view of main elements of a screw compressor. *(Sullair Refrigeration, Inc.)*

compressor. The refrigerant vapor enters one end of the compressor at the top and leaves the other end at the bottom. At the suction position of the compressor a void is created into which the inlet vapor flows. Just before the point where the interlobe space leaves the inlet port, the entire length of the cavity or gully is filled with gas. As the rotation continues, the trapped gas is moved circumferentially around the housing of the compressor. Further rotation results in meshing of the male lobe with the female gully, decreasing the volume in the cavity and compressing the gas. At a certain point into the compression process the discharge port is uncovered and the compressed gas is discharged by further meshing of the lobe and the gully.

The screw compressor was developed in the 1930s and first became popular for refrigeration service in Europe in the 1950s and 1960s. On some early compressors the two rotors were geared to each other and no lubrication was provided between the rotors. Current practice is to drive the female rotor with the male rotor and inject oil between the two rotors for lubrication and sealing. In the package water chiller shown in Fig. 11-16 the screw compressor is just to the right of the control panel, showing its

Figure 11-16 A water-chilling package that uses a screw compressor. *(Dunham-Bush, Inc.)*

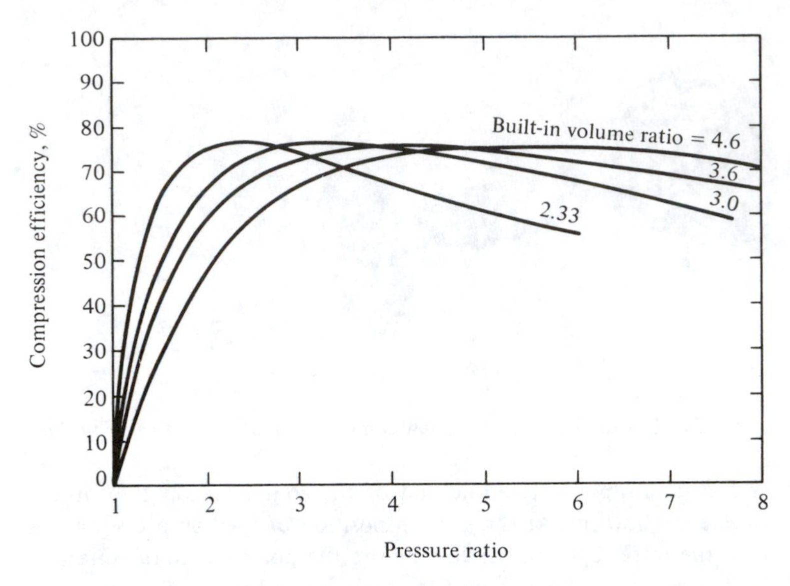

Figure 11-17 Compression efficiencies of screw compressors of various built-in volume and pressure ratios.

relatively small size compared with the condenser (in back) and the water-chilling evaporator at the bottom. The vessel below the control panel is the oil separator, a standard component in a screw-compressor system.

11-17 Performance characteristics of screw compressors The explanation in Section 11-16 stated that at a certain point in the compression process the discharge port is uncovered. This point is a function of the design of the compressor and establishes a built-in volume ratio of the compressor. This ratio has a corresponding built-in pressure ratio associated with it, and any compressor has its best performance at a certain pressure ratio. Figure 11-17 shows curves[3] of compression efficiency for several compressors of different built-in pressure ratios. Normal operation of most refrigeration systems occurs over a range of pressure ratios as the condenser and evaporator conditions change, so that a screw compressor does not always operate at peak efficiency. The peak efficiencies are quite high, however, and there is little sacrifice of efficiency if the pressure ratio does not change radically.

11-18 Capacity control Many screw compressors are equipped with a sliding valve for capacity control. It is in the housing of the compressor and can be moved axially. As the valve is opened, it delays the position at which compression begins. The capacity can be modulated down to about 10 percent of full capacity, although there is loss of efficiency in the capacity reduction.

PART III: VANE COMPRESSORS

11-19 Vane compressors The two basic types of vane compressors are the *roller* or *single-vane type* and the *multiple-vane* type. Vane compressors are used mostly in do-

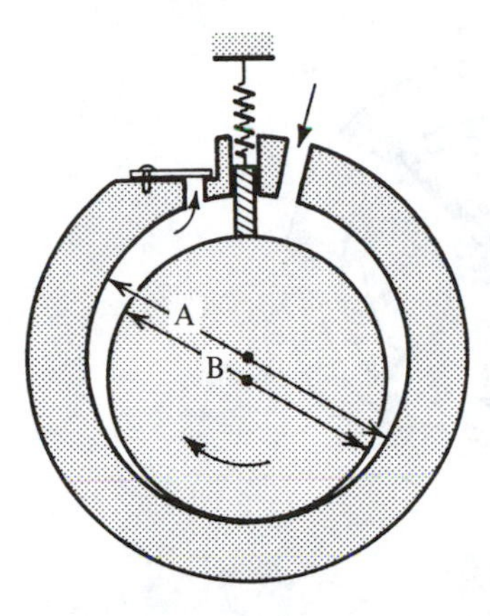

Figure 11-18 Roller-type vane compressor.

mestic refrigerators, freezers, and air conditioners, although they can also be used as booster compressors in the low-pressure portion of large multistage compression systems. In the roller type (Figs. 11-18 and 11-19) the centerline of the shaft is the same as the centerline of the cylinder. The centerline of the shaft, however, is located eccentrically on the rotor, so that as the rotor revolves it makes contact with the cylinder. The roller-type compressor has a spring-loaded divider which separates the suction and discharge chambers.

Figure 11-19 A roller-type vane compressor. *(General Electric Company.)*

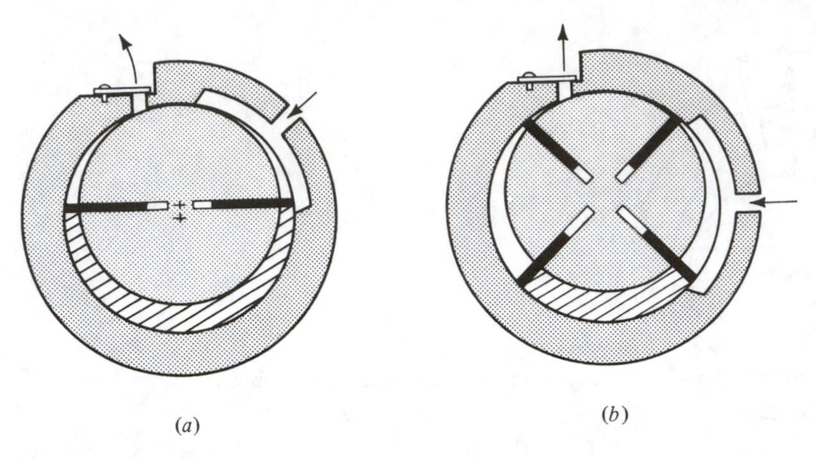

Figure 11-20 Multiple-vane compressors: (*a*) two-vane and (*b*) four-vane.

The formula for the displacement rate D of the roller-type compressor is

$$D = \frac{\pi}{4}(A^2 - B^2)L(\text{rotative speed}) \qquad \text{m}^3/\text{s} \qquad (11\text{-}10)$$

where A = cylinder diameter, m
B = roller diameter, m
L = cylinder length, m
and the rotative speed is in revolutions per second.

In the multiple-vane compressor (Fig. 11-20) the rotor revolves about its own centerline, but the centerlines of the cylinder and the rotor do not coincide. The rotor has two or more sliding vanes, which are held against the cylinder by centrifugal force.

For the two-vane compressor in Fig. 11-20 the displacement per revolution is proportional to twice the crosshatched area. For the four-vane compressor the displacement per revolution is proportional to 4 times the crosshatched area. Up to a certain point, then, the displacement is greatest on the compressor with the largest number of vanes.

In the two types of rotary compressors shown here no suction valves are needed, and since the suction gas enters the compressor continuously, gas pulsation is at a minimum.

PART IV: CENTRIFUGAL COMPRESSORS

11-20 Role of centrifugal compressors The first commercial centrifugal compressor used in refrigeration service was promoted by Willis Carrier in 1920. Since then the centrifugal compressor has become the dominant type of compressor in large installations.

Centrifugal compressors serve refrigeration systems in the range of 200 to 10,000 kW of refrigerating capacity. Evaporating temperatures in multistage machines may extend down to the -50 to -100°C range, although one of the largest uses of the

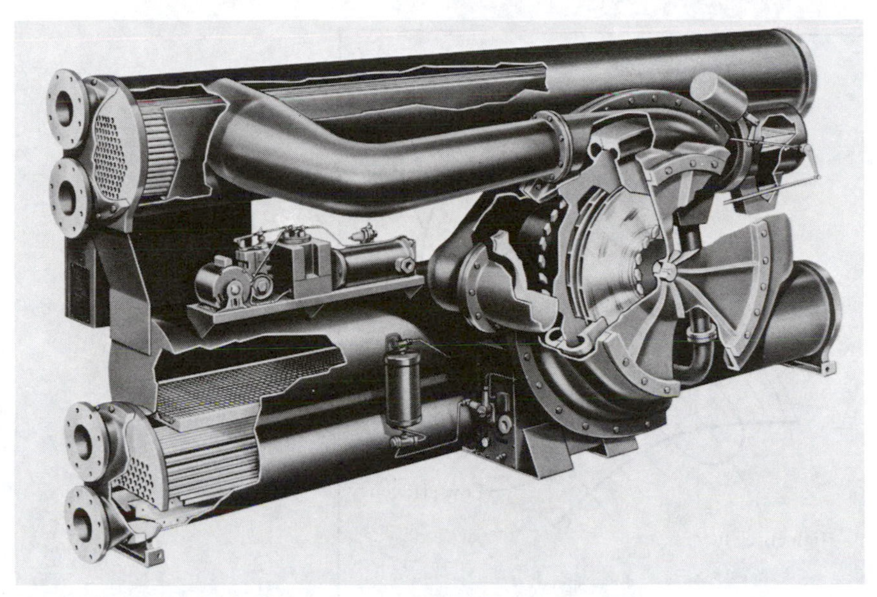

Figure 11-21 A centrifugal compressor system. The condenser is the top vessel, and the water-cooling evaporator is at the bottom. The two impellers of the two-stage compressor are driven by an electric motor in the rear. *(The Trane Company.)*

compressor is for chilling water to about 6 to 8°C in air-conditioning systems. A cutaway view of a complete refrigeration system using a centrifugal compressor is shown in Fig. 11-21.

11-21 Operation Centrifugal compressors are similar in construction to centrifugal pumps in that the incoming fluid enters the eye of the spinning impeller and is thrown by centrifugal force to the periphery of the impeller. Thus the blades of the impeller impart a high velocity to the gas and also build up the pressure. From the impeller the gas flows either into diffuser blades or into a volute, where some of the kinetic energy is converted into pressure. The centrifugal compressor may be manufactured with only one wheel if the pressure ratio is low, although the machines are generally multistage. Centrifugal compressors operate with adiabatic compression efficiencies of 70 to 80 percent.

11-22 Flash-gas removal A centrifugal compressor with two or more stages invites the use of flash-gas removal. Flash gas can be removed by partially expanding the liquid from the condenser, separating the flash gas, and then recompressing the gas instead of dropping its pressure further. Flash-gas removal, discussed further in Sec. 16-2, increases the efficiency of the cycle and is conveniently achieved when two or more stages of compression are available.

11-23 Performance characteristics Impellers in centrifugal compressors are equipped with backward-curved blades. Section 6-16 on fans mentioned backward-curved blades and showed performance characteristics of a fan with forward-curved blades. Admit-

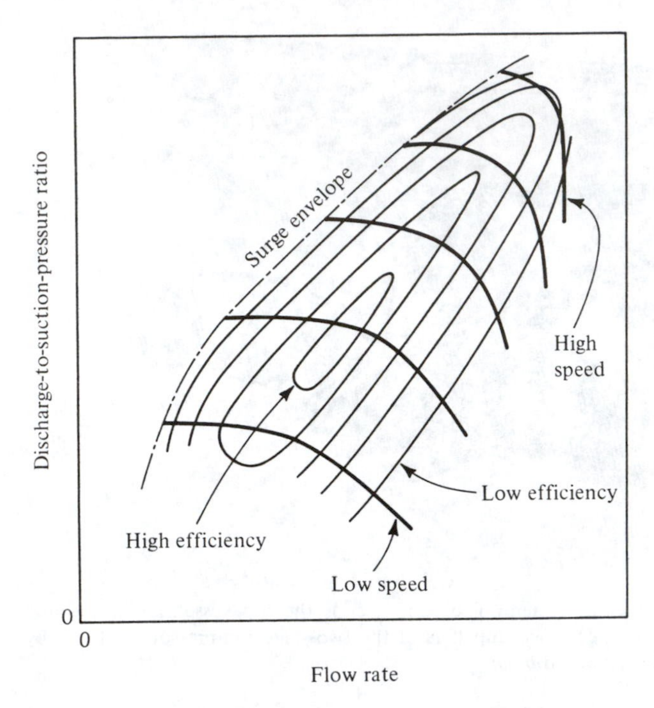

Figure 11-22 Performance of a centrifugal compressor.

tedly the air flowing through a fan was treated as an incompressible fluid while in the centrifugal compressor the refrigerant vapor is clearly compressed. The predominant characteristic prevails in both machines, however, in that for a constant-speed compressor as the flow rate starts at zero and increases, the pressure buildup developed by the compressor starts at some nonzero value, increases for a time, and then progressively drops off.

One choice of coordinates for presenting the characteristics is shown in Fig. 11-22: the discharge-to-suction-pressure ratio versus the flow rate. The graph shows the performance for several different compressor speeds and isoefficiency lines (lines of constant efficiency). No performance curves are shown to the left of the *surge line*; this surge phenomenon will be discussed in Sec. 11-26.

11-24 Tip speed to develop pressure A rough estimate of the tip speed of the impeller can be made by using several fundamental relationships for turbomachinery. The torque the impeller ideally imparts to the gas is

$$T = w(V_{2t}r_2 - V_{1t}r_1) \tag{11-11}$$

where T = torque, N · m
w = mass rate of flow, kg/s
V_{2t} = tangential velocity of refrigerant leaving impeller, m/s
r_2 = radius of exit of impeller, m
V_{1t} = tangential velocity of refrigerant entering impeller, m/s
r_1 = radius of inlet of impeller, m

If the refrigerant enters the impeller in an essentially radial direction, the tangential component of the velocity $V_{1t} = 0$, and so

$$T = wV_{2t}r_2 \tag{11-12}$$

The power required at the shaft is the product of the torque and the rotative speed

$$P = T\omega = wV_{2t}r_2\omega \tag{11-13}$$

where P = power, W
ω = rotative speed, rad/s

At least at very low refrigerant flow rates the tip speed of the impeller and the tangential velocity of the refrigerant are nearly identical; therefore

$$r_2\omega = V_{2t}$$

and
$$P = wV_{2t}^2 \tag{11-14}$$

Another expression for ideal power is the product of the mass rate of flow and the isentropic work of compression,

$$P = w\,\Delta h_i\,(1000\text{ J/kJ}) \tag{11-15}$$

Equating the two expressions for power, Eqs. (11-14) and (11-15), yields

$$V_{2t}^2 = 1000\Delta h_i \tag{11-16}$$

Although Eq. (11-16) is based on some idealizations, it can provide an order-of-magnitude estimate of the tip speed and can also show important comparisons, as in Example 11-2.

Example 11-2 Calculate the speed of the impeller tip in order to compress the following refrigerants from saturated vapor at 10°C to a pressure corresponding to a condensing temperature of 30°C when the refrigerant is (*a*) refrigerant 11 and (*b*) ammonia.

Solution (*a*) In the isentropic compression of refrigerant 11 from saturated vapor at 10°C to a saturated condensing temperature of 30°C

$$\Delta h_i = 406.7 - 393.9 = 12.8\text{ kJ/kg}$$

The tip speed is

$$V_{2t} = \sqrt{1000(12.8)} = 113.1\text{ m/s}$$

(*b*) For ammonia

$$\Delta h_i = 1560 - 1472 = 88\text{ kJ/kg}$$

The tip speed is

$$V_{2t} = \sqrt{1000(88)} = 297\text{ m/s}$$

11-25 Choice of impeller and refrigerant Two crucial impeller dimensions are the wheel diameter and the width between impeller faces. The designer of a centrifugal compressor system must select a combination of these dimensions along with a choice of refrigerant. The magnitude of the wheel diameter is heavily dictated by the discharge pressure that must be achieved, because for a given rotative speed a large wheel diameter will provide a higher tip speed, which results in a higher pressure ratio. The results of Example 11-2 provide some insight into the influence of the refrigerant choice on required tip speed. If a centrifugal compressor is driven by an electric motor operating at 60 r/s, the wheel diameter needed for the 113.1 m/s tip speed (refrigerant 11) is 0.6 m, while for the tip speed of 297 m/s with ammonia the wheel diameter must be 1.58 m. The required wheel diameter for ammonia would probably be impractical. Furthermore, from the standpoint of strength of the wheel, the tip speed for ammonia is nearing the usual limitation[3] of 300 m/s.

The initial conclusion is to choose a refrigerant with properties similar to refrigerant 11 in preference to ammonia. Centrifugal compressors could and do handle ammonia, but additional stages of compression might be required. For example, if the compression in Example 11-2 were executed in two stages, the Δh_i could be cut in half and the tip speeds of both wheels would be 210 m/s.

Another decision of the designer is the width of passage in the impeller. To increase the capacity, increase the width between the faces of the impeller, which, of course, also increases the power requirement. Centrifugal-compressor designers constantly struggle to maintain high efficiencies with machines of small capacity. One reason for the dropoff of efficiency with low capacities is that the impeller width becomes narrow and the friction of the gas on the impeller faces becomes large relative to the flow rate through the impeller. The choice of a low-density refrigerant allows one to maintain a wide impeller width for a given capacity.

Refrigerants 11 and 113 especially meet the requirements described above and are the popular refrigerants used in water-chilling systems with centrifugal compressors. But ammonia, refrigerant 12, and other refrigerants are used successfully with centrifugal compressors.

11-26 Surging Figure 11-22 shows no performance data to the left of the *surge envelope*, although the classic performance of a backward-curved-blade pump, fan, and compressor would be shown by the dashed line in Fig. 11-23. As the refrigeration load drops off and the flow rate decreases from point A, the performance rides up the pressure-flow characteristic to point B. Further decrease in flow rate sends the operation to point C where the pressure-ratio capability of the compressor drops. The drop in pressure ratio is due both to the inherent characteristics of backward-curved-blade turbomachines and to the fact that flow separation begins occurring at the blades.

Although the compressor capacity drops significantly when operation moves to point C the heat load on the evaporator continues to boil off refrigerant, building up the evaporator pressure and decreasing the pressure ratio. The compressor is then momentarily able to shift operation back to point A where the cycle begins to repeat itself. This sequence, called *surging*, is characterized by objectionable noise and wide fluctuations of load on the compressor and motor. The period of the cycle is usually 2 to 5 s, depending upon the size of the installation.

One reason for not showing data to the left of the surge envelope is that steady-state readings cannot be obtained. Centrifugal fans (Chap. 6) have the same pressure-

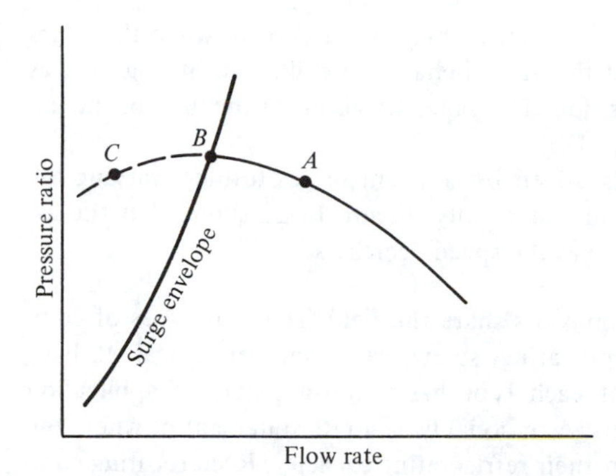

Figure 11-23 Surging in a centrifugal compressor.

flow characteristics as the compressor shown in Fig. 11-23 and are thus subject to the same surge phenomenon. Many fans operate from time to time in the surge region, however, and the only undesirable result is a low rumbling sound. In the case of centrifugal compressors, however, operating in the surge region is definitely objectionable and should be avoided. Some compressors are equipped with a discharge-gas bypass that at low refrigeration loads throttles discharge gas back to the suction line in order to provide a false load on the compressor.

11-27 Capacity control The two most efficient and most widely used methods of capacity control are (1) adjusting prerotation vanes at the impeller inlet and (2) varying the speed. Two methods that are not efficient and not widely used are varying the condenser pressure and bypassing the discharge gas. The latter was mentioned as a means of preventing surge and is sometimes combined with prerotation vanes.

Equation (11-11) for the torque indicates that if a positive component is provided for V_{1t}, the torque will be reduced, which also translates into reduced pumping capability. Prerotation vanes provide a swirl to the gas entering the impeller so that the inlet gas has a tangential velocity in the direction the impeller is rotating. Figure 11-24 shows how the position of the prerotation vanes influences the compressor character-

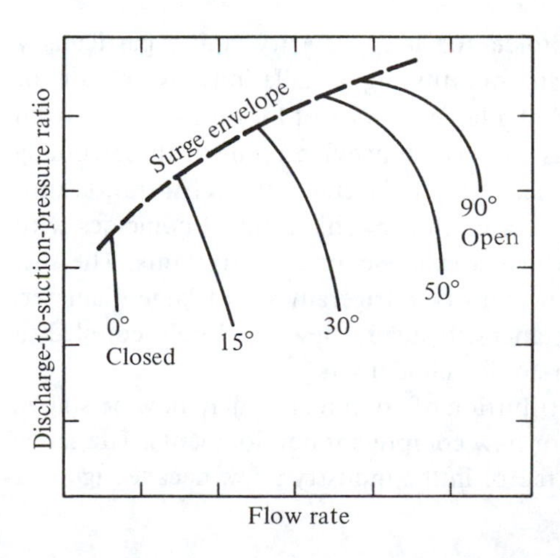

Figure 11-24 Centrifugal-compressor characteristics at various settings of the prerotation vanes.

istics. The use of prerotation vanes is an efficient method of control when the vanes are near their fully open position, but the vanes behave more like a throttling valve as they approach their nearly closed position. The vanes are visible at the inlet of the impeller in the compressor shown in Fig. 11-21.

When a centrifugal compressor is driven by a steam or gas turbine, varying the speed for capacity control can be achieved readily. Figure 11-22 shows that the capacity drops off at a given pressure ratio as the speed decreases.

11-28 How the various types of compressors share the field The four types of compressors studied in this chapter, reciprocating, screw, vane, and centrifugal, all have somewhat different qualities, so that each type has found a sphere of application where it has advantages over the others. A reasonably accurate statement of where the compressors are used can be based on their refrigerating capacity. Reciprocating compressors dominate from very small refrigerating capacities to about 300 kW. Centrifugal compressors are most widely used for units having refrigerating capacities of 500 kW and above. The screw compressor has found a niche in the 300-to-500-kW capacities and competes against large reciprocating compressors and against small centrifugal compressors. The vane compressor competes against the reciprocating compressor primarily in the market for domestic refrigerators and air conditioners.

Probably more reciprocating compressors are manufactured than any other type because they are the choice for smaller refrigeration units in high production. For high-capacity refrigeration systems, the large physical size of the reciprocating compressor shifts the choice in favor of the more compact screw and centrifugal compressors, which battle for the market in the 300- to 500-kW capacities. An uneasy truce sometimes is arrived at in industrial refrigeration plants, where a combination of screw and reciprocating compressors is used. The operating strategy is to use the screw compressor for the base load and bring on reciprocating compressors to accommodate the variations above the base. The reason for this distribution of load is that the screw compressor is efficient when operating near full load, it has fewer moving parts than the reciprocating compressor, and is developing a reputation for long operating life. The reciprocating compressor seems to have better efficiencies at part-load operation than the screw compressor and can accommodate the varying portion of the load more efficiently.

The centrifugal compressor is the choice for large-capacity units, particularly for water-chilling plants used for large air-conditioning installations. A feature of most such installations is that air is cooled at a large number of locations remote from the compressor room. Since using water as the heat-conveying agent in these cooling coils is preferable to the complexities of delivering refrigerant, the refrigeration unit for most large air-conditioning plants is a water chiller. This concept coincides with the characteristics of a centrifugal unit which uses low-density refrigerants. The suction and discharge pipes needed for the low-density refrigerant are of large diameter, and it is impractical to run them large distances. Instead, they are closely coupled to the water-chilling evaporator and the water-cooled condenser.

While it may seem that the market distribution of compressors may now be stable, engineers should be alert to the potential for new compressor developments. The screw compressor, for example, which was not a factor in the industry a few decades ago, has now established itself.

PROBLEMS

11-1 An ammonia compressor has a 5 percent clearance volume and a displacement rate of 80 L/s and pumps against a condensing temperature of 40°C. For the two different evaporating temperatures of -10 and 10°C, compute the refrigerant flow rate assuming that the clearance volumetric efficiency applies. *Ans.* 0.37 kg/s at 10°C

11-2 A refrigerant 22 compressor with a displacement rate of 60 L/s operates in a refrigeration system that maintains a constant condensing temperature of 30°C. Compute and plot the power requirement of this compressor at evaporating temperatures of -20, -10, 0, 10, and 20°C. Use the actual volumetric efficiencies from Fig. 11-12 and the following isentropic works of compression for the five evaporating temperatures, respectively, 39.9, 30.2, 21.5, 13.7, and 6.5 kJ/kg

11-3 The catalog for a refrigerant 22, four-cylinder, hermetic compressor operating at 29 r/s, a condensing temperature of 40°C, and an evaporating temperature of -4°C shows a refrigerating capacity of 115 kW. At this operating point the motor (whose efficiency is 90 percent) draws 34.5 kW. The bore of the cyclinders is 87 mm and the piston stroke is 70 mm. The performance data are based on 8°C of subcooling of the liquid leaving the condenser. Compute (*a*) the actual volumetric efficiency and (*b*) the compression efficiency. *Ans.* (*a*) 77.4%; (*b*) 71%

11-4 An automotive air conditioner using refrigerant 12 experiences a complete blockage of the airflow over the condenser, so that the condenser pressure rises until the volumetric efficiency drops to zero. Extrapolate the actual volumetric-efficiency curve of Fig. 11-12 to zero and estimate the maximum discharge pressure, assuming an evaporating temperature of 0°C. *Ans.* 5300 kPa

11-5 Compute the maximum displacement rate of a two-vane compressor having a cylinder diameter of 190 mm and a rotor 80 mm long with a diameter of 170 mm. The compressor operates at 29 r/s. *Ans.* 22 L/s

11-6 A two-stage centrifugal compressor operating at 60 r/s is to compress refrigerant 11 from an evaporating temperature of 4°C to a condensing temperature of 35°C. If both wheels are to be of the same diameter, what is this diameter? *Ans.* 0.53 m

REFERENCES

1. W. L. McGrath: New Refrigeration System Reduces Electrical Demand of Air Conditioning Equipment, *Refrig. Eng.*, vol. 65, no. 2, p. 52, February 1957.
2. "Trane Reciprocating Refrigeration," The Trane Company, LaCrosse, Wis., 1977.
3. "Handbook and Product Director, Equipment Volume," chap. 12, American Society of Heating, Refrigerating, and Air-Conditioning Engineers, Atlanta, Ga., 1979.

ADDITIONAL READINGS ON COMPRESSORS

Methods of Testing for Rating Positive Displacement Refrigerant Compressors, Standard 23-78, American Society of Heating, Refrigerating, and Air-Conditioning Engineers, Atlanta, Ga., 1978.

W. D. Cooper: Refrigeration Compressor Performance as Affected by Suction Vapor Superheating, *ASHRAE Trans.*, vol. 80, pt. 1, pp. 195–204, 1974.

J. Brown and S. F. Pearson: Piston Leakage in Refrigeration Compressors, J. Refrig., vol. 6, no. 5, p. 104, September-October 1963.

E. H. Jensen: Effect of Compressor Characteristics on Motor Performance, *ASHRAE Trans.*, vol. 66, pp. 194–201, 1960.

O. Jensen: Heat Transfer in a Refrigeration Compressor, *Kulde*, vol. 20, no. 1, p. 1, February 1966.

G. Lorentzen: Influence of Speed on Compressor Volumetric Efficiency, *Refrig. Eng.*, vol. 60, no. 3, p. 272, March 1952.

J. F. T. MacLaren and S. V. Kerr: Analysis of Valve Behavior in Reciprocating Compressors, *12th Int. Cong. Refrig., Madrid, 1967*, pap. 3.39.

U. Ritter, G. Schoberth, and E. Emblik: Development of Oil-Free Compressors, J. Refrig., vol. 2, no. 2., March-April 1959.

D. N. Shaw: Helical Rotary Screw Compressor Heating/Cooling Systems, *ASHRAE Trans.*, vol. 83, Pt. 1, pp. 177–184, 1977.

T. Stillson: Helical Rotary Screw Compressor Applications for Energy Conservation, *ASHRAE Trans.*, vol. 83, pt. 1, pp. 185–201, 1977.

CHAPTER

TWELVE

CONDENSERS AND EVAPORATORS

12-1 Condensers and evaporators as heat exchangers Since both the condenser and evaporator are heat exchangers, they have certain features in common. One classification of condensers and evaporators (Table 12-1) is according to whether the refrigerant is on the inside or outside of the tubes and whether the fluid cooling the condenser or being refrigerated is a gas or a liquid. The gas referred to in Table 12-1 is usually air, and the liquid is usually water, but other substances are used as well.

The most widely used types of condensers and evaporators are shell-and-tube heat exchangers (Fig. 12-1) and finned-coil heat exchangers (Fig. 12-2). Table 12-1 indicates that certain combinations are not frequently used, particularly the configuration where the gas is passed through tubes. The reason is that volume flow rates of gases are high relative to those of liquids and would result in high pressure drops if forced through the tubes.

We shall study evaporators and condensers together to take advantage of the features they have in common. For example, the laws governing the flow of water through the shell and over the tube bundle of a heat exchanger are the same whether the heat exchanger is an evaporator or a condenser. It is important to realize, however, that the mechanisms prevailing when refrigerant boils are quite different from those when refrigerant condenses.

12-2 Overall heat-transfer coefficient The overall heat-transfer coefficient for an evaporator or condenser is the proportionality constant, which, when multiplied by the heat-transfer area and the mean temperature difference between the fluids, yields

Table 12-1 Some types of evaporators and condensers

Component	Refrigerant	Fluid
Condenser	Inside tubes	Gas outside Liquid outside†
	Outside tubes	Gas inside† Liquid inside
Evaporator	Inside tubes	Gas outside Liquid outside
	Outside tubes	Gas inside† Liquid inside

† Seldom used.

Figure 12-1 Shell-and-tube water-cooled condenser. *(ITT Bell & Gossett–Fluid Handling Division.)*

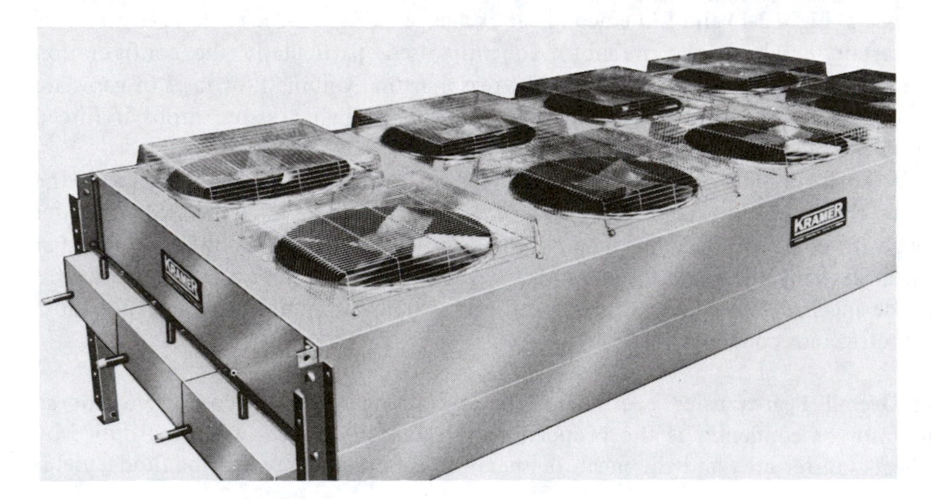

Figure 12-2 An air-cooled condenser. *(Kramer Trenton Co.).*

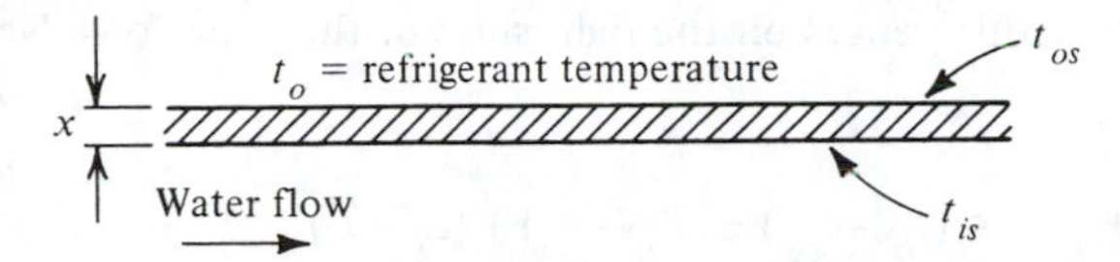

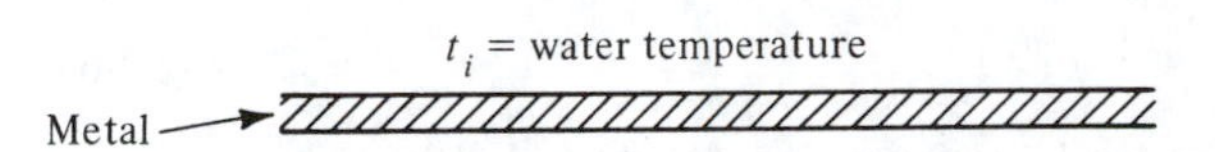

Figure 12-3 Heat transfer between refrigerant and water through a tube.

the rate of heat transfer. If heat flows across a tube, as in Fig. 12-3, between refrigerant on the outside and water on the inside, for example, under steady-state conditions the rate of heat transfer q in watts is the same from the refrigerant to the outside surface of the tube, from the outside to the inside surface of the tube, and from the inside surface of the tube to the water. The expressions for q in each of these transfers are, respectively,

$$q = h_o A_o (t_o - t_{os}) \tag{12-1}$$

$$q = \frac{k}{x} A_m (t_{os} - t_{is}) \tag{12-2}$$

$$q = h_i A_i (t_{is} - t_i) \tag{12-3}$$

where q = rate of heat transfer, W
h_o = heat-transfer coefficient on outside of tube, W/m^2 · K
A_o = outside area of tube, m^2
t_o = refrigerant temperature, °C
t_{os} = temperature of outside surface of tube, °C
k = conductivity of tube metal, W/m · K
x = thickness of tube, m
t_{is} = temperature of inside surface of tube, °C
A_m = mean circumferential area of tube, m^2
h_i = heat-transfer coefficient on inside of tube, W/m^2 · K
A_i = inside area of tube, m^2
t_i = water temperature, °C

To express the overall heat-transfer coefficient the area on which the coefficient is based must be specified. Two acceptable expressions for the overall heat-transfer coefficient are

$$q = U_o A_o (t_o - t_i) \tag{12-4}$$

and

$$q = U_i A_i (t_o - t_i) \tag{12-5}$$

where U_o = overall heat-transfer coefficient based on outside area, W/m^2 · K
U_i = overall heat-transfer coefficient based on inside area, W/m^2 · K

From Eqs. (12-4) and (12-5) it is clear that $U_o A_o = U_i A_i$. The U value is always associated with an area. Knowledge of U_o or U_i facilitates computation of the rate of heat transfer q.

To compute the U value from knowledge of the individual heat-transfer coefficients, first divide Eq. (12-1) by $h_o A_o$, Eq. (12-2) by kA_m/x, and Eq. (12-3) by

h_iA_i, leaving only the temperature differences on the right sides of the equations. Next add the three equations, giving

$$\frac{q}{h_oA_o} + \frac{qx}{kA_m} + \frac{q}{h_iA_i} = (t_o - t_{os}) + (t_{os} - t_{is}) + (t_{is} - t_i)$$

$$= t_o - t_i \tag{12-6}$$

Alternate expressions for $t_o - t_i$ are available from Eqs. (12-4) and (12-5)

$$t_o - t_i = \frac{q}{U_oA_o} = \frac{q}{U_iA_i} \tag{12-7}$$

Equating Eqs. (12-6) and (12-7) and canceling q provides an expression for computing the U values

$$\frac{1}{U_oA_o} = \frac{1}{U_iA_i} = \frac{1}{h_oA_o} + \frac{x}{kA_m} + \frac{1}{h_iA_i} \tag{12-8}$$

The physical interpretation of the terms in Eq. (12-8) is that $1/U_oA_o$ and $1/U_iA_i$ are the total resistances to heat transfer between the refrigerant and water. This total resistance is the sum of the individual resistances

1. From the refrigerant to the outside surface of the tube $1/h_oA_o$
2. Through the tube $x/(kA_m)$
3. From the inside surface of the tube to the water $1/h_iA_i$

Later in this chapter modifications will be made in Eq. (12-8) to account for fouling of the tube and performance of fins.

12-3 Liquid in tubes; heat transfer and pressure drop The expression for the heat-transfer coefficient for fluids flowing inside tubes, as was first shown in Fig. 2-6, is of the form

$$\text{Nu} = C\ \text{Re}^n\ \text{Pr}^m$$

where n and m are exponents. The constant C and exponents in the equation are

$$\frac{hD}{k} = 0.023\left(\frac{VD\rho}{\mu}\right)^{0.8}\left(\frac{c_p\mu}{k}\right)^{0.4} \tag{12-9}$$

where h = convection coefficient, $W/m^2 \cdot K$
D = ID of tube, m
k = thermal conductivity of fluid, $W/m \cdot K$
V = mean velocity of fluid, m/s
ρ = density of fluid, kg/m^3
μ = viscosity of fluid, $Pa \cdot s$
c_p = specific heat of fluid, $J/kg \cdot K$

Equation (12-9) is applicable to turbulent flow, which typically prevails with the velocities and fluid properties experienced in most commercial evaporators and condensers. McAdams[1] proposed the constant of 0.023 in Eq. (12-9), but Katz et al.[2] found that the actual value in condensers is about 15 percent higher because the tubes are relatively short and the effects of higher turbulence due to entrance effects increase the rate of heat transfer slightly.

Example 12-1 Compute the heat-transfer coefficient for water flow inside the tubes (8 mm ID) of an evaporator if the water temperature is 10°C and its velocity is 2 m/s.

Solution The properties of water at 10°C are
$\mu = 0.00131$ Pa · s $\rho = 1000$ kg/m^3 $k = 0.573$ W/m · K $c_p = 4190$ J/kg · K
The Reynolds number is

$$\text{Re} = \frac{(2\ \text{m/s})\,(0.008\ \text{m})\,(1000\ \text{kg/m}^3)}{0.00131\ \text{Pa}\cdot\text{s}} = 12{,}214$$

This value of the Reynolds number indicates that the flow is turbulent, so Eq. (12-9) applies. The Prandtl number is

$$\text{Pr} = \frac{(4190\ \text{J/kg}\cdot\text{K})\,(0.00131\ \text{Pa}\cdot\text{s})}{0.573\ \text{W/m}\cdot\text{K}} = 9.6$$

The Nusselt number can now be computed from Eq. (12-9)

$$\text{Nu} = 0.023(12{,}214^{0.8})\,(9.6^{0.4}) = 106$$

from which the heat-transfer coefficient can be computed as

$$h = \frac{0.573\ \text{W/m}\cdot\text{K}}{0.008\ \text{m}}\,(106) = 7592\ \text{W/m}^2\cdot\text{K}$$

As the fluid flows inside the tubes through a condenser or evaporator, a pressure drop occurs both in the straight tubes and in the U-bends or heads of the heat exchanger. Some drop in pressure is also attributable to entrance and exit losses. The expression for pressure drop of fluid flowing in straight tubes from Chap 7 is

$$\Delta p = f\frac{L}{D}\,\frac{V^2}{2}\,\rho \tag{12-10}$$

where Δp = pressure drop, Pa
f = friction factor, dimensionless
L = length of tube, m
Since the pressure drop in the straight tubes in an evaporator or condenser may represent only 50 to 80 percent of the total pressure drop, experimental or catalog data on the pressure drop as a function of flow rate are desirable. If the pressure drop at one flow rate is known, it is possible to predict the pressure drop at other flow rates. The expression applicable to straight tubes, Eq. (12-10), indicates that the pressure drop is proportional to the square of the velocity and thus the square of the flow rate.

The other contributors to pressure drop resulting from changes in flow area and direction are also almost exactly proportional to the square of the flow rate, so if the pressure drop and flow rate Δp_1 and w_1 are known, the pressure drop Δp_2 at a different flow rate w_2 can be predicted:

$$\Delta p_2 = \Delta p_1 \left(\frac{w_2}{w_1} \right)^2 \tag{12-11}$$

12-4 Liquid in shell; heat transfer and pressure drop In shell-and-tube evaporators, where refrigerant boils inside tubes, the liquid being cooled flows in the shell across bundles of tubes, as shown schematically in Fig. 12-4. The liquid is directed by baffles so that it flows across the tube bundle many times and does not short-circuit from the inlet to the outlet. The analytical prediction of the heat-transfer coefficient of liquid flowing normal to a tube is complicated in itself, and the complex flow pattern over a bundle of tubes makes the prediction even more difficult. In order to proceed with the business of designing heat exchangers, engineers resort to correlations that relate the Nusselt, Reynolds, and Prandtl numbers to the geometric configuration of the tubes and baffles. Such an equation by Emerson[3] can be modified to the form

$$\frac{hD}{k} = \text{(terms controlled by geometry)} (\text{Re}^{0.6}) (\text{Pr}^{0.3}) \left(\frac{\mu}{\mu_w} \right)^{0.14} \tag{12-12}$$

where μ = viscosity of fluid at bulk temperature, Pa • s
μ_w = viscosity of fluid at tube-wall temperature, Pa • s

The Reynolds number in this equation is GD/μ, where G is the mass velocity or mass rate of flow divided by a characteristic flow area.

Although in this text we shall delve no deeper into the complexities of designing a shell-and-tube heat exchanger, one important but simple realization emerges from Eq. (12-12): for a given evaporator or condenser when water flows in the shell outside the tubes

$$\text{Water-side heat-transfer coefficient} = \text{(const) (flow rate)}^{0.6} \tag{12-13}$$

The convection coefficient varies as the 0.6 power of the flow rate compared with the 0.8 power for flow inside tubes, as indicated by Eq. (12-9).

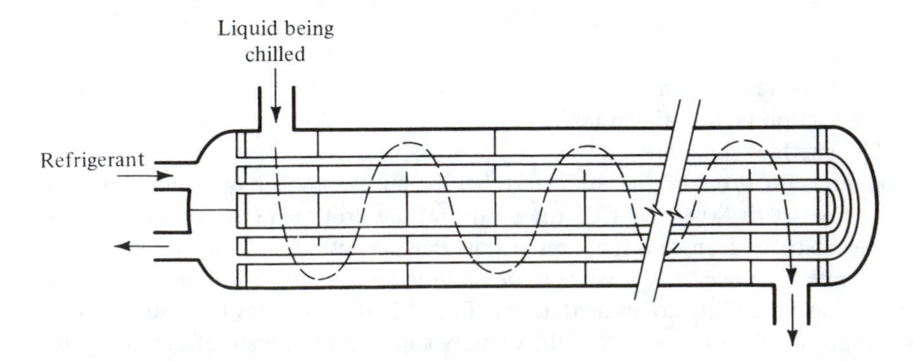

Figure 12-4 Shell flow of liquid across tube bundles.

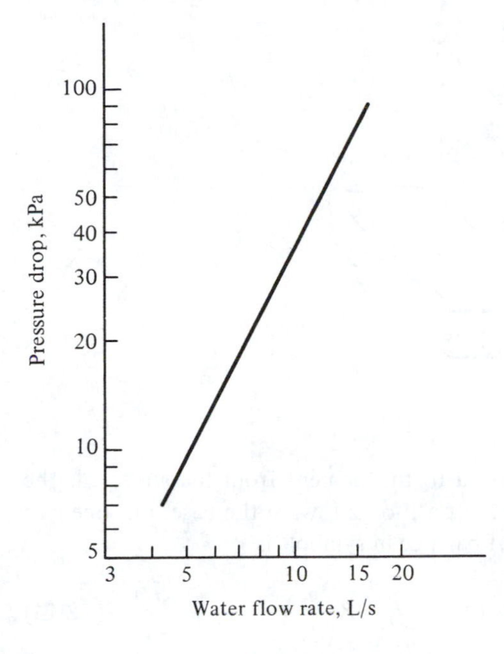

Figure 12-5 Pressure drop of water flowing in the shell of an evaporator. *(York Division of Borg Warner.)*

The pressure drop of liquid flowing through the shell across tube bundles is also difficult to predict analytically, but when an experimental point is available for one flow rate, predictions of the pressure drop at other flow rates can be made quite accurately. Figure 12-5 shows the water pressure drop taken from catalog data of a water-chilling evaporator. The applicable exponent in the pressure-drop–flow-rate relationship here is 1.9.

12-5 Extended surface; fins Equation (12-8) expresses the resistances a heat exchanger encounters when transferring heat from one fluid to another. Suppose that in Eq. (12-8) $1/h_oA_o$ is 80 percent of the total resistance to heat transfer. Efforts to improve the U value by increasing h_i provide only modest benefits. If, for example, h_i were doubled so that $1/h_iA_i$ is cut in half, the decrease in the total resistance could at best be reduced by 10 percent. The resistance on the outside of the tube, $1/h_oA_o$, is said to be the *controlling resistance.*

When one of the fluids in a condenser or evaporator is a gas (hereafter considered to be air), the properties of the air compared with those of the liquid, such as water, result in heat-transfer coefficients of the order of one-tenth to one-twentieth that of the water. The air-side resistance in a configuration such as shown in Fig. 12-2 would provide the controlling resistance. In order to decrease $1/hA$, the area A is usually increased by using fins.

The bar fin, shown in Fig. 12-6 is a rudimentary fin whose performance can be predicted analytically and will be used to illustrate some important characteristics. The fins are of length L and thickness $2y$ m. The conductivity of the metal is k W/m · K, and the air-side coefficient is h_f W/m^2 · K. To solve for the temperature distribution through the fin, a heat balance can be written about an element of thickness dx m. The heat balance states that the rate of heat flow entering the element at position 1

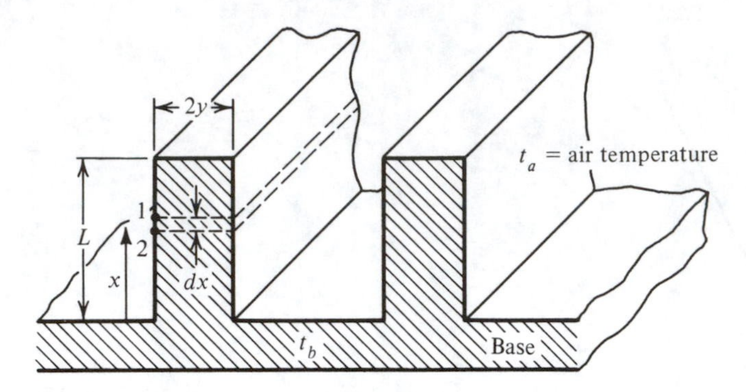

Figure 12-6 Bar fin.

from the end of the fin plus that transferred to the element from the air equals the rate of heat transferred out of the element at position 2 toward the base. For one-half a fin width and a fin depth of Z m, the heat balance in symbols is

$$kyZ\left(\frac{dt}{dx}\right)_1 + Z\,dx\,h_f(t_a - t) = kyZ\left(\frac{dt}{dx}\right)_2 \tag{12-14}$$

where t_a = temperature of air
t = temperature of fin

Canceling Z and factoring gives

$$ky\left[\left(\frac{dt}{dx}\right)_2 - \left(\frac{dt}{dx}\right)_1\right] = dx\,h_f(t_a - t) \tag{12-15}$$

For the differential length dx the change in the temperature gradient is

$$\left(\frac{dt}{dx}\right)_1 - \left(\frac{dt}{dx}\right)_2 = \frac{d}{dx}\left(\frac{dt}{dx}\right)dx = \frac{d^2t}{dx^2}\,dx \tag{12-16}$$

Substituting into Eq. 12-15, we get

$$\frac{d^2t}{dx^2} = \frac{h_f(t_a - t)}{ky} \tag{12-17}$$

By solving the second-order differential equation (12-17) the temperature distribution throughout the fin can be shown to be

$$\frac{t - t_b}{t_a - t_b} = \frac{\cosh M(L - x)}{\cosh ML} \tag{12-18}$$

where t_b = temperature of base of fin, °C

$$M = \sqrt{\frac{h_f}{ky}}$$

When a finned coil cools air, points in the fin farther away from the base are higher

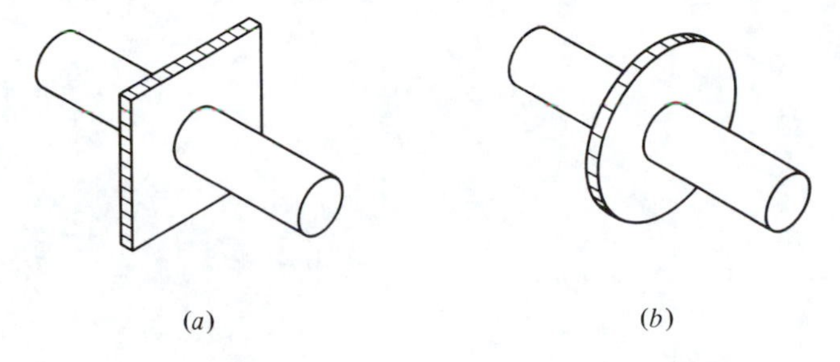

Figure 12-7 Determining fin effectiveness of a rectangular plate fin (*a*) by treating it as an (*b*) annular fin of the same area.

in temperature than points close to the base. The net result of the higher temperature of most of the fin is that less heat is transferred than if the entire fin were at temperature t_b. The ratio of the actual rate of heat transfer to that which would be transferred if the fin were at temperature t_b is called the *fin effectiveness*

$$\text{Fin effectiveness} = \eta = \frac{\text{actual } q}{q \text{ if fin were at base temperature}} \tag{12-19}$$

Harper and Brown[4] found that the fin effectiveness for the bar fin at Fig. 12-6 can be represented by

$$\eta = \frac{\tanh ML}{ML}$$

The bar fin is not a common shape but the dominant type of finned surface is the rectangular plate fin mounted on cylindrical tubes. The net result is a rectangular or square fin mounted on a circular base, one section of which is shown in Fig. 12-7*a*. The fin effectiveness of the rectangular plate fin is often calculated by using properties of the corresponding annular fin (Fig. 12-7*b*), for which a graph of the fin effectiveness is available, as in Fig. 12-8. The corresponding annular fin has the same area and thickness as the plate fin it represents.

Example 12-2 What is the fin effectiveness of a rectangular plate fin made of aluminum 0.3 mm thick mounted on a 16-mm-OD tube if the vertical tube spacing is 50 mm and the horizontal spacing is 40 mm? The air-side heat-transfer coefficient is 65 W/m^2 · K, and the conductivity of aluminum is 202 W/m · K.

Solution The annular fin having the same area as the plate fin (Fig. 12-9) has an external radius of 25.2 mm. The half-thickness of the fin $y = 0.15$ mm

$$M = \sqrt{\frac{65}{202(0.00015)}} = 46.3 \text{ m}^{-1}$$

$$(r_e - r_i)M = (0.0252 - 0.008)(46.3) = 0.8$$

From Fig. 12-8 for $(r_e - r_i)M = 0.8$ and $r_e/r_i = 25.2/8 = 3.15$ the fin effectiveness η is 0.72.

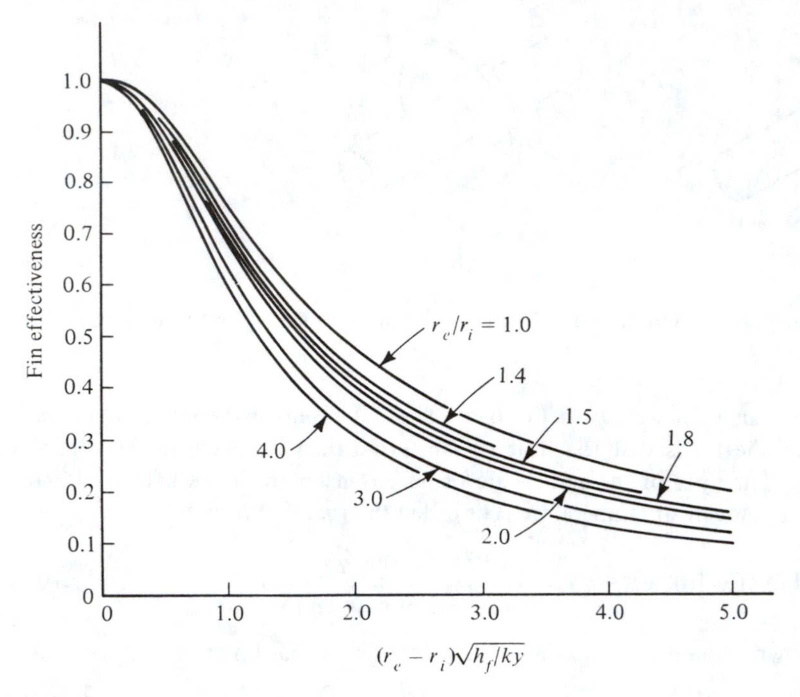

Figure 12-8 Fin effectiveness of an annular fin.[5] The external radius of the fin is r_e m and the internal radius is r_im.

The air-side area of a finned condenser or evaporator is composed of two portions, the prime area and the extended area. The *prime* area A_p is that of the tube between the fins, and the *extended* area A_e is that of the fin. Since the prime area is at the base temperature, it has a fin effectiveness of 1.0. It is to the extended surface that the fin effectiveness less than 1.0 applies. Equation (12-8) for the overall heat-transfer coefficient can be revised to read

$$\frac{1}{U_o A_o} = \frac{1}{U_i A_i} = \frac{1}{h_f(A_p + \eta A_e)} + \frac{x}{kA_m} + \frac{1}{h_i A_i} \tag{12-20}$$

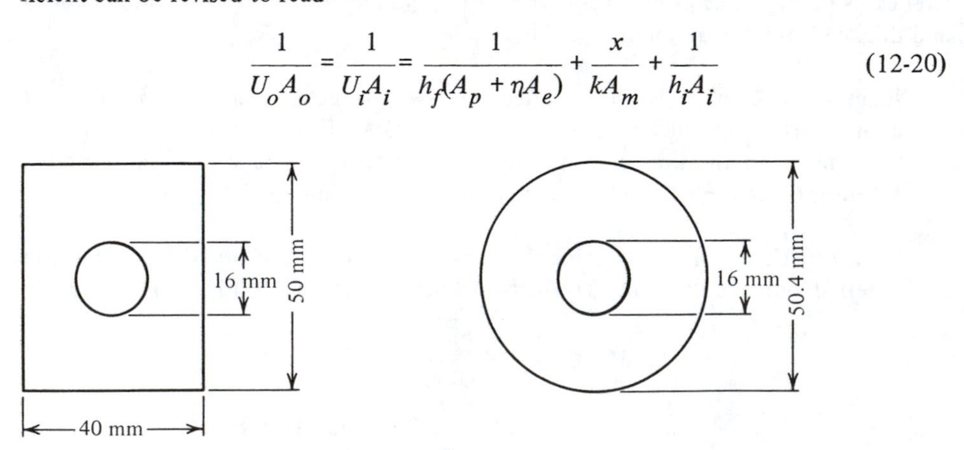

Figure 12-9 Annular fin of same area as rectangular plate fin.

12-6 Gas flowing over finned tubes; heat transfer and pressure drop A precise prediction of the air-side heat-transfer coefficient when the air flows over finned tubes is complicated because the value is a function of geometric factors, e.g., the fin spacing, the spacing and diameter of the tubes, and the number of rows of tubes deep. Usually the coefficient varies approximately as the square root of the face velocity of the air. A rough estimate of the air-side coefficient h_f can be computed from the equation derived from illustrative data in the ARI standard[6]

$$h_f = 38V^{0.5} \tag{12-21}$$

where V is the face velocity in meters per second.

Rich[7] conducted tests of coils of various fin spacings and correlated the dimensionless heat-transfer numbers with specially defined Reynolds numbers.

The drop in pressure of the air flowing through a finned coil is also dependent upon the geometry of the coil. Figure 12-10 shows the pressure drop of a commercial cooling coil when the finned surfaces are dry. As expected, the pressure drop is higher

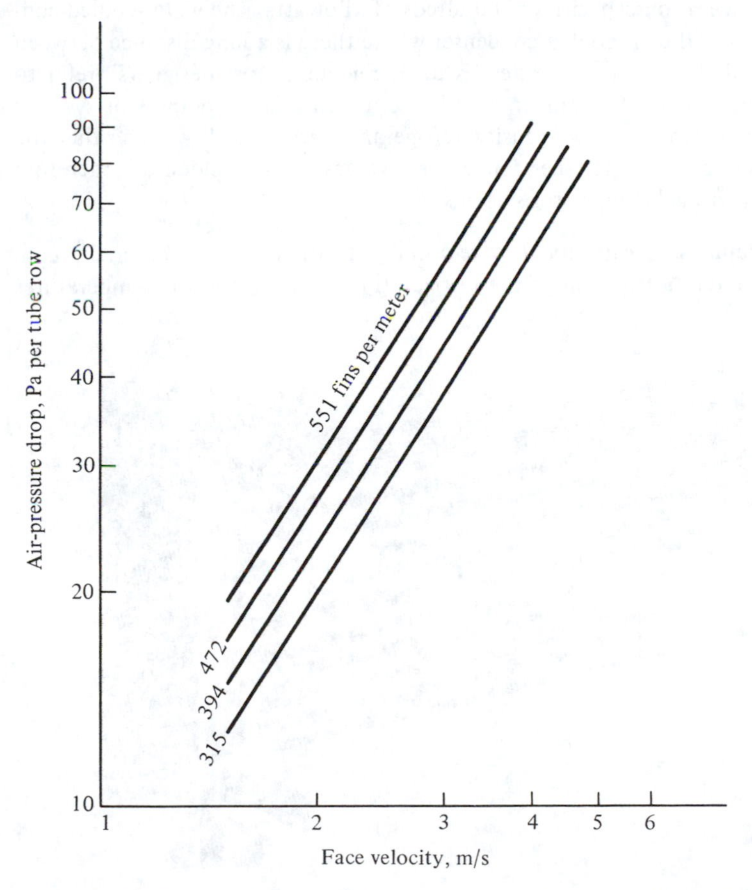

Figure 12-10 Pressure drop of air flowing through a finned coil *(Bohn Heat Transfer Division of Gulf & Western Manufacturing Company.)*

for coils with a large number of fins per meter of tube length. The ordinate is the pressure drop per number of rows of tubes deep, so the values would be multiplied by 6 for a six-row coil, for example.

For the coil series whose pressure drops are shown in Fig. 12-10 the pressure drop for a given coil varies as the face velocity to the 1.56 power. That exponent is fairly typical of commercial plate-fin coils.

12-7 Condensers The previous sections presented tools for computing heat-transfer coefficients and pressure drops of the fluid exchanging heat with the refrigerant in a condenser or evaporator. For the condenser the fluid to which heat is rejected is usually either air or water. Air-cooled condensers are shown in Fig. 12-2 and a shell-and-tube condenser in Fig. 12-1. Another type of water-cooled condenser has cleanable tubes (Fig. 12-11). When the condenser is water-cooled, the water is sent to a cooling tower (Chap. 19) for ultimate rejection of the heat to the atmosphere. Some years ago air-cooled condensers were used only in small refrigeration systems (less than 100 kW refrigerating capacity), but now individual air-cooled condensers are manufactured in sizes matching refrigeration capacities of hundreds of kilowatts. The water-cooled condenser is favored over the air-cooled condenser where there is a long distance between the compressor and the point where heat is to be rejected. Most designers prefer to convey water rather than refrigerant in long lines. In centrifugal-compressor systems large pipes are needed for the low-density refrigerants (see Sec. 11-25), so that the compressor is close-coupled to the condenser. Water-cooled condensers therefore predominate in centrifugal-compressor systems.

12-8 Required condensing capacity The required rate of heat transfer in the condenser is predominately a function of the refrigerating capacity and the temperatures

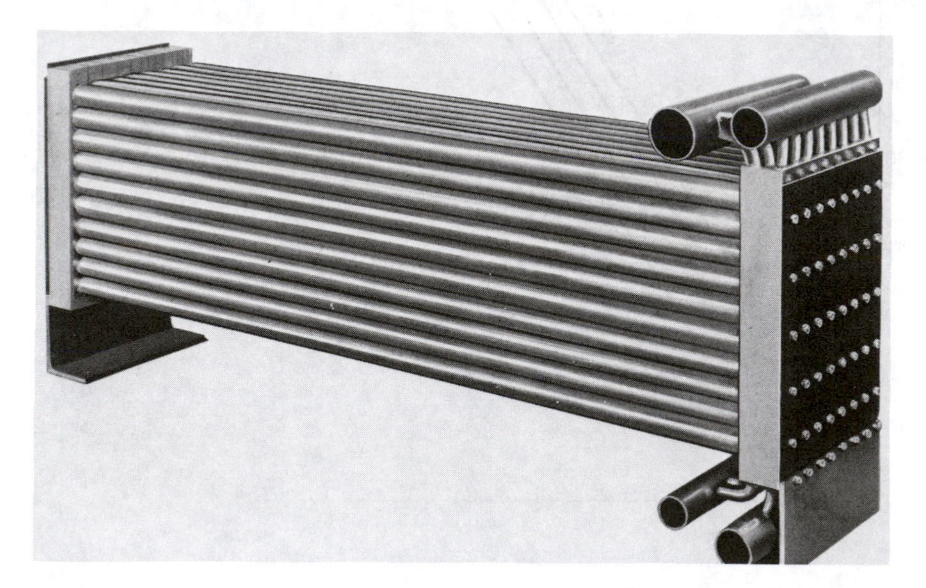

Figure 12-11 Water-cooled condenser with cleanable tubes. *(Halstead and Mitchell, a Division of Halstead Industries, Inc.)*

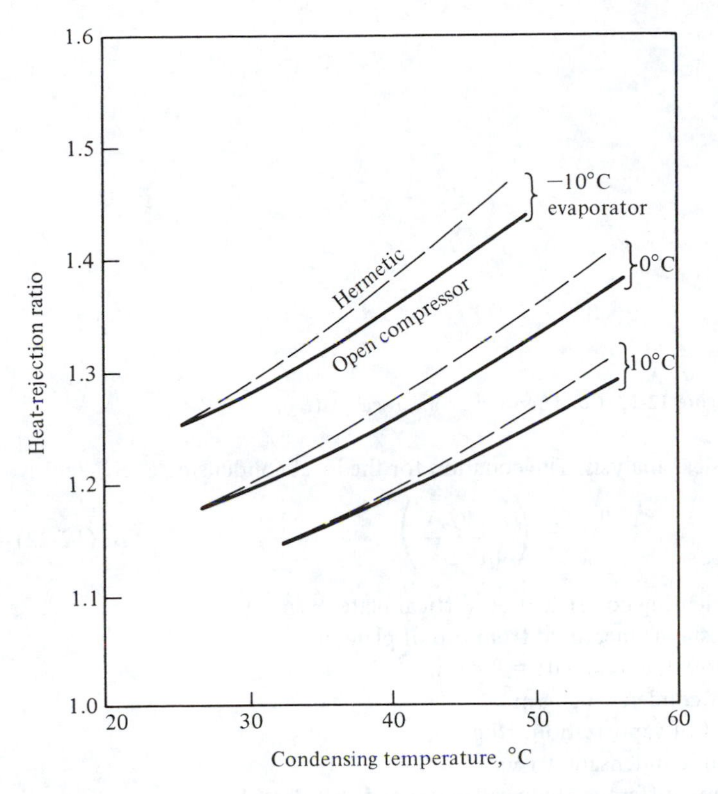

Figure 12-12 Typical values of the ratio of the heat rejected at the condenser to the refrigerating capacity for refrigerants 12 and 22.

of evaporation and condensation. The condenser must reject both the energy absorbed by the evaporator and the heat of compression added by the compressor. A term often used to relate the rate of heat flow at the condenser to that of the evaporator is the *heat-rejection ratio*

$$\text{Heat-rejection ratio} = \frac{\text{rate of heat rejected at condenser, kW}}{\text{rate of heat absorbed at evaporator, kW}}$$

Theoretical calculations of the condenser heat rejection can be made from the standard vapor-compression cycle (Sec. 10-14), but they do not take into consideration the additional heat added by inefficiencies in the compressor. A graph of typical values of heat-rejection ratios is shown in Fig. 12-12. When the motor driving the compressor is hermetically sealed, some of the heat associated with inefficiencies of the electric motor is added to the refrigerant stream and must ultimately be removed at the condenser. The heat-rejection ratios of the hermetically sealed compressors are usually slightly higher than those of the open-type compressor.

12-9 Condensing coefficient The basic equation for calculating the local coefficient of heat transfer of vapor condensing on a vertical plate (Fig. 12-13) was developed by

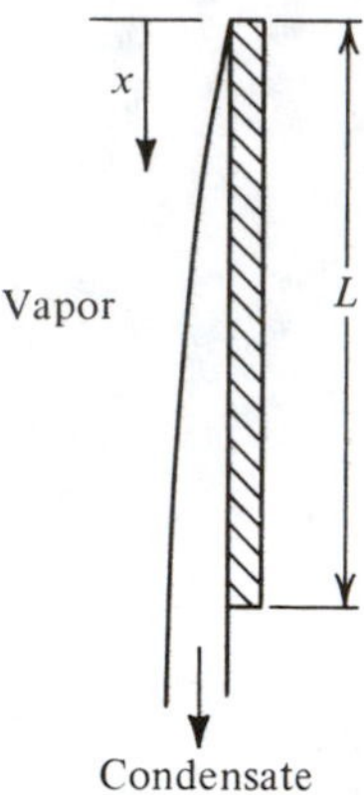

Figure 12-13 Condensation on a vertical plate.

Nusselt[8] by pure physical analysis. The equation for the local condensing coefficient is

$$\frac{h_{cv}x}{k} = \left(\frac{g\rho^2 h_{fg} x^3}{4\mu k\, \Delta t}\right)^{1/4} \tag{12-22}$$

where h_{cv} = local condensing coefficient on vertical plate, $W/m^2 \cdot K$
x = vertical distance measured from top of plate, m
g = acceleration due to gravity = 9.81 m/s^2
ρ = density of condensate, kg/m^3
h_{fg} = latent heat of vaporization, J/kg
μ = viscosity of condensate, Pa $\cdot$ s
Δt = temperature difference between vapor and the plate, K

The mean condensing coefficient over the total height of the plate L is

$$\overline{h_{cv}} = \frac{\int_0^L h_{cv}\, dx}{L} = 0.943 \left(\frac{g\rho^2 h_{fg} k^3}{\mu\, \Delta t\, L}\right)^{1/4} \quad W/m^2 \cdot K \tag{12-23}$$

The equation for the mean condensing coefficient for vapor condensing on the outside of horizontal tubes is

$$h_{ct} = 0.725 \left(\frac{g\rho^2 h_{fg} k^3}{\mu\, \Delta t\, ND}\right)^{1/4} \quad W/m^2 \cdot K \tag{12-24}$$

where N = number of tubes in vertical row
D = OD of tube, m

Several investigators have found that the constant 0.725 in Eq. (12-24) agrees closely with experimental results. White[9] found the constant to be 0.63, and Goto et al.[10] found it to be approximately 0.65.

Equations (12-22) and (12-24) are beautiful examples of how equations of motion and energy are combined. They are expressions of the coefficient of heat transfer across the film of liquid which is continuously condensing on the surface and continuously draining away. A qualitative examination of Eq. (12-24) shows that an increase in k increases h_{ct} because a greater rate of heat flow can be transferred across a given thickness of liquid film. A high value of density ρ or a low value of viscosity μ results in more rapid draining of the condensed liquid, which decreases the film thickness and increases h_{ct}. A high value of h_{fg} means that for each joule transferred a small mass of vapor condenses, which keeps the film thinner and increases h_{ct}. Finally, a high Δt indicates a high rate of condensation with a resultant thick film, which decreases h_{ct}.

Equations (12-23) and (12-24) apply to *film* condensation, one of the two types of condensation. The other type of condensation is *dropwise*. In film condensation the condensed liquid spreads over the entire condensing surface, whereas in dropwise condensation the condensed liquid gathers in globules, leaving some of the vapor in direct contact with the surface. Dropwise condensation provides a higher coefficient of heat transfer, but it can occur only on clean surfaces. To be safe, therefore, performance of condensers is predicted on the basis of film condensation.

Equations (12-23) and (12-24) apply to condensation on the outside of tubes, where there is little or no vapor velocity. When calculating the condensing coefficient on the inside of tubes[11] allowance must be made for the reduction in condensing area due to liquid collecting in the bottom of the tube before draining out the end. When subjected to a vapor velocity, the drag of the vapor on the condensate may accelerate or retard draining, depending upon the relative directions of flow of the condensate and vapor.

12-10 Fouling factor After a water-cooled condenser has been in service for some time its U value usually degrades somewhat because of the increased resistance to heat transfer on the water side due to fouling by the impurities in the water from the cooling tower. The new condenser must therefore have a higher U value in anticipation of the reduction that will occur in service. The higher capacity with new equipment is provided by specifying a *fouling factor* $1/h_{ff}$ m^2 · K/W. This term expands Eq. (12-8) for the U value into

$$\frac{1}{U_o} = \frac{1}{h_o} + \frac{xA_o}{kA_m} + \frac{A_o}{h_{ff}A_i} + \frac{A_o}{h_i A_i} \tag{12-25}$$

Several different agencies have established standards for the fouling factor to be used. One trade association[12] specifies 0.000176 m^2 · K/W, which means that the condenser should leave the factory with a $1/U_o$ value 0.000176 A_o/A_i less than the minimum required to meet the quoted capacity of the condenser.

12-11 Desuperheating Even when the refrigerant condenses at a constant pressure, its temperature is constant only in the condensing portion. Because the vapor coming from the compressor is usually superheated, the distribution of temperature will be as shown in Fig. 12-14. Because of the distortion in the temperature profile caused by the desuperheating process, the temperature difference between the refrigerant and the cooling fluid is no longer correctly represented by the LMTD

$$\text{LMTD} = \frac{(t_c - t_i) - (t_c - t_o)}{\ln\,[(t_c - t_i)/(t_c - t_o)]} \tag{12-26}$$

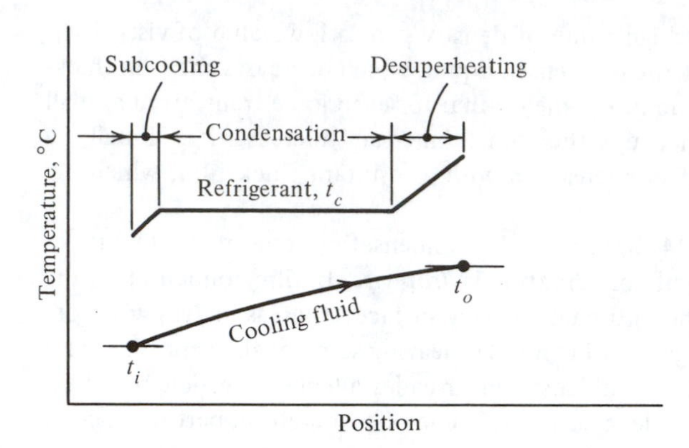

Figure 12-14 Temperature distributions in a condenser.

It is common practice to use Eq. (12-26) anyway with the following justification. Although the temperature difference between the refrigerant and cooling fluid is higher in the desuperheating section than calculated from Eq. (12-26), the convection coefficient in this section is normally lower than the condensing coefficient. The two errors compensate somewhat for each other, and the application of Eq. (12-26) along with the condensing coefficient over the entire condenser area usually provides reasonably accurate results.

12-12 Condenser design An example will illustrate how some of the principles described in the previous sections are combined in designing a condenser.

Example 12-3 The condensing area is to be specified for a refrigerant 22 condenser of a refrigerating system that provides a capacity of 80 kW for air conditioning. The evaporating temperature is 5°C, and the condensing temperature is 45°C at design conditions. Water from a cooling tower enters the condenser at 30°C and leaves at 35°C.

A two-pass condenser with 42 tubes, arranged as shown in Fig. 12-15, will be used, and the length of tubes is to be specified to provide the necessary area. The tubes are copper and are 14 mm ID and 16 mm OD.

Solution The steps in the solution of this design are as follows: calculate the re-

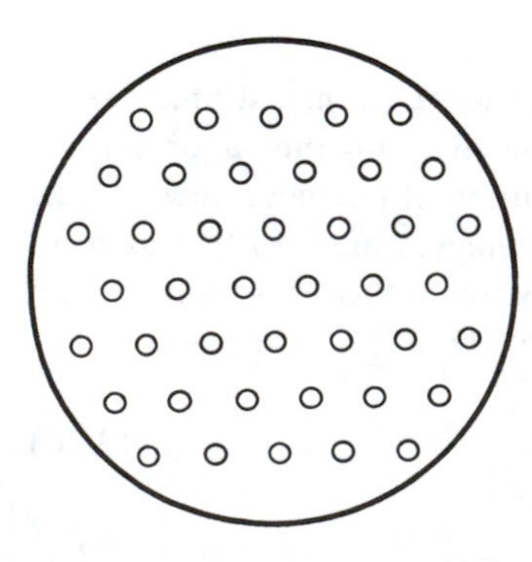

Figure 12-15 Tube arrangement of condenser in Example 12-3.

quired rate of heat transfer, the individual heat-transfer coefficients, and the overall heat-transfer coefficient; then compute the required area and tube length.

Rate of heat transfer From Fig. 12-12, assuming the compressor to be a hermetic one, the heat-rejection ratio at a condensing temperature of 45°C and an evaporating temperature of 5°C is 1.27. The rate of heat rejected at the condenser q is

$$q = (80\ \text{kW})\ (1.27) = 101.6\ \text{kW}$$

Condensing coefficient From Eq. (12-24)

$$h_{cond} = 0.725 \left(\frac{g\rho^2 h_{fg} k^3}{\mu\, \Delta t\, ND}\right)^{1/4}$$

The density ρ and latent heat of vaporization h_{fg} at 45°C are available from Table A-6

$$\rho = \frac{1}{0.90203\ \text{L/kg}} = 1.109\ \text{kg/L} = 1109\ \text{kg/m}^3$$

$$h_{fg} = 160{,}900\ \text{J/kg}$$

The conductivity k and viscosity μ of the liquid refrigerant at 45°C are available from Table 15-5

$$k = 0.0779\ \text{W/m} \cdot \text{K} \qquad \mu = 0.000180\ \text{Pa} \cdot \text{s}$$

The average number of tubes in a vertical row N is

$$N = \frac{2+3+4+3+4+3+4+3+4+3+4+3+2}{13} = 3.23$$

The temperature difference between the vapor and the tube is unknown at this point; therefore Δt will be assumed to be 5 K and the value adjusted later if necessary.

$$h_{cond} = 0.725 \left[\frac{9.81(1109^2)\ (160{,}900)\ (0.0779^3)}{0.000180(5)\ (3.23)\ (0.016)}\right]^{1/4}$$

$$= 1528\ \text{W/m}^2 \cdot \text{K}$$

Resistance of metal The conductivity of copper is 390 W/m · K, and the resistance of the tube is

$$\frac{xA_o}{kA_m} = \frac{(0.016 - 0.014)/2}{390}\ \frac{16}{(14+16)/2} = 0.000002735\ \text{m}^2 \cdot \text{K/W}$$

a value that will prove to be negligible in comparison to the other resistances.

Fouling factor From Sec. 12-10

$$\frac{1}{h_{ff}} = 0.000176\ \text{m}^2 \cdot \text{K/W}$$

Water-side coefficient The flow rate of water needed to carry the heat away from the condenser with a temperature rise from 30 to 35°C is

$$\frac{101.6 \text{ kW}}{(4.19 \text{ kJ/kg} \cdot \text{K})(35.0 - 30.0)} = 4.85 \text{ kg/s}$$

and the volume flow rate is

$$\frac{4.85 \text{ kg/s}}{1000 \text{ kg/m}^3} = 0.00485 \text{ m}^3\text{/s}$$

The water velocity through the tubes V is

$$V = \frac{0.00485 \text{ m}^3\text{/s}}{(21 \text{ tubes per pass})(\pi/4)(0.014^2)} = 1.5 \text{ m/s}$$

Equation (12-9) can be used to calculate the water-side heat-transfer coefficient h_w using the water properties at 32°C

$$\rho = 995 \text{ kg/m}^3 \qquad \mu = 0.000773 \text{ Pa} \cdot \text{s}$$
$$c_p = 4190 \text{ J/kg} \cdot \text{K} \quad k = 0.617 \text{ W/m} \cdot \text{K}$$

$$h_w = \frac{0.617(0.023)}{0.014}\left[\frac{1.5(0.014)(995)}{0.000773}\right]^{0.8}\left[\frac{4190(0.00773)}{0.617}\right]^{0.4}$$

$$h_w = 1.014(27030^{0.8})(5.25^{0.4}) = 6910 \text{ W/m}^2 \cdot \text{K}$$

Required area

$$\frac{1}{U_o} = \frac{1}{1528} + 0.000002735 + \frac{0.016}{0.014}(0.000176) + \frac{0.016}{0.014}\frac{1}{6910} = 0.001023$$

$$U_o = 977 \text{ W/m}^2 \cdot \text{K}$$

The LMTD is

$$\text{LMTD} = \frac{(45 - 30) - (45 - 35)}{\ln \dfrac{(45 - 30)}{(45 - 35)}} = 12.33\text{°C}$$

$$A_o = \frac{101{,}600 \text{ W}}{977(12.33)} = 8.43 \text{ m}^2$$

Length of tubes

$$\text{Length} = \frac{8.43 \text{ m}^2}{(42 \text{ tubes})(0.016\pi)} = 4.0 \text{ m}$$

A recheck should now be made of the assumption of the 5-K temperature difference used in calculating the condensing coefficient. A recalculation may be necessary.

The designer would also check the water-pressure drop in the condenser to see that it does not exceed a reasonable value (perhaps of the order of 70 kPa.)

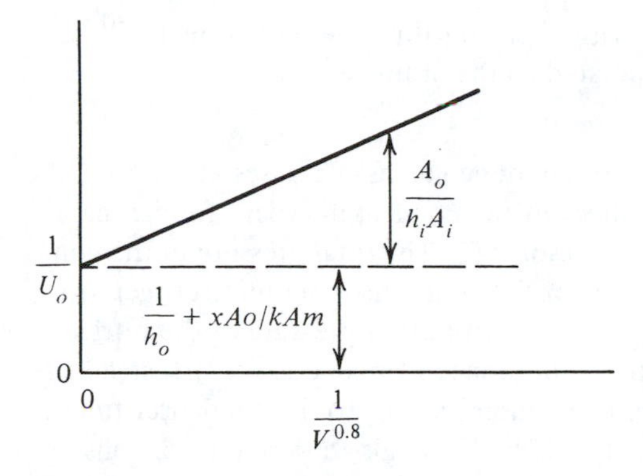

Figure 12-16 A Wilson plot to determine the individual heat-transfer coefficients of a condenser.

12-13 Wilson plots Constructing a Wilson plot is a technique of processing heat-transfer data to determine the individual heat-transfer coefficients in a heat exchanger. The concept was introduced by Wilson[13] and is often applied to condensers and evaporators to determine the condensing or evaporating heat-transfer coefficient along with the air- or water-side coefficient.

If it is a water-cooled condenser that is being analyzed, for example, a series of heat-transfer tests is run and the U value determined for various flow rates of cooling water. If the condenser tubes are clean, Eq. (12-8) applies and h_o is the condensing-side coefficient and h_i the water-side coefficient.

$$\frac{1}{U_o} = \frac{1}{h_o} + \frac{xA_o}{kA_m} + \frac{A_o}{h_i A_i} \tag{12-27}$$

The properties of the cooling water are primarily a function of temperature, and if the temperature range throughout the tests is not large, the properties may be assumed constant. Equation (12-9) can then be simplified to

$$h_i = (\text{const})\,(V^{0.8}) \tag{12-28}$$

The Wilson plot for this heat exchanger is a graph of $1/U_o$ versus $1/V^{0.8}$, as shown in Fig. 12-16. The intercept on the ordinate is $1/h_o + xA_o/kA_m$. The resistance of the tube, xA_o/kA_m, can be calculated and then subtracted from the intercept to yield the reciprocal of the condensing coefficient, $1/h_o$. At any velocity the value of A_o/h_iA_i is also available from which the value of h_i can be extracted. The significance of the intercept is that when $1/V^{0.8}$ is zero, V is infinite. An infinite water velocity would wipe out the water-side heat-transfer resistance completely, leaving only the other two resistances.

Drawing a straight line on Fig. 12-16 is possible only when the value of h_o remains constant throughout the tests. The loading, or rate of heat removal, of the condenser has some effect upon h_o since it controls the temperature difference across the condensing film.

The Wilson plot is applicable to other types of heat exchangers in addition to the shell-and-tube condenser. If a Wilson plot were applied to an air-cooled condenser where air flows over the outside of finned coils and refrigerant condenses inside the tubes, the tests would be run at various air velocities. Equation (12-21) indicates that the heat-transfer coefficient for air flowing over finned surfaces varies as the square root of the velocity. The abscissa of the Wilson plot in this case would be $1/V^{0.5}$ in order to achieve a straight line to be extrapolated to the ordinate.

12-14 Air and noncondensables If air or other noncondensable gases enter the refrigeration system, they will ultimately collect in the condenser, where foreign gases reduce the efficiency of the system for two reasons: (1) The total pressure in the condenser is elevated, which requires more power for the compressor per unit refrigeration capacity. The condenser pressure is raised over the saturation pressure of the refrigerant by the amount of the partial pressure of the noncondensable gas. (2) Instead of diffusing throughout the condenser, the noncondensables cling to the condenser tubes. Thus the condensing-surface area is reduced, which also tends to raise the condensing pressure.

Noncondensables can be removed from the condenser by purging. The purging operation consists of drawing a mixture of refrigerant vapor and noncondensables from the condenser, separating the refrigerant, and discharging the noncondensables. Ammonia systems are commonly equipped with purgers, but the only other systems equipped with purgers are the centrifugal-compressor systems, which use low-pressure refrigerants such as refrigerants 11 and 113.

12-15 Evaporators In most refrigerating evaporators the refrigerant boils in the tubes and cools the fluid that passes over the outside of the tubes. Evaporators that boil refrigerant in the tubes are often called *direct-expansion evaporators*, and Fig. 12-17 shows an air-cooling evaporator and Fig. 12-18 a liquid cooler. The tubes in the liquid

Figure 12-17 Air-cooling evaporator. The device on the left end is a refrigerant distributor to feed the several circuits uniformly. *(McQuay Group, McQuay-Perfex Inc.)*

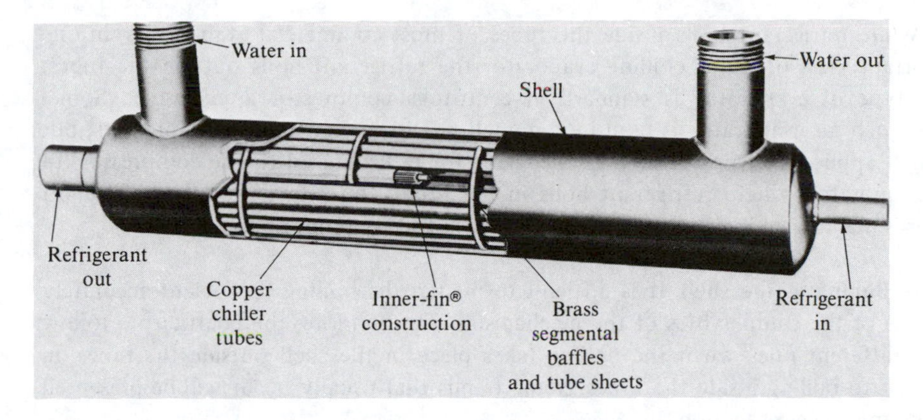

Figure 12-18 A liquid chilling evaporator in which refrigerant boils inside finned tubes. *(Dunham-Bush, Inc.)*

chiller in Fig. 12-18 have fins inside the tubes in order to increase the conductance on the refrigerant side.

Direct-expansion evaporators used for air-conditioning applications are usually fed by an expansion valve that regulates the flow of liquid so that the refrigerant vapor leaves the evaporator with some superheat, as shown in Fig. 12-19*a*. Another concept is the liquid-recirculation or liquid-overfeed evaporator in Fig. 12-19*b*, in which excess liquid at low pressure and temperature is pumped to the evaporator (see Chap. 16). Some liquid boils in the evaporator, and the remainder floods out of the outlet. The liquid from the evaporator is separated out, and the vapor flows on to the compressor. Low-temperature industrial refrigeration systems often use this type of evaporator, which has the advantage of wetting all the interior surfaces of the evaporator and maintaining a high coefficient of heat transfer.

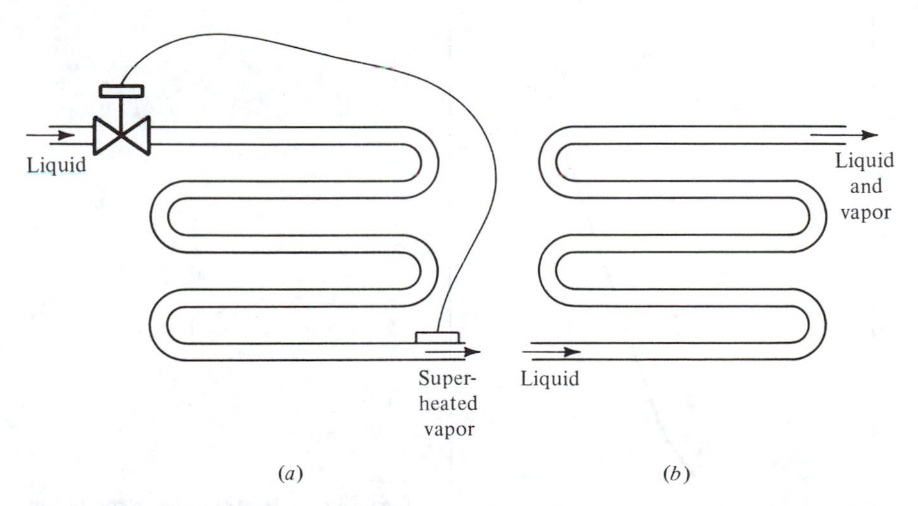

Figure 12-19 (*a*) Air-conditioning evaporator with refrigerant leaving in a superheated state, (*b*) liquid-recirculation evaporator with liquid refrigerant carried out of the evaporator.

While refrigerant boils inside the tubes of most commercial evaporators, in one important class of liquid-chilling evaporator the refrigerant boils outside the tubes. This type of evaporator is standard in centrifugal-compressor applications. Sometimes such an evaporator is used in conjunction with reciprocating compressors, but in such applications provision must be made for returning oil to the compressor. In the evaporators where refrigerant boils in the tubes, the velocity of the refrigerant vapor is maintained high enough to carry oil back to the compressor.

12-16 Boiling in the shell It is difficult to predict the boiling coefficient accurately because of the complexities of the mechanisms. Furthermore, the coefficients follow some different rules when the boiling takes place in the shell outside the tubes, in contrast to boiling inside the tubes. Some trends that usually occur will be presented in this and the next section.

The classic prediction for the heat-transfer coefficient for pool boiling of water at atmospheric pressure is shown in Fig. 12-20. The tests were conducted by immersing a heated wire in a container of water. In the boiling regime AB the boiling is called *nucleate boiling*, where bubbles form on the surface and rise through the pool. The equation of the curve is approximately

$$\frac{q}{A} = C\,\Delta t^{3 \text{ to } 4}$$

where q = rate of heat transfer, W
A = heat-transfer area, m^2
C = constant
Δt = difference in temperature between metal surface and boiling fluid, K

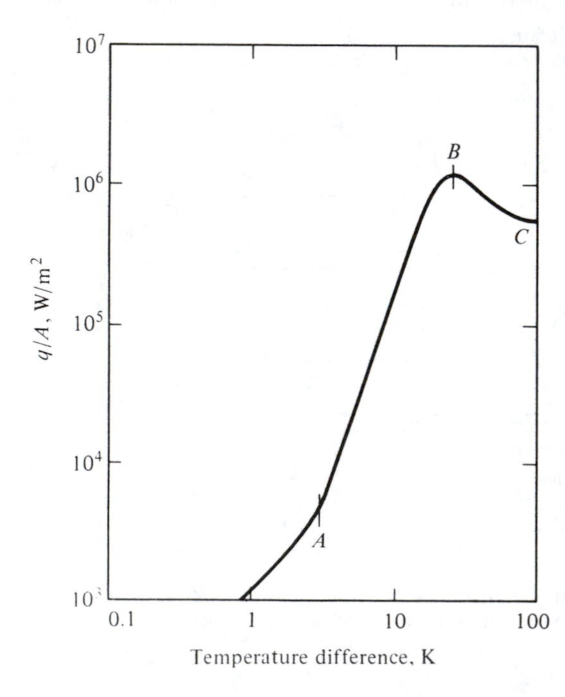

Figure 12-20 Heat-transfer coefficient for pool boiling of water. *(From W. H. McAdams, "Heat Transmission," 3d ed., p. 370, McGraw-Hill, New York, 1954.)*

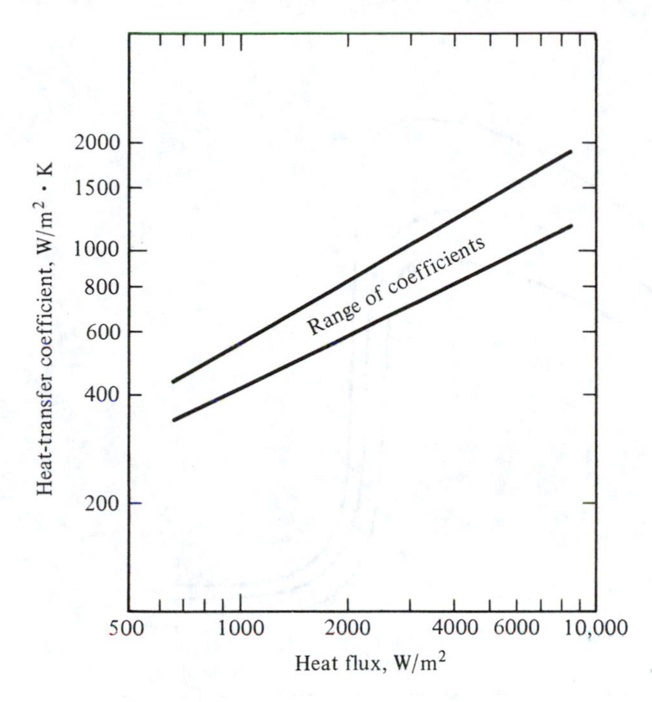

Figure 12-21 Heat-transfer coefficients for refrigerants 12 and 22 boiling outside of tube bundles.

To write the equation in another form divide both sides by Δt,

$$\frac{q}{A\ \Delta t} = h_r = C\ \Delta t^{2\text{ to }3}$$

where h_r is the boiling coefficient, W/(m^2 · K). The value of h_r increases as the temperature difference increases, which physically is due to the greater agitation. The disturbance frees the bubbles of vapor from the metal surface sooner and allows the liquid to come into contact with the metal.

The rate of evaporation can increase to a peak, point B, where so much vapor covers the metal surface that the liquid can no longer intimately contact the metal. A further increase in the temperature difference decreases the rate of heat transfer.

The graph in Fig. 12-20 is useful in predicting the trends for heat-transfer coefficients for boiling outside tube bundles. Hoffmann[14] summarized the work of several investigators to provide the band shown in Fig. 12-21.

12-17 Boiling inside tubes When refrigerant boils inside the tubes, the heat-transfer coefficient changes progressively as the refrigerant flows through the tube. The refrigerant enters the evaporator tube with a low fraction of vapor. As the refrigerant proceeds through the tube, the fraction of vapor increases, intensifying the agitation and increasing the heat-transfer coefficient. When the refrigerant is nearly all vaporized, the coefficient drops off to the magnitude applicable to vapor transferring heat by forced convection. Figure 12-22 shows local coefficients throughout a tube for three different levels of temperature. The heat-transfer coefficient is highest for the high evaporating temperature, probably because at high evaporating temperatures and

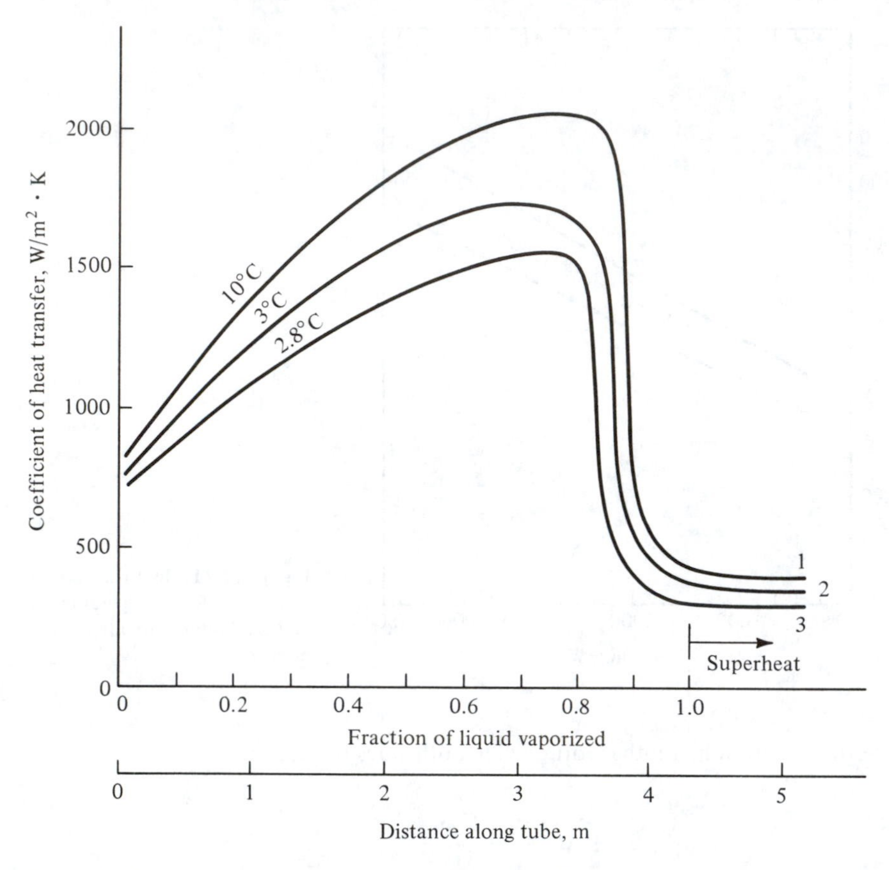

Figure 12-22 Heat-transfer coefficients of refrigerant 22 boiling inside tubes. Curve 1 at 10°C, curve 2 at 3°C, and curve 3 at 2.8°C temperatures of evaporation.[15]

pressures the vapor density is high, permitting a greater fraction of the metal to be wetted with liquid.

12-18 Evaporator performance From the discussion of boiling heat-transfer coefficients in Secs. 12-16 and 12-17 the coefficient will be expected to increase with an increase in loading. This assumption is borne out by the performance of commercial evaporators. We encounter the performance of evaporators again in Chap.14, and Fig. 14-8 shows the performance of a water-chilling evaporator where the refrigerant boils inside the tubes. For a given temperature of entering water the lines on the capacity-versus-evaporating-temperature graph would be straight if the U value remained constant. Instead, the lines are curved upward, indicating an increase in U value at more intense loadings due to the improved boiling heat-transfer coefficient.

12-19 Pressure drop in tubes The pressure of the refrigerant drops as it flows through tube-type evaporators. The effect of pressure drop on system performance is that the compressor must pump from a lower suction pressure, which increases the power requirement. On the other hand a high refrigerant velocity can be achieved if more pressure drop is permitted, and this high velocity improves the heat-transfer coefficient. Typical pressure drops for air-conditioning evaporators are 15 to 30 kPa.

12-20 Frost When the surface temperatures of an air-cooling evaporator fall below 0°C frost will form. Frost is detrimental to the operation of the refrigeration system for two reasons:[16] (1) thick layers of frost act as insulation, and (2) in forced-convection coils the frost reduces the airflow rate. With a reduced airflow rate the U value of the coil drops, and the mean temperature difference between the air and refrigerant must increase in order to transfer the same rate of heat flow. Both these factors penalize the system by requiring a lower evaporating temperature.

Numerous methods of defrosting are available, and probably the most popular ones are hot-gas defrost and water defrost. In hot-gas defrost, discharge gas from the compressor is sent directly to the evaporator and the evaporator performs temporarily as a condenser. The heat of condensation melts off the frost, which drains away. In water defrost, a stream of water is directed over the coil until all the frost is melted.

PROBLEMS

12-1 An air-cooled condenser is to reject 70 kW of heat from a condensing refrigerant to air. The condenser has an air-side area of 210 m^2 and a U value based on this area of 0.037 $kW/m^2 \cdot K$; it is supplied with 6.6 m^3/s of air, which has a density of 1.15 kg/m^3. If the condensing temperature is to be limited to 55°C, what is the maximum allowable temperature of inlet air? *Ans.* 40.6°C

12-2 An air-cooled condenser has an expected U value of 30 $W/m^2 \cdot K$ based on the air-side area. The condenser is to transfer 60 kW with an airflow rate of 15 kg/s entering at 35°C. If the condensing temperature is to be 48°C, what is the required air-side area? *Ans.* 184 m^2

12-3 A refrigerant 22 condenser has four water passes and a total of 60 copper tubes that are 14 mm ID and have 2 mm wall thickness. The conductivity of copper is 390 $W/m \cdot K$. The outside of the tubes is finned so that the ratio of outside to inside area is 1.7. The cooling-water flow through the condenser tubes is 3.8 L/s

(*a*) Calculate the water-side coefficient if the water is at an average temperature of 30°C, at which temperature $k = 0.614$ $W/m \cdot K$, $\rho = 996$ kg/m^3, and $\mu = 0.000803$ $Pa \cdot s$

(*b*) Using a mean condensing coefficient of 1420 $W/m^2 \cdot K$, calculate the overall heat-transfer coefficient based on the condensing area. *Ans.* 1067 $W/m^2 \cdot K$

12-4 A shell-and-tube condenser has a U value of 800 $W/m^2 \cdot K$ based on the water-side area and a water pressure drop of 50 kPa. Under this operating condition 40 percent of the heat-transfer resistance is on the water side. If the water-flow rate is doubled, what will the new U value and the new pressure drop be? *Ans.* 964 $W/m^2 \cdot K$, 200 kPa

12-5 (*a*) Compute the fin effectiveness of a bar fin made of aluminum that is 0.12 mm thick and 20 mm long when $h_f = 28$ $W/m^2 \cdot K$, the base temperature is 4°C, and the air temperature is 20°C. *Ans.* 0.775

(*b*) If you are permitted to use twice as much metal for the fin as originally specified in part (*a*) and you can either double the thickness or double the length, which choice would be preferable in order to transfer the highest rate of heat flow? Why?

12-6 Compute the fin effectiveness of an aluminum rectangular plate fin of a finned air-cooling evaporator if the fins are 0.18 mm thick and mounted on 16-mm-OD tubes. The tube spacing is 40 mm in the direction of airflow and 45 mm vertically. The air-side coefficient is 55 $W/m^2 \cdot K$. *Ans.* 0.68

12-7 What is the *UA* value of a direct-expansion finned coil evaporator having the following areas: refrigerant side, 15 m^2; air-side prime, 13.5 m^2; and air-side extended 144 m^2? The refrigerant-side heat-transfer coefficient is 1300 $W/m^2 \cdot K$, and the air-side coefficient is 48 $W/m^2 \cdot K$. The fin effectiveness is 0.64. *Ans.* 4027 W/K

12-8 A refrigerant 22 system having a refrigerating capacity of 55 kW operates with an evaporating temperature of 5°C and rejects heat to a water-cooled condenser. The compressor is hermetically sealed. The condenser has a U value of 450 $W/m^2 \cdot K$ and a heat-transfer area of 18 m^2 and receives a flow rate of cooling water of 3.2 kg/s at a temperature of 30°C. What is the condensing temperature? *Ans.* 41.2°C

12-9 Calculate the mean condensing heat-transfer coefficient when refrigerant 12 condenses on the outside of the horizontal tubes in a shell-and-tube condenser. The outside diameter of the tubes is 19 mm, and in the vertical rows of tubes there are, respectively, two, three, four, three, and two tubes. The refrigerant is condensing at a temperature of 52°C, and the temperature of the tubes is 44°C. *Ans.* 1066 $W/m^2 \cdot K$

12-10 A condenser manufacturer guarantees the U value under operating conditions to be 990 $W/m^2 \cdot K$ based on the water-side area. In order to allow for fouling of the tubes, what is the U value required when the condenser leaves the factory? *Ans.* 1200 $W/m^2 \cdot K$

12-11 In Example 12-3 the temperature difference between the refrigerant vapor and tube was initially assumed to be 5 K in order to compute the condensing coefficient. Check the validity of this assumption. *Ans.* Δt from 8.2 to 12.3°C

12-12 (*a*) A Wilson plot is to be constructed for a finned air-cooled condenser by varying the rate of airflow. What should the abscissa of the plot be?

(*b*) A Wilson plot is to be constructed for a shell-and-tube water chiller in which refrigerant evaporates in the tubes. The rate of water flow is to be varied for the Wilson plot. What should the abscissa of the plot be?

12-13 The following values were measured[17] on an ammonia condenser:

U_o, $W/m^2 \cdot K$	2300	2070	1930	1760	1570	1360	1130	865
V, m/s	1.22	0.975	0.853	0.731	0.610	0.488	0.366	0.244

Water flowed inside the tubes, and the tubes were 51 mm OD and 46 mm ID and had a conductivity of 60 $W/m \cdot K$. Using a Wilson plot, determine the condensing coefficient. *Ans.* 8600 $W/m^2 \cdot K$

12-14 Develop Eq. (12-23) from Eq. (12-22).

12-15 From Fig. 12-21, determine C and b in the equation $h = C\,\Delta t^b$ applicable to values in the middle of the typical range. *Ans.* $h = 222\Delta t^{1.06}$

12-16 Section 12-18 makes the statement that on a graph of the performance of a water-chilling evaporator with the coordinates of Fig. 12-23 a curve for a given enter-

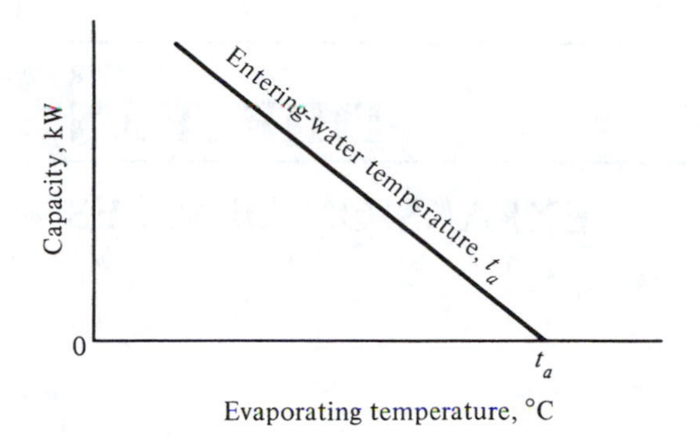

Figure 12-23 Evaporator performance curve.

ing water temperature is a straight line if the heat-transfer coefficients are constant. Prove this statement.

REFERENCES

1. W. H. McAdams: "Heat Transmission," 3d ed., McGraw-Hill, New York, 1954.
2. D. L. Katz, R. E. Hope, S. C. Datski, and D. B. Robinson: Condensation of F-12 with Finned Tubes, *Refrig. Eng.*, vol. 53, no. 4, p. 211, March 1947.
3. W. H. Emerson: Shell-Side Pressure Drop and Heat Transfer with Turbulent Flow in Segmentally Baffled Shell-and-Tube Heat Exchangers, *Int. J. Heat Mass Transfer*, vol. 6, no. 8, pp. 649–668, August 1963.
4. D. R. Harper III and W. P. Brown: Mathematical Equations for Heat Conduction in the Fins of Air Cooled Engines, *NACA Tech. Rep. 158*, p. 677, 1922.
5. K. A. Gardner: Efficiency of Extended Surfaces, *Trans. ASME*, vol. 67, pp. 621–631, 1945.
6. Standard for Forced-Circulation Air-Cooling and Air-Heating Coils, Standard 410, Air-Conditioning and Refrigeration Institute, Arlington, Va., 1972.
7. D. G. Rich: The Effect of Fin Spacing on the Heat Transfer and Friction Performance of Multi-Row, Smooth Plate Fin-and-Tube Heat Exchangers, *ASHRAE Trans.*, vol. 79, pt. 2, pp. 137–145, 1974.
8. W. Z. Nusselt: Die Oberflaechenkondensation des Wasserdampfes, *Ver. Dtsch. Ing.*, vol. 60, pp. 541–569, July 1916.
9. R. E. White: Condensation of Refrigerant Vapors: Apparatus and Film Coefficients for F-12, *Refrig. Eng.*, vol. 55, no. 5, p. 375, April 1948.
10. M. Goto, H. Hotta, and S. Tezuka: Film Condensation of Refrigerant Vapors on a Horizontal Tube, *15th Int. Congr. Refrig., Venice, 1979, Pap.* B1-20.
11. J. C. Chato: Laminar Condensation Inside Horizontal and Inclined Tubes, *ASHRAE J.*, vol. 4, no. 2, p. 52, 1962.
12. "Standards of the Tubular Exchanger Manufacturers Association," 5th ed., New York, 1970.
13. E. E. Wilson: A Basis for Rational Design of Heat-Transfer Apparatus, *Trans. ASME*, vol. 37, p. 47, 1915.
14. E. Hoffmann: Waermeuebergangzahlen verdampfender Kaeltemittel, *Kaeltetechnik*, vol. 9, no. 1, pp. 7-12, January 1957.
15. S. G. Kuvshinov, I. F. Yatsunov, and N. I. Frolova: Performance of Shell-and-Tube Evaporators with Refrigerant Boiling inside Tubes, *Kholod. Tekh.*, vol. 50, no. 9, pp. 39–43, 1973.
16. W. F. Stoecker: How Frost Formation on Coils Affects Refrigeration Systems, *Refrig. Eng.*, vol. 65, no. 2, p. 42, February 1957.
17. A. P. Kratz, H. J. Macintire, and R. E. Gould: Heat Transfer in Ammonia Condensers, pt. III, *Univ. Ill. Eng. Exp. Stn. Bull.* 209, June 17, 1930.

CHAPTER

THIRTEEN

EXPANSION DEVICES

13-1 Purpose and types of expansion devices The last of the basic elements in the vapor-compression cycle, after the compressor, condenser, and evaporator, is the expansion device. The purpose of the expansion device is twofold: it must reduce the pressure of the liquid refrigerant, and it must regulate the flow of refrigerant to the evaporator.

This chapter explains the operation of the common types of expansion devices, the *capillary tube,* the *superheat-controlled expansion valve,* the *float valve,* and the *constant-pressure expansion valve.* Operation of a refrigeration system using these devices will be discussed, with special emphasis on balanced and unbalanced flow conditions occurring between the expansion device and the compressor. The two most commonly used expansion devices, the capillary tube and the superheat-controlled expansion valve, are singled out for a more thorough study of their operating characteristics.

13-2 Capillary tubes The capillary tube serves almost all small refrigeration systems, and its application extends up to refrigerating capacities of the order of 10 kW. A capillary tube is 1 to 6 m long with an inside diameter generally from 0.5 to 2 mm. The name is a misnomer, since the bore is too large to permit capillary action. Liquid refrigerant enters the capillary tube, and as it flows through the tube, the pressure drops because of friction and acceleration of the refrigerant. Some of the liquid flashes into vapor as the refrigerant flows through the tube.

Numerous combinations of bore and length are available to obtain the desired restriction. Once the capillary tube has been selected and installed, however, the tube cannot adjust to variations in discharge pressure, suction pressure, or load. The

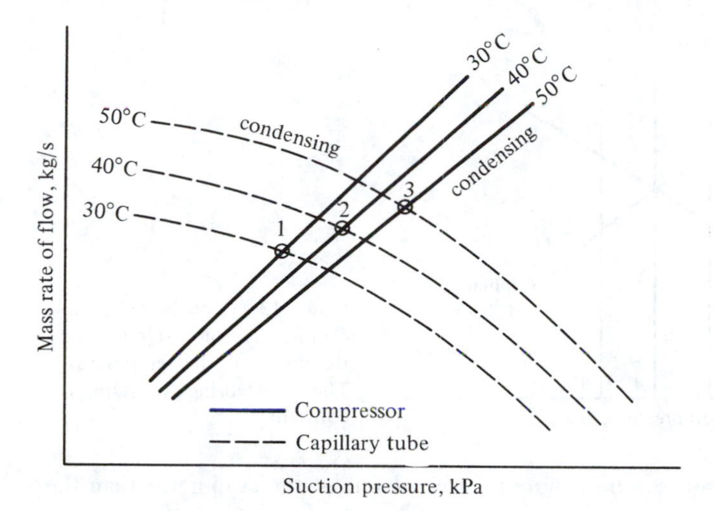

Figure 13-1 Balance points with a reciprocating compressor and capillary tube.

compressor and expansion device must arrive at suction and discharge conditions which allow the compressor to pump from the evaporator the same flow rate of refrigerant that the expansion device feeds to the evaporator. A condition of unbalanced flow between these two components must necessarily be temporary.

For a closer look at balance points the mass rate of flow fed by the capillary tube can be plotted on the same graph as the mass rate of flow pumped by the compressor. Figure 13-1 is such a plot with the flow through the capillary tube shown in dashed lines and the pumping capacity of a reciprocating compressor shown in solid lines. At high condensing pressures the capillary tube feeds more refrigerant to the evaporator than it does at low condensing pressures because of the increase in pressure difference across the tube. The compressor-capacity curves are the same as those explained in Chap. 11 in the study of compressors. At a 30°C condensing temperature, for example, the compressor and capillary tube must search for a suction pressure which allows them both to pass equal mass rates of flow. This suction pressure is found at point 1, which is the balance point at a 30°C condensing temperature. Points 2 and 3 are the balance points at 40°C and 50°C condensing temperatures, respectively.

The compressor and capillary tube do not have complete liberty to fix the suction pressure because the heat-transfer relationships of the evaporator must also be satisfied. If the evaporator heat transfer is not satisfied at the compressor–capillary-tube balance point, an unbalanced condition results which can starve the evaporator or overfeed the evaporator.

Starving the evaporator results when the suction pressure rises and the capillary does not feed sufficient refrigerant to refrigerate the evaporator surfaces adequately. Figure 13-2 shows a balance point for a constant condensing pressure between the compressor and the capillary tube at suction pressure *A*. A heavy heat load, when received at the evaporator, manifests itself as a high temperature of the fluid to be chilled. The suction temperature and pressure will rise to some point *B*. At suction

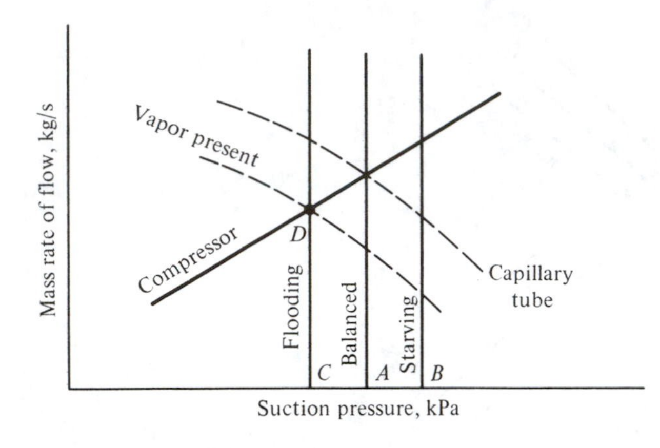

Figure 13-2 Unbalanced conditions causing starving or flooding of the evaporator. The condensing pressure is constant.

pressure B the compressor can draw more refrigerant out of the evaporator than the capillary tube can supply, so the evaporator soon becomes short of refrigerant. Since the evaporator cannot be emptied indefinitely, something must happen to restore the balance. The corrective condition on most units without a receiver (vessel that stores liquid between the condenser and expansion device) is that liquid backs up into the condenser. The condensing area is thereby reduced and the condenser pressure raised. With the elevated condenser pressure, the compressor capacity is reduced and the capillary-tube rate of feed is increased until balance is restored. Another possibility for regaining a balanced flow rate is that the heat-transfer coefficient in the starved evaporator decreases. A greater temperature difference must develop between the fluid being chilled and the refrigerant in the evaporator, which occurs by means of the suction pressure dropping back to pressure A and restoring balanced flow.

An opposite unbalanced condition results if the refrigeration load falls off to less than the refrigeration capacity at the balance point. If the refrigeration load drops off, the suction temperature and pressure drop to some point C. At suction pressure C the capillary tube can feed more refrigerant to the evaporator than the compressor can draw out. The evaporator fills with liquid and would spill over into the compressor with disastrous results were it not prevented. Slugging the compressor with liquid can be prevented by limiting the charge of refrigerant in the system. The charge is carefully measured so that there is enough refrigerant to fill the evaporator but no more. Balance of flow is restored when some gas enters the capillary tube, reducing the feed rate of the capillary tube[1] because of the high specific volume of the vapor. A new balance point is at point D in Fig. 13.2.

Although point D represents balanced flow, it is not a satisfactory condition. The state of the refrigerant entering the capillary tube shown on the pressure-enthalpy diagram in Fig. 13-3 is in the mixture region, which reduces the refrigerating effect compared with that when saturated or subcooled liquid enters the capillary tube. Each kilogram of refrigerant provides a reduced refrigerating effect in Fig. 13-3, but the work per kilogram remains unchanged.

Many capillary tubes are installed so that they become a part of a heat exchanger. The heat exchanger is constructed by attaching the suction line to the capillary tube.

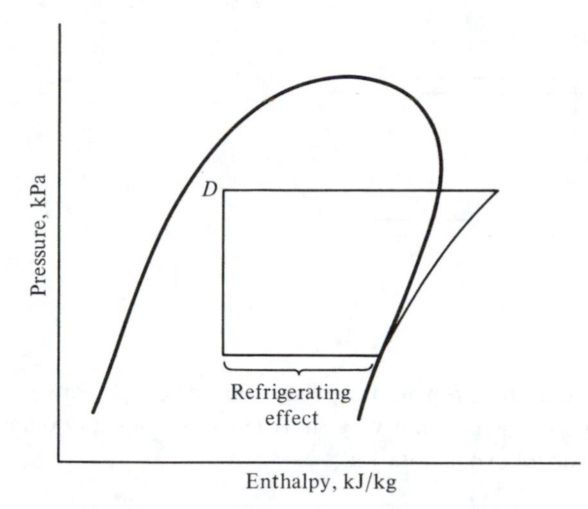

Figure 13-3 Reduction in refrigerating effect when some vapor enters the capillary tube.

The cold suction gas from the evaporator retards the flashing of the liquid flowing through the capillary tube.

Capillary tubes have certain advantages and disadvantages. Their advantages are predominant enough to give them universal acceptance in factory-sealed systems. They are simple, have no moving parts, and are inexpensive. They also allow the pressures in the system to equalize during the off cycle. The motor driving the compressor can then be one of low starting torque.

The disadvantages of capillary tubes are that they are not adjustable to changing load conditions, are susceptible to clogging by foreign matter, and require the mass of refrigerant charge to be held within close limits. This last feature has dictated that the capillary tube be used only on hermetically sealed systems, where there is less likelihood of the refrigerant leaking out. The capillary tube is designed for one set of operating conditions, and any change in the applied heat load or condensing temperature from design conditions represents a decrease in operating efficiency.

13-3 Selection of a capillary tube The designer of a new refrigeration unit employing a capillary tube must select the bore and length of the tube so that the compressor and tube fix a balance point at the desired evaporating temperature. Final adjustment of the length is most often "cut and try." A longer tube than desired is first installed in the system with the probable result that the balance point will occur at too low an evaporating temperature. The tube is shortened until the desired balance point is reached.

The method of calculating the necessary bore and length of a capillary tube using graphs[2] will be summarized in Sec. 13-7. As a preparation for that method, a purely analytical technique based on fundamental laws will be described in Secs. 13-4 and 13-5. The method and illustration in Example 13-1 are influenced by proposals of Hopkins[3] and Cooper et al.[4] but modified to take advantage of a digital computer to assist in the computations.

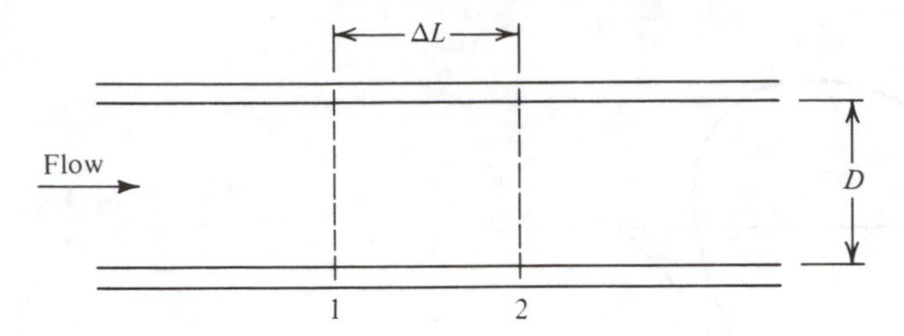

Figure 13-4 Incremental length of capillary tube.

13-4 Analytical computation of pressure drop in a capillary tube The equations relating states and conditions at points 1 and 2 in a very short length of capillary tube in Fig. 13-4 will be written using the following notation:

A = cross-sectional area of inside of tube, m^2
D = ID of tube, m
f = friction factor, dimensionless
h = enthalpy, kJ/kg
h_f = enthalpy of saturated liquid, kJ/kg
h_g = enthalpy of saturated vapor, kJ/kg
ΔL = length of increment, m
p = pressure, Pa
Re = Reynolds number = $VD/\nu\mu$
ν = specific volume, m^3/kg
ν_f = specific volume of saturated liquid, m^3/kg
ν_g = specific volume of saturated vapor, m^3/kg
V = velocity of refrigerant, m/s
w = mass rate of flow, kg/s
x = fraction of vapor in mixture of liquid and vapor
μ = viscosity, Pa • s
μ_f = viscosity of saturated liquid, Pa • s
μ_g = viscosity of saturated vapor, Pa • s

The fundamental equations applicable to the control volume bounded by points 1 and 2 in Fig. 13-4 are (1) conservation of mass, (2) conservation of energy, and (3) conservation of momentum.

The equation for conservation of mass states that

$$w = \frac{V_1 A}{\nu_1} = \frac{V_2 A}{\nu_2} \tag{13-1}$$

or

$$\frac{w}{A} = \frac{V_1}{\nu_1} = \frac{V_2}{\nu_2} \tag{13-2}$$

and w/A will be constant throughout the length of the capillary tube.

The statement of conservation of energy is

$$1000h_1 + \frac{V_1^2}{2} = 1000h_2 + \frac{V_2^2}{2} \tag{13-3}$$

which assumes negligible heat transfer in and out of the tube.

The momentum equation in words states that the difference in forces applied to the element because of drag and pressure difference on opposite ends of the element equals that needed to accelerate the fluid

$$\left[(p_1 - p_2) - f\frac{\Delta L}{D}\frac{V^2}{2v}\right]A = w(V_2 - V_1) \tag{13-4}$$

As the refrigerant flows through the capillary tube, its pressure and saturation temperature progressively drop and the fraction of vapor x continuously increases. At any point

$$h = h_f(1 - x) + h_g x \tag{13-5}$$

and

$$v = v_f(1 - x) + v_g x \tag{13-6}$$

In Eq. (13-4) V, v, and f all change as the refrigerant flows from point 1 to point 2, but some simplification results from Eq. (13-2), which shows that V/v is constant so that

$$f\frac{\Delta L}{D}\frac{V^2}{2v} = f\frac{\Delta L}{D}\frac{V}{2}\frac{w}{A} \tag{13-7}$$

In the calculation to follow in Example 13-1 the V used in Eq. (13-7) will be the mean velocity

$$V_m = \frac{V_1 + V_2}{2} \tag{13-8}$$

Since expressing the friction factor f for the two-phase flow is complex, we shall use an approximation and later compare the calculation with experimental results as a check on the validity of this approximation as well as of any other approximation built into the method.

For Reynolds numbers in the lower range of the turbulent region an applicable equation for the friction factor f is

$$f = \frac{0.33}{\text{Re}^{0.25}} = \frac{0.33}{(VD/\mu v)^{0.25}} \tag{13-9}$$

The viscosity of the two-phase refrigerant at a given position in the tube is a function of the vapor fraction x

$$\mu = \mu_f(1 - x) + \mu_g x \tag{13-10}$$

The mean friction factor f_m applicable to the increment of length 1-2 is

$$f_m = \frac{f_1 + f_2}{2} = \frac{0.33/\text{Re}_1^{0.25} + 0.33/\text{Re}_2^{0.25}}{2} \tag{13-11}$$

13-5 Calculating the length of an increment The essence of the analytical calculation method is to determine the length of the increment 1-2 in Fig. 13-4 for a given reduction in saturation temperature of the refrigerant. The flow rate and all the conditions at point 1 are known, and for an arbitrarily selected temperature at point 2 the remaining conditions at point 2 and the ΔL will be computed in the following specific steps:

1. Select t_2.
2. Compute $p_2, h_{f2}, h_{g2}, v_{f2}$, and v_{g2}, all of which are functions of t_2.
3. Combine the continuity equation (13-2) and the energy equation (13-3)

$$1000h_2 + \frac{v_2^2}{2}\left(\frac{w}{A}\right)^2 = 1000h_1 + \frac{V_1^2}{2} \tag{13-12}$$

   Substitute Eqs. (13-5) and (13-6) into Eq. (13-12)

$$1000h_{f2} + 1000(h_{g2} - h_{f2})x + \frac{[v_{f2} + (v_{g2} - v_{f2})x]^2}{2}\left(\frac{w}{A}\right)^2 = 1000h_1 + \frac{V_1^2}{2} \tag{13-13}$$

   Everything in Eq. (13-13) is known except x, which can be solved by the quadratic equation

$$x = \frac{-b \pm \sqrt{b^2 - 4ac}}{2a} \tag{13-14}$$

where

$$a = (v_{g2} - v_{f2})^2\left(\frac{w}{A}\right)^2\frac{1}{2}$$

$$b = 1000(h_{g2} - h_{f2}) + v_{f2}(v_{g2} - v_{f2})\left(\frac{w}{A}\right)^2$$

$$c = 1000(h_{f2} - h_1) + \left(\frac{w}{A}\right)^2\frac{1}{2}v_{f2}^2 - \frac{V_1^2}{2}$$

4. With the value of x known, h_2, v_2, and V_2 can be computed.
5. Compute the Reynolds number at point 2 using the viscosity from Eq. (13-10), the friction factor at point 2 from Eq. (13-9), and the mean friction factor for the increment from Eq. (13-11).
6. Finally, substitute Eqs. (13-7) and (13-8) into Eq. (13-4) to solve for ΔL.

Example 13-1 What length of capillary tube (ID = 1.63 mm) will drop the pressure of saturated liquid refrigerant 22 at 40°C to the saturation temperature of the evaporator of 5°C? The flow rate is 0.010 kg/s.

Solution Since the calculation will be performed on a computer, to facilitate the computation equations for properties of saturated refrigerant 22 applicable to a temperature range of -20 to 50°C will be used.

$$\ln\left(\frac{p}{1000}\right) = 15.06 - \frac{2418.4}{t + 273.15} \tag{13-15}$$

$$v_f = \frac{v_f}{1000} = \frac{0.777 + 0.002062t + 0.00001608t^2}{1000} \tag{13-16}$$

$$v_g = \frac{-4.26 + 94050(t + 273.15)/p}{1000} \tag{13-17}$$

$$h_f = 200.0 + 1.172t + 0.001854t^2 \tag{13-18}$$

$$h_g = 405.5 + 0.3636t - 0.002273t^2 \tag{13-19}$$

$$\mu_f = 0.0002367 - 1.715 \times 10^{-6}t + 8.869 \times 10^{-9}t^2 \tag{13-20}$$

$$\mu_g = 11.945 \times 10^{-6} + 50.06 \times 10^{-9}t + 0.2560 \times 10^{-9}t^2 \tag{13-21}$$

Conditions at entrance to capillary tube, point 1 The entering refrigerant is saturated liquid at 40°C, and with $x = 0$ the properties from Eqs. (13-15) to (13-21) are

$p_1 = 1{,}536{,}000$ Pa $\quad v_1 = v_{f1} = 0.000885$ m^3/kg
$h_1 = h_{f1} = 249.9$ kJ/kg $\quad \mu = \mu_{f1} = 0.0001823$ Pa • s

$$\frac{w}{A} = \frac{0.010}{\pi(0.00163^2)/4} = 4792.2 \text{ kg/s} \cdot \text{m}^2$$

$$V_1 = \frac{w}{A} v_1 = 4.242 \text{m/s}$$

$$\text{Re}_1 = 42{,}850 \qquad f_1 = \frac{0.33}{\text{Re}_1^{0.25}} = 0.0229$$

Conditions at point 2 Arbitrarily select $t_2 = 39$°C. Then

$p_2 = 1{,}498{,}800$ Pa $\quad h_{f2} = 248.5$ kJ/kg $\quad h_{g2} = 416.2$ kJ/kg
$v_{f2} = 0.000882$ m^3/kg $\quad v_{g2} = 0.01533$ m^3/kg
$\mu_{f2} = 0.0001833$ Pa • s $\quad \mu_{g2} = 0.00001429$ Pa • s

From Eq. (13-14)

$x = 0.008$

From Eqs. (13-5) and (13-6) and using an equation of the same form for viscosity, we get

$h_2 = 249.84$ kJ/kg $\quad v_2 = 0.0009952$ m^3/kg
$\mu_2 = 0.0001820$ Pa • s

The following terms can now be calculated:

$$V_2 = \frac{w}{A} v_2 = 4.769 \text{ m/s} \qquad \text{Re}_2 = 42{,}923$$

$$f_2 = \frac{0.33}{42{,}923^{0.25}} = 0.0229$$

$$f_m = \frac{0.0229 + 0.0229}{2} = 0.0229$$

$$V_m = \frac{4.242 + 4.769}{2} = 4.506$$

From Eq. (13-4) the magnitude of the expression

$$f_m \frac{\Delta L}{D} \frac{V_m}{2} \frac{V}{v}$$

is found to be 34,964, and when the known values are substituted,

$$\Delta L_{1-2} = 0.2306 \text{ m}$$

Continuation to succeeding increments The conditions at point 2 that have just been computed are the entering conditions to the next increment of length, in which the saturation temperature drops to 38°C. Table 13-1 presents a summary of the calculations near the entrance to the tube and as the temperature approaches the evaporating temperature of 5°C. The cumulative length of the capillary tube required for the specified reduction in pressure is 2.118 m.

13-6 Choked flow Table 13-1 shows that near the end of the capillary tube the increments of length needed to drop the saturation temperature 1°C become progressively

Table 13-1 Capillary-tube calculations in Example 13-1

Position	Temperature, °C	Pressure, kPa	x	Specific volume, m^3/kg	Enthalpy, kJ/kg	Velocity, m/s	Increment length, m	Cumulative length, m
1	40	1536.4	0.000	0.000885	249.85	4.242		
2	39	1498.8	0.008	0.000995	249.84	4.769	0.2306	0.231
3	38	1461.9	0.016	0.001110	249.84	5.320	0.2013	0.432
4	37	1425.8	0.023	0.001230	249.84	5.895	0.1770	0.609
5	36	1390.3	0.031	0.001355	249.83	6.496	0.1565	0.765
6–31	...	...	...	...	...	...	...	...
32	9	657.65	0.194	0.007660	249.18	36.71	0.0097	2.089
33	8	637.90	0.199	0.008048	249.11	38.57	0.0085	2.098
34	7	618.61	0.204	0.008452	249.03	40.51	0.0075	2.105
35	6	599.78	0.209	0.008873	248.95	42.52	0.0066	2.112
36	5	581.38	0.213	0.009309	248.86	44.61	0.0049	2.118

Table 13-2 Continuation of capillary-tube calculation

Position	Temperature, °C	Pressure, kPa	x	Specific volume, m^3/kg	Enthalpy, kJ/kg	Velocity, m/s	Increment length, m	Cumulative length, m
42	-1	479.97	0.239	0.01231	248.11	59.00	0.0017	2.137
43	-2	464.50	0.243	0.01288	247.95	61.73	0.0012	2.138
44	-3	449.41	0.247	0.01347	247.77	64.56	0.0007	2.139
45	-4	434.71	0.250	0.01409	247.58	67.50	0.0003	2.139
46	-5	420.38	0.254	0.01472	247.37	70.55	-0.0001	

smaller. It might be interesting to ask what happens if the evaporating temperature is lowered, say, to -10°C. Table 13-2 shows a continuation of Table 13-1. To drop the pressure from 434.7 to 420.4 kPa (saturation temperature from -4 to -5°C), a negative length of capillary tube is required. Clearly an impossible situation has developed, called *choked flow*. The phenomenon is similar to flow in a converging nozzle when the outlet pressure has been reduced until sonic velocity occurs at the throat. Further reductions in the discharge pressure fail to increase the flow rate through the nozzle. An occurrence of choked flow more closely related to the flow in a capillary tube is in long gas pipelines where because of the pressure drop due to friction the specific volume and velocity increase until sonic velocity occurs. The thermodynamic conditions are represented by the *Fanno line*[6] shown on the enthalpy-entropy diagram of Fig. 13-5, where the enthalpy decreases while the entropy increases as the fluid flows through the tube. At the sonic velocity the Fanno line demands a decrease in entropy, which is forbidden by the second law of thermodynamics for this adiabatic process. Nature adjusts the flow rate so that the sonic velocity occurs at the very exit of the tube.

The implication of choked flow in a capillary tube of a refrigeration system is that at some suction pressure the flow-rate curves of the capillary tube shown in Fig. 13-1 may reach a plateau, as shown in Fig. 13-6, where further decreases in suction pressure do not increase the rate of flow through the capillary tube. For system per-

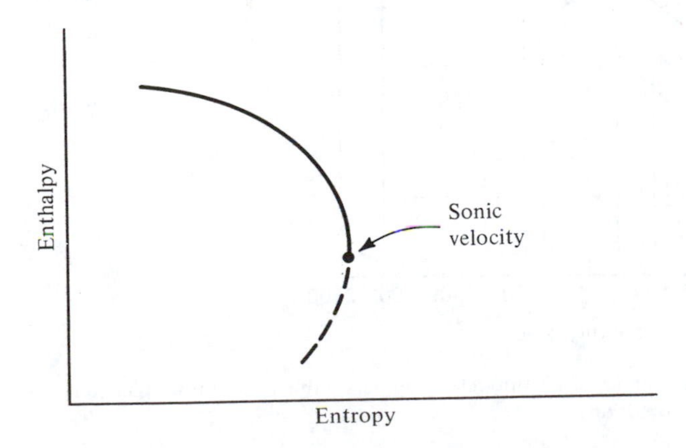

Figure 13-5 Fanno line showing choked-flow conditions.

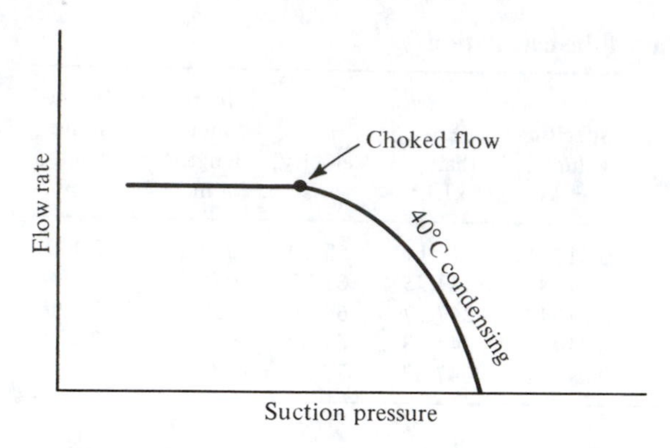

Figure 13-6 Flow through a capillary tube.

formance this means that to achieve a flow balance between the compressor and capillary tube the suction pressure must drop until the flow rate through the compressor matches the flow rate through the capillary tube. The result is a penalty in operating efficiency. Many appliance manufacturers bond the suction line from the evaporator to the capillary tube to remove heat from the refrigerant in the capillary tube, lowering the specific volume and thus retarding choking.

13-7 Graphical method of capillary-tube selection Graphs[2] to facilitate the selection of capillary tubes are based on data by Hopkins[3] and revised with data by Whitesel.[7,8] The first graph (Fig. 13-7) presents the refrigerant flow rate as a function of

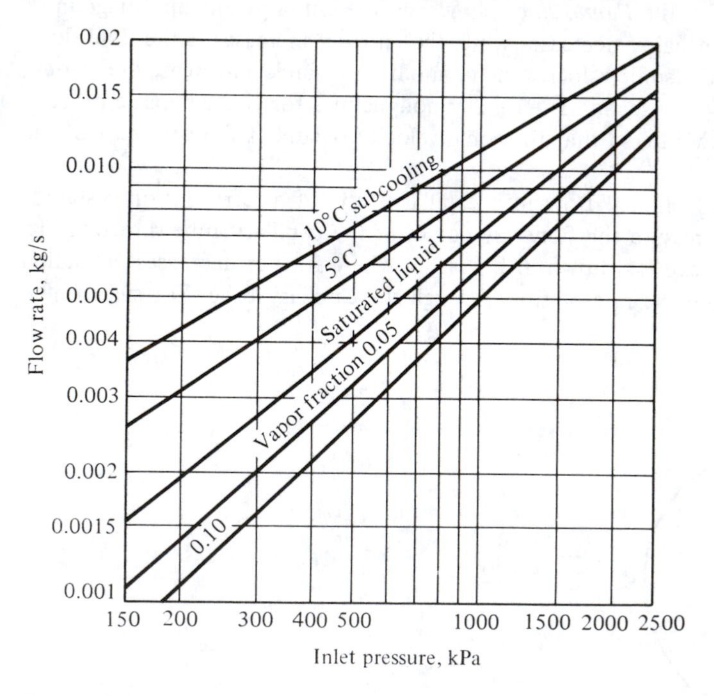

Figure 13-7 Flow rate of refrigerant 12 or 22 through a capillary tube 1.63 mm in diameter and 2.03 m long under choked-flow conditions.

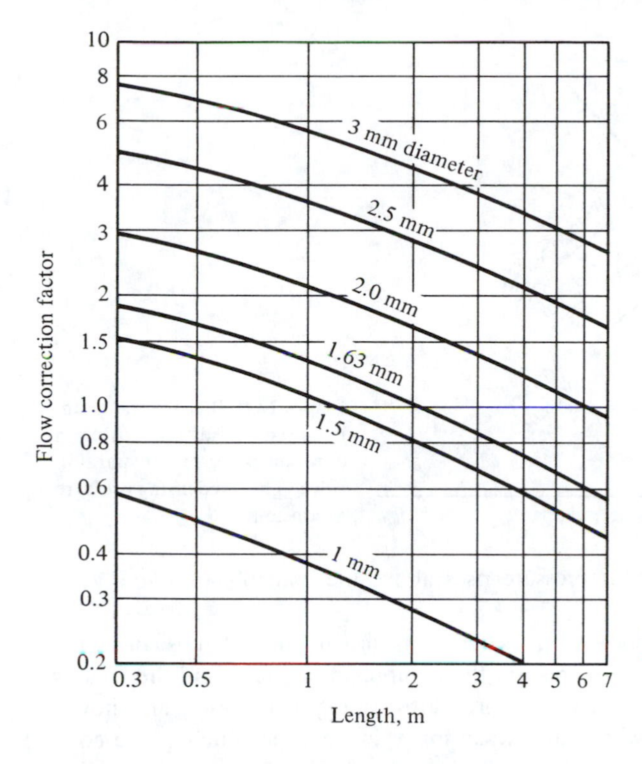

Figure 13-8 Correction factors to be applied to Fig. 13-7 for other diameters and lengths of capillary tubes.

the entering pressure to the capillary tube for a tube that is 1.63 mm in diameter and 2.03 m long. The various curves in Fig. 13-7 represent performance at a variety of inlet conditions–magnitudes of subcooling and fractions of vapor. The companion graph to Fig. 13-7 is the one in Fig. 13-8, presenting correction factors to the flow rate of Fig. 13-7 for other lengths and diameters.

Both Figs. 13-7 and 13-8 apply to choked-flow conditions, and the ASHRAE Handbook[2] contains further graphs to correct for non-choked-flow conditions. The length chosen that is based on choked-flow conditions may be only slightly different from that needed for the nonchoked condition. Example 13-1 indicates that for the evaporatoring temperature of 5°C the required length of capillary tube is 2.118 m, while if the expansion proceeds to the choked condition at –4°C, the length is 2.139 m.

The data in Fig. 13-7 also provide a check on the accuracy of the calculation method used in Example 13-1. The diameter of the tube in Example 13-1 is the same as that in Fig. 13-7, and the flow rate of saturated liquid entering at a pressure of 1536 kPa is 0.01 kg/s in both Example 13-1 and Fig. 13-7. The comparison arises in the length of the capillary tube for choked-flow conditions: 2.03 m in Fig. 13-7 and 2.14 m in Example 13-1; therefore the calculation method shows a length 5 percent greater than the graph.

13-8 Constant-pressure expansion valve The *contant-pressure expansion valve* maintains a constant pressure at its outlet, the entrance to the evaporator. It senses the evaporator pressure, and when that pressure drops below the control point, the valve

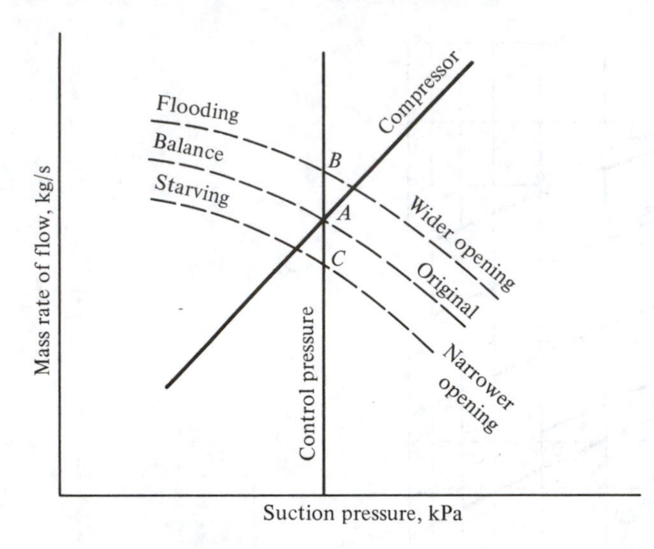

Figure 13-9 Balanced and unbalanced conditions using a constant-pressure expansion valve. The condensing pressure is constant.

opens wider. When the evaporator pressure rises above the control point, the valve partially closes.

The effect of the valve operation on the performance of the system is charted in Fig. 13-9. At a constant condensing pressure the compressor capacity and the feeding capacity of the expansion valve at several degrees of opening of the valve are shown. Point *A* is the balance point, where the expansion valve feeds as much as the compressor pumps from the evaporator. If the refrigeration load drops off, the suction temperature and pressure attempt to drop but the valve resists the drop in pressure by opening wider. Under the new condition the compressor capacity remains at *A*, but the feed rate of the valve changes to *B*. The evaporator will then flood under this unbalanced-flow condition. Starving of the evaporator can occur if the refrigeration load increased and the valve operated at point *C*.

Use of the constant-pressure expansion valve has been limited to systems of refrigerating capacity less than about 30 kW, in which a critical charge of refrigerant is feasible to prevent liquid from flooding out of the evaporator. Its primary use is where the evaporating temperature should be maintained at a certain point to control humidity or to prevent freezing in water coolers. The pressure-limiting characteristic can be used to advantage when protection is required against overload of the compressor due to high suction pressure.

13-9 Float valves The float valve is a type of expansion valve which maintains the liquid at a constant level in a vessel or an evaporator. A float switch which opens completely when the liquid level drops below the control point and closes completely when the level reaches the control point will give the same net performance as a modulating type of float control.

By maintaining a constant liquid level in the evaporator the float valve always establishes balanced conditions of flow between the compressor and itself. Figure 13-10 shows an original balance point at *A*. If the refrigeration load should increase, the evaporating temperature and pressure rise, which momentarily allows the compressor

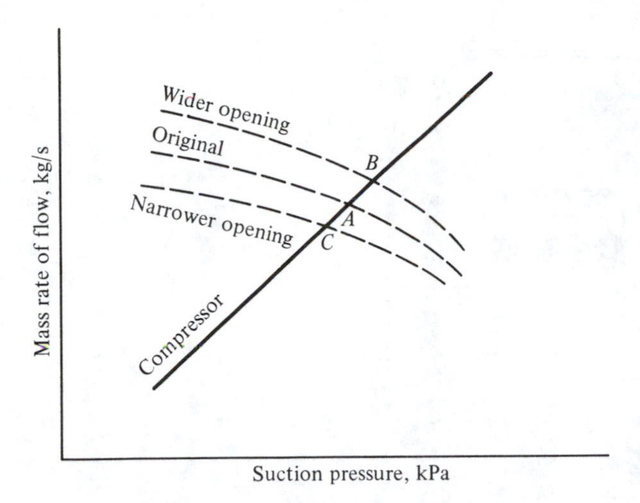

Figure 13-10 Balance points with various load conditions using a float valve. The condensing pressure is constant.

to pump a greater rate of flow than the valve is feeding. The valve reacts to keep the level constant by widening its average opening. A new balance point occurs at point *B*. If the refrigeration load decreases, the suction pressure drops and the level rises, prompting the valve to close somewhat and give a balance point at *C*.

Float valves and float-switch-solenoid combinations are used primarily in large installations. They can regulate the flow to flooded evaporators in response to the level of liquid refrigerant in the shell of the evaporator or in a chamber connected to the evaporator. They should not be used in continuous-tube evaporators, where it is impossible to establish a level of liquid refrigerant by which they can be controlled.

13-10 Superheat-controlled (thermostatic) expansion valve The most popular type of expansion device for moderate-sized refrigeration systems is the superheat-controlled valve, usually called a *thermostatic expansion valve.* The name may be misleading because control is actuated not by the temperature in the evaporator but by the magnitude of superheat of the suction gas leaving the evaporator. The superheat expansion valve regulates the rate of flow of liquid refrigerant in proportion to the rate of evaporation in the evaporator. The balances of the flow rate between the compressor and superheat-controlled expansion valve are therefore practically identical to those shown for the float valve in Fig. 13-10. Figure 13-11 is a photograph of a thermostatic expansion valve.

The superheat of the suction gas operates the thermostatic expansion valve as follows. A feeler bulb (Fig. 13-12) is partially filled with liquid of the same refrigerant as that used in the system. The fluid in the bulb is called the *power fluid.* The feeler bulb is clamped to the outlet of the evaporator so that the bulb and the power fluid closely assume the temperature of the suction gas. The pressure of the power fluid bears on the top of the diaphragm, and the evaporator pressure pushes on the bottom of the diaphragm. A slight force exerted by the spring on the valve stem keeps the valve closed until the pressure above the diaphragm overcomes the spring force plus the force of the evaporator pressure. For the pressure above the diaphragm to be higher than the pressure below the diaphragm, the power fluid must be at a tempera-

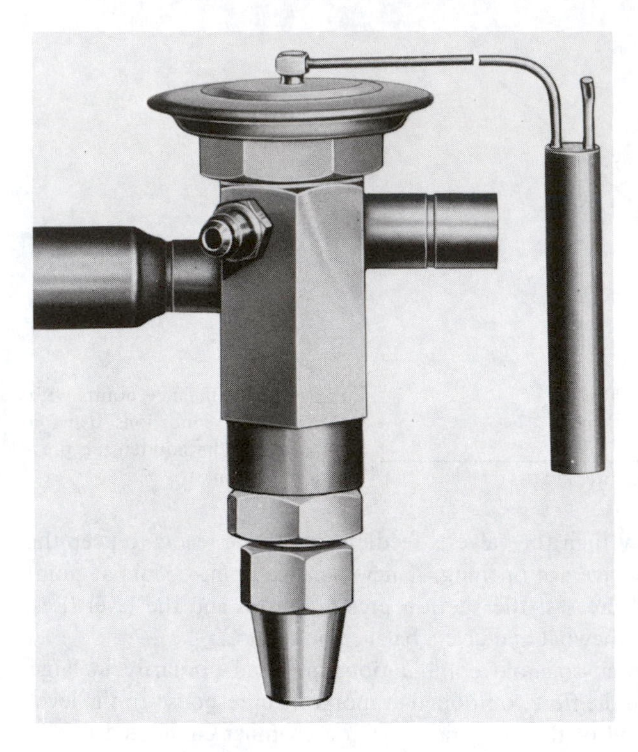

Figure 13-11 A thermostatic expansion valve. *(Sporlan Valve Co.)*

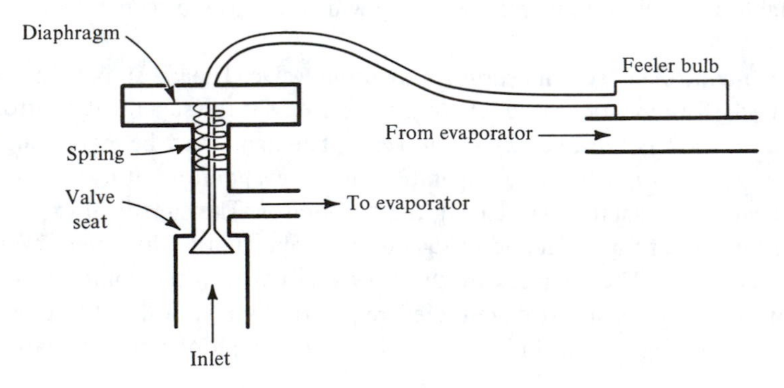

Figure 13-12 A schematic diagram of the basic superheat-controlled expansion valve.

ture higher than the saturation temperature in the evaporator. The suction gas must therefore be superheated in order to bring the power fluid up to the pressure which opens the valve.

The expansion valve experiences a throttling range like that in the pneumatic controls discussed in Sec. 9-8. As shown in Fig. 13-13, a certain amount of suction superheat is required to begin opening the valve (4°C in Fig. 13-13). To overcome the force of the spring a progressively greater force must be supplied by the power fluid

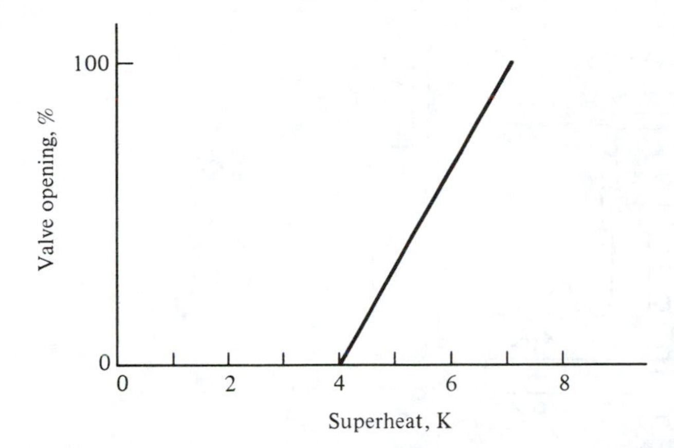

Figure 13-13 Throttling range of superheat-controlled valve.

to open the valve farther. This higher force is developed by increasing the superheat, and Fig. 13-13 shows 7 K of superheat needed to open the valve completely.

The operation of the valve maintains an approximately constant quantity of liquid in the evaporator because if the amount of liquid decreases, more evaporator surface is exposed for superheating the refrigerant, opening the valve farther.

The type of thermostatic expansion valve shown in Fig. 13-12 is of the internal-equalizer type, in which the evaporator pressure at the valve outlet is admitted internally to the underneath side of the diaphragm. Some refrigerating systems have an appreciable pressure drop in the evaporator or employ multiple refrigerant circuits in the evaporator, necessitating a distributor (Fig. 12-17) with a resultant pressure drop. With the internal equalizer, therefore, a higher refrigerant pressure is applied to the underside of the diaphragm than actually exists in the suction line at the point where the bulb is attached. This condition requires an increased amount of superheat to open the valve and reduces the effectiveness of the evaporator. To eliminate this situation, an *external equalizer* is employed, which applies beneath the diaphragm the pressure of the refrigerant at the outlet of the evaporator. This process is accomplished by connecting a small tube from the suction line to the chamber beneath the diaphragm, as shown in the cutaway view of Fig. 13-14.

13-11 Manufacturers' ratings of thermostatic expansion valves The catalogs of manufacturers of expansion valves usually show the refrigerating capacity associated with the flow rate of which the valve is capable. In order to provide some reserve capacity, most manufacturers show the refrigerating capacity at perhaps 75 percent of the full flow rate of the valve. The flow rate through the valve is a function of the pressure difference across the valve, and the velocity through the fully opened valve can be computed from the hydraulic formula

$$\text{Velocity} = C\sqrt{2(\text{pressure difference})} \quad \text{m/s} \tag{13-22}$$

where C is an experimentally determined constant and the pressure difference is in kilopascals. Although the refrigerant following the throttling process in the valve is a mixture of vapor and liquid, Equation (13-22) applies to liquid because the vaporiza-

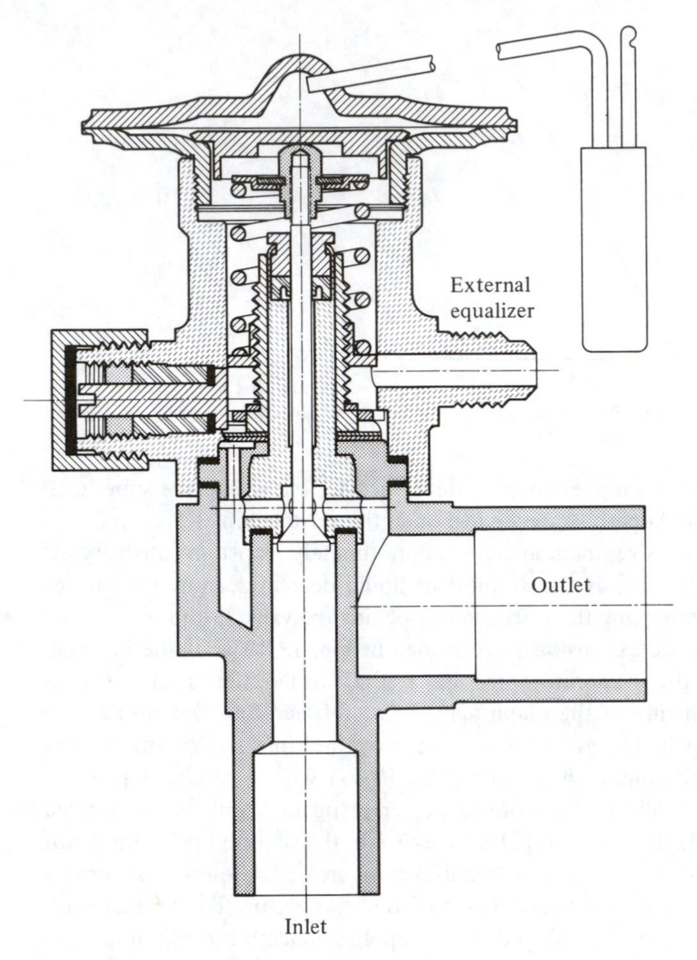

Figure 13-14 A sectional view of a superheat-controlled expansion valve showing an external equalizer. *(Alco Controls Division of Emerson Electric Company.)*

tion does not occur until after the fluid has passed through the valve. The liquid is momentarily in a metastable condition.[9]

The thermostatic expansion valve is frequently called upon to operate over a wide range of evaporator temperatures. An expansion valve for a low-temperature system, for example, not only must control the refrigerant flow at the design temperature but must also be able to feed the evaporator properly during the pulldown operation. A characteristic of the thermostatic expansion valve feeding a low-temperature evaporator is shown in Fig. 13-15. If a pressure difference of 100 kPa between the power-fluid and the evaporator pressure is required to open the valve fully, the superheat of the suction gas must be 5 K when the evaporator is operating at 5°C. If the same valve feeds an evaporator operating at -30°C, the superheat required to provide the 100-kPa pressure difference is 12 K. At the low-temperature condition, therefore, a considerable portion of the evaporator is not effective because a large area is required

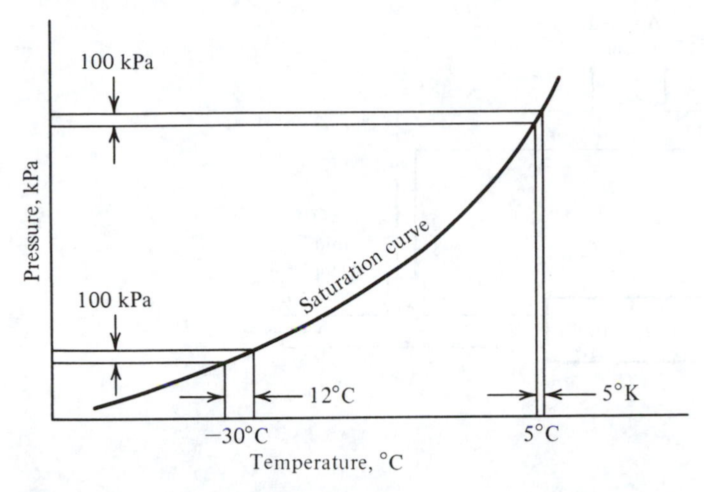

Figure 13-15 Pressure-temperature characteristic of a refrigerant 22 power fluid which results in greater superheat at low-temperature operation.

for superheating the gas. Decreasing the superheat adjustment on the valve may correct the low-temperature condition but may also reduce the superheat at the high operating temperature to the point where liquid floods out of the evaporator into the compressor during pulldown of temperature.

One solution to the operating problems at low temperature is to use a valve with a *cross charge*, i.e., one having a power fluid different from the refrigerant in the system. The power fluid is selected so that its properties are closely related to those of the refrigerant, as shown in Fig. 13-16. The power-fluid characteristics are chosen so that the superheat required to open the valve is nearly constant over the entire operating range.

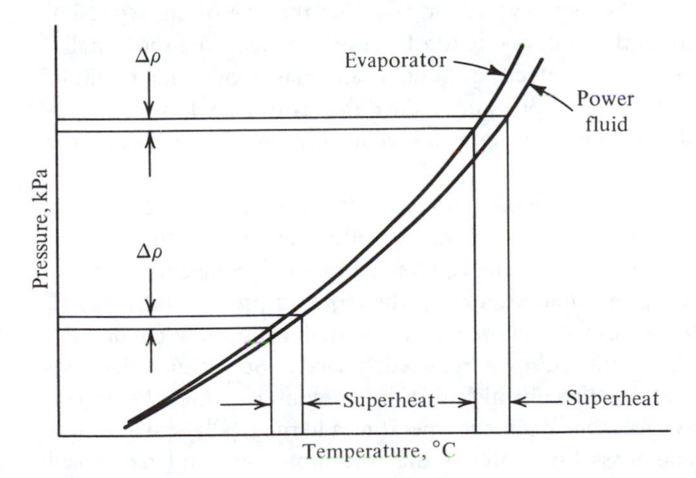

Figure 13-16 Evaporator and power-fluid temperatures in a thermostatic expansion valve with a cross charge.

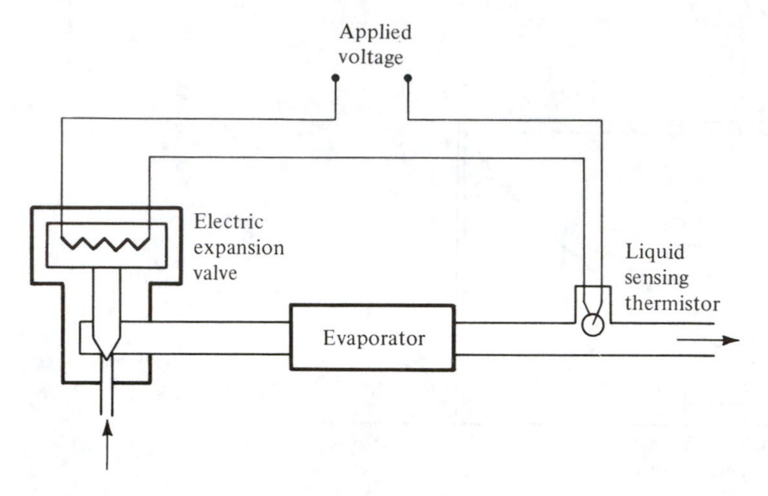

Figure 13-17 An electric expansion valve.

13-12 Electric expansion valves The electric expansion valve, shown schematically in Fig. 13-17, uses a thermistor to sense the presence of liquid in the outlet stream of the evaporator. When no liquid is present, the temperature of the thermistor increases, which drops its resistance and permits a greater current flow through the heater at the valve. The valve is thereby opened, allowing an increased refrigerant flow rate. One of the applications of the electric expansion valve is for heat pumps (Chap. 18), where the flow rate of refrigerant is reversed in order to change from heating to cooling. Since its control is independent of refrigerant pressures, the electric expansion valve can function with flow through the valve in either direction.

13-13 Application While there is some latitude in selecting the size of an expansion device, faulty operation results if the device is too large or too small for the installation. A valve that is too large often overfeeds or "hunts" and may allow some liquid to flood out of the evaporator to the compressor. A valve that is too small passes insufficient liquid so that a balance point occurs at a low suction pressure, reducing the capacity of the system.

During winter operation the condensing pressure drops on systems that use a cooling tower, an evaporative condenser, or an air-cooled condenser, and the pressure differential across the valve may become inadequate. To establish a balance of flow, the suction pressure must drop in order to develop the required pressure differential. Instead of achieving the higher capacity normally associated with a low condensing pressure, the low suction pressure causes a reduced capacity. Sometimes the condensing pressure must be kept artificially high during the winter so that the expansion valve can feed properly. Another danger in operating a hermetically sealed motor at low suction pressures and mass flow rates is that the motor will not be cooled sufficiently and will burn out.

In a not uncommon operating problem the valve passes insufficient refrigerant because vapor is mixed with the liquid entering the valve. The high specific volume of the vapor compared with that of the liquid means that the valve can pass as a vapor only a fraction of the mass flow of refrigerant it can pass as a liquid. Two frequent causes of losing the liquid seal to the valve are (1) an insufficient charge of refrigerant and (2) a high elevation of the expansion valve above the condenser or receiver. If the valve is higher than the condenser or receiver, the difference in static head may make the pressure low enough at the valve to flash some liquid into vapor. A liquid-to-suction heat exchanger (Sec. 10-15) sometimes corrects this situation.

PROBLEMS

13-1 Using the method described in Sec. 13-5 and the entering conditions given in Table 13-1 for Example 13-1 at position 4, compute the length of tube needed to drop the temperature to 36°C. Use property values from refrigerant 22 tables when possible.

13-2 A capillary tube is to be selected to throttle 0.011 kg/s of refrigerant 12 from a condensing pressure of 960 kPa and a temperature of 35°C to an evaporator operating at -20°C.

(*a*) Using Figs. 13-7 and 13-8, select the bore and length of a capillary tube for this assignment. *Ans.* For example, 1.63 mm diameter and 1.2 m long.

(*b*) If the evaporating temperature had been 5°C rather than -20°C, would the selection of part (*a*) be suitable? Discuss assumptions that have been made.

13-3 A refrigerant 22 refrigerating system operates with a condensing temperature of 35°C and an evaporating temperature of -10°C. If the vapor leaves the evaporator saturated and is compressed isentropically, what is the COP of the cycle (*a*) if saturated liquid enters the expansion device and (*b*) if the refrigerant entering the expansion device is 10 percent vapor as in Fig. 13-3? *Ans.* (*a*) 4.71, (*b*) 4.20

13-4 Refrigerant 22 at a pressure of 1500 kPa leaves the condenser and rises vertically 10 m to the expansion valve. The pressure drop due to friction in the liquid line is 20 kPa. In order to have no vapor in the refrigerant entering the expansion valve, what is the maximum allowable temperature at that point? *Ans.* 35.3°C

13-5 A superheat-controlled expansion valve in a refrigerant 22 system is not equipped with an external equalizer. The valve supplies refrigerant to an evaporator coil and comes from the factory with a setting that requires 5 K superheat in order to open the valve at an evaporator temperature of 0°C.

(*a*) What difference in pressure on opposite sides of the diaphragm is required to open the valve?

(*b*) When the pressure at the entrance of the evaporator is 600 kPa, how much superheat is required to open the valve if the pressure drop of the refrigerant through the coil is 55 kPa? *Ans.* 7.4 K

13-6 The catalog of an expansion-valve manufacturer specifies a refrigerating capacity of 45 kW for a certain valve when the pressure difference across the valve is 500 kPa. The catalog ratings apply when vapor-free liquid at 37.8°C enters the expansion valve and the evaporating temperature is 4.4°C. What is the expected rating of the valve when the pressure difference across it is 1200 kPa? *Ans.* Catalog specifies 69.4 kW

REFERENCES

1. L. A. Staebler: Theory and Use of a Capillary Tube for Liquid Refrigerant Control, *Refrig. Eng.*, vol. 55, no. 1, p. 55, January 1948.
2. "ASHRAE Handbook and Product Directory, Equipment Volume," American Society of Heating, Refrigerating, and Air-Conditioning Engineers, Atlanta, Ga., 1979.
3. N. E. Hopkins: Rating the Restrictor Tube, *Refrig. Eng.*, vol. 58, no. 11, p. 1087, November 1950.
4. L. Cooper, C. K. Chu, and W. R. Brisken: Simple Selection Method for Capillaries Derived from Physical Flow Conditions, *Refrig. Eng.*, vol. 65, no. 7, p. 37, July 1957.
5. M. M. Bolstad and R. C. Jordan: Theory and Use of the Capillary Tube Expansion Device, *Refrig. Eng.*, vol. 56, no. 6, p. 519, December 1948.
6. A. H. Shapiro: "The Dynamics and Thermodynamics of Compressible Fluid Flow," Ronald, New York, 1953.
7. H. A. Whitesel: Capillary Two-Phase Flow, *Refrig. Eng.*, vol. 65, no. 4, p. 42, April 1957.
8. H. A. Whitesel: Capillary Two-Phase Flow, pt. II, *Refrig. Eng.*, vol. 65, no. 9, p. 35, September 1957.
9. P. F. Pasqua: Metastable Flow of Freon 12, *Refrig. Eng.*, vol. 61, no. 10, p. 1084, October 1953.

CHAPTER

FOURTEEN

VAPOR-COMPRESSION-SYSTEM ANALYSIS

14-1 Balance points and system simulation The performance characteristics of individual components making up the vapor-compression system have been explored in Chap. 11 for compressors, Chap. 12 for condensers and evaporators, and Chap. 13 for expansion devices. These components never work in isolation but are combined into a system, so that their behavior is interdependent. It is the purpose of this chapter to predict the performance of the entire system when the characteristics of the individual components are known. A further function of the techniques to be explained in this chapter is to analyze the influence of externally imposed conditions. For example, system analysis can predict the influence on refrigeration capacity of a change in ambient temperature of the air serving the condenser.

A traditional method of system analysis used by engineers has been through the determination of *balance points*. In this process the performance characteristics of two interrelated components are expressed in terms of the same variables and plotted on a graph. The intersection of corresponding curves indicates the conditions at which the performance characteristics of both components are satisfied, and it is at this point that the system composed of these two components will operate.

In recent years another approach to system analysis has emerged, called *system simulation*[1] which is performed by mathematical rather than graphical procedures. The intersection of two curves, which determines a balance point, suggests the mathematical counterpart of the simultaneous solution of two equations. System simulation is indeed the simultaneous solution of the equations representing the performance characteristics of all components in the system as well as appropriate equations for energy and mass balances and equations of state. To simulate steady-state performance, which is our interest, all the equations are algebraic; the simulation of dynamic performance of systems must include differential equations.

Systems with a small number of components such as the vapor-compression system can be simulated either graphically or mathematically. Both techniques will be illustrated in this chapter. The technique chosen for the mathematical simulation will be the method of *successive substitution*, which is one of the most straightforward technique of simultaneously solving the performance equations.

The chapter first presents performance data in both graphic and mathematical form for a reciprocating compressor and an air-cooled condenser. The first simulation is that of combining those two components into a *condensing unit*. Next, typical performance of an evaporator is presented, and then this component is combined with the condensing unit to provide the performance of a complete system. The influence of the expansion device is then shown qualitatively. The final topic is that of *sensitivity analysis*, where the degree of influence of each of the components–compressor, condenser, and evaporator–on the refrigeration capacity is explored.

14-2 Reciprocating compressor The expected trends of refrigeration capacity and power requirements as functions of the evaporating and condensing temperatures were presented in Chap. 11. Indeed the trends also prevail for actual compressors, as shown by the plot of catalog data for a given compressor in Fig. 14-1. The upper family of

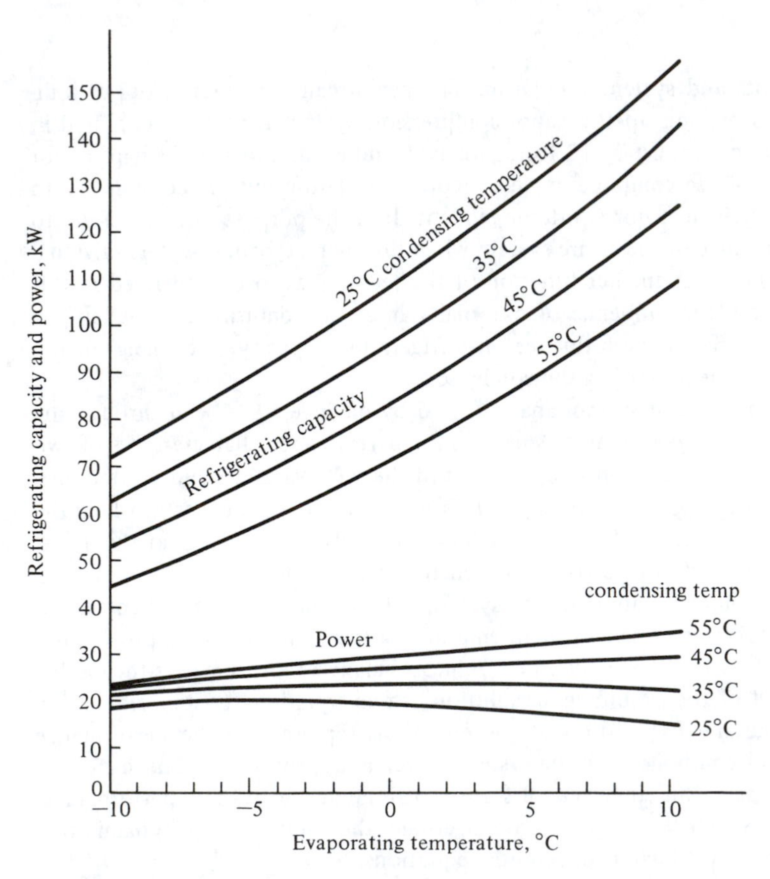

Figure 14-1 Refrigerating capacity and power requirement of a York (Division of Borg-Warner) hermetic reciprocating H62SP-22E, refrigerant 22, 1750 r/min compressor.

curves shows the refrigerating capacity, it being understood that the compressor does not possess refrigerating capacity of itself but is capable of compressing a flow rate of refrigerant that provides the given refrigerating capacity at the evaporator. An increase in evaporating temperature or a decrease in condensing temperature results in increased refrigerating capacity.

The lower set of curves in Fig. 14-1 displays the power required by the compressor.

One choice of the form of the mathematical equations that represent the performance data of Fig. 14-1 is

$$q_e = c_1 + c_2 t_e + c_3 t_e^2 + c_4 t_c + c_5 t_c^2 + c_6 t_e t_c + c_7 t_e^2 t_c + c_8 t_e t_c^2 + c_9 t_e^2 t_c^2 \tag{14-1}$$

and

$$P = d_1 + d_2 t_e + d_3 t_e^2 + d_4 t_c + d_5 t_c^2 + d_6 t_e t_c + d_7 t_e^2 t_c + d_8 t_e t_c^2 + d_9 t_e^2 t_c^2 \tag{14-2}$$

where q_e = refrigerating capacity, kW
P = power required by compressor, kW
t_e = evaporating temperature, °C
t_c = condensing temperature, °C

The constants applicable to Eqs. (14-1) and (14-2) for the compressor in Fig. 14-1 are determined by equation-fitting procedures, e.g., the method of least squares or selecting nine points off the graph and substituting into Eq. (14-1) or (14-2) to develop a set of nine simultaneous equations for the c or d constants. The numerical values are shown in Table 14-1.

In addition to the refrigerating capacity and the power requirement of the compressor, another quantity of interest is the rate of heat rejection required at the condenser. Some compressor catalogs show this quantity, and usually it is simply the sum of the refrigerating capacity and compressor power at a given combination of evaporating and condensing temperatures

$$q_c = q_e + P \tag{14-3}$$

where q_c is the rate of heat rejection at the condenser in kilowatts. A graph of the heat rejection rate for the compressor of Fig. 14-1 is shown in Fig. 14-2. The abscissa

Table 14-1 Constants in Eqs. (14-1) and (14-2)

c_1 = 137.402	d_1 = 1.00618
c_2 = 4.60437	d_2 = −0.893222
c_3 = 0.061652	d_3 = −0.01426
c_4 = −1.118157	d_4 = 0.870024
c_5 = −0.001525	d_5 = −0.0063397
c_6 = −0.0109119	d_6 = 0.033889
c_7 = −0.00040148	d_7 = −0.00023875
c_8 = −0.00026682	d_8 = −0.00014746
c_9 = 0.000003873	d_9 = 0.0000067962

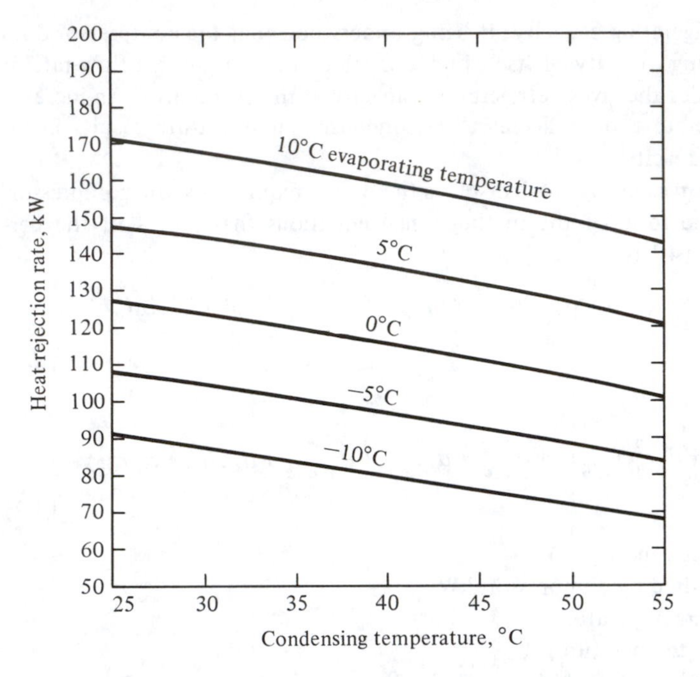

Figure 14-2 Heat-rejection rate of a York (Division of Borg-Warner) hermetic reciprocating compressor, H62SP-22E, refrigerant 22, 1750 r/min.

has been chosen as the condensing temperature for future convenience, and each curve in the family applies to a different evaporating temperature.

14-3 Condenser performance The precise representation of the heat-transfer performance of a condenser can be quite complex, because the refrigerant vapor enters the condenser superheated and following the onset of condensation in the tube the fraction of liquid and vapor changes constantly through the condenser. A satisfactory representation of air-cooled condenser performance for most engineering calculations is available, however, through an assumption of a constant heat-exchanger effectiveness for the condenser, namely

$$q_c = F(t_c - t_{amb}) \tag{14-4}$$

where F = capacity per unit temperature difference, kW/K

t_{amb} = ambient temperature, °C

Figure 14-3 shows the catalog performance for a certain air-cooled condenser for which F in Eq. (14-4) is 9.39 kW/K.

14-4 Condensing-unit subsystem; graphic analysis The first system—or more correctly subsystem—to be analyzed is that of a condensing unit. It consists of a compressor and condenser (Fig. 14-4) and performs the function of drawing low-pressure vapor from the evaporator, compressing and condensing the refrigerant, and supplying high-pressure liquid to the expansion device. Condensing units can be purchased as a pack-

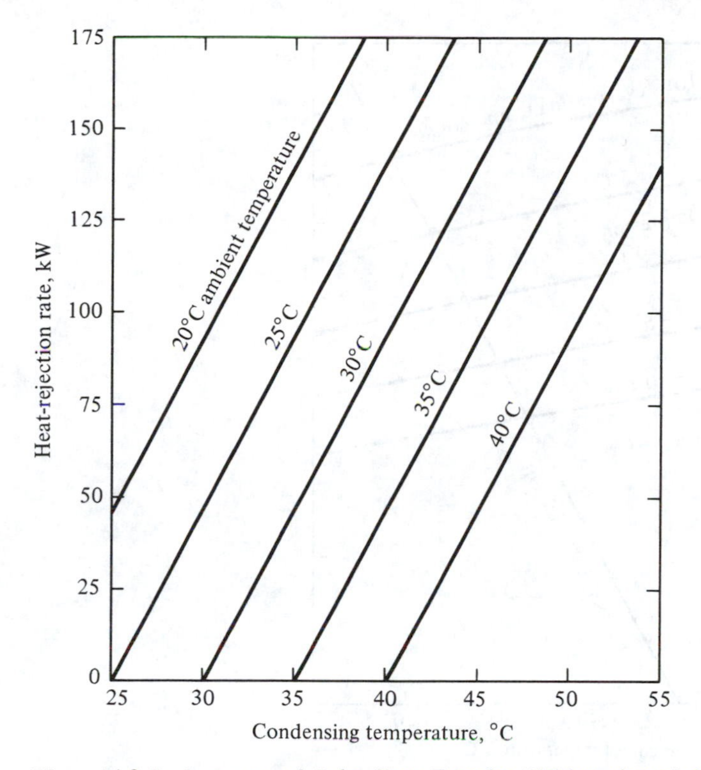

Figure 14-3 Performance of Bohn Heat Transfer Division air-cooled condenser, refrigerant 22, model no. 36.

age and may be installed outdoors to serve an air- or liquid-chilling evaporator located inside a building.

The behavior of the condensing unit is influenced by the evaporating temperature (and thus the suction pressure) of the vapor received from the evaporator. When t_e changes, the pumping capability and therefore the refrigerating capacity and t_c change. The superposition of Figs. 14-2 and 14-3, as shown in Fig. 14-5, can quantify the behavior of the condensing unit formed by the combination of that particular compres-

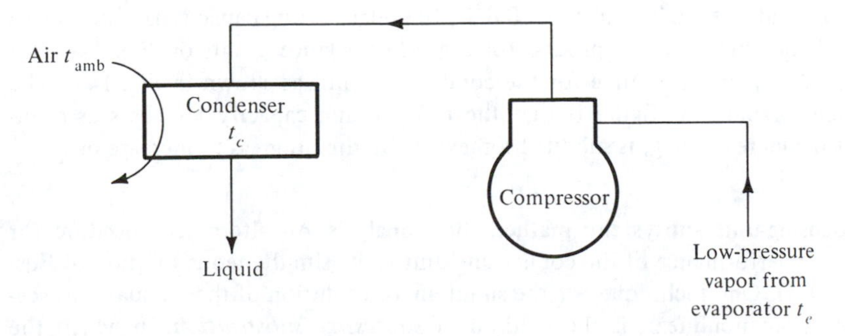

Figure 14-4 Condensing unit.

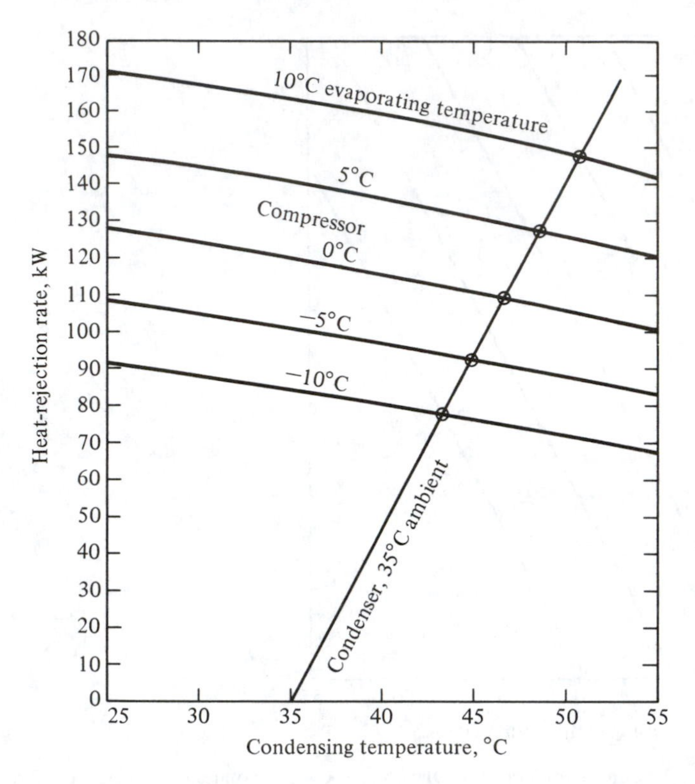

Figure 14-5 Balancing points of compressor and condenser that indicate performance of condensing unit.

sor and condenser. For a given ambient temperature, say, 35°C, Fig. 14-5 shows the balance points where the heat-rejection rates and condensing temperatures of both the compressor and condenser are simultaneously satisfied.

The refrigerating capacity provided by the condensing unit at various evaporating temperatures would be important information to have for the condensing unit. Such data can be extracted from a combination of Figs. 14-5 and 14-1. From Fig. 14-5, for example, one balance point is at t_e = 10°C and a condensing temperature t_c = 50.8°C. From Fig. 14-1 at t_e = 10°C and t_c = 50.8°C, the refrigerating capacity is found to be 115.5 kW. Following a similar process for the other balance points on Fig. 14-5, we can develop a performance curve for the condensing unit, as shown in Fig. 14-6. The trend apparent from that figure is that the refrigerating capacity increases as t_e increases but the increase in q_e is subdued somewhat by the progressive increase in t_c.

14-5 Condensing-unit subsystem; mathematical analysis An alternate procedure for simulating the performance of the condensing unit is the simultaneous solution of Eqs. (14-1) to (14-4). One technique for the simultaneous solution of these equations, several of which are nonlinear, is the method of *successive substitution*, in which the calculation sequence is set up and trial values are introduced for certain variables in

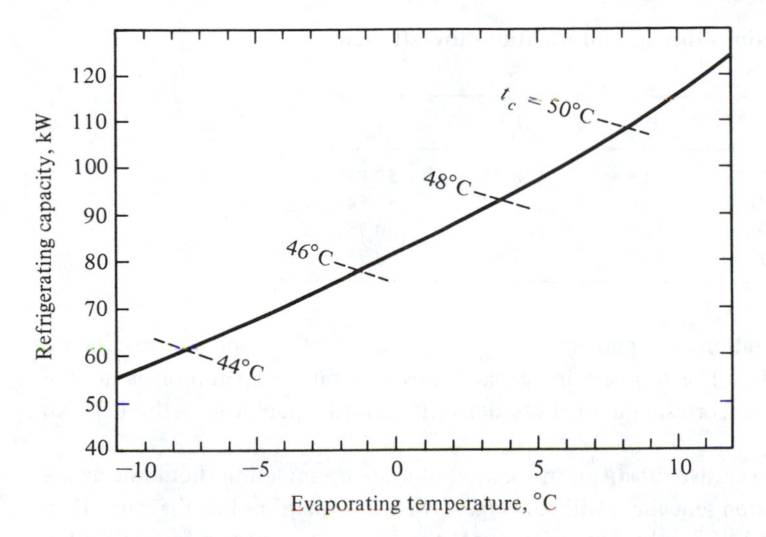

Figure 14-6 Performance of condensing unit consisting of the compressor of Fig. 14-1 and the condenser of Fig. 14-3. The temperature of ambient air for the condenser is 35°C.

order to get started. The values of the variables are updated each time the calculation proceeds through the loop.

A calculation loop or *information-flow diagram* for simulating the condensing unit is shown in Fig. 14-7 for an ambient temperature of 35°C and an evaporating temperature t_e = 10°C. A trial value for t_c of 50°C is arbitrarily selected and entered into the calculation in order to get started. The values of q_e, P, and q_c are calculated and then from Eq. (14-4) a new value of t_c is computed which replaces the trial value of 50°C for the next loop. The values of the variables computed through the several cycles starting with the trial value of t_c = 50°C are shown in Table 14-2. The converged values correspond to the balance point on Fig. 14-5, where the condenser-performance line intersects the 10°C evaporating-temperature line of the compressor performance.

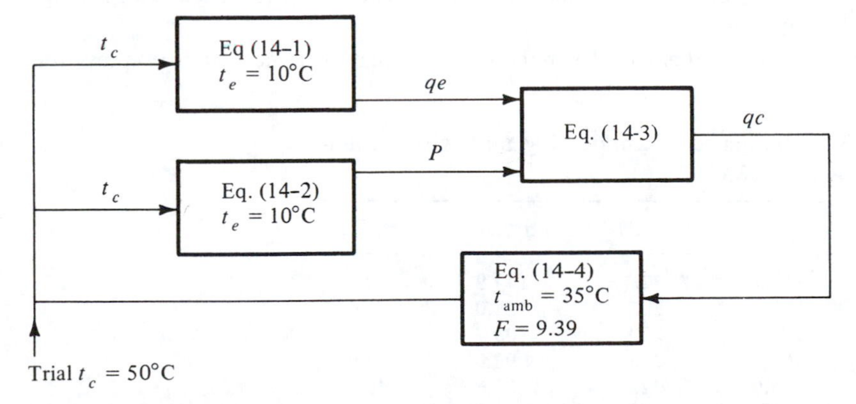

Figure 14-7 Information-flow diagram for condensing unit with t_{amb} = 35°C and t_e = 10°C.

Table 14-2 Calculations through information-flow diagram of Fig. 14-7

Cycle	q_e	P	q_c	t_c
1	116.71	32.06	148.77	50.84
2	115.32	32.46	147.77	50.74
3	115.49	32.41	147.90	50.75
4	115.47	32.41	147.88	50.75

Similar computations are performed for other values of t_e, and the results are shown in Table 14-3. The refrigerating capacity, evaporating temperatures, and condensing temperatures correspond to those derived from the graphical method shown in Fig. 14-6.

Users of the successive-substitution method of system simulation should be aware that not all calculation sequences will converge. Other information-flow diagrams than the one used in Fig. 14-7 can be devised to relate the four equations and four variables. Some calculation sequences will converge, as Table 14-2 illustrates, and some will diverge. If the sequence diverges, a different arrangement should be tried.

14-6 Evaporator performance Chapter 12 has explained the influence of individual heat-transfer coefficients, particularly the boiling coefficient, on the performance of a refrigerant evaporator. For system simulation we are concerned with the overall performance of the evaporator; one of the forms in which the behavior can be displayed is shown in Fig. 14-8 for a specific evaporator. The general trends evident from Fig. 14-8 are (1) that the capacity increases with a reduction in evaporating temperature and/or an increase in the temperature of entering water and (2) that the capacity is reduced when the rate of water flow is decreased at a given inlet temperature.

If the U value of the evaporator remained constant over the full range of operation shown in Fig. 14-8, the lines representing a given temperature of entering water would be straight (see Prob. 12-16). Instead they are curved upward slightly, indicating that the U value increases as the refrigeration capacity increases. This trend can be explained by the increase in the boiling heat-transfer coefficient (as discussed in Chap. 12) as the heat flux increases.

For subsequent mathematical simulation, an equation is needed to express the

Table 14-3 Performance of condensing unit with an ambient temperature of 35°C

t_e, °C	q_e, kW	P, kW	q_c, kW	t_c, °C
10	115.5	32.4	147.9	50.8
5	97.9	30.1	128.0	48.6
0	81.9	27.8	109.7	46.7
−5	67.5	25.4	92.9	44.9
−10	55.0	22.6	77.6	43.3

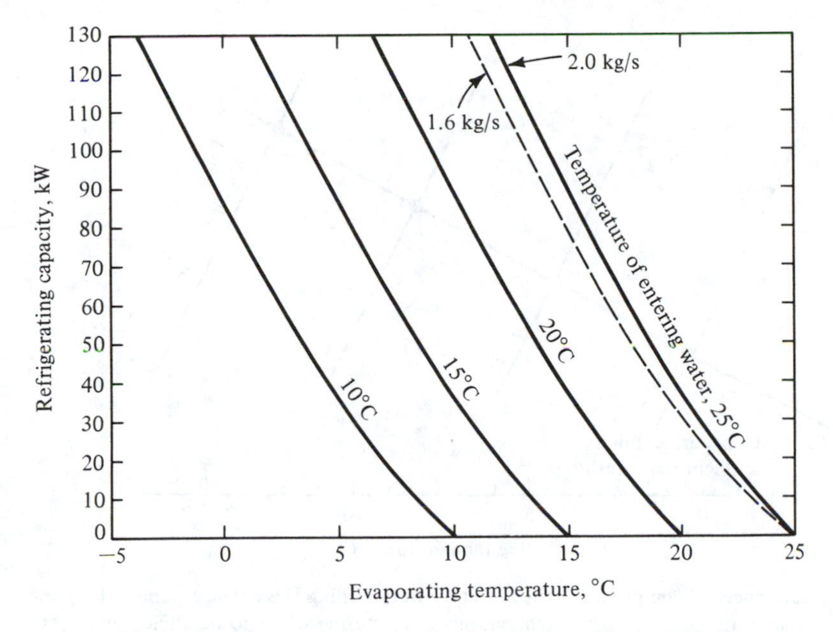

Figure 14-8 Refrigerating capacity of a Dunham-Bush, refrigerant 22, direct-expansion, inner-fin liquid chiller CH660B. The solid lines show performance with 2 kg/s water flow and the dashed line with 1.6 kg/s.

evaporator capacity shown in Fig. 14-8. An adequate equation could originate from

$$q_e = G(t_{wi} - t_e)$$

where t_{wi} = temperature of entering water, °C
G = proportionality factor, kW/K

If the U value were constant and the lines on Fig. 14-8 were straight, G would be a constant. Instead G increases with the temperature difference $t_{wi} - t_e$, and as an approximation G can be proposed as a linear function of the temperature difference. For the evaporator in Fig. 14-8 with a water flow rate of 2 kg/s

$$G = 6.0\,[1 + 0.046(t_{wi} - t_e)] \tag{14-5}$$

thus

$$q_e = 6.0[1 + 0.046(t_{wi} - t_e)]\,(t_{wi} - t_e) \tag{14-6}$$

14-7 Performance of complete system; graphic analysis The complete system consists of the compressor, condenser, and evaporator and the performance of the combination of two of these components (the compressor and condenser) has already been predicted in the balance-point determination that resulted in Fig. 14-6. By the superposition of Figs. 14-6 for the condensing unit and Fig. 14-8 for the evaporator the performance of the complete system can be predicted. Figure 14-9 shows this combination and the balance points of the system that occur at various temperatures of the return chilled water.

In summary, a graphic simulation of the vapor-compression system can be performed by first establishing the balance points for the condensing unit and then com-

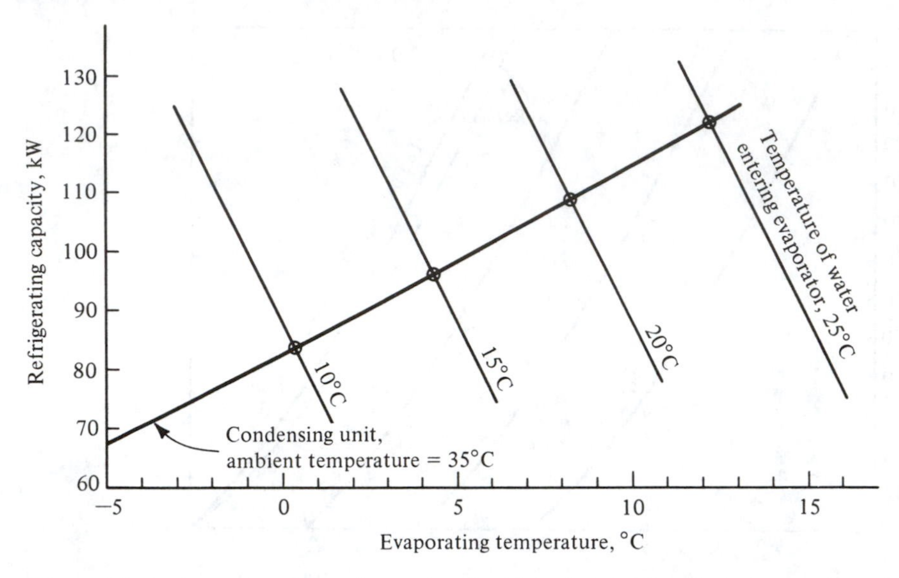

Figure 14-9 Performance of complete system found by determining the balance points of a condensing unit and an evaporator at various temperatures of entering water to be chilled and 35°C ambient temperature.

bining the evaporator performance with the condensing-unit performance to find the balance points of the entire system.

14-8 Simulation of complete system; mathematical analysis In the mathematical simulation it is not necessary to combine the components in pairs; instead the three components can be simulated simultaneously. The sequence of the calculation is shown by the information-flow diagram in Fig. 14-10. Starting with trial values of

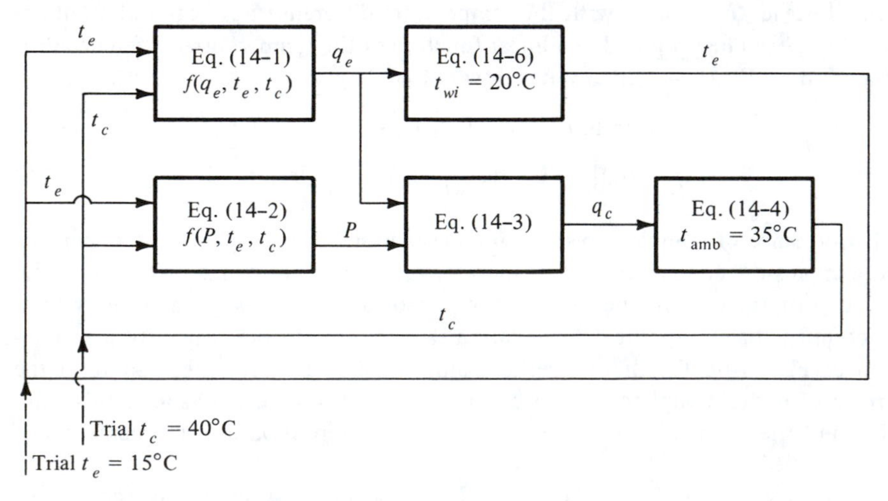

Figure 14-10 Information-flow diagram for simulating the complete vapor-compression system with $t_{wi} = 20°C$ and $t_{amb} = 35°C$.

Table 14-4 Simulation using information-flow diagram of Fig. 14-10 for ambient temperature of 35°C and entering-water temperature of 20°C

t_e, °C	t_c, °C	q_e, kW	P, kW	q_c, kW
15.0†	40.0†	158.0	26.2	184.2
4.6	54.6	87.6	31.8	119.4
10.0	47.7	120.4	31.0	151.4
7.3	51.1	103.5	31.8	135.3
8.7	49.4	111.8	31.5	143.3
8.0	50.3	107.7	31.6	140.3
8.3	49.8	109.7	31.6	140.8
8.2	50.0	108.7	31.6	140.3
8.2	49.9	109.2	31.6	140.8
8.2	50.0	109.0	31.6	140.6
8.2	50.0	109.1	31.6	140.7
8.2	50.0	109.0	31.6	140.6

† Trial

t_e = 15°C and t_c = 40°C, Table 14-4 shows the values of the operating variables as iterations through the calculation loops proceed. The converged values of capacity and evaporating temperature, 109.0 kW and 8.2°C, respectively, from Table 14-4 check with the balance point for 20°C entering-water temperature from Fig. 14-9. In addition, the mathematical simulation shows t_c = 50.0°C, which checks with the condensing temperature shown in Fig. 14-6 at the evaporating temperature of 8.2°C.

14-9 Some performance trends The results from the simulation of the complete vapor-compression system for various temperatures of entering water to be chilled are summarized in Table 14-5. Each time the temperature of entering water drops 5K the evaporating temperature drops too, but by less than 5K. The refrigerating capacity progressively decreases, due primarily to the compressor characteristics, which result in a reduced pumping capacity as t_e drops. As the refrigerating capacity falls off, the rate of heat rejection at the condenser also diminishes, permitting the condensing temperature to decline for a given ambient temperature.

The power required by the compressor is greatest at the highest temperature of entering water. The operating range in which to be most concerned about overloading

Table 14-5 Operating variables at various temperatures of entering water to be chilled

Ambient temperature = 35°C

t_{wi}, °C	t_e, °C	t_c, °C	q_e, kW	P, kW	q_c, kW
25	12.1	51.7	123.3	33.4	156.7
20	8.2	50.0	109.0	31.6	140.6
15	4.3	48.4	95.6	29.8	125.4
10	0.4	46.8	83.1	28.0	111.1

the motor driving the compressor is at high temperatures of entering chilled water. When the temperature of entering water is 10°C, the evaporating temperature is 0.4°C, which is drawing close to the freezing temperature of water. Catalog data would probably not be shown for a chiller at lower temperatures than these, or if the data are shown, there would be a reminder to the user to protect the water from freezing by adding an antifreeze.

14-10 The expansion device Up to now in this chapter the complete vapor-compression system has been described as consisting of three components, the compressor, condenser, and evaporator, and no mention has been made of the expansion device. In predicting system performance so far it has tacitly been assumed that the expansion device is able to regulate the flow of refrigerant into the evaporator so that the heat-transfer surfaces on the refrigerant side of the evaporator are wetted with liquid refrigerant. Chapter 13 on expansion devices explained that the capillary tube achieves this goal in only certain combinations of condensing and evaporating pressures. The superheat-controlled expansion valve operates on the principle of maintaining a small amount of superheat in the evaporator but does provide most of the evaporator surfaces with liquid throughout a wide range of condensing and evaporating pressures.

The consequence for system performance of the expansion device starving the evaporator is illustrated in Fig. 14-11, showing the balance points between the condensing unit and the evaporator. When the evaporator is starved, the overall heat-transfer coefficient of the evaporator drops and the balance point shifts to a lower evaporating temperature and refrigerating capacity.

Even when using the controlling type of expansion device, such as the superheat-controlled valve, there may be conditions that result in a starved evaporator, e.g., (1) the expansion valve is too small, (2) some vapor is present in the liquid entering the expansion valve, or (3) the pressure difference across the valve is too small. Condition 2 is likely to occur if the refrigerant charge in the system is too small, the pressure drop in the liquid line is high due to friction, or the valve and evaporator are located at a higher elevation than the condenser. Condition 3 occurs many times in systems with air-cooled condensers when the ambient temperature is low. In such cases the condens-

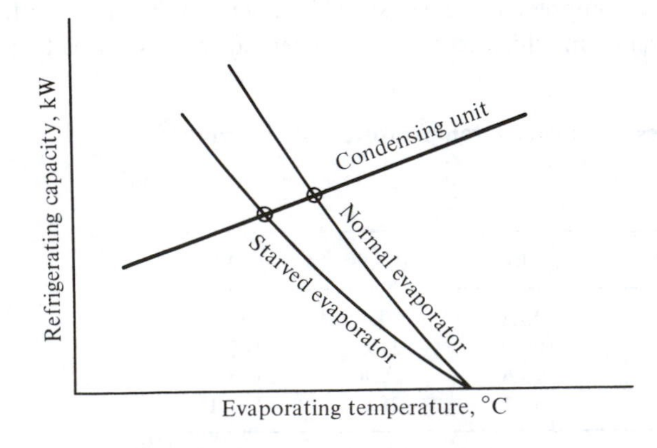

Figure 14-11 Reduction in capacity and evaporating temperature due to feeding insufficient refrigerant to the evaporator.

Table 14-6 Sensitivity study of vapor-compression system†

Ratio of component capacity to base capacity			Refrigerating capacity, kW	Percent increase
Compressor	Condenser	Evaporator		
1.0	1.0	1.0	95.6	
1.1	1.0	1.0	101.6	6.3
1.0	1.1	1.0	96.8	1.3
1.0	1.0	1.1	97.6	2.1
1.1	1.1	1.1	105.2	10.0

† The study analyzes the effect on the refrigerating capacity of the system caused by 10 percent increases in component capacities. The ambient temperature is 35°C, and the temperature of entering water to be chilled is 15°C.

ing temperature drops so low that there is insufficient pressure difference across the valve. The extreme consequence of this condition is that the evaporating temperature and pressure drop so low that the flow rate of refrigerant diminishes considerably. With hermetically sealed compressors and motors, the motor is cooled by the flow of refrigerant across it, and if the flow rate of refrigerant becomes too low, the motor may burn out.

14-11 Sensitivity analysis In engineering terminology *sensitivity analysis* means the process of examining how one variable in a system is affected by a change in another variable or parameter of the system. Of special applicability to the vapor-compression system is the exploration of the influence on refrigerating capacity of changes in capacity of each of the components—the compressor, condenser, and evaporator.[2] A simple extension of the computer runs of the sequence of Fig. 14-10 permits such an analysis. Table 14-6 summarizes the effects of increasing the capacity of each component, one at a time, by 10 percent. A 10 percent increase in compressor capacity means a 10 percent increase in refrigerating capacity and power for given evaporating and condensing pressures. A 10 percent increase in condenser capacity means a 10 percent increase in F in Eq. (14-4), and a 10 percent increase in evaporator capacity means a 10 percent increase in q_e in Eq. (14-6) for given values of t_{wi} and t_e. Table 14-6 shows that the compressor capacity has a dominant influence on the system capacity and that the evaporator is next in importance. These influences agree approximately with those of Kaufman,[3] whose results were derived from a graphical analysis. The relative influence shown in Table 14-6 will not apply precisely to all vapor-compression systems because the influences are also affected by the choice of the base condition. For example, if the base condition has an extremely large condenser, a 10 percent increase in this large condenser will not show as much increase in refrigerating capacity as the 1.3 percent indicated by Table 14-6.

An important advantage to a designer of having information like that shown in Table 14-6 is that the data can be used for optimization. By combining the data from Table 14-6 with knowledge of the costs of increasing (or the savings of decreasing) component capacity the designer can decide how to reduce the first cost of the system that produces a given refrigerating capacity.

PROBLEMS

14-1 Either graphically or by using the computer, for an ambient temperature of 30°C develop the performance characteristics of a condensing unit (of the form of Fig. 14-6 or Table 14-3) if the compressor has performance shown by Fig. 14-1 [or Eqs. (14-1) and (14-2)] and the condenser has characteristics shown by Fig. 14-3 [or Eq. (14-4)].
Ans.

q_e, kW	122.8	104.4	87.6	72.5	59.3
t_e, °C	10	5.0	0	-5.0	-10.0

14-2 Combine the condensing unit of Prob. 14-1 (using answers provided) with the evaporator of Fig. 14-8 to form a complete system. The water flow rate to the evaporator is 2 kg/s, and the temperature of water to be chilled is 10°C

(*a*) What are the refrigerating capacity and power requirement of this system? *Ans.* q_e = 87.6 kW, P = 26.4 kW

(*b*) This system pumps heat between 10°C and an ambient temperature of 30°C, which is the same temperature difference as from 15 to 35°C, for which information is available in Table 14-4. Explain why the refrigerating capacity and power requirement are less at the lower temperature level.

14-3 Section 14-11 suggests that the influences of the several components shown in Table 14-6 are dependent upon relative sizes of the components at the base condition. If the base system is the same as that tabulated in Table 14-6 except that the condenser is twice as large [F = 18.78 kW/K in Eq. (14-4)], what is the increase in system capacity of a 10 percent increase in condenser capacity above this new base condition? The ambient temperature is 35°C, and the entering temperature of water to be chilled is 15°C. *Ans.* 0.62%

14-4 For the components of the complete system described in Secs. 14-7, 14-8, and 14-11 the following costs (or savings) are applicable to a 1 percent change in component capacity. An optimization is now to proceed by increasing or decreasing sizes of components in order to reduce the first cost of the system. What relative changes in component sizes should be made in order to reduce the first cost of the system but maintain a fixed refrigerating capacity? *Ans.* Decrease evaporator capacity 3 times the increase in compressor capacity

Component	Increase (saving) in first cost for 1% increase (decrease) in component capacity
Compressor	\$2.80
Condenser	0.67
Evaporator	1.40

REFERENCES

1. Stoecker, W. F.: "Design of Thermal Systems," 2d ed., McGraw-Hill, New York, 1980.
2. Backstrom, M.: The Use of Influence Numbers in Calculating Refrigeration Plant, *Kyltek. Tidskr.*, vol. 22, no. 3, pp. 43–48, June 1963.
3. Kaufman, B.: Graphical Analysis for Air Conditioning System Performance, *Refrig. Eng.*, vol. 67, no. 7, p. 52, July 1956.

CHAPTER

FIFTEEN

REFRIGERANTS

15-1 Primary and secondary refrigerants Most of this chapter is devoted to *primary refrigerants* by which is meant the refrigerants used in vapor-compression systems. Section 15-11 describes the properties of *secondary refrigerants,* which are liquids used for transporting low-temperature heat energy from one location to another. Other names for secondary refrigerants are antifreezes and brines. Two substances form the refrigerant combination in absorption refrigeration systems, but their characteristics will be presented later in Chap. 17.

The most common refrigerants are the fluorinated hydrocarbons, but numerous other substances also function well as refrigerants, including many inorganic compounds and hydrocarbons. This chapter will describe and present characteristics of only the most widely used refrigerants.

15-2 Halocarbon compounds The halocarbon group includes refrigerants which contain one or more of the three halogens chlorine, fluorine, and bromine. The numerical designation, the chemical name, and the chemical formula of some of the commercially available members of this group are shown in Table 15-1.

The numbering system in the halocarbon group follows this pattern: the first digit on the right is the number of fluorine atoms in the compound; the second digit from the right is one more than the number of hydrogen atoms in the compound; and the third digit from the right is one less than the number of carbon atoms. When the third digit is zero, it is omitted.

Table 15-1 Some halocarbon refrigerants

Numerical designation	Chemical name	Chemical formula
11	Trichloromonofluoromethane	CCl_3F
12	Dichlorodifluoromethane	CCl_2F_2
13	Monochlorotrifluoromethane	$CClF_3$
22	Monochlorodifluoromethane	$CHClF_2$
40	Methyl chloride	CH_3Cl
113	Trichlorotrifluoroethane	CCl_2FCClF_2
114	Dichlorotetrafluoroethane	$CClF_2CClF_2$

15-3 Inorganic compounds Many of the early refrigerants were inorganic compounds, and some have maintained their prominence to this day. These compounds are listed in Table 15-2.

15-4 Hydrocarbons Many hydrocarbons are suitable as refrigerants, especially for service in the petroleum and petrochemical industry. Several such refrigerants are listed in Table 15-3.

15-5 Azeotropes An azeotropic mixture of two substances is one which cannot be separated into its components by distillation. An azeotrope evaporates and condenses as a single substance with properties that are different from those of either constituent. The most popular azeotrope is refrigerant 502, which is a mixture of 48.8 percent refrigerant 22 and 51.2 percent refrigerant 115. The properties of saturated refrigerant 502 are given in Table A-8, and the properties of superheated vapor are shown in Fig. A-5.

15-6 Thermodynamic comparison of some common refrigerants Some thermodynamic and efficiency characteristics of several common refrigerants are presented in Table 15-4. The pressures, refrigerating effect, volume flow per unit refrigeration capacity, and the coefficient of performance (COP) are based on the standard vapor-compression cycle (Sec. 10-12) with an evaporating temperature of $-15^\circ C$ and a condensing temperature of $30^\circ C$.

Table 15-2 Some inorganic refrigerants

Numerical designation†	Chemical name	Chemical formula
717	Ammonia	NH_3
718	Water	H_2O
729	Air	
744	Carbon dioxide	CO_2
764	Sulfur dioxide	SO_2

† The last two digits are the molecular weight.

Table 15-3 Hydrocarbon refrigerants

Numerical designation†	Chemical name	Chemical formula
50	Methane	CH_4
170	Ethane	C_2H_6
290	Propane	C_3H_8

† Follows same principle as the halocarbon scheme.

The operating pressures should be low enough for lightweight vessels and pipes to contain the refrigerant. On the other hand, pressures below atmospheric pressure, which occur in refrigerant 11 evaporators, have the disadvantage of drawing air into the evaporator if there are any leaks. Systems using this refrigerant should be equipped with a purger to eliminate any air that leaks in. A low pressure ratio is desirable from the standpoint of any type of compressor—reciprocating, screw, or centrifugal.

The refrigerating effect would at first seem to be a good indicator of the cycle efficiency, but this property must be considered in combination with the work of compression. Refrigerant 717 (ammonia), for example, has a refrigerating effect much larger than the other refrigerants, but the work of compression of ammonia is also high, so that its COP is of the same order of magnitude as that of the other refrigerants shown.

The flow rate of suction vapor per kilowatt of refrigeration influences the pumping rate and/or the type of compressor. Refrigerants 22, 502, and 717 show comparable values of this term, while refrigerant 12 is less dense than those three and requires a higher vapor flow rate. The high volume flow rate of refrigerant 11 indicates why it is used in centrifugal compressors (Sec. 11-25).

The coefficient of performance of a standard refrigeration cycle using refrigerant 11 is higher than the others, but this advantage is not sufficient to permit its application in any but systems using centrifugal compressors. The COPs of refrigerants 12, 22, and 717 are nearly the same, and while that of refrigerant 502 is the lowest shown,

Table 15-4 Thermodynamic characteristics of several refrigerants

Operation on a standard vapor-compression cycle with an evaporating temperature of -15°C and a condensing temperature of 30°C

Refrigerant	Evaporating pressure, kPa	Condensing pressure, kPa	Pressure ratio	Refrigerating effect, kJ/kg	Suction vapor flow per kW of refrigeration, L/s	COP
11	20.4	125.5	6.15	155.4	4.90	5.03
12	182.7	744.6	4.08	116.3	0.782	4.70
22	295.8	1192.1	4.03	162.8	0.476	4.66
502	349.6	1308.6	3.74	106.2	0.484	4.37
717	236.5	1166.6	4.93	1103.4	0.462	4.76

this refrigerant has some other advantages that will be discussed in the next section. As a basis of comparison the COP of a Carnot cycle operating at the evaporating and condensing temperatures of Table 15-4 is 5.74.

15-7 Physical and chemical comparison Two important characteristics of refrigerants from a safety standpoint are its flammability and toxicity.[1] Of the refrigerants listed in Table 15-4, ammonia is listed as flammable with 16 to 25 percent ammonia by volume in air, while the others are considered nonflammable. With respect to toxicity, refrigerant 12 is considered nontoxic in concentrations up to 20 percent by volume for an exposure period of less than 2 h, while ammonia is assigned to a group of refrigerants considered injurious or lethal in concentrations of $\frac{1}{2}$ to 1 percent for exposures of $\frac{1}{2}$ h duration. Refrigerants 11, 22, and 502 are in a class slightly more toxic than refrigerant 12.

How the refrigerant combines with oil in the system is a factor in its selection. No chemical reaction between the refrigerant and the lubricating oil of the compressor is anticipated, but the miscibility of the oil and the refrigerant is of concern. In reciprocating and screw compressors some oil carries out of the compressor with the refrigerant discharge gas. This oil passes through the condenser and on to the evaporator, where the refrigerant vaporizes off, leaving the oil to reduce the heat-transfer effectiveness of the evaporator.

Several procedures are available to prevent oil from reaching the evaporator or to remove it after it collects. An oil separator placed in the discharge-gas line removes oil continuously and returns it to the compressor, where it belongs. Refrigerant 12 and oil are miscible, whereas refrigerant 22 is partially miscible and ammonia is not miscible with oil. Oil in the evaporator of a refrigerant 12 system is not nearly so detrimental to heat transfer as in an ammonia system, where it separates. Oil can be drained from ammonia evaporators, but in refrigerant 12 systems the velocity in the suction line must be kept high enough to carry oil back to the compressor. The popularity of several refrigerants rises and falls over the years: refrigerant 12 was most popular and was then supplanted in popularity by refrigerant 22 because of its lower volume flow rate per unit capacity. Refrigerant 502 has become popular lately, because it has comparable volume flow rates to refrigerant 22 but oil is more miscible in it and because it has lower discharge temperatures (see Fig. 11-13).

When a leak occurs in a refrigeration system, the refrigerant may come in contact with the product, such as food. The halocarbons are generally considered to have negligible effect on foods, furs, or fabrics for short exposures. Prolonged exposure to ammonia could result in food tasting or smelling of ammonia, although it should be remembered that a small amount of ammonia (0.01 to 0.1 percent) is present in foods naturally.[2]

Reaction of a refrigerant with a material of construction used in the piping, vessels, and compressor does not usually influence the selection of the refrigerant, but the refrigerant used does frequently dictate the material employed in the system. Certain metals may be attacked by refrigerants. Ammonia, for example, reacts with copper, brass, or other cuprous alloys in the presence of water. Iron and steel are therefore used in ammonia systems. The halocarbons may react with zinc but not copper, aluminum, iron, or steel. In the presence of a small quantity of water, however, the halocarbons form acids which attack most metals. The halocarbons attack natural rubber; therefore synthetic material should be used as gaskets and for other sealing purposes.

15-8 Thermal conductivity and viscosity of refrigerants Occasionally the engineer will need viscosity and thermal-conductivity data for making heat transfer and/or flow calculations; Table 15-5 presents a summary of these properties.

15-9 Ozone depletion An alarm was sounded[3] in the mid-1970s that the chlorine from halogenated hydrocarbons released to the environment was using up ozone in the stratosphere. A reduction in the ozone composition of the stratosphere would permit

Table 15-5 Thermal conductivities and viscosities of saturated refrigerant liquid and vapor[1]

		Viscosity, Pa • s		Conductivity, W/m • K	
Refrigerant	t, °C	Liquid	Vapor	Liquid	Vapor
11	-40	0.000922		0.106	
	-20	0.000694		0.100	
	0	0.000546		0.0943	
	20	0.000441	0.0000103	0.0890	
	40	0.000367	0.0000119	0.0832	0.00841
	60	0.000312	0.0000127	0.0777	0.0093
12	-40	0.000409		0.0931	
	-20	0.000325	0.0000108	0.0857	0.00734
	0	0.000267	0.0000118	0.0784	0.00838
	20	0.000225	0.0000126	0.0711	0.00938
	40	0.000194	0.0000135	0.0637	0.0105
	60	0.000169	0.0000148	0.0564	0.0118
22	-40	0.000330	0.0000101	0.120	0.0069
	-20	0.000275	0.0000110	0.110	0.00817
	0	0.000237	0.0000120	0.100	0.00942
	20	0.000206	0.0000130	0.090	0.0107
	40	0.000182	0.0000144	0.0805	0.0119
	60	0.000162	0.0000160	0.0704	0.0133
502	-40	0.000356	0.0000100	0.0898	0.00796
	-20	0.000284	0.0000111	0.0820	0.00907
	0	0.000233	0.0000120	0.0742	0.0102
	20	0.000193	0.0000132	0.0665	0.0114
	40	0.000153	0.0000146	0.0585	0.0124
	60	0.000117	0.0000161	0.0486	0.0144
717	-40			0.632	
	-20	0.000236	0.0000097	0.585	0.0204
	0	0.000190	0.0000104	0.540	0.0218
	20	0.000152	0.0000112	0.493	0.0267
	40	0.000122	0.0000120	0.447	0.0318
	60	0.000098	0.0000129	0.400	0.0381

more ultraviolet radiation to reach the earth and might cause cancer. The initial reactions to the warnings were to stop using the offending halocarbons as the propellent in aerosol containers and to reduce their use in foam insulation. Although the application of halocarbons as refrigerants is only one of the uses of these substances, on a worldwide basis considerable quantities of halocarbons are used for refrigeration purposes, and the immediate response of the refrigerant industry[4] has been to tighten up procedures for preventing spills into the atmosphere. Refrigerants 11 and 12, with their three and two chlorine atoms, respectively, have a greater impact than refrigerant 22 which has one chlorine atom. During the coming years the situation will be monitored carefully and the need for protection of the environment may demand that still more care be exercised and/or new refrigerants be employed.

15-10 Basis of choice of refrigerant The characteristics of refrigerants presented in this chapter are dominant factors in the choice. The following is a brief and rough review of the principal applications of some refrigerants.

Air. The major use of air as a refrigerant is in aircraft, where the light weight of an air system compensates for its low COP.

Ammonia. Large industrial low-temperature installations are the applications where ammonia is most frequently used. Many new ammonia systems come into operation each year.

Carbon dioxide. This refrigerant is sometimes used for direct-contact freezing of food. Its high condensing pressure usually limits its application to the low-temperature side of a cascade system where a different refrigerant operates in the high-temperature section.

Refrigerant 11. Along with refrigerant 113 this refrigerant is popular for centrifugal compressor systems.

Refrigerant 12. This refrigerant is used primarily with reciprocating compressors for service in domestic refrigeration appliances and in automotive air conditioners.

Refrigerant 22. Because a smaller and lower-cost compressor can be used with refrigerant 22 than with refrigerant 12, this refrigerant has taken over many air-conditioning applications from refrigerant 12.

Refrigerant 502. This is one of the newer refrigerants, with some of the advantages of refrigerant 22 but with the further advantage of better behavior with oil and lower compressor discharge temperatures than refrigerant 22.

15-11 Secondary refrigerants Secondary refrigerants are fluids that carry heat from a substance being cooled to the evaporator of a refrigeration system. The secondary refrigerant experiences a change in temperature when it absorbs the heat and liberates it at the evaporator, and the secondary refrigerant does not change phase. Technically speaking, water could be a secondary refrigerant, but the substances we particularly wish to explore are brines and antifreezes, which are solutions with freezing temperatures below 0°C. Several of the most widely used antifreezes are solutions of water and ethylene glycol, propylene glycol, or calcium chloride. Propylene glycol has the unique feature of being safe in contact with foods. The properties[5,6] of these antifreezes are similar; those of ethylene glycol will be presented so that some quantitative implications can be derived.

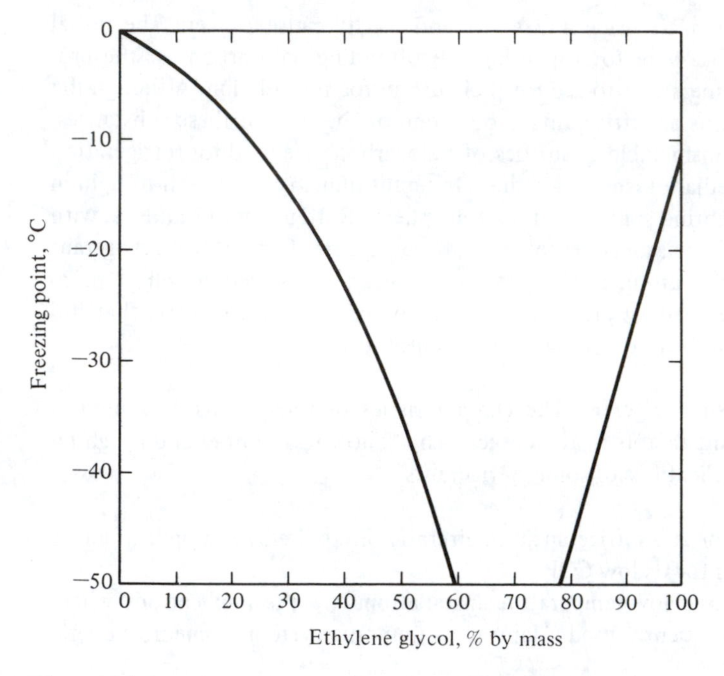

Figure 15-1 Freezing point of ethylene glycol solutions.

One of the most important properties of antifreeze solutions is the freezing point, shown in Fig. 15-1. The freezing points form the classical phase diagram shown in skeleton form in Fig. 15-2. The curves of the freezing points show that the solution of the two constituents has a lower freezing point than either substance individually. Figure 15-2 shows possible phases and mixtures that can exist at various concentrations

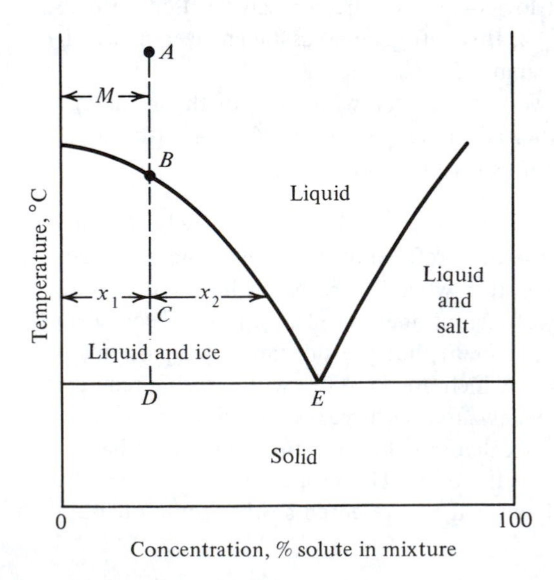

Figure 15-2 Phase diagram of an antifreeze.

and temperatures. If the antifreeze at temperature A has a concentration M, what will the behavior of the antifreeze be as it progressively cools? The antifreeze remains a liquid until the temperature drops to B. Further cooling to C results in a slush which is a mixture of ice and liquid. The antifreeze at C has concentrated itself by freezing out some of its water into ice. The percent of ice in the mixture at C is given by

$$\text{Percent ice} = \frac{x_2}{x_1 + x_2}(100) \tag{15-1}$$

and the percent liquid is given

$$\text{Percent liquid} = \frac{x_1}{x_1 + x_2}(100) \tag{15-2}$$

Cooling the solution below D solidifies the entire mixture.

Point E, called the *eutectic point*, represents the concentration at which the lowest temperature can be reached with no solidification. Strengthening the solution beyond the eutectic concentration is fruitless, because the freezing temperature rises.

Further properties of ethylene glycol solutions are presented in Fig. 15-3 (specific gravity), Fig. 15-4 (thermal conductivity), Fig. 15-5 (viscosity), and Fig. 15-6 (specific heat).

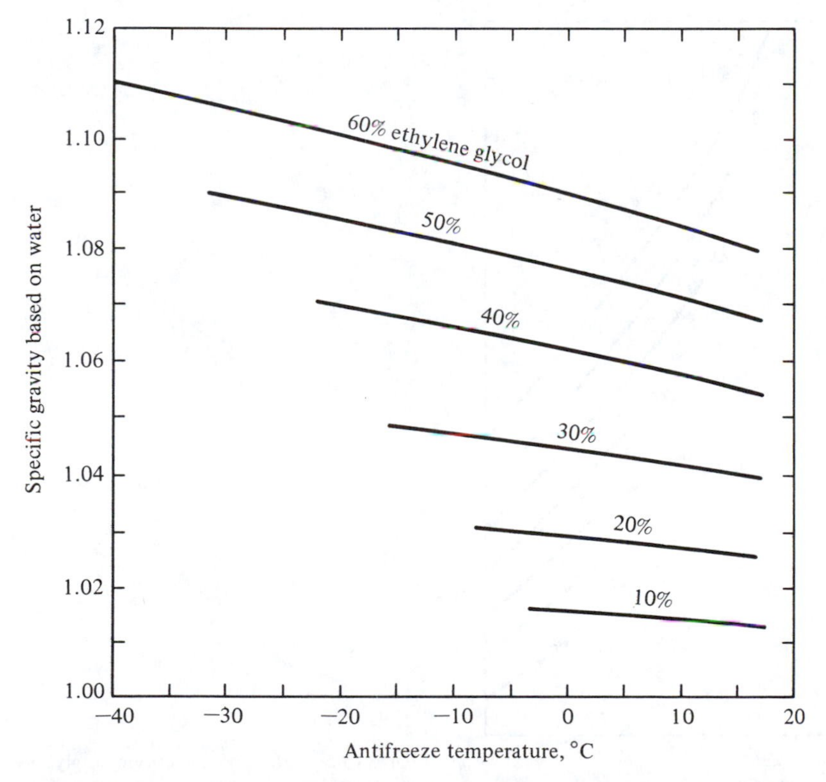

Figure 15-3 Specific gravity of ethylene glycol-water solutions based on water at 4°C (density = 1000 kg/m^3).

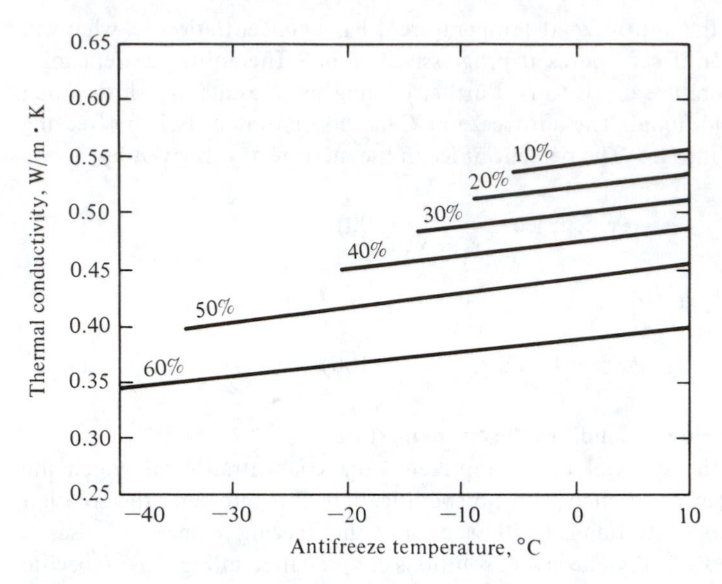

Figure 15-4 Thermal conductivity of ethylene glycol-water solutions.

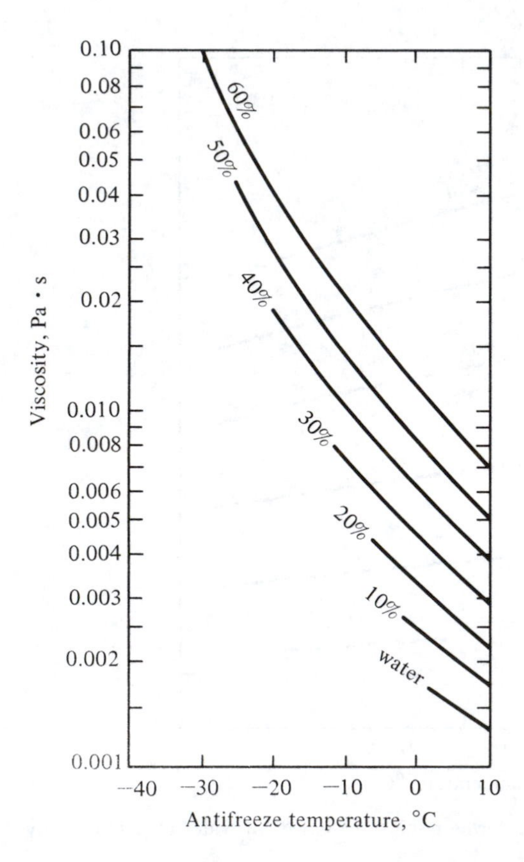

Figure 15-5 Viscosity of ethylene glycol-water solutions.

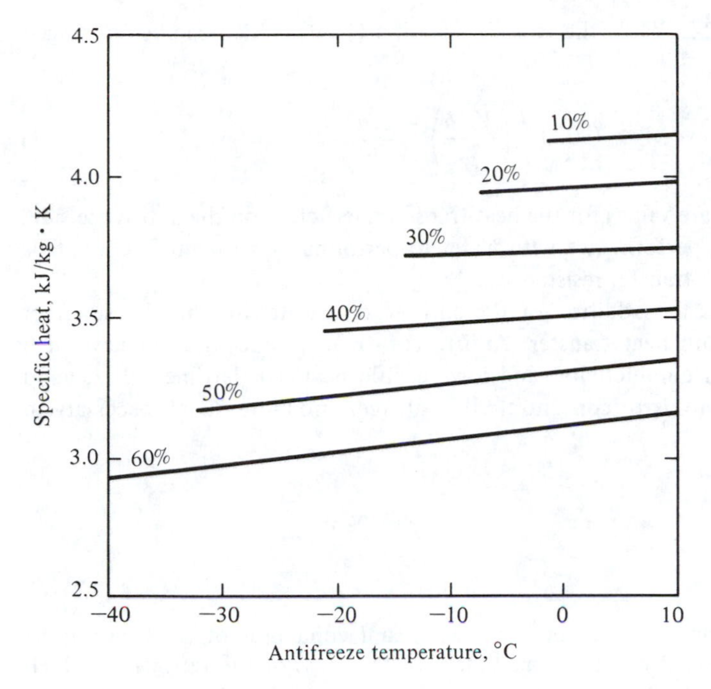

Figure 15-6 Specific heat of ethylene glycol-water solutions.

In designing a refrigeration system, it may be necessary to convert a manufacturer's data given for water into data for an antifreeze. For pressure drop in straight tubes the following ratio holds

$$\frac{\Delta p_a}{\Delta p_w} = \frac{f_a \dfrac{L_a}{D_a} \dfrac{V_a^2}{2} \rho_a}{f_w \dfrac{L_w}{D_w} \dfrac{V_w^2}{2} \rho_w} \tag{15-3}$$

where the a subscript refers to antifreeze and the w subscript refers to water. The velocity is V and the density is ρ. For a given heat exchanger the length L and diameter D for the two fluids cancel. An expression for the friction in the low Reynolds number range of turbulent flow was presented as Eq. (13-9)

$$f = \frac{0.33}{\mathrm{Re}^{0.25}} \tag{15-4}$$

If the properties and velocities of water and the antifreeze are known, pressure-drop data for antifreeze flow can be estimated from the knowledge of the pressure drop when using water.

For a conversion of heat-transfer data from water to antifreeze the task is more difficult. The overall U value of a heat exchanger depends upon the heat-transfer coefficient of both the antifreeze and the other fluid flowing through the exchanger, as discussed in Sec. 12-2. While the Nusselt equation (12-9), which can be rearranged as

$$h = 0.023 \frac{k}{D}\left(\frac{VD\rho}{\mu}\right)^{0.8}\left(\frac{c_p\mu}{k}\right)^{0.4} \tag{15-5}$$

gives reasonably accurate values for the heat-transfer coefficient on the antifreeze side, a conversion from the performance with water to performance with antifreeze entails a separation of the heat-transfer resistances.

For all antifreezes the addition of the antifreeze to water has an adverse effect upon pressure drop and heat transfer. Antifreezes of high concentration have high viscosity, low thermal conductivity, and low specific heat—all detrimental. A good operating rule, therefore, is to concentrate the antifreeze no more than is necessary to prevent its freezing.

PROBLEMS

15-1 The machine room housing the compressor and condenser of a refrigerant 12 system has dimensions 5 by 4 by 3 m. Calculate the mass of the refrigerant which would have to escape into the space to cause a toxic concentration for a 2-h exposure. *Ans.* 60 kg

15-2 Using data from Table 15-4 for the standard vapor-compression cycle operating with an evaporating temperature of -15°C and a condensing temperature of 30°C, calculate the mass flow rate of refrigerant per kilowatt of refrigeration and the work of compression for (*a*) refrigerant 22 and (*b*) ammonia. *Ans.* (*b*) 0.000906 kg/s, 0.210 kW

15-3 A 20% ethylene glycol solution in water is gradually cooled.

(*a*) At what temperature does crystallization begin?

(*b*) If the antifreeze is cooled to -25°C, what percent will have frozen into ice? *Ans.* 51%

15-4 A solution of ethylene glycol and water is to be prepared for a minimum temperature of -30°C. If the antifreeze is mixed at 15°C, what is the required specific gravity of the antifreeze solution at this temperature? *Ans.* 1.06

15-5 For a refrigeration capacity of 30 kW, how many liters per second of 30% solution of ethylene glycol–water must be circulated if the antifreeze enters the liquid chiller at -5°C and leaves at -10°C? *Ans.* 1.54 L/s

15-6 A manufacturer's catalog gives the pressure drop through the tubes of a heat exchanger as 70 kPa for a given flow rate of water at 15°C. If a 40% ethylene glycol-water solution at -20°C flows through the heat exchanger at the same mass flow rate as the water, what will the pressure drop be? Assume turbulent flow. At 15°C the viscosity of water is 0.00116 Pa • s. *Ans.* 131 kPa

15-7 Compute the convection heat-transfer coefficient for liquid flowing through a 20-mm-ID tube when the velocity is 2.5 m/s if the liquid is (*a*) water at 15°C, which has a viscosity of 0.00116 Pa • s and a thermal conductivity of 0.584 W/m • K; (*b*) 40% solution of ethylene glycol at -20°C. *Ans.* (*b*) 2182 $W/m^2 \cdot k$

REFERENCES

1. "ASHRAE Handbook and Product Directory, Fundamentals Volume," American Society of Heating, Refrigerating, and Air-Conditioning Engineers, Atlanta, Ga., 1981.
2. Effect of Ammonia, *Refrig. Res. Found. Inform. Bull.* p. 4, Washington, D.C., January 1979.
3. M. J. Molina and F. S. Rowland: Stratospheric Sink for Chlorofluoromethanes: Chlorine Atom-Catalyzed Destruction of Ozone, *Nature,* no. 249, p. 810, 1974.
4. B. A. Thrush: The Halocarbon Contamination Problem, pt 1: Atmospheric Effects of Halocarbons, *15th Int. Cong. Refrig., Venice, September 1979,* plenary pap.
5. "Glycols," Chemicals and Plastics Division, Union Carbide Corporation, 1978.
6. C. S. Cragoe: Properties of Ethylene Glycol and Its Aqueous Solutions, *Nat. Bur. Std. Rep.* 4268, 1955.

CHAPTER

SIXTEEN

MULTIPRESSURE SYSTEMS

16-1 Multipressure systems in industrial refrigeration A multipressure system is a refrigeration system that has two or more low-side pressures. The low-side pressure is the pressure of the refrigerant between the expansion valve and the intake of the compressor. A multipressure system is distinguished from the single-pressure system, which has but one low-side pressure. A multipressure system may be found, for example, in a dairy where one evaporator operates at -35°C to harden ice cream while another evaporator operates at 2°C to cool milk. Another typical application might be in a process industry where a two- or three-stage compression arrangement serves an evaporator operating at a low temperature of -20°C or lower.

This chapter considers only multipressure systems having two low-side pressures, but the principles developed here will apply to more than two low-side pressures. Two functions often integral to multipressure systems are the removal of flash gas and intercooling. They will be discussed first. Then several combinations of multiple evaporators and compressors will be analyzed.

16-2 Removal of flash gas A saving in the power requirement of a refrigeration system results if the flash gas that develops in the throttling process between the condenser and evaporator is removed and recompressed before complete expansion. When saturated liquid expands through an expansion valve, the fraction of vapor or flash gas progressively increases. The expansion process shown on the pressure-enthalpy diagram in Fig. 16-1 takes place from 1 to 2. The state point, as the expansion proceeds, moves into a region of a greater fraction of vapor.

The end point of the expansion, 2, could have been achieved by interrupting the expansion at 3 and separating the liquid and vapor phases, which are 4 and 6, respectively. The expansion could then continue by expanding the liquid at 4 and the vapor

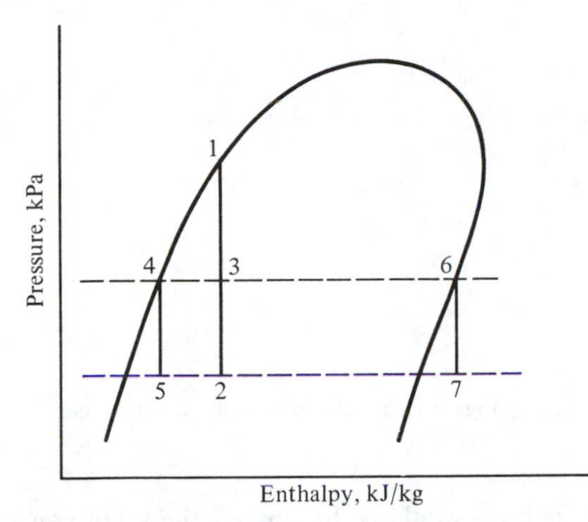

Figure 16-1 Expansion process showing replacement of process 3-2 with the combination of 4-5 and 6-7.

at 6 to the final pressure, giving 5 and 7, respectively. The combination of refrigerant at states 5 and 7 gives point 2.

Inspection of the expansion from 6 to 7 confirms that it is wasteful. In the first place, the refrigerant at 7 can do no refrigerating; in the second place, work will be required to compress the vapor back to the pressure it had at 6. Why not perform part of the expansion, separate the liquid from the vapor, continue expanding the liquid, and recompress the vapor without further expansion? The equipment to achieve this separation is called a *flash tank* (see Fig. 16-2). The expansion from 1 to 3 takes place through a float valve, which serves the further purpose of maintaining a constant level in the flash tank. To recompress the vapor at 6, a compressor must be available with a suction pressure of 6. Thus two compressors are needed in the system.

The flash tank must separate liquid refrigerant from vapor. The separation occurs when the upward velocity of the vapor is low enough for the liquid particles to drop back into the tank. Normally vapor velocities less than 1 m/s will provide adequate separation. This velocity is found by dividing the volume flow of the vapor by the surface area of the liquid.

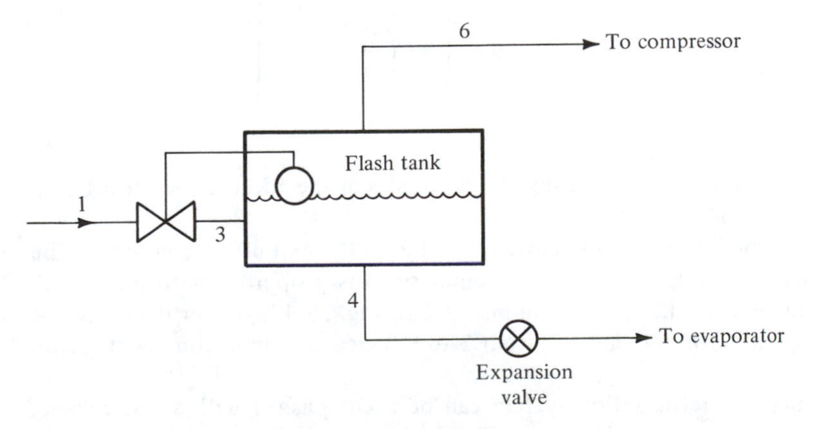

Figure 16-2 Flash tank for removing flash gas during expansion process.

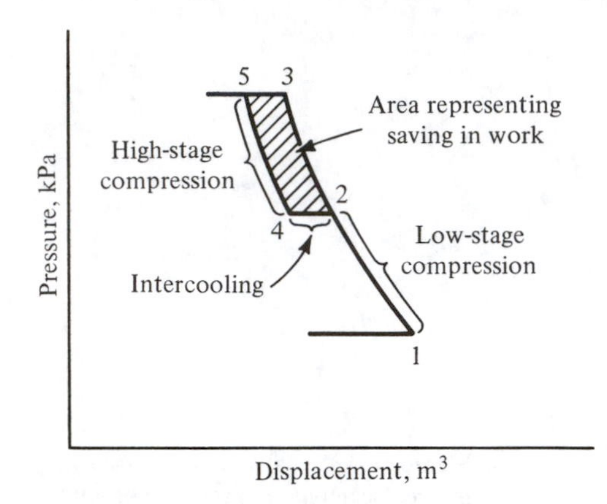

Figure 16-3 Intercooling in two-stage compression.

The most efficient way to remove flash gas would be to separate the vapor continuously as it forms and to recompress it immediately. No practical means has yet been developed to accomplish this.

16-3 Intercooling Intercooling between two stages of compression reduces the work of compression per kilogram of vapor. In two-stage compression of air, for example, an intercooling from point 2 to 4 on the pressure-displacement diagram of Fig. 16-3 saves some work. If the processes are reversible, the saving is represented by the shaded area in Fig. 16-3.

Figure 16-4 shows how compression with intercooling appears on the pressure-enthalpy diagram of a refrigerant. Processes 1-2-3 and 4-5 are on lines of constant entropy, but process 2-3 falls on a flatter curve than process 4-5. Between the same two pressures, therefore, process 4-5 shows a smaller increase in enthalpy, which indicates that less work is required than in 2-3.

Another way of showing that the work of compression increases when the process moves out farther into the superheat region is to examine the equation for work in a reversible polytropic compression of a perfect gas

$$W = -\int v\,dp = \frac{n}{n-1} p_1 v_1 \left[1 - \left(\frac{p_2}{p_1} \right)^{(n-1)/n} \right]$$

where p = pressure, Pa

v = specific volume, m^3/kg

n = polytropic exponent relating the pressure and specific volume during compression, pv^n = const

and where subscript 1 refers to the entrance and 2 to the exit of the compressor. Between two given pressures, the work of compression is proportional to the specific volume of entering gas. The specific volume at 2 in Fig. 16-4 is greater than it is at 4; so the work required for compressing from 2 to 3 is greater than in compressing from 4 to 5.

Intercooling in a refrigeration system can be accomplished with a water-cooled heat exchanger or by using refrigerant. (Fig. 16-5*a* and *b*). The water-cooled intercooler may be satisfactory for two-stage air compression, but for refrigerant compres-

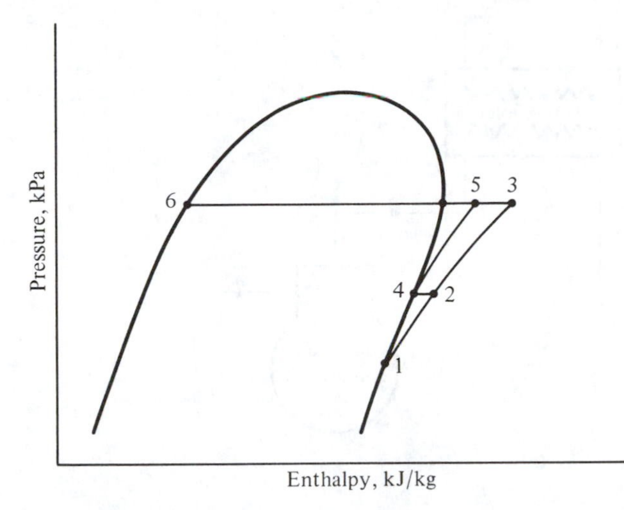

Figure 16-4 Intercooling of a refrigerant in two-stage compression.

sion the water is usually not cold enough. The alternate method of Fig. 16-5*b* uses liquid refrigerant from the condenser to do the intercooling. Discharge gas from the low-stage compressor bubbles through the liquid in the intercooler. Refrigerant leaves the intercooler at 4 as saturated vapor.

Intercooling with liquid refrigerant will usually decrease the total power requirements when ammonia is the refrigerant but not when refrigerant 12 or 22 is used, as illustrated in Examples 16-1 and 16-2. In the examples of this chapter, assume that liquid leaves the condenser saturated, vapor leaves the evaporator saturated, and the compressions are isentropic.

Example 16-1 Calculate the power needed to compress 1.2 kg/s of ammonia from saturated vapor at 80 kPa to 1000 kPa (*a*) by single-stage compression and (*b*) by two-stage compression with intercooling by liquid refrigerant at 300 kPa.

Solution Table 16-1 shows the summary of the calculations with the subscripts referring to state points in Fig. 16-4.

The high-stage compressor in the intercooled system must compress 1.2 kg/s plus the flow rate of refrigerant that evaporates to desuperheat the gas at 2. The flow rate of ammonia compressed in the high stage can be calculated by making a heat and a mass balance about the intercooler, as shown in Fig. 16-6.

Heat balance:

$$w_6(316 \text{ kJ/kg}) + (1.2 \text{ kg/s})(1588 \text{ kJ/kg}) = w_4(1450 \text{ kJ/kg})$$

Mass balance:

$$w_6 + 1.2 = w_4$$

Solving gives

$$w_4 = 1.346 \text{ kg/s}$$

Intercooling the ammonia with liquid refrigerant reduced the power requirement from 468 to 453.2 kW.

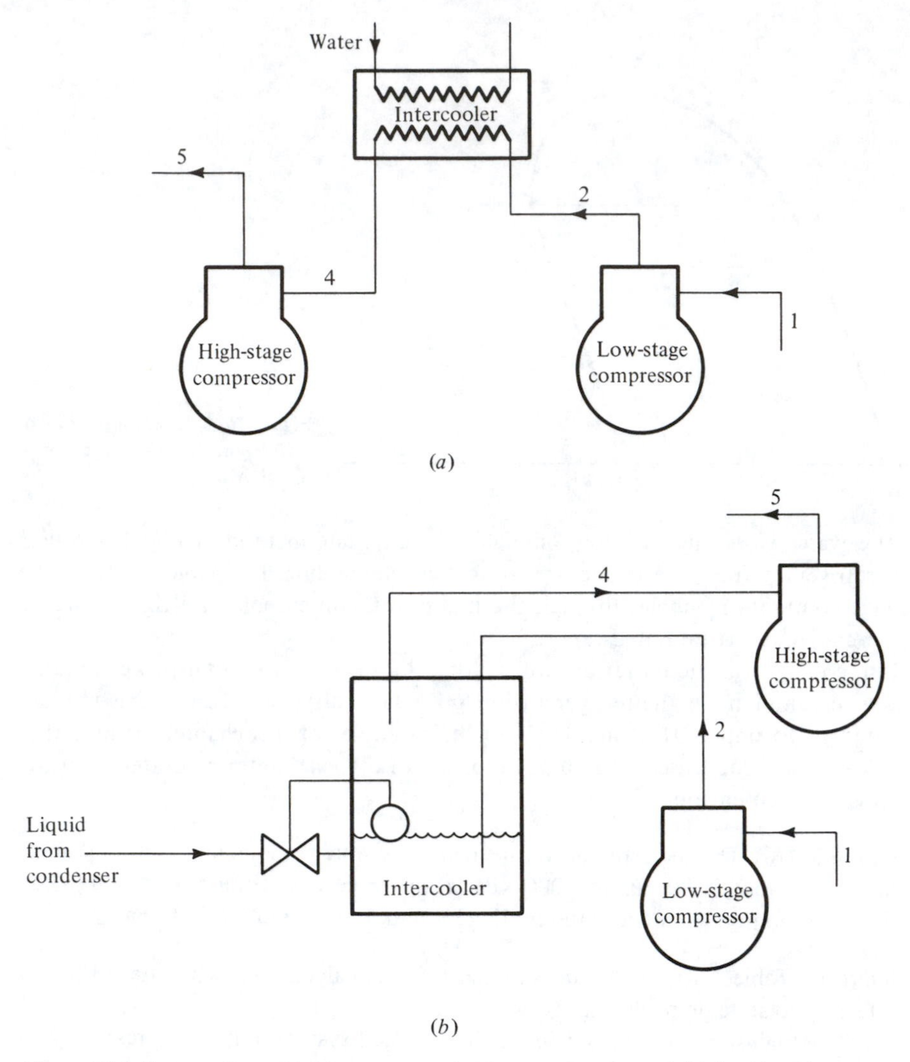

Figure 16-5 Intercooling with (*a*) a water-cooled heat exchanger, and (*b*) liquid refrigerant.

A further benefit of intercooling the ammonia is that the discharge temperature from the high-stage compressor will be reduced from a value of 146°C, the temperature at 3 based on isentropic compression, to 77°C. The lower discharge temperature permits better lubrication and results in longer life of the compressor.

Example 16-2 Compare a compression of 3.5 kg/s of refrigerant 22 from saturated vapor at 100 kPa to a condensing pressure of 1000 kPa (*a*) by single-stage compression and (*b*) by two-stage compression with intercooling at 300 kPa, using liquid refrigerant.

Solution Table 16-2 shows the summary of the calculations with the subscripts referring to state points in Fig. 16-4.

Table 16-1 Comparison of ammonia compression with and without intercooling

	Without intercooling, processes 1-2 and 2-3	With intercooling, processes 1-2, 2-4, and 4-5
$h_2 - h_1$, kJ/kg	1588 - 1410	1588 - 1410
$h_3 - h_2$, kJ/kg	1800 - 1588	
$h_5 - h_4$, kJ/kg	....	1628 - 1450
Flow rate, kg/s, 1 to 2	1.2	1.2
2 to 3	1.2	
4 to 5	....	1.346
Power required, kW, 1 to 2	213.6	213.6
2 to 3	254.4	
4 to 5	....	239.6
Total power, kW	468.0	453.2

Table 16-2 Comparison of refrigerant 22 compression with and without intercooling

	Without intercooling, processes 1-2 and 2-3	With intercooling, processes 1-2, 2-4, and 4-5
$h_2 - h_1$, kJ/kg	416 - 387	416 - 387
$h_3 - h_2$, kJ/kg	449 - 416	
$h_5 - h_4$, kJ/kg	....	430 - 399
Flow rate, kg/s, 1 to 2	3.5	3.5
2 to 3	3.5	
4 to 5	....	3.74
Power required, kW, 1 to 2	101.5	101.5
2 to 3	115.5	
4 to 5	....	115.9
Total power, kW	217.0	217.4

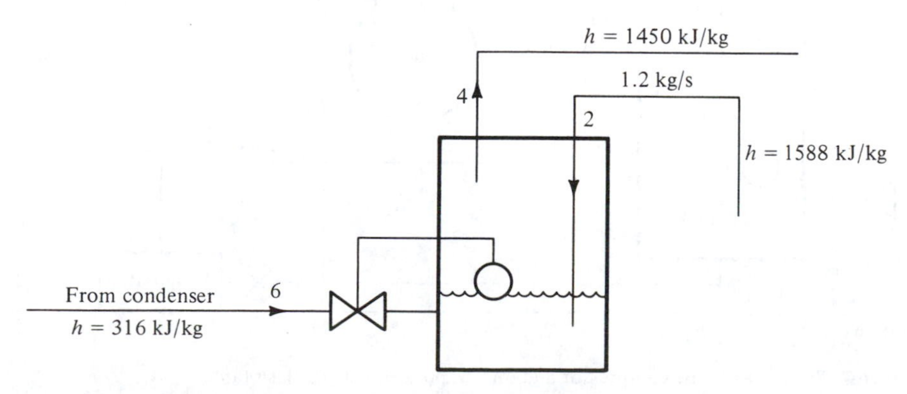

Figure 16-6 Heat and mass balance around intercooler in Example 16-1.

For refrigerant 22 intercooling with liquid refrigerant midway during the compression is ineffective. The dissimilarity during intercooling of ammonia and refrigerant 22 systems results from the difference of their properties. The lines of constant entropy for ammonia become more flat in the superheat region than those of refrigerant 22. In a refrigerant 22 system, therefore, the saving in work per kilogram by performing the compression close to the saturated-vapor line does not compensate for the increased flow rate which must be pumped by the high-stage compressor.

There is an optimum pressure at which the intercooling should take place in an ammonia system. In the compression of air, where the intercooling is achieved by rejecting heat to the ambient or to cooling water, that intermediate pressure for minimum total power is

$$p_i = \sqrt{p_s p_d} \tag{16-1}$$

where p_i = intercooler pressure, kPa
p_s = suction pressure of low-stage compressor, kPa
p_d = discharge pressure of high-stage compressor, kPa

The development of the equation does not consider the additional refrigerant compressed by the high-stage compressor, but it does provide an approximate guideline for the optimal intermediate pressure.

16-4 One evaporator and one compressor The flash tank and intercooler appear in most multipressure systems and will now be examined in various compressor-evaporator combinations.

With one compressor and one evaporator the flash tank may function as shown schematically in Fig. 16-7. A pressure-reducing valve throttles the flash gas from the intermediate pressure to the evaporator pressure. The throttling is necessary because

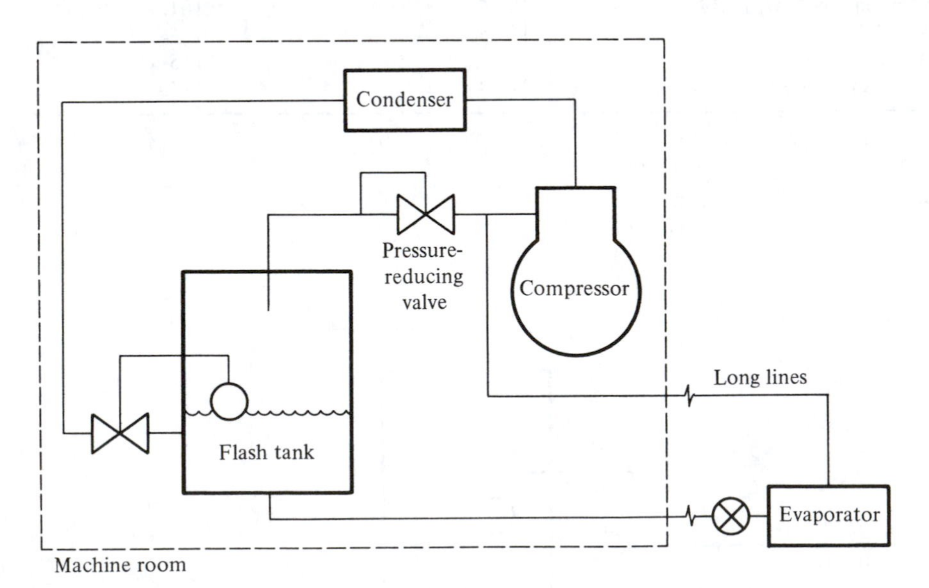

Figure 16-7 System with one compressor and one evaporator using a flash tank.

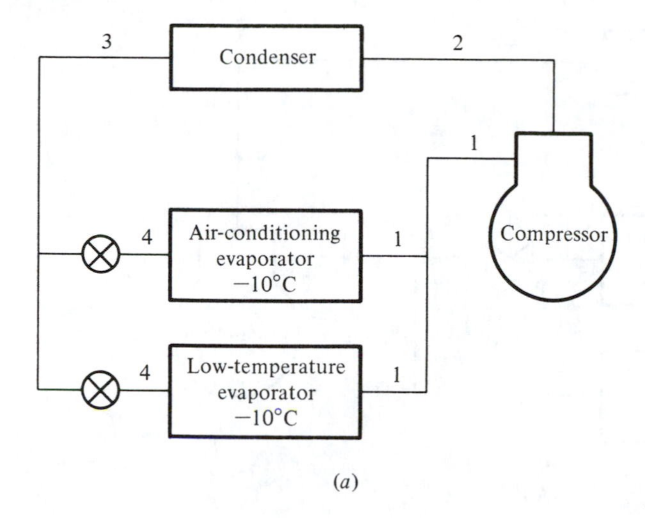

(*a*)

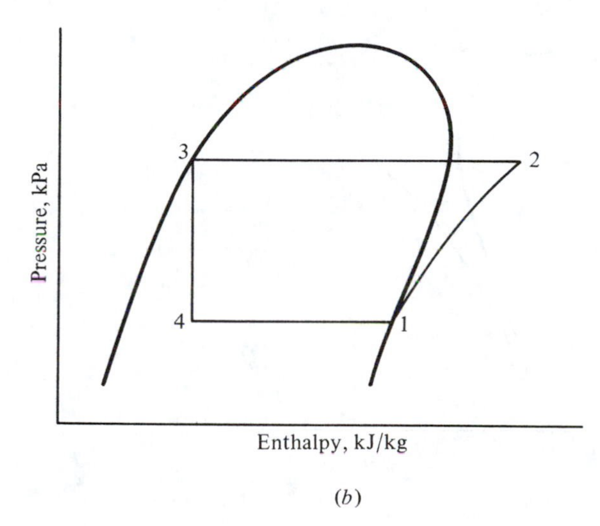

(*b*)

Figure 16-8 (*a*) One compressor and two evaporators with the air-conditioning evaporator operating at -10°C. (*b*) Pressure-enthalpy diagram for system of (*a*).

there is no compressor available with a high suction pressure. Calculations would show that the flash tank does not improve the performance of the system. The only reason for using the flash tank would be to keep the flash gas in the machine room rather than sending it to the evaporator. The flash gas in the evaporator tubes and long suction line does no refrigeration but does increase the pressure drop. This system is used infrequently.

16-5 Two evaporators and one compressor In many situations one compressor serves two evaporators having different temperature requirements. An example is an industry which needs low-temperature refrigeration for a process and which must also provide air conditioning for some offices. Figure 16-8*a* shows one method of arranging this system, and Fig. 16-8*b* shows the corresponding pressure-enthalpy diagram. In Fig. 16-8*a*

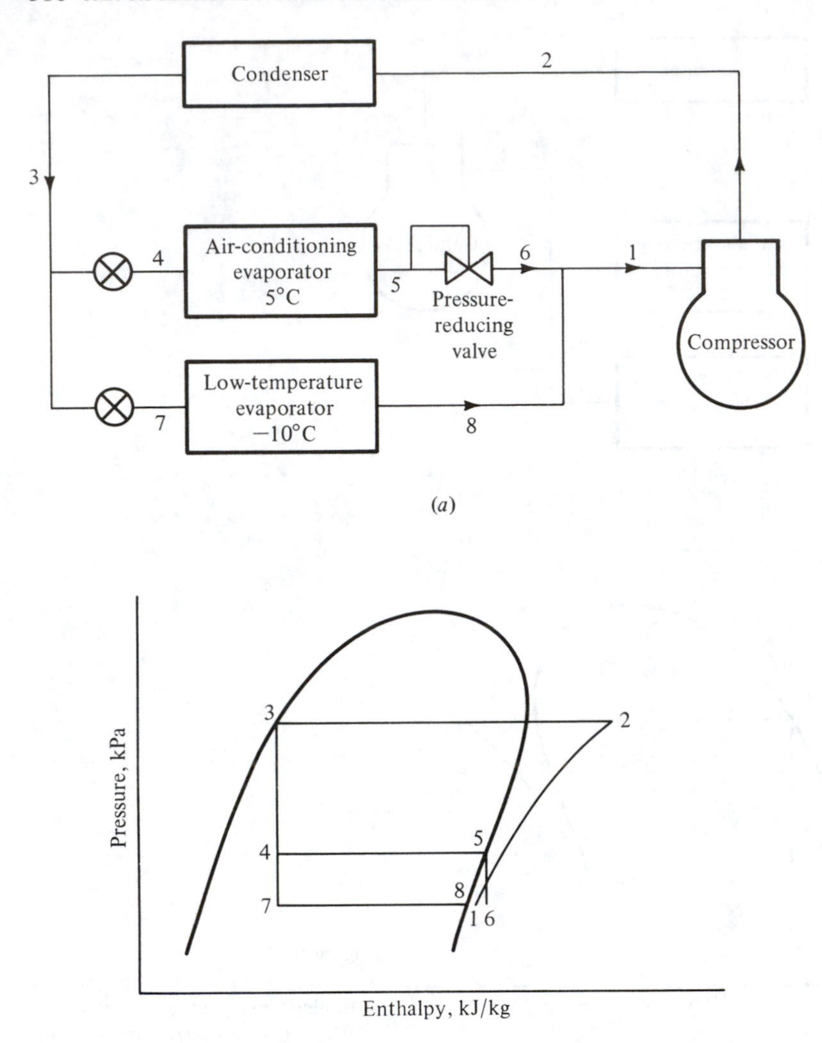

Figure 16-9 (*a*) One compressor and two evaporators with a pressure-reducing valve to maintain a high temperature in the air-conditioning evaporator. (*b*) Pressure-enthalpy diagram for system of (*a*).

the air-conditioning evaporator operates at -10°C even though a higher temperature in this evaporator would cool the air sufficiently. Furthermore, difficulties may arise when an evaporator operates at an unnecessarily low temperature: an evaporator which cools air for air conditioning may collect frost, which blocks the flow of air; an evaporator which chills a liquid may freeze the liquid; and an evaporator which cools air for a room where meat or produce is stored may dehumidify the air so much that the products will be dehydrated.

To overcome the drawbacks of the system in Fig. 16-8*a*, a revision may be made as shown in Fig. 16-9*a*. A pressure-reducing valve installed after the high-temperature evaporator regulates the pressure and maintains a temperature in the air-conditioning evaporator of 5°C, for example. Figure 16-9*b* shows the corresponding pressure-enthalpy diagram. Differences in performance between the systems in Figs. 16-8*a* and 16-9*a* are as follows. In the system of Fig. 16-9*a*, the refrigerating effect in the high-temperature evaporator is greater than it is in the system of Fig. 16-8*a*. This is an advantage for the system of Fig. 16-9*a*. To counterbalance this advantage, the compression in Fig. 16-9*b* occurs farther out in the superheat region than in Fig. 16-8*b*. The system of Fig. 16-9*a* therefore demands more work per kilogram of refrigerant.

From a power standpoint, the systems are practically a standoff, but for proper operation of the high-temperature evaporator the system of Fig. 16-9*a* is preferred.

16-6 Two compressors and one evaporator Two-stage compression with intercooling and removal of flash gas is often the ideal way to serve one low-temperature evaporator. This system requires less power than with a single compressor, and often the saving in power will justify the cost of the extra equipment.

Example 16-3 Calculate the power required by the two compressors in an ammonia system which serves a 250-kW evaporator at -25°C. The system uses two-stage compression with intercooling and removal of flash gas. The condensing temperature is 35°C.

Solution First sketch the schematic diagram of the system (Fig. 16-10*a*) and the corresponding pressure-enthalpy diagram (Fig. 16-10*b*). The functions of the intercooler and flash tank are combined in one vessel.

The intermediate pressure for optimum economy can be calculated from Eq. (16-1):

p_s = saturation pressure at -25°C = 152 kPa
p_d = saturation pressure at 35°C = 1352 kPa
$p_i = \sqrt{152(1352)} = 453$ kPa

The enthalpies at all points can now be determined from Table A-3 and Fig. A-1:

$h_1 = h_g$ at -25°C = 1430 kJ/kg
$h_2 = h$ at 453 kPa after isentropic compression = 1573
$h_3 = h_g$ at 453 kPa = 1463
$h_4 = h$ at 1352 kPa after isentropic compression = 1620
$h_5 = h_f$ at 35°C = 366 $\quad h_6 = h_5 = 366$
$h_7 = h_f$ at 453 kPa = 202 $\quad h_8 = h_7 = 202$

Next, the mass rates of flow through the compressors can be calculated by means of heat and mass balances.
Heat balance about the evaporator:

$$w_1 = \frac{250 \text{ kW}}{1430 - 202} = 0.204 \text{ kg/s}$$

$$w_1 = w_2 = w_7 = w_8 = 0.204 \text{ kg/s}$$

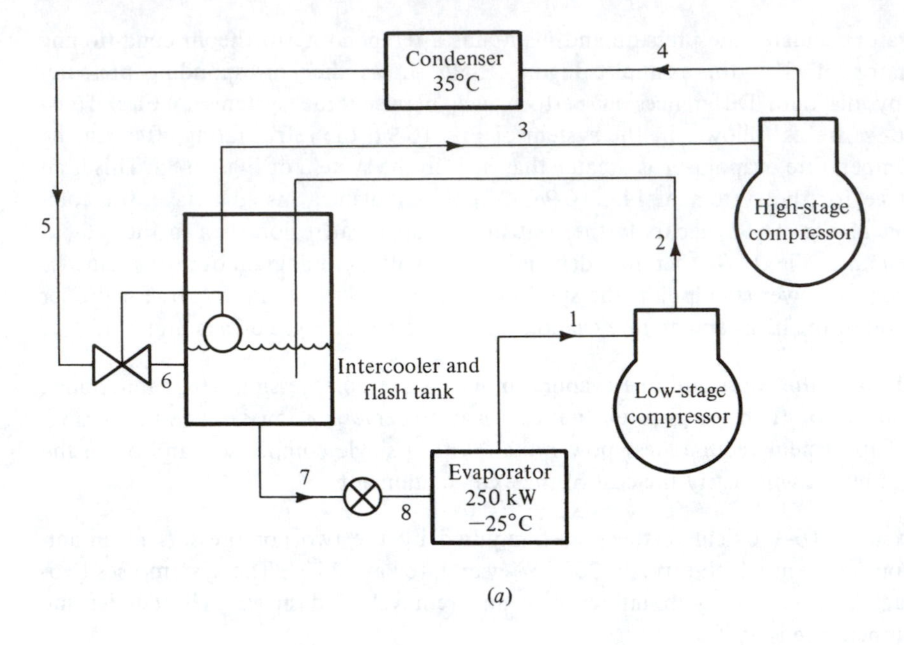

(a)

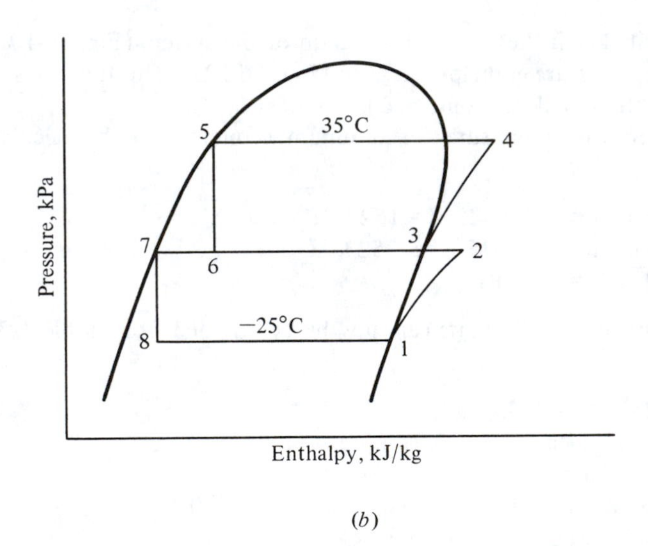

(b)

Figure 16-10 (a) Two compressors and one evaporator in Example 16-3. (b) Pressure-enthalpy diagram for system in (a).

Heat and mass balance about the intercooler:

$$w_2 h_2 + w_6 h_6 = w_7 h_7 + w_3 h_3$$

$$w_6 = w_3 \qquad \text{and} \qquad w_7 = w_2$$

$$0.204(1573) + w_3(366) = 0.204(202) + w_3(1463)$$

$$w_3 = 0.255 \text{ kg/s}$$

Low-stage power: (0.204 kg/s) (1573 - 1430 kJ/kg) = 29.2 kW
High-stage power: (0.255 kg/s) (1620 - 1463 kJ/kg) = 40.0 kW
Total power: 29.2 + 40.0 = 69.2 kW

This power requirement can be compared with that of a single-compressor system developing 250 kW of refrigeration at -25°C with a condensing temperature of 35°C. The pressure-enthalpy diagram is shown in Fig. 16-11. The enthalpies are:

$h_1 = 1430$ kJ/kg $\quad h_2 = 1765 \quad h_3 = h_4 = 366$

$$w_1 = \frac{250 \text{ kW}}{1430 - 366} = 0.235 \text{ kg/s}$$

Power = 0.235(1765 - 1430) = 78.7 kW

The two-stage compressor system requires 69.2 kW, or 12 percent less power than the single-compressor system.

16-7 Two compressors and two evaporators The system which has two evaporators operating at different temperatures is common in industrial refrigeration. A dairy cooling milk and manufacturing ice cream has been mentioned. A frozen-food plant may require two evaporators at different temperatures, one at -40°C to quick-freeze the food and the other at -25°C to hold the food after it is frozen. Process and chemical industries often require different temperatures of refrigeration in various sections of the plant. Evaporators at two different temperatures can be handled efficiently by a two-stage system which employs intercooling and removal of flash gas.

Example 16-4 In an ammonia system one evaporator is to provide 180 kW of refrigeration at -30°C and another evaporator is to provide 200 kW at 5°C. The system uses two-stage compression with intercooling and is arranged as in Fig. 16-12*a*. The condensing temperature is 40°C. Calculate the power required by the compressors.

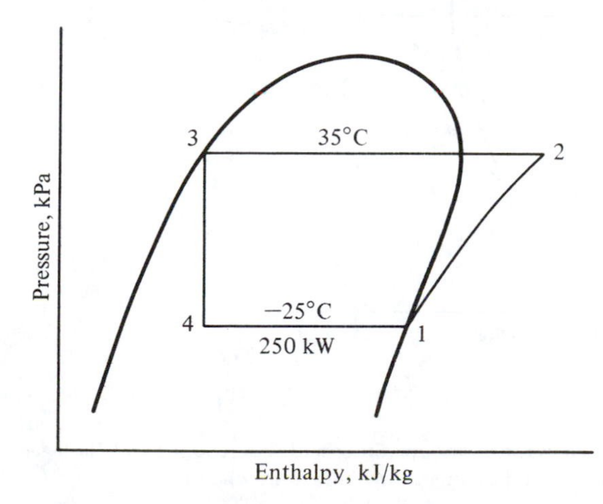

Figure 16-11 Pressure-enthalpy diagram for single-compressor system for conditions in Example 16-3.

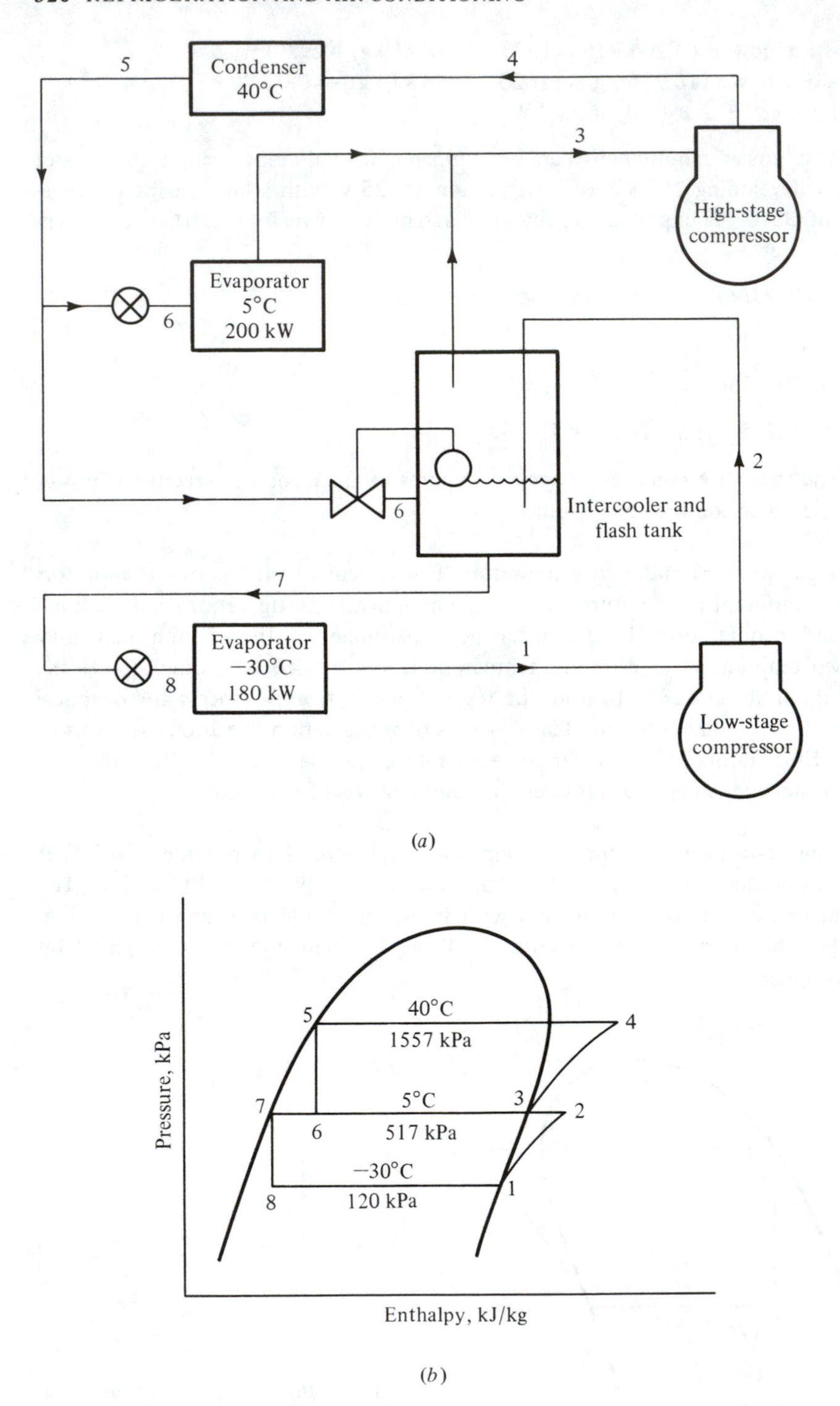

Figure 16-12 (*a*) Two compressors and two evaporators operating with intercooling and flash-gas removal. (*b*) The pressure-enthalpy diagram corresponding to the system in (*a*).

Solution Sketch the pressure-enthalpy diagram of the cycle as in Fig. 16-12*b*. The discharge pressure of the low-stage compressor and the suction pressure of the high-stage compressor are the same as the pressure in the 5°C evaporator.

Next determine the enthalpies at the state points.

$h_1 = h_g$ at -30°C = 1423 kJ/kg
$h_2 = h$ at 517 kPa after isentropic compression = 1630
$h_3 = h_g$ at 5°C = 1467
$h_4 = h$ at 1557 kPa after isentropic compression = 1625
$h_5 = h_f$ at 40°C = 390.6 $\quad h_6 = h_5 = 390.6$
$h_7 = h_f$ at 5°C = 223 $\quad h_8 = h_7 = 223$

The mass rates of flow are

$$w_1 = \frac{180 \text{ kW}}{1423 - 223} = 0.150 \text{ kg/s}$$

$$w_7 = w_8 = w_2 = w_1 = 0.150 \text{ kg/s}$$

Probably the simplest way to calculate the mass rate of flow handled by the high-stage compressor is to make a heat and mass balance about both the high-temperature evaporator and the intercooler, as shown in Fig. 16-13.

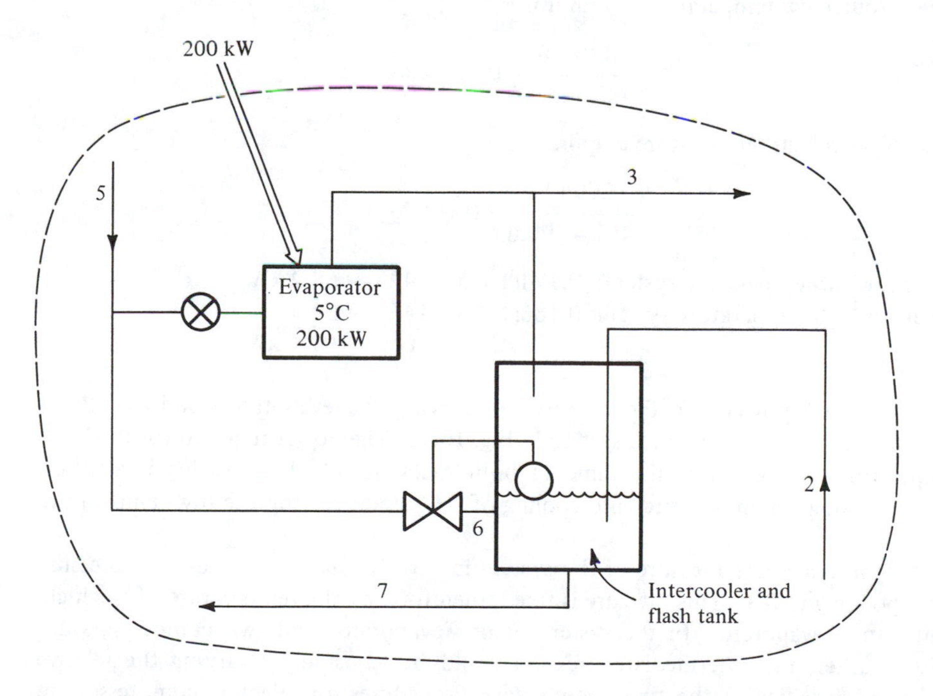

Figure 16-13 Heat and mass balance about high-temperature evaporator and intercooler in Example 16-4.

Heat balance:

$$w_5h_5 + 200\text{ kW} + w_2h_2 = w_3h_3 + w_7h_7$$

Mass balance:

$$w_2 = w_7 = 0.150\text{ kg/s}$$

Therefore

$$w_5 = w_3$$

Combining gives

$$390.6w_3 + 200 + 0.150(1630) = 1467w_3 + 0.150(223)$$

Solving leads to

$$w_3 = 0.382\text{ kg/s}$$

The power required by the compressors can now be calculated:

Low-stage power: 0.150(1630 - 1423) = 31.1 kW
High-stage power: 0.382(1625 - 1467) = 60.4
Total 91.5 kW

If one compressor served each evaporator in single-stage compression, the power requirements of the two compressors would have been as follows:
Flow through low-temperature evaporator:

$$\frac{180\text{ kW}}{1423 - 390.6} = 0.174\text{ kg/s}$$

Flow through high-temperature evaporator:

$$\frac{200\text{ kW}}{1467 - 390.6} = 0.186\text{ kg/s}$$

Power for low-temperature system: 0.174(1815 - 1423) = 68.2 kW
Power for high-temperature system: 0.186(1625 - 1467) = 29.4
Total 97.6 kW

The combined power for the compressors serving the evaporators individually is greater than with the combined system in Fig. 16-12. The power required for the high-temperature evaporator is the same in both cases, so all of the saving is attributable to flash-gas removal and intercooling of refrigerant serving the low-temperature evaporator.

The intermediate pressure of the system in Fig. 16-12*a* is the saturation pressure corresponding to 5°C. This pressure is fixed, therefore, by the temperature of the high-temperature evaporator. In the system of one evaporator and two compressors discussed earlier, the intermediate pressure could be adjusted by varying the relative pumping capacities of the two compressors. Procedures for selecting compressors to

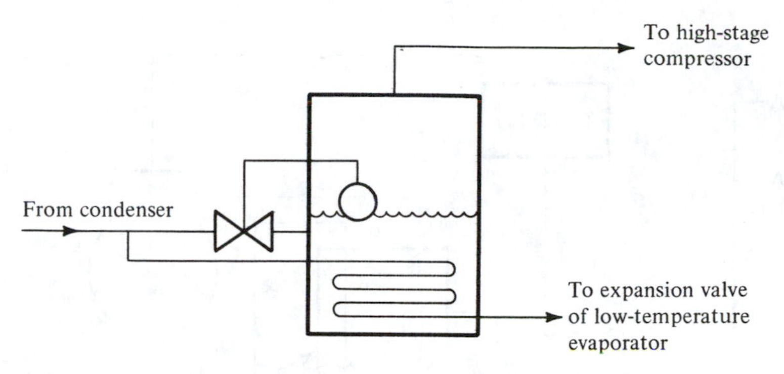

Figure 16-14 A liquid subcooler.

give a specified intermediate pressure are extensions of the techniques of system simulation discussed in Chap. 14 (see Prob. 16-6).

16-8 Auxiliary equipment Modifications are sometimes made in the equipment for intercooling and removal of flash gas. If the temperature of the discharge gas from the low-stage compressor is sufficiently high, a water-cooled heat exchanger may remove some of the heat from the discharge gas before it flows into the main intercooler.

A device which gives the same result as flash-gas removal is called a *liquid subcooler* (see Fig. 16-14). It cools the liquid refrigerant by evaporating a small fraction of the liquid. Compared with the direct-contact flash tank, the subcooler cannot cool the liquid to quite as low a temperature. On the other hand, the subcooler maintains the liquid at a high pressure. If the liquid must flow through a long line before it reaches the expansion valve, there is less possibility that the pressure drop in the line will flash the liquid to vapor and thus restrict the flow through the expansion valve.

16-9 Compound compressors In the systems shown in this chapter, where there are two levels of compression, the flow diagrams have shown two compressors. Single compressors are available that accept both high and low suction pressures and can thus serve the purpose of the two compressors shown in Fig. 16-10. In compound reciprocating compressors, four cylinders of a six-cylinder compressor might perform the low-stage compression and the remaining two cylinders would accomplish the high-stage compression. In a compound screw compressor, the entrance of gas at the intermediate pressure takes place part way along the compression process.

16-10 Liquid-recirculation systems Section 12-15 on evaporators referred to liquid-recirculation systems. A schematic diagram of a liquid-recirculation system is shown in Fig. 16-15. A liquid pump delivers the low-temperature liquid to the evaporators, which can be regulated by thermostatically controlled solenoid valves. Since the pump supplies several times as much refrigerant to the evaporator as can be evaporated, some liquid washes out of the evaporator and returns to the separator. Liquid-recirculation

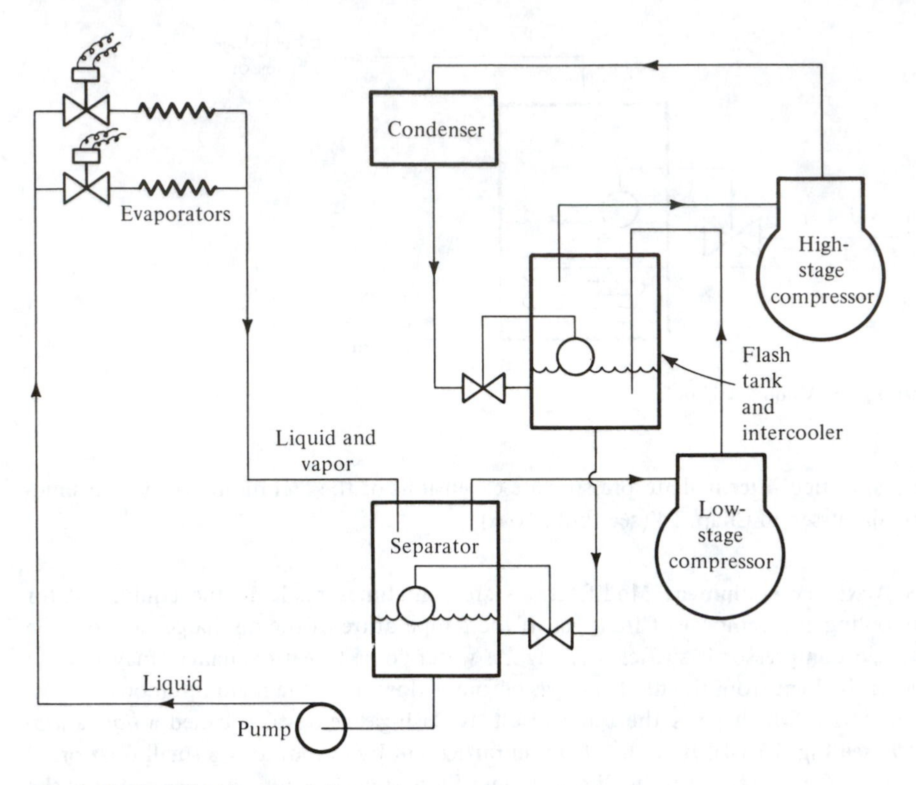

Figure 16-15 A liquid-recirculation system.

systems usually achieve good feeding of the evaporators and provide high heat-transfer coefficients on the refrigerant side. It is not essential that a liquid-recirculation system be served by a multistage compression system, but they are usually associated because the liquid-recirculation system is particularly advantageous in the low-temperature systems that multipressure compression usually serves.

16-11 Summary In multipressure systems the removal and recompression of flash gas before complete expansion decrease the power required by the compressors. Intercooling between stages of compression reduces the power requirements, at least when ammonia is the refrigerant. Intercooling decreases the discharge temperature of the refrigerant from the high-stage compressor. High discharge temperatures cause oil carbonization, sticky compressor valves, and lubrication difficulties in reciprocating compressors.

Any decision to use multiple-stage systems should be based on an economic analysis. The savings in power must be compared with the additional cost of equipment to determine whether the added investment is warranted. Factors such as the refrigerant used, the type of compressor (whether reciprocating or screw), and the size of the system also have an influence. Using ammonia as an example, the practical minimum evaporating temperatures for reciprocating compressors are approximately -30°C for a single-stage, -50°C for a two-stage, and -70°C for a three-stage system. Screw compressors are capable of operating against larger pressure ratios than reciprocating compressors. A further advantage of multistaging is that it decreases the pressure difference across which the compressor works, thus reducing wear on bearing surfaces.

PROBLEMS

In the following problems, liquid leaves the condensers saturated, vapor leaves the evaporators saturated, and compressions are isentropic.

16-1 A cylindrical tank 2 m long mounted with its axis horizontal is to separate liquid ammonia from ammonia vapor. The ammonia vapor bubbles through the liquid and 1.2 m^3/s leaves the surface of the liquid. If the velocity of the vapor is limited to 1 m/s and the vessel is to operate with the liquid level two-thirds of the diameter from the bottom, what must the diameter of the tank be? *Ans.* 0.636 m

16-2 A liquid subcooler as shown in Fig. 16-14 receives liquid ammonia at 30°C and subcools 0.6 kg/s to 5°C. Saturated vapor leaves the subcooler for the high-stage compressor at −1°C. Calculate the flow rate of ammonia that evaporates to cool the liquid. *Ans.* 0.0575 kg/s

16-3 In a refrigerant 22 refrigeration system the capacity is 180 kW at a temperature of −30°C. The vapor from the evaporator is pumped by one compressor to the condensing pressure of 1500 kPa. Later the system is revised to a two-stage compression operating on the cycle shown in Fig. 16-16 with intercooling but no removal of flash gas at 600 kPa.

(*a*) Calculate the power required by the single compressor in the original system.

(*b*) Calculate the total power required by the two compressors in the revised system. *Ans.* 70.9 kW

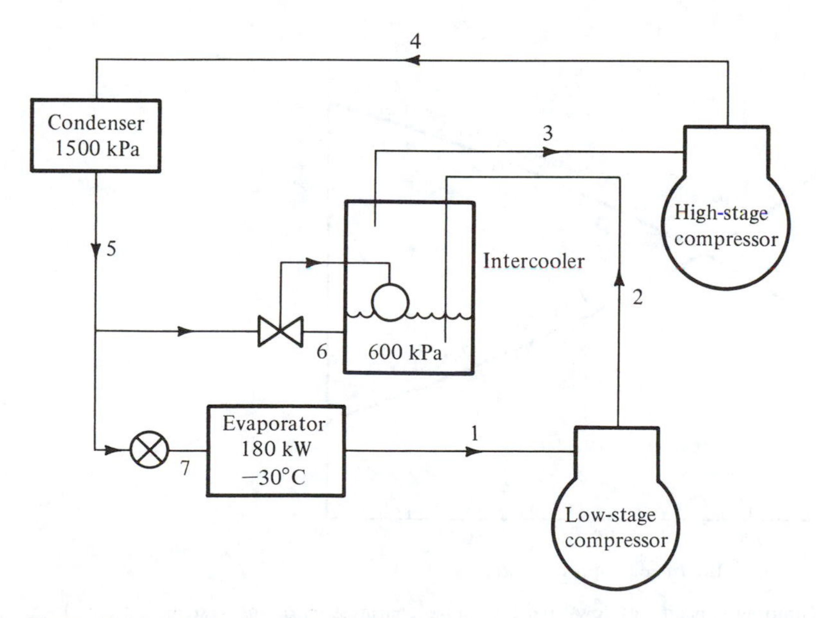

Figure 16-16 Intercooling system in Prob. 16-3.

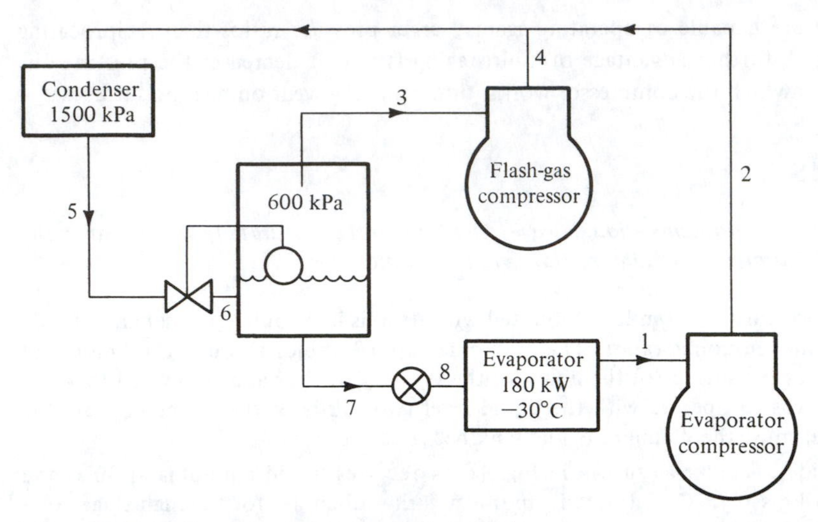

Figure 16-17 Flash-gas removal system in Prob. 16-4.

16-4 A refrigerant 22 system has a capacity of 180 kW at an evaporating temperature of −30°C when the condensing pressure is 1500 kPa.

(*a*) Compute the power requirement for a system with a single compressor.

(*b*) Compute the total power required by the two compressors in the system shown in Fig. 16-17 where there is no intercooling but there is flash-gas removal at 600 kPa. *Ans.* 60.7 kW

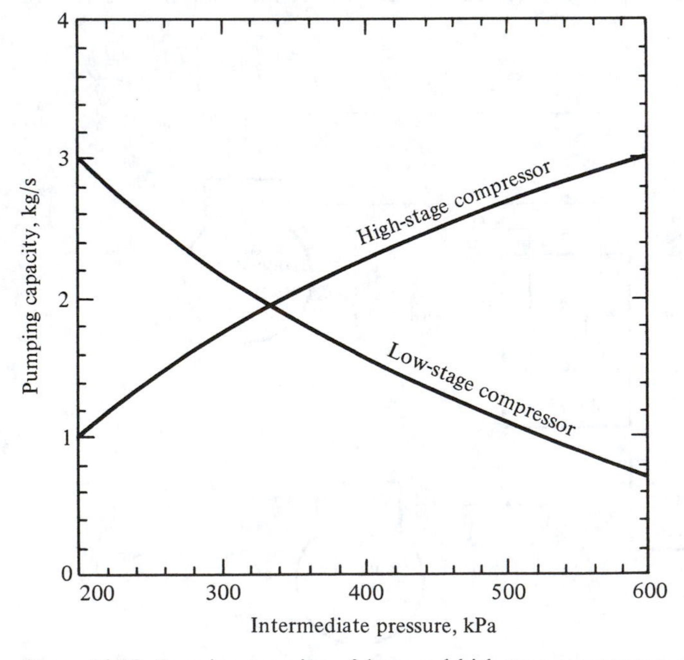

Figure 16-18 Pumping capacity of low- and high-stage compressors of the system in Prob. 16-6. The suction pressure of the low-stage compressor is 105 kPa (−40°C), and the discharge pressure of the high-stage compressor is 1192 kPa (30°C).

16-5 A two-stage ammonia system using flash-gas removal and intercooling operates on the cycle shown in Fig. 16-12*a*. The condensing temperature is 35°C. The saturation temperature of the intermediate-temperature evaporator is 0°C, and its capacity is 150 kW. The saturation temperature of the low-temperature evaporator is −40°C, and its capacity is 250 kW. What is the rate of refrigerant compressed by the high-stage compressor? *Ans.* 0.411 kg/s

16-6 A two-stage refrigerant 22 system that uses flash-gas removal and intercooling serves a single low-temperature evaporator, as in Fig. 16-10*a*. The evaporating temperature is −40°C, and the condensing temperature is 30°C. The pumping capacities of the high- and low-stage compressors are shown in Fig. 16-18. What is (*a*) the refrigerating capacity of the system and (*b*) the intermediate pressure? *Ans.* (*a*) 318 kW, (*b*) 390 kPa

CHAPTER

SEVENTEEN

ABSORPTION REFRIGERATION

17-1 Relation of the absorption to the vapor-compression cycle Ferdinand Carré, a Frenchman, invented the absorption system and took out a United States patent in 1860. The first use of the system in the United States was probably made by the Confederate States during the Civil War after the supply of natural ice had been cut off from the North.

The absorption cycle is similar in certain respects to the vapor-compression cycle. A refrigeration cycle will operate with the condenser, expansion valve, and evaporator shown in Fig. 17-1 if the low-pressure vapor from the evaporator can be transformed into high-pressure vapor and delivered to the condenser. The vapor-compression system uses a compressor for this task. The absorption system first absorbs the low-pressure vapor in an appropriate absorbing liquid. Embodied in the absorption process is the conversion of vapor into liquid; since this process is akin to condensation, heat must be rejected during the process. The next step is to elevate the pressure of the liquid with a pump, and the final step releases the vapor from the absorbing liquid by adding heat.

The vapor-compression cycle is described as a *work-operated cycle* because the elevation of pressure of the refrigerant is accomplished by a compressor that requires work. The absorption cycle, on the other hand, is referred to as a *heat-operated cycle* because most of the operating cost is associated with providing the heat that drives off the vapor from the high-pressure liquid. Indeed there is a requirement for some work in the absorption cycle to drive the pump, but the amount of work for a given quantity of refrigeration is minor compared with that needed in the vapor-compression cycle.

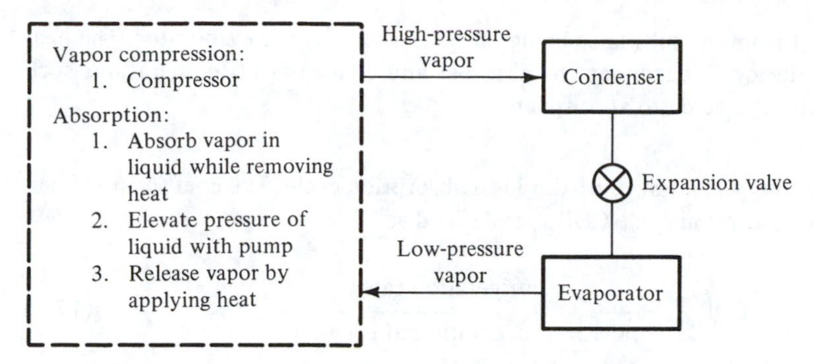

Figure 17-1 Methods of transforming low-pressure vapor into high-pressure vapor in a refrigeration system.

17-2 The absorption cycle The basic absorption cycle is shown in Fig . 17-2. The condenser and evaporator are as shown in Fig. 17-1, and the compression operation is provided by the assembly in the left half of the diagram. Low-pressure vapor from the evaporator is absorbed by the liquid solution in the absorber. If this absorption process were executed adiabatically, the temperature of the solution would rise and eventually the absorption of vapor would cease. To perpetuate the absorption process the absorber is cooled by water or air that ultimately rejects this heat to the atmosphere. The pump receives low-pressure liquid from the absorber, elevates the pressure of the liquid, and delivers the liquid to the generator. In the generator, heat from a high-temperature source drives off the vapor that had been absorbed by the solution. The liquid solution returns to the absorber through a throttling valve whose purpose is to provide a pressure drop to maintain the pressure difference between the generator and absorber.

The pattern for the flow of heat to and from the four heat-exchange components in the absorption cycle is that high-temperature heat enters the generator while low-

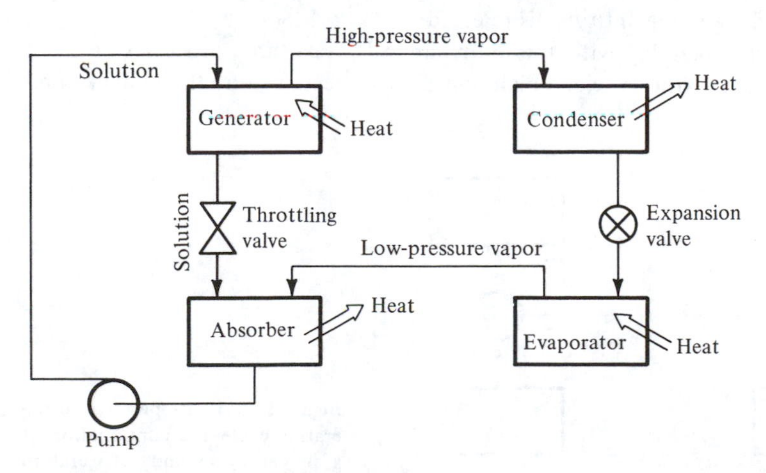

Figure 17-2 The basic absorption unit.

temperature heat from the substance being refrigerated enters the evaporator. The heat rejection from the cycle occurs at the absorber and condenser at temperatures such that the heat can be rejected to atmosphere.

17-3 Coefficient of performance of the ideal absorption cycle The coefficient of performance of the absorption cycle COP_{abs} is defined as

$$COP_{abs} = \frac{\text{refrigeration rate}}{\text{rate of heat addition at generator}} \tag{17-1}$$

In certain respects applying the term COP to the absorption system is unfortunate because the value is appreciably lower than that of the vapor-compression cycle (0.6 versus 3, for example). The comparatively low value of COP_{abs} should not be considered prejudicial to the absorption system, because the COPs of the two cycles are defined differently. The COP of the vapor-compression cycle is the ratio of the refrigeration rate to the power in the form of *work* supplied to operate the cycle. Energy in the form of work is normally much more valuable and expensive than energy in the form of heat.

Further insight into the distinction of the effectiveness of the absorption and vapor-compression cycles is provided by the exercise of determining the COP of the ideal absorption cycle. Stated more precisely, the COP of an ideal *heat-operated refrigeration cycle* will be evaluated. Figure 17-3 suggests how to proceed with the analysis, because the processes in the box on the left consist of a power cycle that develops work needed to perform the compression of the vapor from the evaporator to the condenser. These two cycles are shown schematically in Fig. 17-3. The power cycle receives energy in the form of heat q_g at an absolute temperature T_s, delivers some energy W in the form of work to the refrigeration cycle, and rejects a quantity of energy q_a in the form of heat at a temperature T_a. The refrigeration cycle receives the work W and with it pumps heat q_e at the refrigerating temperature of T_r to a temperature T_a, where the quantity q_c is rejected.

The ideal cycle operating with thermodynamically reversible processes between two temperatures is a Carnot cycle, which appears as a rectangle on the temperature-

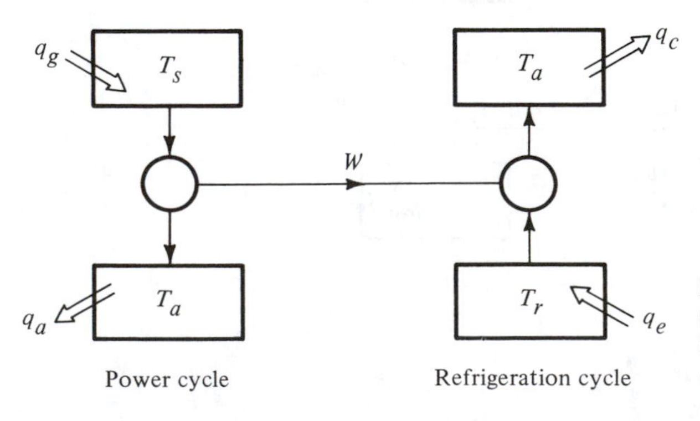

Figure 17-3 Heat-operated refrigeration cycle as a combination of a power cycle and refrigeration cycle.

entropy diagram. For the power cycle on the left side of Fig. 17-3

$$\frac{q_g}{W} = \frac{T_s}{T_s - T_a} \tag{17-2}$$

and for the refrigeration cycle on the right side of Fig. 17-3

$$\frac{q_e}{W} = \frac{T_r}{T_a - T_r} \tag{17-3}$$

The refrigeration rate in Eq. (17-1) is q_e, and the rate of heat addition at the generator is q_g. Using the expressions for q_g and q_e from Eqs. (17-2) and (17-3), respectively, the COP is

$$\text{COP} = \frac{q_e}{q_g} = \frac{WT_r}{T_a - T_r}\,\frac{T_s - T_a}{WT_s} = \frac{T_r(T_s - T_a)}{T_s(T_a - T_r)} \tag{17-4}$$

Example 17-1 What is the COP of an ideal heat-operated refrigeration system that has a source temperature of heat of 100°C, a refrigerating temperature of 5°C, and an ambient temperature of 30°C?

Solution

$$\text{COP} = \frac{(5 + 273.15)(100 - 30)}{(100 + 273.15)(30 - 5)} = 2.09$$

Several trends are detectable from Eq. (17-4):

1. As T_s increases, the COP increases.
2. As T_r increases, the COP increases.
3. As T_a increases, the COP decreases.

17-4 Temperature-pressure-concentration properties of LiBr–water solutions Lithium bromide is a solid salt crystal; in the presence of water vapor it will absorb the vapor and become a liquid solution. The liquid solution exerts a water-vapor pressure that is a function of the solution temperature and the concentration of the solution. If two vessels were connected as in Fig. 17-4, one vessel containing LiBr–water solution and

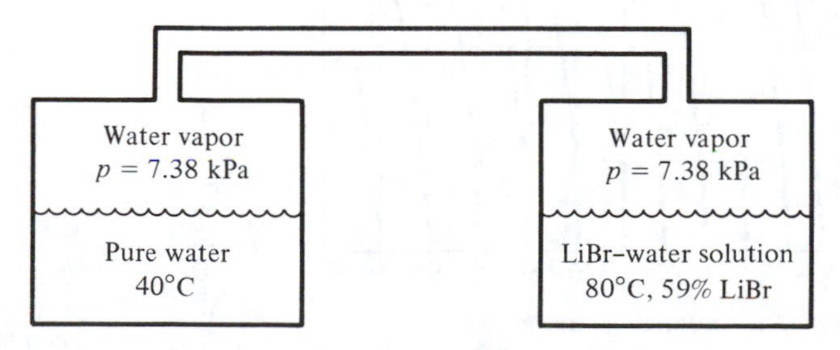

Figure 17-4 Equilibrium water–vapor pressure.

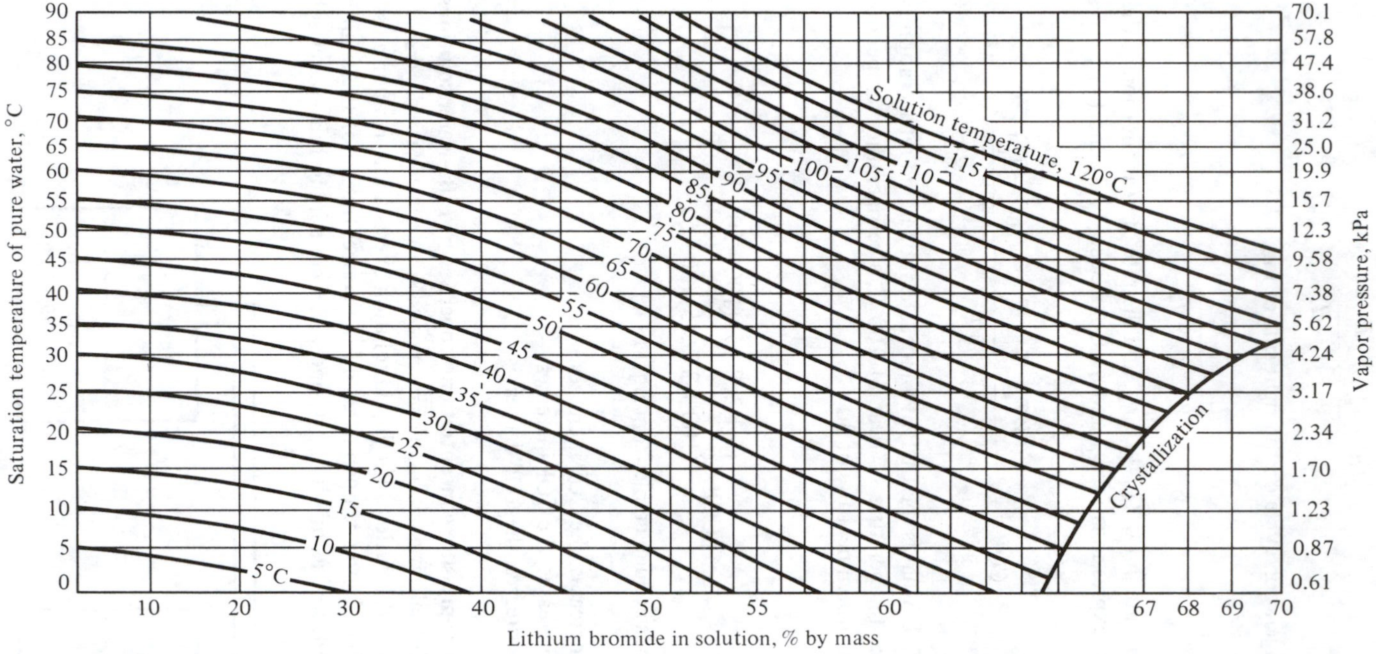

Figure 17-5 Temperature-pressure-concentration diagram of saturated LiBr–water solutions, developed from data in Ref. 1.

the other pure water, each liquid would exert a water-vapor pressure. At equilibrium the water-vapor pressures exerted by the two liquids would be equal. An example of one equilibrium condition is noted in Fig. 17-4. If the temperature of pure water is 40°C, its vapor pressure is 7.38 kPa. That same vapor pressure would be developed by a LiBr-water solution at a temperature of 80°C and a concentration $x = 59\%$ LiBr on a mass basis. Many other combinations of temperatures and concentrations of solution also provide a vapor pressure of 7.38 kPa.

Figure 17-5 is a temperature-pressure-concentration diagram for LiBr-water solutions. Concentration is the abscissa of the graph and water-vapor pressure could be considered the ordinate, as shown on the vertical scale on the right. For convenience the saturation temperature of pure water corresponding to these vapor pressures is shown as the ordinate on the left. The chart applies to saturated conditions where the solution is in equilibrium with the water vapor, as in the vessel on the right in Fig. 17-4.

The pressures, temperatures, and concentrations chosen as example conditions in Fig. 17-4 can now be verified. If the temperature of pure water is 40°C, the vapor pressure the liquid exerts is 7.38 kPa, which can be determined from the opposite vertical scale in Fig. 17-5. A LiBr-water solution with a concentration x of 59% and a temperature of 80°C also develops a water-vapor pressure of 7.38 kPa. If the solution had a concentration x of 54% and temperature of 70°C, the water-vapor pressure would likewise be 7.38 kPa.

17-5 Calculation of mass flow rates in the absorption cycle The first stage in analyzing the simple LiBr-water absorption refrigeration cycle can now be performed by using the property data presented in Fig. 17-5.

Example 17-2 Compute the rate flow of refrigerant (water) through the condenser and evaporator in the cycle shown in Fig. 17-6 if the pump delivers 0.6 kg/s and the following temperatures prevail: generator, 100°C; condenser, 40°C; evaporator, 10°C; and absorber, 30°C.

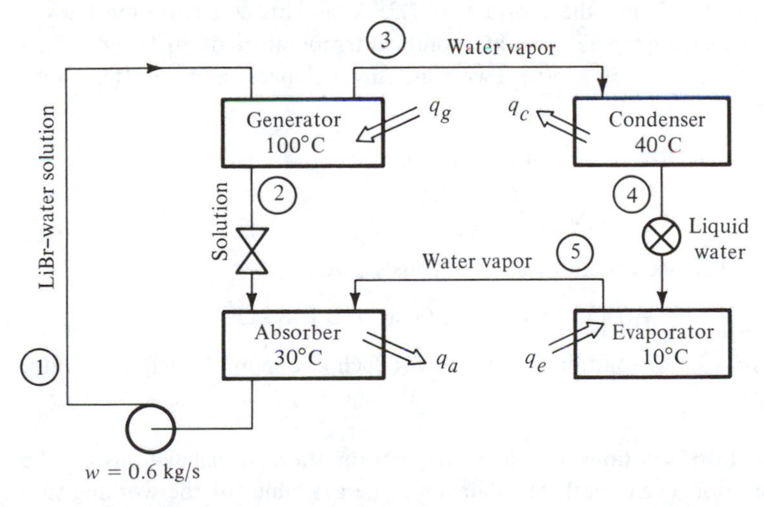

Figure 17-6 Absorption cycle in Example 17-2

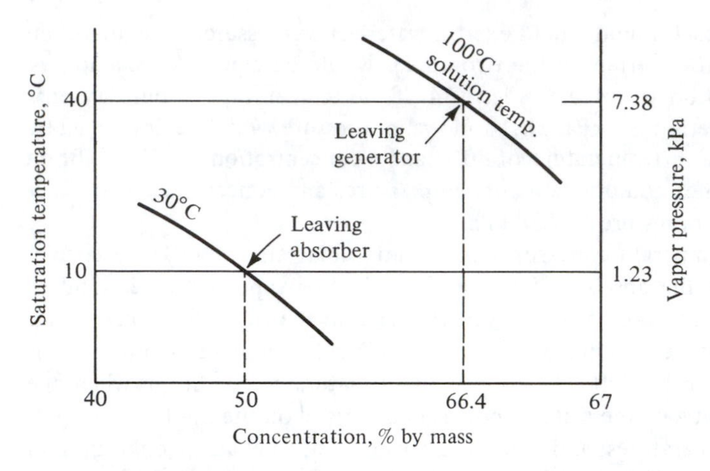

Figure 17-7 Conditions of solution in Example 17-2.

Solution The computation of the mass flow rate incorporates material balances using applicable concentrations of the LiBr in the solution. Two different pressures exist in the system: a high pressure prevails in the generator and condenser, while the low pressure prevails in the absorber and evaporator. Since a saturated condition of pure water prevails in the condenser due to simultaneous existence of liquid and vapor, the condensing temperature of 40°C fixes the pressure in the condenser (and thus in the generator) of 7.38 kPa. From similar reasoning, the evaporator temperature of 10°C establishes the low pressure at 1.23 kPa. Figure 17-7 is a skeleton *p-x-t* diagram extracted from Fig. 17-5 to display the state points of the LiBr solution. The solution leaving a component is representative of the solution in the component, so the state point of the solution at point 2 leaving the generator is found from Fig. 17-7 at the intersection of the solution temperature of 100°C and the pressure of 7.38 kPa. This concentration is $x_2 = 0.664 = 66.4\%$. Leaving the absorber at a solution temperature of 30°C and a pressure of 1.23 kPa, $x_1 = 0.50 = 50\%$. Two mass-flow balances can be written about the generator:

Total mass-flow balance: $w_2 + w_3 = w_1 = 0.6$

LiBr balance: $w_1 x_1 = w_2 x_2$

$$0.6(0.50) = w_2(0.664)$$

Solving the two balance equations simultaneously gives

$$w_2 = 0.452 \text{ kg/s} \quad \text{and} \quad w_3 = 0.148 \text{ kg/s}$$

Approximately 4 kg of solution is pumped for each kilogram of refrigerant water vapor developed.

17-6 Enthalpy of LiBr solutions In order to perform thermal calculations on the absorption refrigeration cycle, enthalpy data must be available for the working substances at all crucial positions in the cycle. Water in liquid or vapor form flows in and out of the condenser and evaporator, so enthalpies at these points can be determined from a table of properties of water. In the generator and absorber, LiBr-water solutions exist for which the enthalpy is a function of both the temperature of the solution and the concentration. Figure 17-8 presents enthalpy data for LiBr-water solutions. The data are applicable to saturated or subcooled solutions and are based on a zero enthalpy of liquid water at 0°C and a zero enthalpy of solid LiBr at 25°C. Since the zero enthalpy for the water in the solution is the same as that in conventional tables of properties of water, the water-property tables can be used in conjunction with Fig.

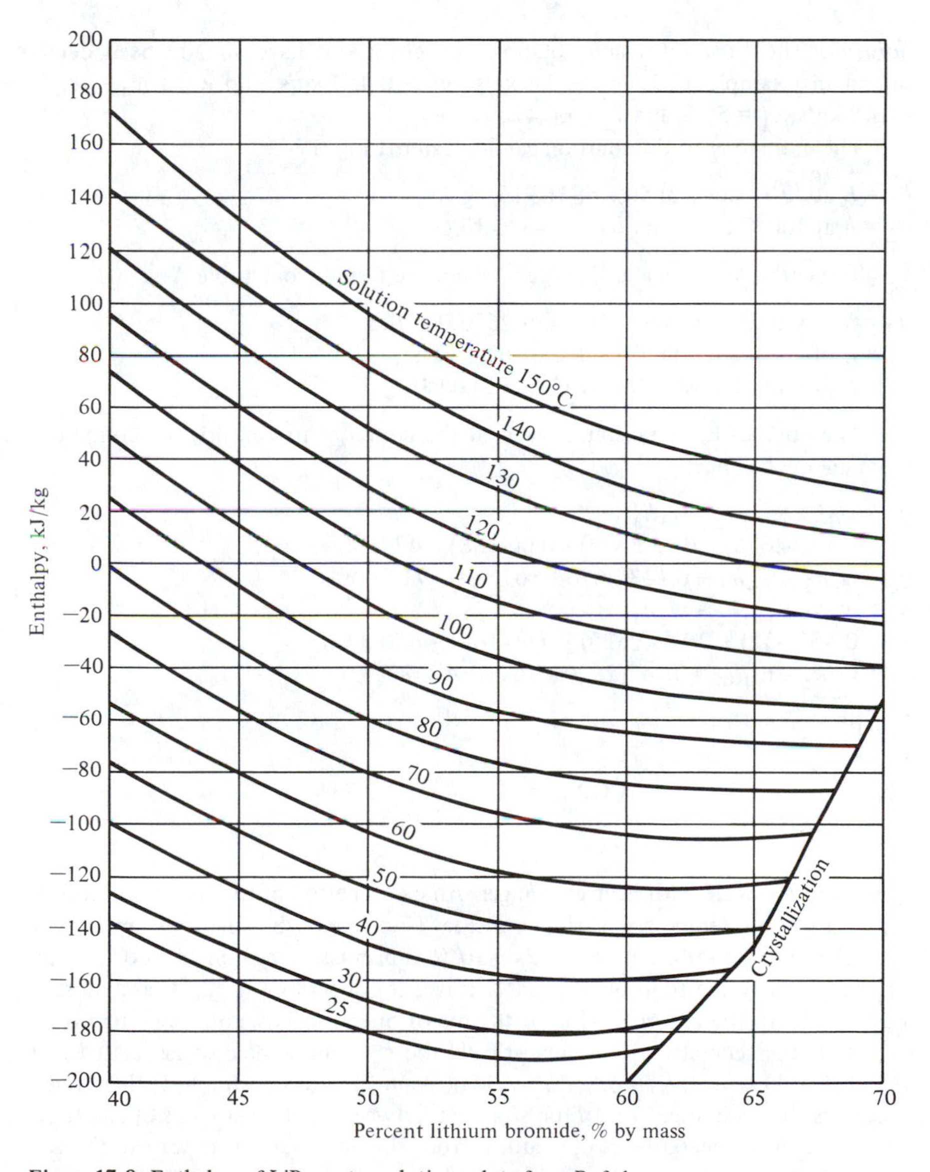

Figure 17-8 Enthalpy of LiBr–water solutions; data from Ref. 1.

17-8. The enthalpy values shown in Ref. 4 are based on a different datum plane for the solid LiBr, but the end results of calculations are essentially the same.

17-7 Thermal analysis of simple absorption system

Example 17-3 For the absorption system of Example 17-2 shown in Fig. 17-6 compute q_g, q_a, q_c, q_e, and the COP.

Solution The flow rates and solution concentrations have already been determined in Example 17-2: $w_1 = 0.6$ kg/s, $w_2 = 0.452$ kg/s, and $w_3 = w_4 = w_5 = 0.148$ kg/s; $x_1 = 50\%$, and $x_2 = 66.4\%$.

The enthalpies of the solution can be read off Fig. 17-8:

$h_1 = h$ at 30°C and x of 50% = -168 kJ/kg
$h_2 = h$ at 100°C and x of 66.4% = -52 kJ/kg

The enthalpies of water liquid and vapor are found from Table A-1:

$h_3 = h$ of saturated vapor at 100°C = 2676.0 kJ/kg
$h_4 = h$ of saturated liquid at 40°C = 167.5 kJ/kg
$h_5 = h$ of saturated vapor at 10°C = 2520.0 kJ/kg

The rates of heat transfer at each of the components can now be computed from energy balances:

$$q_g = w_3h_3 + w_2h_2 - w_1h_1$$
$$= 0.148(2676) + 0.452(-52) - 0.6(-168) = 473.3 \text{ kW}$$
$$q_c = w_ch_3 - w_4h_4 = 0.148(2676 - 167.5) = 371.2 \text{ kW}$$
$$q_a = w_2h_2 + w_5h_5 - w_1h_1$$
$$= 0.452(-52) + 0.148(2520) - 0.6(-168) = 450.3 \text{ kW}$$
$$q_e = w_5h_5 - w_4h_4 = 0.148(2520 - 167.5) = 348.2 \text{ kW}$$

Finally,

$$\text{COP}_{\text{abs}} = \frac{q_e}{q_g} = \frac{348.2}{476.6} = 0.736$$

17-8 Absorption cycle with heat exchanger An examination of the simple absorption cycle and operating temperatures shown in Fig. 17-6 reveals that the solution at point 1 leaves the absorber at a temperature of 30°C and must be heated to 100°C in the generator. Similarly the solution at point 2 leaves the generator at 100°C and must be cooled to 30°C in the absorber. One of the major operating costs of the system is the heat added in the generator q_g, and realistically there will be some cost associated with the removal of heat in the absorber q_a. A logical addition to the simple cycle is a heat exchanger as shown in Fig. 17-9 to transfer heat between the two streams of solutions. This heat exchanger heats the cool solution from the absorber on its way to the generator and cools the solution returning from the generator to the absorber.

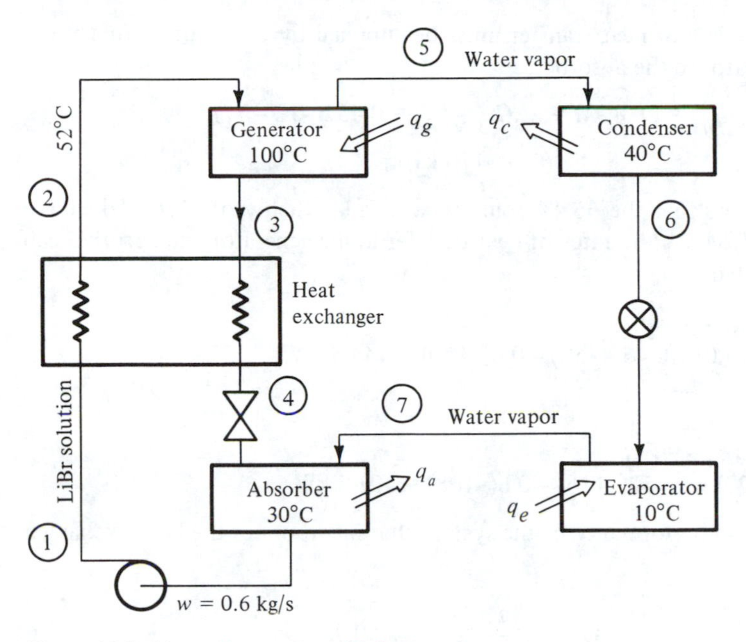

Figure 17-9 Absorption system with heat exchanger.

Example 17-4 The simple cycle operating at the temperatures shown in Fig. 17-6 is modified by the insertion of a heat exchanger, as shown in Fig. 17-9, such that the temperature at point 2 is 52°C. The mass rate of flow delivered by the solution pump is 0.6 kg/s. What are the rates of energy transfer at each of the components and the COP_{abs} of this cycle?

Solution Certain quantities are unchanged from Examples 17-2 and 17-3:

$$w_1 = w_2 = 0.6 \text{ kg/s} \qquad w_3 = w_4 = 0.452 \text{ kg/s}$$

and

$$w_5 = w_6 = w_7 = 0.148 \text{ kg/s}$$

The enthalpies that remain unchanged are

$h_1 = -168$ kJ/kg $\quad h_3 = -52$ kJ/kg
$h_5 = 2676.0$ kJ/kg $\quad h_6 = 167.5$ kJ/kg $\quad h_7 = 2520.0$ kJ/kg

The heat-transfer rates at the condenser and evaporator remain unchanged

$$q_c = 371.2 \text{ kW} \qquad \text{and} \qquad q_e = 348.2 \text{ kW}$$

The temperature of the 50% solution leaving the heat exchanger at point 2 is 52°C, and the solution at that condition has an enthalpy of -120 kJ/kg, as indicated by Fig. 17-8. The rate of heat absorbed by the solution passing from the absorber to the generator q_{hx} is

$$q_{hx} = w_1(h_2 - h_1) = 0.6[-120 - (-168)] = 28.8 \text{ kW}$$

Since this same rate of heat transfer must be supplied by the solution that flows from the generator to the absorber,

$$q_{hx} = 28.8 \text{ kW} = w_3(h_3 - h_4) = 0.452(-52 - h_4)$$

and so

$$h_4 = -116 \text{ kJ/kg}$$

Figure 17-8 shows that the 66.4% solution with an enthalpy of -116 kJ/kg has a temperature of 64°C. The rates of heat transfer in the generator and absorber can now be computed

$$\begin{aligned} q_g &= w_5 h_5 + w_3 h_3 - w_2 h_2 \\ &= 0.148(2676.0) + 0.452(-52) - 0.6(-120) = 444.5 \text{ kW} \end{aligned}$$

and

$$\begin{aligned} q_a &= w_7 h_7 + w_4 h_4 - w_1 h_1 \\ &= 0.148(2520) + 0.452(-116) - 0.6(-168) = 421.3 \text{ kW} \end{aligned}$$

The coefficient of performance of the system that incorporates the heat exchanger is

$$\text{COP}_{\text{abs}} = \frac{q_e}{q_g} = \frac{348.2}{444.5} = 0.783$$

This COP is an improvement over the value of 0.736 applicable to the simple system without a heat exchanger.

Another comparison of interest is that of the absorption cycle to the ideal heat-operated cycle whose COP is expressed by Eq. (17-4). Equation (17-4) expects just one temperature of heat rejection, T_a, while in the absorption cycles just analyzed there are two, 30 and 40°C. Choosing the mean of those two temperatures as the heat-rejection temperature gives

$$\text{COP}_{\text{ideal}} = \frac{(10 + 273.15)(100 - 35)}{(100 + 273.15)(35 - 10)} = 1.97$$

Therefore the absorption COP at these conditions is less than half of that of an ideal heat-powered refrigerating unit.

17-9 Configuration of commercial absorption units The construction of commercial absorption plants, taking advantage of the fact that the condenser and generator operate at the same pressure, combines these components in one vessel. Similarly, since the evaporator and absorber operate at the same pressure, these components likewise can be installed in the same vessel, as Fig. 17-10 shows. In the high-pressure vessel the water vapor from the generator drifts to the condenser, where it is liquefied, while in the low-pressure vessel the water vapor released at the evaporator flows downward to the absorber. To enhance the heat-transfer rate at the evaporator a circulating pump sprays the evaporating water over the evaporator tubes to chill the water from the refrigeration load. Note that the chilled water serving the refrigeration load is a

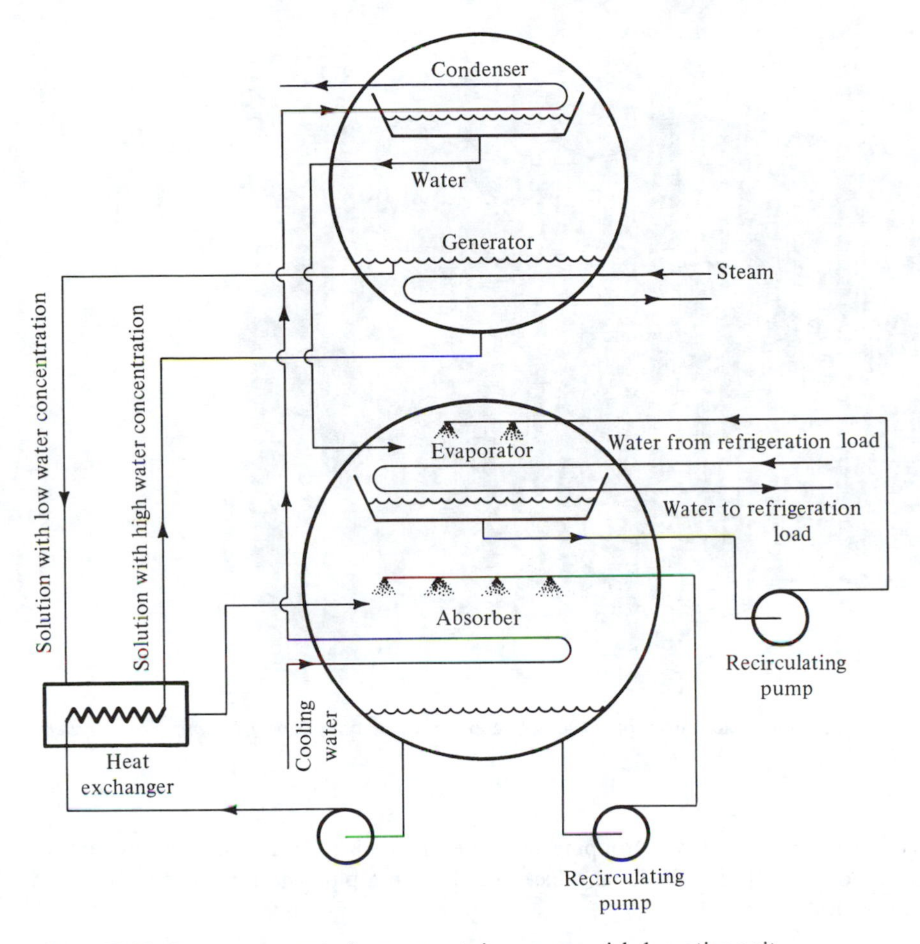

Figure 17-10 One arrangement of components in a commercial absorption unit.

separate circuit from the water serving as refrigerant in the absorption unit. Maintaining separate water circuits helps maintain better purity in the absorption unit and permits the water serving the refrigeration load to operate at pressures above atmospheric. Another feature shown in Fig. 17-10 is that the cooling water from the cooling tower passes in series through the absorber and condenser, extracting heat in both components.

In the photograph of the absorption unit in Fig. 17-11 the high- and low-pressure vessels can be distinguished. It is also possible to combine all the components into one vessel with an internal separator between the high- and low-pressure chambers.

17-10 Crystallization On the two property charts for LiBr–water solutions (Figs. 17-5 and 17-8) *crystallization lines* appear in the lower right section. The region to the right and below those lines indicates a solidification of the LiBr. The process is similar to that of the solidification of an antifreeze (Chap. 15) in that as the LiBr solidifies, it dilutes the liquid solution so that the state of the solution continues to be represented

Figure 17-11 An absorption unit of 1200 kW cooling capacity. *(York Division, Borg-Warner Corporation.)*

by the crystallization curve. Dropping into the crystallization region thus indicates the formation of a slush, which can block the flow in a pipe and interrupt the operation of the absorption unit.

Example 17-5 In the system shown in Fig. 17-9, the ambient wet-bulb temperature decreases so that the temperature of cooling water drops, which also reduces the condensing temperature to 34°C. All other temperatures specified on Fig. 17-9 remain unchanged. Is there a danger of crystallization?

Solution The crucial component from the standpoint of crystallization is the heat exchanger, shown in Fig. 17-12. The reduction in condensing temperature drops the high-side pressure so that the concentration of solution leaving the generator at point 3 is 69%. If the mass flow rate delivered by the solution pump remains constant at 0.6 kg/s, new mass rates of flow apply at points 3 and 5

$$w_3 = \frac{w_1 x_1}{x_3} = \frac{0.60(0.50)}{0.69} = 0.435 \text{ kg/s}$$

and

$$w_5 = 0.60 - 0.435 = 0.165 \text{ kg/s}$$

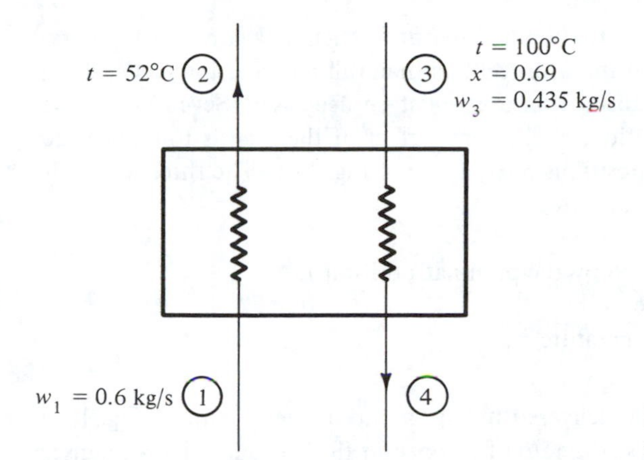

Figure 17-12 Heat exchanger in Example 17-5.

The enthalpies surrounding the heat exchanger that are known at this stage are

h_1 = h at 30°C and x of 50% = −168 kJ/kg
h_2 = h at 52°C and x of 50% = −120 kJ/kg
h_3 = h at 100°C and x of 69% = −54 kJ/kg

An energy balance about the heat exchanger yields

$$h_4 = h_3 - \frac{w_1(h_2 - h_1)}{w_3} = -54 - \frac{0.6(66)}{0.435} = -120 \text{ kJ/kg}$$

From Fig. 17-8 the condition of solution at point 4, which is h = -120 kJ/kg, x = 0.69, is found to be crystallized. Some of the solution has thus solidified, and there is danger of blocking the flow and causing refrigeration to cease.

Example 17-5 illustrates two facts: (1) the position in the system where crystallization is most likely to occur is where the solution from the generator leaves the heat exchanger, and (2) an operating condition conducive to crystallization is at low condensing pressures. Commercial LiBr–water absorption units have controls that avoid crystallization, one element of them usually being to maintain a condensing pressure artificially high even when low-temperature cooling water is available for the condenser.

17-11 Capacity control The true meaning of *capacity control* is "capacity reduction," since operation without capacity control yields maximum refrigeration capacity. The need for capacity control arises when the refrigeration load drops off, as reflected in a reduction in chilled-water temperature returning to the absorption unit (assuming a constant-rate flow of chilled water). With no capacity control the temperature of the chilled water leaving the evaporator would decrease, as would the pressure on the low-pressure side of the absorption unit. The low-side pressure could reduce to the point where the refrigerant water would freeze.

Most control systems on absorption units attempt to regulate a constant temperature of chilled water leaving the evaporator. At less than full refrigeration loads, then, the refrigerating capacity of the absorption unit must be decreased. Several methods are available to achieve this reduction, but the net effect of them all is to reduce the flow rate of refrigerant water at positions 5, 6, and 7 in Fig. 17-9. The three methods for reducing the refrigerant-water flow are:

1. Reducing the flow rate delivered by the pump at position 1
2. Reducing the generator temperature
3. Increasing the condensing temperature

Method 1 If the mass rate of flow delivered by the pump in the system of Fig. 17-9 were reduced from 0.6 to 0.4 kg/s, the rate of refrigerant flow through the condenser and evaporator would also be reduced by one-third, resulting in a corresponding reduction in refrigerating capacity. This method is efficient since the rate of heat addition at the generator is reduced by the same proportion as the refrigerating capacity.

The statement that the mass flow rate of refrigerant changes in the same proportion as that of the flow rate delivered by the pump is correct provided that the concentrations of solution remain unchanged. The concentrations remain unchanged only if the operating temperatures in the components also remain fixed, and such is not the usual situation. In the condenser, for example, if the supply temperature and flow rate of condenser cooling water remains constant, the reduction in flow rate of refrigerant being condensed causes the condensing temperature to drop. Similarly in the generator if the supply conditions of steam or hot water remain constant, the generator temperature increases. These particular changes in temperatures result in higher LiBr concentrations leaving the generator at point 3, which results in higher COP_{abs}, but they also are the conditions that could induce crystallization. On some commercial absorption units the adjustment of flow rate is the primary method of capacity control but is combined with method 2 or method 3 when there is danger of crystallization.

Method 2 A reduction in the generator temperature will reduce the refrigerating capacity and can be achieved by throttling the pressure of the steam entering the generator or reducing the flow rate of hot water, depending upon which is the heat source of the generator.

Example 17-6 The absorption cycle shown in Fig. 17-9 and analyzed in Example 17-4 is equipped with capacity control to throttle the steam providing heat to the generator and thus reduces the steam pressure and the generator temperature. If the generator temperature is reduced to 95°C while all other temperatures and the flow rate through the pump noted in Fig. 17-9 remain constant, determine the new refrigerating capacity, the rate of heat addition at the generator, and COP_{abs}.

Solution The reduction in generator temperature reduces x_3 from 66.4 to 65% and h_3, which is the enthalpy of solution at point 3, from -52 to -59 kJ/kg.

The revised mass flow rates are

$$w_3 = w_2 \frac{x_2}{x_3} = 0.6 \frac{0.5}{0.65} = 0.462 \text{ kg/s}$$

and
$$w_5 = w_6 = w_7 = 0.6 - 0.462 = 0.138 \text{ kg/s}$$

From an energy balance about the generator and evaporator, respectively,

$$\begin{aligned} q_{gen} &= w_5 h_5 + w_3 h_3 - w_2 h_2 \\ &= 0.138(2676) + 0.462(-59) - 0.6(-120) = 414.0 \text{ kW} \end{aligned}$$

and

$$q_{ev} = w_7(h_7 - h_6) = 0.138(2520 - 167.5) = 324.6 \text{ kW}$$

The coefficient of performance is

$$\text{COP}_{abs} = \frac{324.6}{414.0} = 0.784$$

An evaluation of Example 17-6 shows that the reduction of the generator temperature does reduce the refrigeration capacity–from 348.2 to 324.6 kW. The rate of heat addition at the generator also drops so that the COP_{abs} remains essentially unchanged (0.784 versus 0.783).

Example 17-6 presents only a portion of the picture of the behavior during capacity control. As emphasized in Chap. 14 on vapor-compression-system analysis, a change at one component is likely to affect the conditions at other components. With the change in generator temperature made in Example 17-6, the flow rate of refrigerant water passing through the condenser and evaporator drops, so with a constant flow rate and entering temperature of condenser cooling water the condensing temperature will drop. Similarly with a constant flow rate and entering temperature of water to be chilled, the evaporator temperature will increase. The changes in those two temperatures will affect the high- and low-side pressure, respectively, and thus influence the concentrations leaving the absorber and generator.

Method 3 A further method for reducing the refrigerating capacity of an absorption unit is to increase the condensing temperature; this can be done conveniently by increasing the temperature of cooling water supplied to the condenser, which in turn can be achieved by bypassing a fraction of the water around the cooling tower. The effect on the cycle performance of increasing the condensing temperature is the same as that of decreasing the generator temperature, namely, reducing the LiBr concentration of solution returning from the generator to the absorber as shown in the skeleton *p-x-t* diagram in Fig. 17-13. If point *A* is the original operating condition, an increase in condensing temperature and pressure moves the condition of the solution along a line of constant generator temperature to point *B*, so that the LiBr concentration drops. For a given rate of flow of solution handled by the pump, the flow rate of refrigerant circulating to the condenser and evaporator decreases.

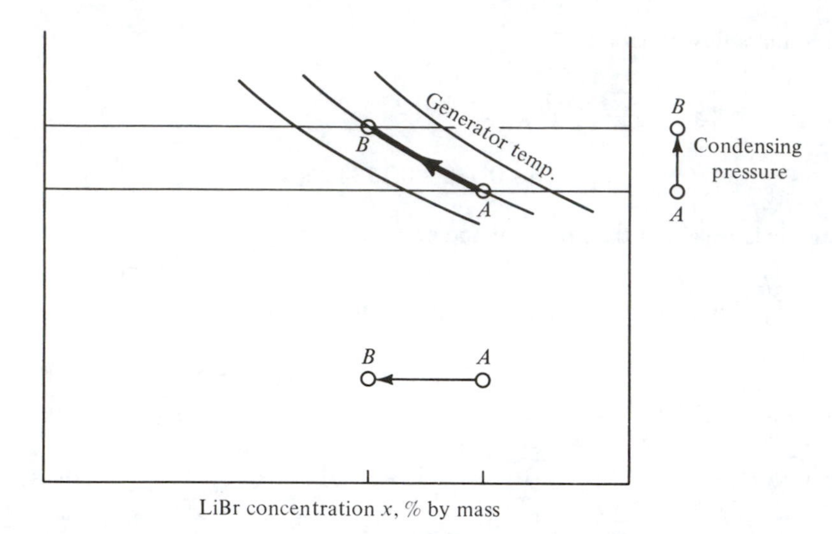

Figure 17-13 Reducing the refrigerating capacity by increasing the condensing temperature and reducing the concentration of LiBr leaving the generator.

17-12 Double-effect system A LiBr absorption unit with improved efficiency is the double-effect system shown schematically in Fig. 17-14. The major distinguishing feature of the double-effect system is that it incorporates a second generator, generator II, which uses the condensing water vapor from generator I to provide its supply of heat. There are three different levels of pressure in each of the vessels shown in Fig. 17-14, and medium-pressure steam (of the order of 1000 kPa) is supplied to generator I instead of the low-pressure steam (approximately 120 kPa) supplied to the generator in a single-stage unit. The LiBr solution from generator I passes through a heat exchanger, where it transfers heat to weak LiBr solution on the way to generator I. The solution that passes on to generator II is heated there by condensing water vapor that was driven off in generator I. The solution next passes through a restriction which drops the pressure to that of vessel 2. Following reduction of pressure some of the water in the solution flashes to vapor, which is liquefied at the condenser.

The double-effect absorption unit operates with higher COPs than the single-stage, as shown in Fig. 17-15.

17-13 Steam-driven combination with vapor compression Some large-capacity water chilling installations use an energy source of high-pressure steam in a system that combines a vapor-compression and an absorption system. The high-pressure steam, as shown in Fig. 17-16, first expands through a turbine, which provides the power for driving the compressor of a vapor-compression system. The exhaust steam from the turbine passes to the generator of the absorption system. The water to be chilled passes in series through the evaporators of the two refrigerating plants. This combination is the counterpart of the type of power plant where some of the energy of high-pressure steam is used to generate power and the condensation of the steam is used for heating or process purposes.

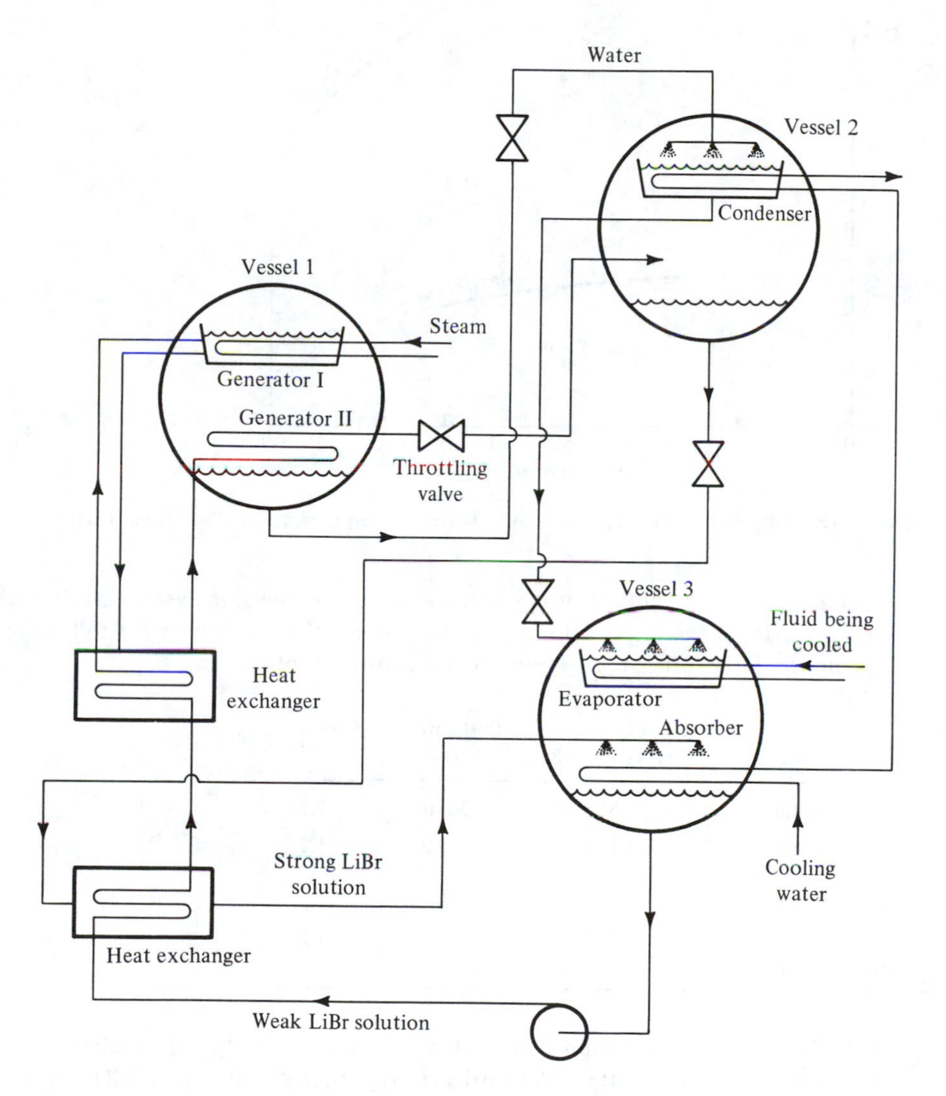

Figure 17-14 Double-effect absorption unit.

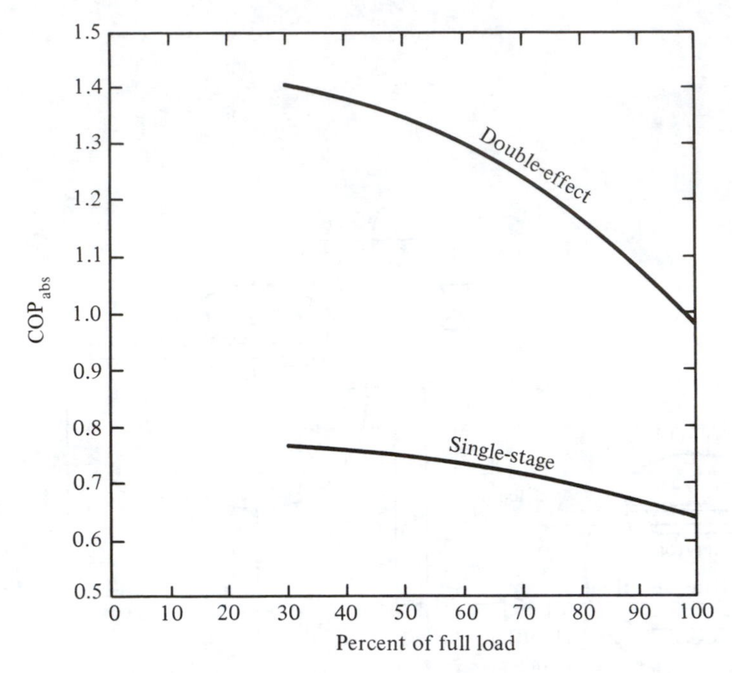

Figure 17-15 COP_{abs} of single-stage and double-effect absorption units. *(The Trane Company.)*

Example 17-7 A combined absorption and vapor-compression system, as shown in Fig. 17-16, is energized with high-pressure steam that undergoes the following conditions through the steam turbine and absorption unit:

Position	Pressure, kPa	Enthalpy, kJ/kg	Flow rate, kg/s
Entering turbine	1500	3080	1.2
Leaving turbine and entering absorption unit	100	2675	1.2
Condensate leaving absorption unit	100	419	1.2

The COP of the vapor-compression unit is 3.6, and the COP_{abs} of the absorption unit is 0.7. What is (*a*) the total refrigerating capacity and (*b*) the COP of the combined system?

Solution (*a*) If heat loss from the steam turbine is neglected, the power delivered by the turbine P equals that extracted from the steam as it flows through the turbine

$$P = (1.2 \text{ kg/s})(3080 - 2675 \text{ kJ/kg}) = 486 \text{ kW}$$

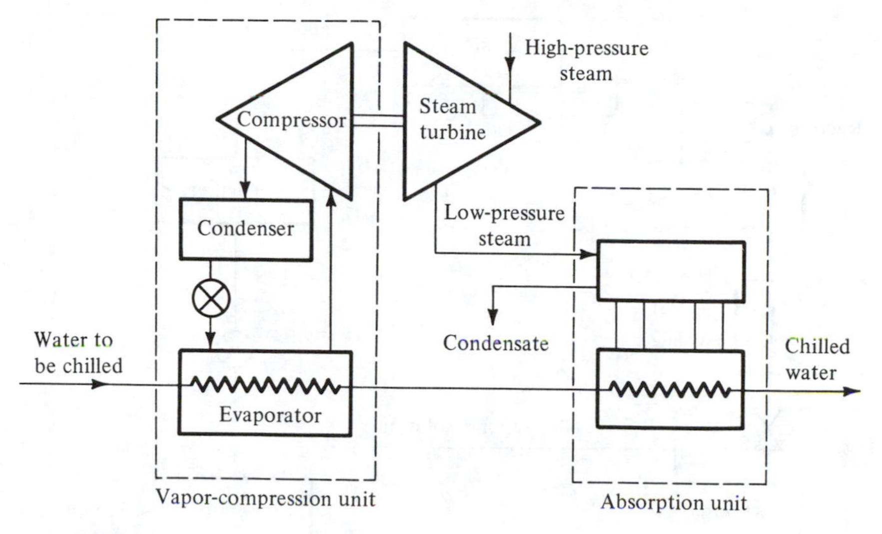

Figure 17-16 Combined absorption and vapor-compression system.

The refrigerating capacity of the vapor-compression system q_{vc} is

$$q_{vc} = (P)\,(\text{COP}) = (486\text{ kW})\,(3.6) = 1750\text{ kW}$$

The rate of heat addition to the absorption unit is

$$(1.2\text{ kg/s})\,(2675 - 419\text{ kJ/kg}) = 2707\text{ kW}$$

and the refrigerating capacity of the absorption unit q_{abs} is

$$q_{abs} = (2707\text{ kW})\,(\text{COP} = 0.7) = 1895\text{ kW}$$

The total refrigerating capacity q_{tot} is therefore

$$q_{tot} = q_{vc} + q_{abs} = 1750 + 1895 = 3645\text{ kW}$$

(*b*) The COP of the combined system, which is a heat-operated refrigerating unit, is the quotient of q_{tot} and the total rate of heat supplied

$$\text{COP}_{comb} = \frac{3645\text{ kW}}{1.2(3080 - 419)} = 1.14$$

which is a favorable COP for a heat-operated unit.

17-14 Aqua-ammonia system This chapter has concentrated so far on absorption systems that use LiBr as the absorbent and water as the refrigerant. Other pairs of substances can also function as absorbent and refrigerants, e.g., water as the absorbent and ammonia as the refrigerant. This combination, called *aqua-ammonia,* was used in absorption systems years before the LiBr-water combination became popular. The aqua-ammonia system, shown schematically in Fig. 17-17, consists of all the components previously described–generator, absorber, condenser, evaporator, and solution heat exchanger–plus a *rectifier* and *analyzer.* The need for them is occasioned by

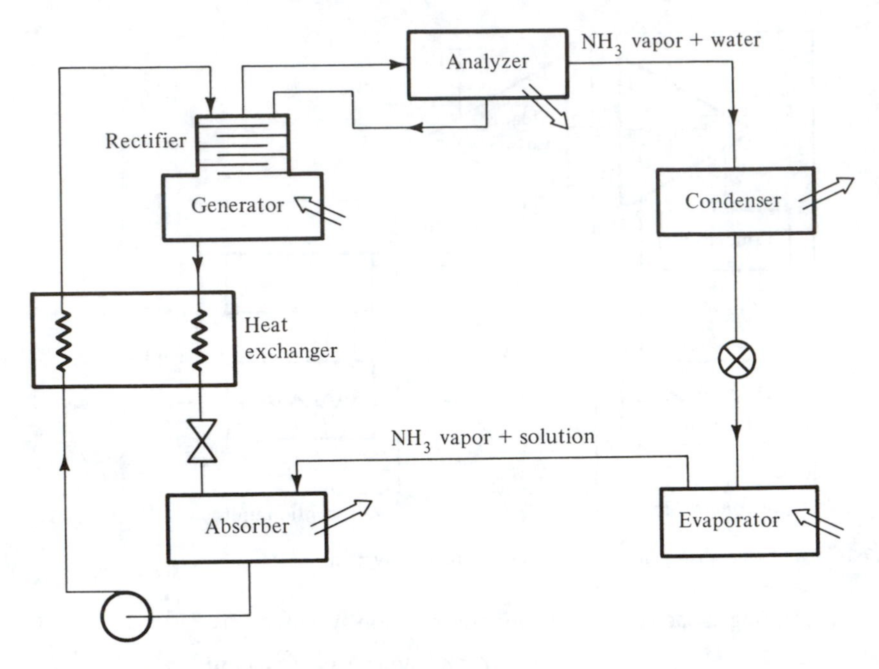

Figure 17-17 Aqua-ammonia absorption system.

the fact that the refrigerant vapor released at the generator (the ammonia) contains water vapor as well. When this water finds its way to the evaporator, it elevates the temperature there. To remove as much water vapor as possible, the vapor driven off at the generator first flows countercurrent to the incoming solution in the rectifier. Next, the solution passes through the analyzer, a water-cooled heat exchanger condensing some water-rich liquid, which drains back to the rectifier. A small amount of water vapor escapes the analyzer and must ultimately be passed as liquid from the evaporator to the absorber.

A comparison of the aqua-ammonia and LiBr–water systems follows. The two systems have comparable COP_{abs}. The aqua-ammonia system is capable of achieving evaporating temperatures below 0°C, but the LiBr–water system is limited in commercial units to no lower than about 3°C. The aqua-ammonia system has the disadvantage of requiring extra components but does have the advantage of operating at pressures above atmospheric. The LiBr–water system operates at pressures below atmospheric, resulting in unavoidable leakage of air into the system, which must be purged periodically. Special inhibitors must be incorporated in the LiBr–water system to retard corrosion.

17-15 Role of the absorption unit in refrigeration practice Absorption systems have experienced many ups and downs. The absorption system was the predecessor of the vapor-compression system in the nineteenth century, and aqua-ammonia systems enjoyed wide application in domestic refrigerators and large industrial installations in the chemical and process industries. The LiBr–water system was commercialized in the

1940s and 1950s as water chillers for large-building air conditioning. They were energized by steam or hot water generated from natural gas and oil-fired boilers. In the 1970s the shift from direct burning of oil and natural gas struck a blow at the application of absorption units but at the same time opened up other opportunities, such as using heat derived from solar collectors to energize the absorption unit. Also, because of the rapidly rising cost of energy, low-temperature-level heat (in the 90 to 110°C range) formerly rejected to the atmosphere in chemical and process plants is now often used to operate absorption systems providing refrigeration that is advantageous elsewhere in the plant. The combination of absorption systems with vapor compression, described in Sec. 17-13, is another application of absorption units that remains attractive.

PROBLEMS

17-1 What is the COP of an ideal heat-operated refrigeration cycle that receives the energizing heat from a solar collector at a temperature of 70°C, performs refrigeration at 15°C, and rejects heat to atmosphere at a temperature of 35°C? *Ans.* 1.47

17-2 The LiBr–water absorption cycle shown in Fig. 17-2 operates at the following temperatures: generator, 105°C; condenser, 35°C; evaporator, 5°C; and absorber, 30°C. The flow rate of solution delivered by the pump is 0.4 kg/s.

(*a*) What are the mass flow rates of solution returning from the generator to the absorber and of the refrigerant? *Ans.* Refrigerant flow rate = 0.093 kg/s

(*b*) What are the rates of heat transfer at each of the components, and the COP_{abs}? *Ans.* Refrigerating capacity = 220 kW

17-3 In the absorption cycle shown in Fig. 17-9 the solution temperature leaving the heat exchanger and entering the generator is 48°C. All other temperatures and the flow rate are as shown in Fig. 17-9. What are the rates of heat transfer at the generator and the temperature at point 4? *Ans.* t_4 = 70°C

17-4 The solution leaving the heat exchanger and returning to the absorber is at a temperature of 60°C. The generator temperature is 95°C. What is the minimum condensing temperature permitted in order to prevent crystallization in the system? *Ans.* 37°C

17-5 One of the methods of capacity control described in Sec. 17-11 is to reduce the flow rate of solution delivered by the pump. The first-order approximation is that the refrigerating capacity will be reduced by the same percentage as the solution flow rate. There are secondary effects also, because if the mean temperatures of the heating medium in the generator, the cooling water in the absorber and condenser, and the water being chilled in the evaporator all remain constant, the temperatures in these components will change when the heat-transfer rate decreases.

(*a*) Fill out each block in the Table 17-1 with either "increases" or "decreases" to indicate qualitative influence of the secondary effect. *Ans.* Capacity decreases by less than reduction in solution flow rate

(*b*) Use the expression for an ideal heat-operated cycle to evaluate the effects of temperature on the COP_{abs}. *Ans.* See Fig. 17-15

17-6 In the double-effect absorption unit shown in Fig. 17-14, LiBr–water solution leaves generator I with a concentration of 67%, passes to the heat exchanger and then

Table 17-1 Influences of reduction in solution flow rate at pump

Component	Temperature	Solution concentration		Refrig. capac.	COP_{abs}
		x_{gen}	x_{abs}		
Generator			X		
Absorber		X			
Condenser			X		
Evaporator		X			

to generator II, where its temperature is elevated to 130°C. Next the solution passes through the throttling valve, where its pressure is reduced to that in the condenser, which is 5.62 kPa. In the process of the pressure reduction, some water vapor flashes off from this solution. For each kilogram of solution flowing through generator II, (*a*) how much mass flashes to vapor, and (*b*) what is the concentration of LiBr–solution that drops into the condenser vessel? *Ans.* (*b*) 68.4%

17-7 The combined absorption and vapor-compression system shown in Fig. 17-16 is to be provided with a capacity control scheme that maintains a constant temperature of the leaving chilled water as the temperature of the return water to be chilled varies. This control scheme is essentially one of reducing the refrigerating capacity. The refrigerant compressor is equipped with inlet vanes (see Chap. 11), the speed of the turbine-compressor can be varied so long as it remains less than the maximum value of 180 r/s, and the control possibilities of the absorption unit are as described in Sec. 17-11. The characteristics of the steam turbine are that both its speed and power diminish if the pressure of the supply steam decreases or the exhaust pressure increases. With constant inlet and exhaust pressures the speed of the turbine increases if the load is reduced. Devise a control scheme and describe the behavior of the entire system as the required refrigerating load decreases.

17-8 The operating cost of an absorption system is to be compared with an electric-driven vapor-compression unit. The cost of natural gas on a heating value basis is $4.20 per gigajoule; when used as fuel in a boiler it has a combustion efficiency of 75 percent. An absorption unit using steam from this boiler has a COP_{abs} of 0.73. If a vapor-compression unit is selected, the COP would be 3.4, and the electric-motor efficiency is 85 percent. At what cost of electricity are the operating costs equal? *Ans.* 8.0 cents/kWh

REFERENCES

1. The Absorption Cooling Process, *Res. Bull.* 14, Institute of Gas Technology, Chicago, 1957.
2. "ASHRAE Handbook, Fundamentals Volume," chap. 1, American Society of Heating, Refrigerating, and Air-Conditioning Engineers, Atlanta, Ga., 1981.
3. Plank, R. (ed.): "Handbuch der Kaeltetechnik," vol. 7, by W. Niebergall, Springer, Berlin, 1959.
4. L. A. McNeely: Thermodynamic Properties of Aqueous Solutions of Lithium Bromide, *ASHRAE Trans.*, vol. 85, pt. 1, pp. 413–434, 1979.

CHAPTER

EIGHTEEN

HEAT PUMPS

18-1 Types of heat pumps All refrigeration systems are heat pumps, because they absorb heat energy at a low temperature level and discharge it to a high temperature level. The designation *heat pump*, however, has developed around the application of a refrigeration system where the heat rejected at the condenser is used instead of simply being dissipated to the atmosphere. There are certain applications and occasions where the heat pump can simultaneously perform useful cooling and useful heat rejection, and this is clearly an advantageous situation.

Heat pumps can be and often are applied in a variety of contexts. Four important classifications this chapter explores are (1) package heat pumps with a reversible cycle, (2) decentralized heat pumps for air-conditioning moderate- and large-sized buildings, (3) heat pumps with a double-bundle condenser, and (4) industrial heat pumps. Common threads run through all four categories, but each group also responds to a unique opportunity or need.

18-2 Package type, reversible cycle This classification especially includes residential and small commercial units that are capable of heating a space in cold weather and cooling it in warm weather. The several different sources and sinks from which the heat pump can draw its heat or reject it—air, water, earth—will be discussed in Sec. 18-3. For purposes of explanation air will be initially assumed to be the source.

The reversible heat pump operates with the flow diagram shown in Fig. 18-1. During heating operation the four-way valve (Fig. 18-2) positions itself so that the high-pressure discharge gas from the compressor flows first to the heat exchanger in the

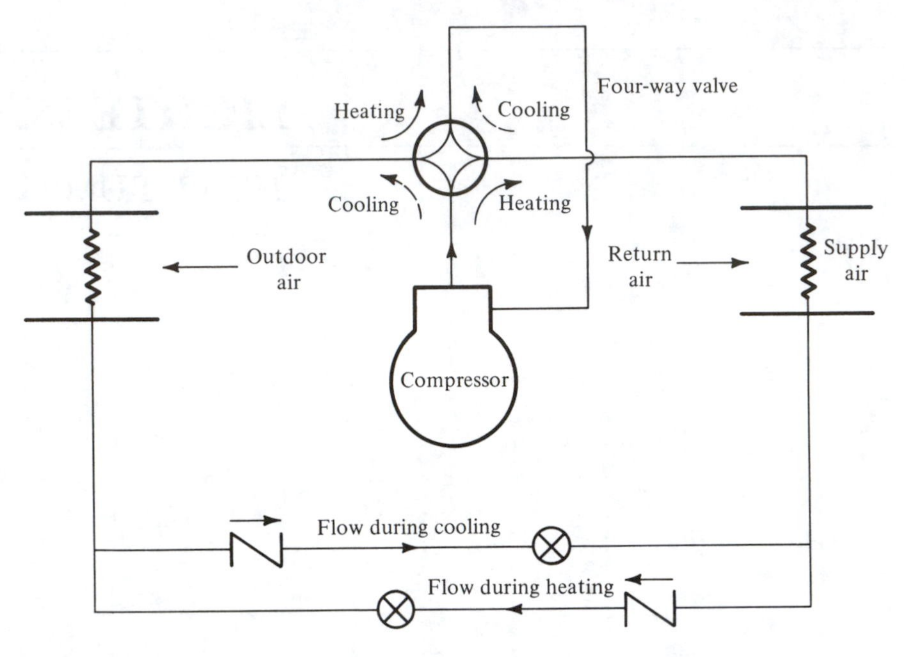

Figure 18-1 Reversible cycle air-source heat pump.

conditioned airstream. In its condensing process the refrigerant rejects heat, warming the air. Liquid refrigerant flows on to the expansion-device section, where the check valve in the upper line prevents flow through its branch and instead the liquid refrigerant flows through the expansion device in the lower branch. The cold low-pressure refrigerant then extracts heat from the outdoor air while it vaporizes. Refrigerant vapor returns to the four-way valve to be directed to the suction side of the compressor.

To convert from heating to cooling operation the four-way valve shifts to its opposite position so that discharge gas from the compressor first flows to the outdoor coil, where the refrigerant rejects heat during condensation. After passing through the

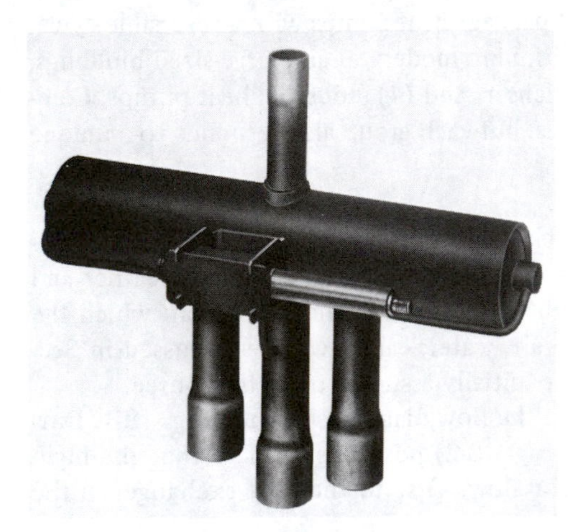

Figure 18-2 Four-way reversing valve. *(Ranco, Inc.)*

expansion device in the upper branch of Fig. 18-1, the low-pressure low-temperature refrigerant evaporates in the heat exchanger that cools air from the conditioned space.

Two branches for the expansion device are needed in Fig. 18-1, because a conventional superheat-controlled expansion valve would perform properly with flow only in one direction. It might seem that a capillary tube would work satisfactorily, because its performance is the same regardless of direction of flow, but the pressure difference across the capillary tube is much higher during winter heating operation than during summer cooling. Thus a capillary tube sized for one season is improperly sized for the other. The electric expansion valve can operate with refrigerant flow in either direction.

18-3 Heat sources and sinks for package-type reversible heat pumps The principal sources and sinks for heat in residential and commercial heat pumps are air, water, and earth. Air is most widely used because it permits a manufacturer to design a product whose capacity and efficiency can be guaranteed for given climatic conditions. Water-source heat pumps using water from wells have performance advantages over the air-source heat pumps, because the water temperature is fairly uniform throughout the year. Air-source heat pumps are subject to a much wider range of source temperatures and experience the penalties of low air-source temperatures when heating capacity is most needed (see Sec. 18-4). In the first wave of interest in heat pumps that occurred in the United States in the 1950s, some earth-source heat pumps were built. They generally had a refrigeration coil buried in the earth. This design proved unsatisfactory due to the expense of installation, the possibility of leaks developing, and the disruption of a large area of earth. The ground source is under consideration once again, particularly in Europe, where heating-only heat pumps are of most interest[1,2] and antifreeze usually circulates through the ground coil. Another approach is to sink a 100-m-long tube 150 mm in diameter into the earth and deliver water to the bottom of the large tube through a small inner tube.

Another heat source under consideration is a solar collector[3] in the *solar-assisted heat pump*, one arrangement of which is shown schematically in Fig. 18-3. The source

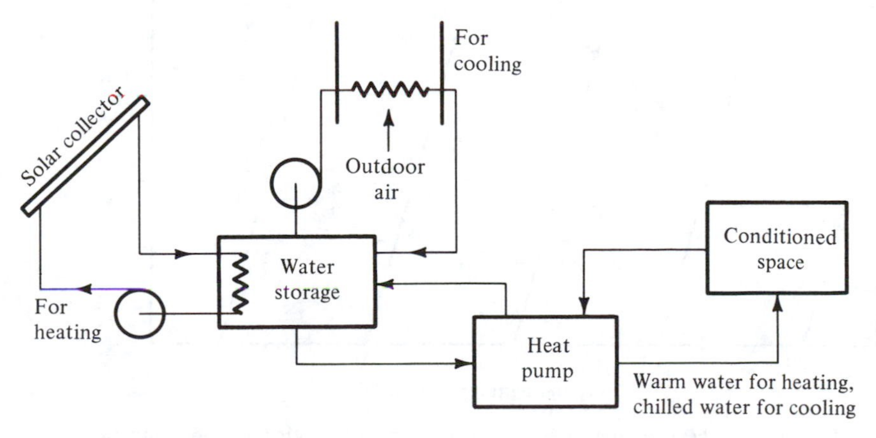

Figure 18-3 Solar-assisted heat pump.

of heat during heating operation, the water in the storage vessel, is capable of serving as a heat source even when its temperature drops to perhaps 5°C. The benefit of this low water temperature is that the solar collector can operate at a much lower temperature, thereby increasing its rate of heat absorption. Also, the range of ambient conditions during which heat can be extracted from the collector can be expanded by the solar-assisted heat pump.

18-4 Heating performance of an air-source heat pump The heat pump is more efficient in converting electric energy to heating than a resistance heater is. Section 10-7 defined the performance factor as a measure of the effectiveness of heating efficiency of a heat pump. The performance factor is the ratio of heat rejected for heating to the electric energy used to drive the compressor. In resistance heating the performance factor is 1.0 since the power derived from heating is the same as the electric power provided to the heater. In the heat pump, with no external losses, the performance factor is

$$\frac{\text{Power to compressor motor + power derived from heat source}}{\text{Power to compressor motor}}$$

The performance factor of a heat pump is therefore always greater than 1.0. The requirement that there be no external losses is an important one, because in practical heat pumps with the motor and compressor located in an outdoor compartment there can be heat loss to the ambient. These heat losses combined with the influence of low

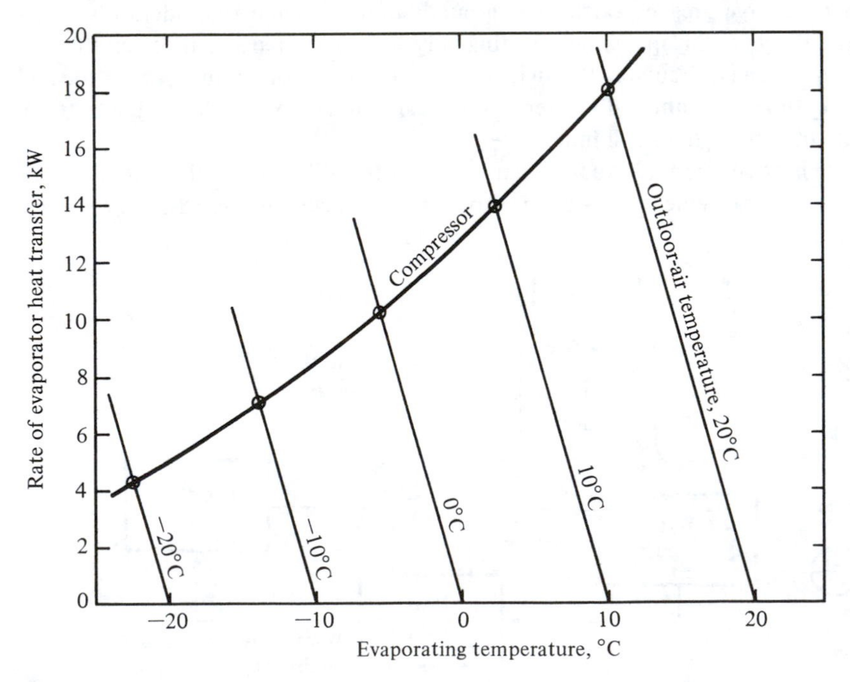

Figure 18-4 Evaporator heat-transfer rates of a heat pump with varying outdoor temperature. The condensing temperature is constant at 40°C.

evaporator temperatures result in performance factors only slightly above 1.0 at extremely low outdoor temperatures.

The preceding paragraphs discussed the efficiency of the air-source heat pump, but another important characteristic is the heating capacity. It decreases as the outdoor temperature drops, as can be demonstrated by Figs. 18-4 and 18-5. The techniques explained in Secs. 14-6 and 14-7 apply. The series of balance points between the compressor and evaporator is shown in Fig. 18-4. The condenser rejects heat to the return air from the conditioned space, whose temperature remains essentially constant, so that a constant condensing temperature of 40°C is assumed. As the outdoor-air temperature drops, the rate of evaporator heat transfer also drops. The heating capacity is the sum of the evaporator heat-transfer rate and the power delivered to the compressor. Figure 18-5 shows the evaporator capacity, compressor power, and heating capacity as a function of outdoor temperature. Values of evaporator heat-transfer rate in Fig. 18-5 are found by transferring data from Fig. 18-4. The compressor power is controlled by the evaporating and condensing temperatures, and the heating capacity is the sum of the evaporator and compressor energy flow rates. The important characteristic to derive from Fig. 18-5 is that the heating capacity drops off as the outdoor-air temperature decreases.

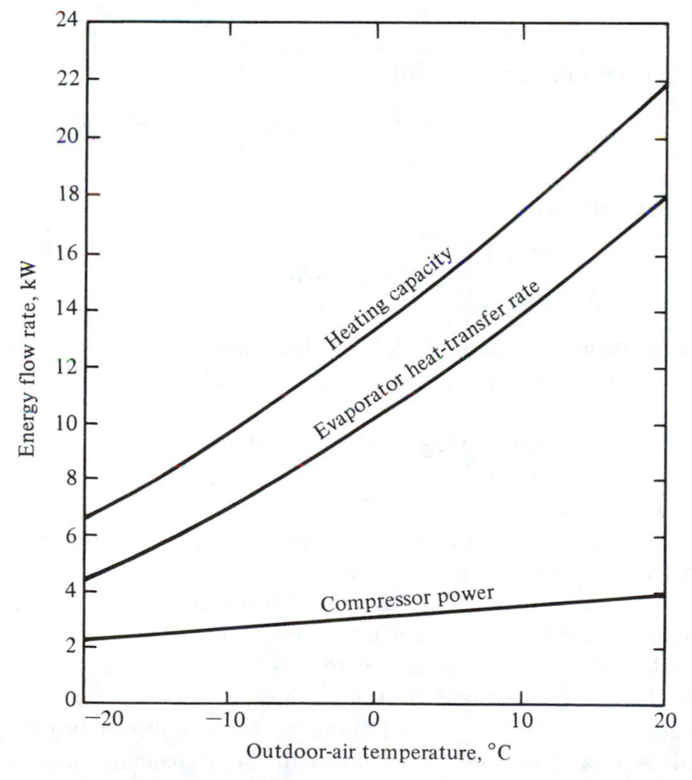

Figure 18-5 Heating capacity, evaporator heat-transfer rate, and compressor power of an air-source heat pump as a function of outdoor-air temperature.

18-5 Comparative heating costs The heat pump strives to supply heat at a lower cost than could be obtained by burning fossil fuel, such as gas or oil. The comparative costs of electricity and fuel decide which method will be the most economical in operation.

Example 18-1 A natural-gas furnace operates with an efficiency of 75 percent, and the cost of gas is \$3.80 per gigajoule. What is the maximum allowable cost of electricity if the heat pump whose characteristics are shown in Fig. 18-5 is to have an operating cost equal to that of the furnace? The average outdoor temperature is 5°C. The motor which drives the compressor has an efficiency of 80 percent, and the motor which drives the outdoor-air fan uses 25 kWh per gigajoule of heating capacity at the 5°C outdoor temperature.

Solution

$$\text{Cost of heating by gas} = \frac{\$3.80/\text{GJ}}{0.75} = \$5.07/\text{GJ}$$

At an outdoor temperature of 5°C the performance factor of the heat pump in Fig. 18-5 is 15.5/3.3 = 4.70.

The electric energy to the compressor motor for 1 GJ of heat is

$$\frac{10^9 \text{ J}}{4.70(\text{motor effic.} = 0.80)} = 266{,}000 \text{ kJ}$$

and

$$266{,}000 \text{ kJ} = \frac{266{,}000}{3600 \text{ s/h}} = 73.9 \text{ kWh}$$

For equal costs the electricity rate would be

$$\frac{\$5.07/\text{GJ}}{73.9 \text{ kWh/GJ}} = \$0.0686/\text{kWh}$$

An electric rate lower than 6.86 cents per kilowatthour makes heating by heat pump more attractive than heating by gas.

18-6 Matching heating capacity to the heating load The heating capacity of an air-source heat pump depends upon the outdoor temperature, as demonstrated in Fig. 18-5. The heating load also depends upon the outdoor-air temperature, which for a residence is approximately proportional to the indoor-outdoor temperature difference. When the heating capacity and heating load are shown on the same graph, as in Fig. 18-6, their intersection is the balance point, which in Fig. 18-6 occurs at a temperature of -4°C. At outdoor temperatures higher than -4°C the heat pump has greater capacity than needed and cycles on and off as necessary to match the load. At outdoor temperatures below the balance point the capacity of the heat pump is less than needed and the temperature of the building would fall unless some additional heating capacity were provided. A typical method of providing the supplementary heating capacity is by using resistance heaters. If 4 kW of capacity in the form of resistance heaters is available, the additional capacity will shift the new balance point in Fig. 18-6 to -8.5°C.

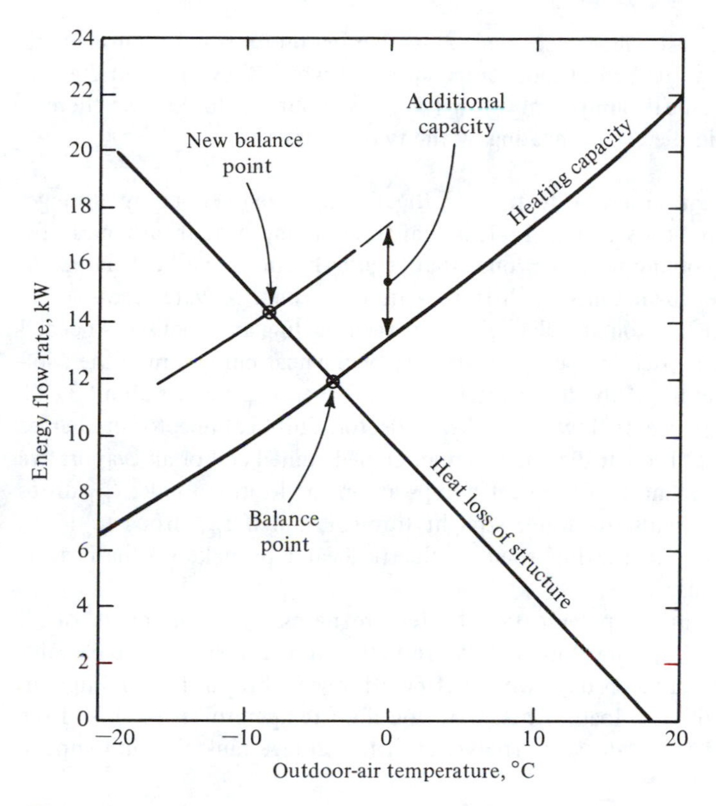

Figure 18-6 Balance points of heating capacity and heating load.

If resistance heaters are used to provide supplementary heat, the performance factor can be only 1.0, and one of the basic purposes of the heat pump is to use electricity to provide heat with a performance factor greater than 1.0. Figure 18-5 shows that the electric motor driving the compressor is not heavily loaded at low outdoor temperatures, which is the time that most heating capacity is needed. Designers continue to explore techniques of utilizing the available motor capacity by such means as bringing an additional cylinder into service or increasing the motor and compressor speed at these low outdoor temperatures.

18-7 Sizing the heat pump Ideally the capacity of the package-type reversible heat pump should match the cooling load of the structure during hot weather and the heating load during cold weather. There will be locations where the combination of climate and thermal characteristics of the structure permit this fortunate combination, but in general this is not the case. In climates warmer than where the summer-winter match is perfect the heat pump is usually sized to match the cooling load, and there is simply excess heating capacity in winter. In colder climates the heat pump is often chosen so that its heating capacity is short of the heating demand, and the deficiency is provided by supplementary electric resistance heaters. The logic of this strategy is that the investment cost per kilowatt of heating capacity is much less in the form of resistance heaters than in the form of heat pump, so that for a few hours of the year a penalty in operating cost will be accepted. A further reason for supplementing with resistance heaters is that at the outdoor temperatures at which they are used the performance factor of the heat pump is also low (perhaps about 1.5 to 2.0), so there is not a great difference in the cost of heating by the two methods.

18-8 Decentralized heat pump A feature of the decentralized-heat-pump arrangement, as shown schematically in Fig. 18-7, is that it can pump heat from zones of a building that require cooling to other zones that require heating. The heat pumps in this concept are water-to-air units, each serving its own zone. A water loop serves these heat pumps, which automatically switch between heating and cooling as needed to maintain the desired space temperature. If most of the heat pumps are in the cooling mode, the temperature of the loop water rises and when t_{ret} reaches about 32°C, three-way valve I diverts water flow to the heat rejector. This heat rejector discharges heat to the atmosphere through the use of an air-cooled finned coil or an evaporative cooler (see Chap. 19). If most of the heat pumps are in the heating mode, t_{ret} drops and three-way valve I sends the water straight through; and if t_{ret} drops to 15°C, three-way valve II opens to the fuel-fired or electric heater to make up the heating deficiency of the system.

The heat rejector and supplementary heater are necessary components of all decentralized-heat-pump systems, but the storage tank and solar collector are options. The storage tank is effective on days when net heat is rejected from the building during the day (because of solar load, lights, warm outdoor temperatures, etc.), and the system has a heat deficiency at night. The water in the storage tank rises in tempera-

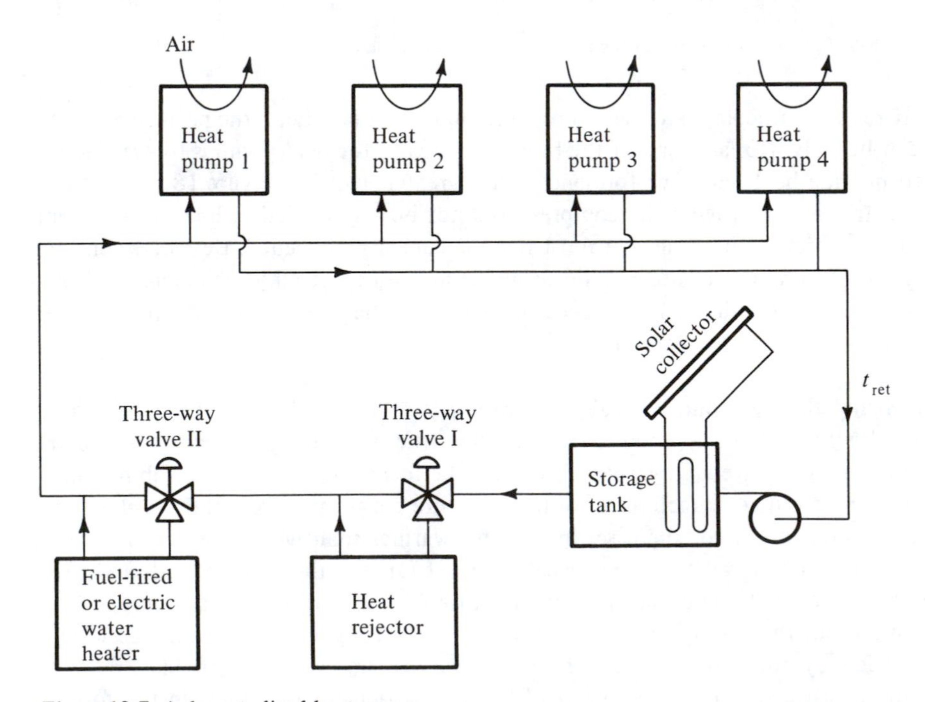

Figure 18-7 A decentralized heat pump.

Figure 18-8 Cutaway of room console decentralized heat pump. *(American Air Filter Co., Inc.)*

ture during the day and provides a source of heat for nighttime operation. The incorporation of the solar collector converts the system into a solar-assisted heat pump, as first shown in Fig. 18-3.

The heat-pump units are available in shapes adaptable to ceiling spaces, small equipment rooms, or as room consoles. A cutaway of a room console is shown in Fig. 18-8. Some console models have provisions for supplying outdoor ventilation air obtained through openings in the wall.

18-9 Double-bundle condenser During cold weather large buildings may require heat at the perimeter zones although the interior zones are unaffected by the outdoor conditions and always require cooling. One type of internal-source heat pump that pumps heat from the interior zones to the perimeter zones is the heat pump with a double-bundle condenser. One arrangement of this system is shown in Fig. 18-9, where a cooling tower cools water for one of the bundles and the water for the heating coils in the perimeter zones flows through the other bundle.

The strategy of operation is that the compressor (usually of the centrifugal type in these systems) have its capacity regulated to maintain t_1 at a constant value, say, 6°C. The controller of the hot-water-supply temperature modulates valve V1 to divert more water to the cooling tower if t_2 rises too high. As t_2 begins to drop, V1 first closes off the water flow to the cooling tower. Upon a continued drop in t_2 electric heaters in the hot-water line are brought into service.

The cooling coil serves an air system (variable-air volume, for example) which may supply both the interior and perimeter zones. The supply-air temperature t_3 could be held constant at 13°C by modulating valve V2. The net result of the operation is that heat removed from the air being cooled is supplied to heating needs. When excess heat is available, it is rejected through the cooling tower. A deficiency of heat for the heating coils is compensated for through the use of electric heaters.

At low outdoor temperatures it may be advantageous to regulate the mix tempera-

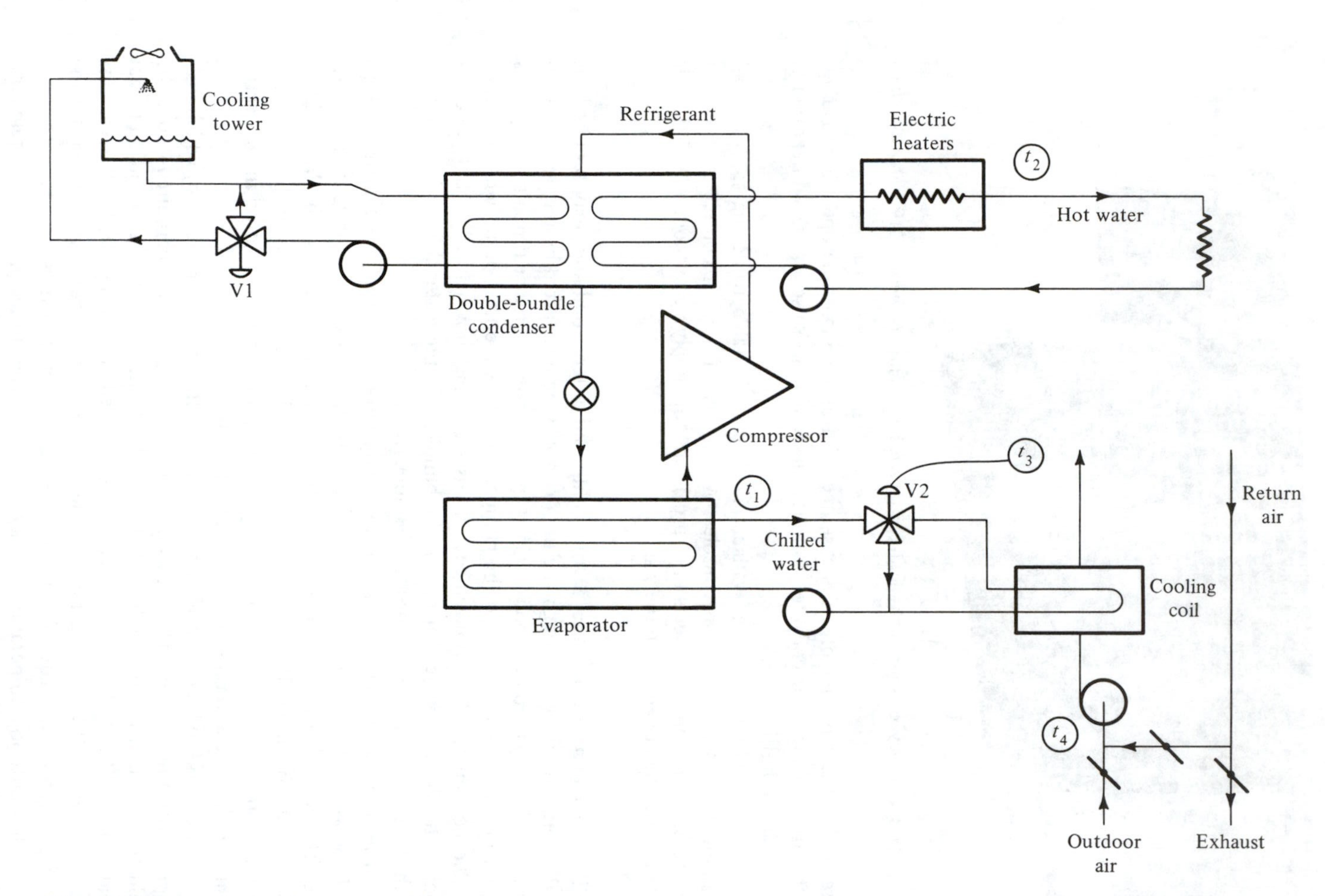

Figure 18-9 Internal source heat pump with double-bundle condenser.

ture t_4 to a higher value than desired for t_3. If there is a shortage of heat at the condenser that would cause the electric heaters to be activated, it would be preferable to elevate t_4 to limit the rate of heat rejected in the exhaust air.

Example 18-2 At one particular moment the demand of the heating coils in Fig. 18-9 is 250 kW when the flow rate of supply air to the interior zones is 36 kg/s with a supply temperature t_3 = 13°C. If the COP of the refrigeration unit is 3.2, what should the mixed-air temperature t_4 be, assuming only sensible cooling in the coil, if the system must provide the heating demand without using supplementary electric heat?

Solution Because the refrigeration unit has a COP of 3.2, the rate of heat transfer at the evaporator must be

$$(250 \text{ kW})(3.2)/(1 + 3.2) = 190.5 \text{ kW}$$

This energy flow rate must be derived from the air being cooled

$$190.5 = (1.0 \text{ kJ/kg} \cdot \text{K})(36 \text{ kg/s})(t_4 - 13°\text{C})$$

$$t_4 = 13 + 5.3 = 18.3°\text{C}$$

18-10 Industrial heat pumps The foregoing applications of heat pumps have been directed toward heating and cooling of buildings. There are certain attractive industrial applications of the heat pump also. One example is a fruit-juice concentrator,[4] shown schematically in Fig. 18-10. The juice, which must be concentrated at low

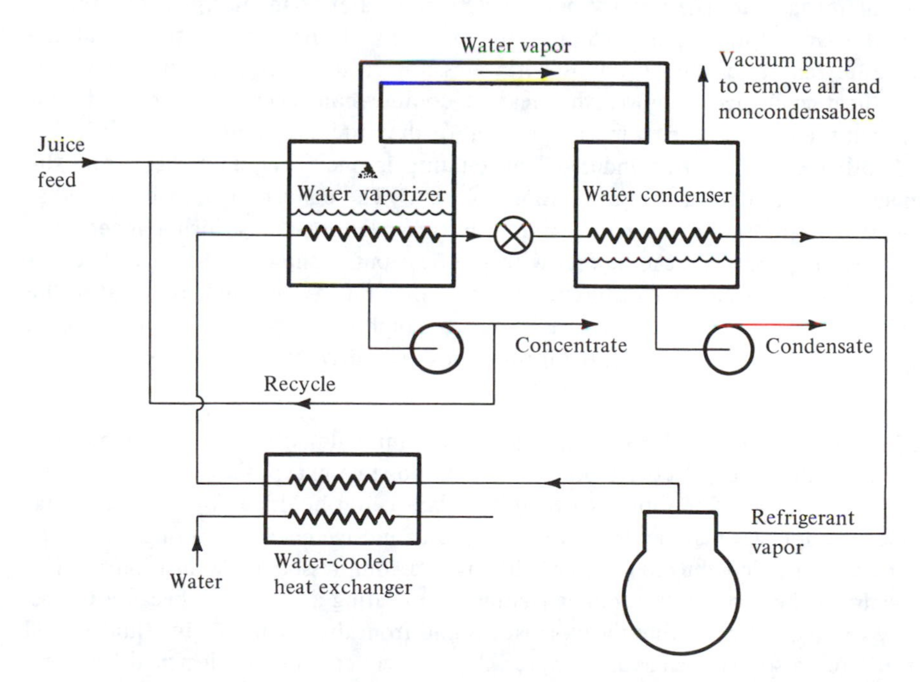

Figure 18-10 Heat pump for concentrating fruit juice.

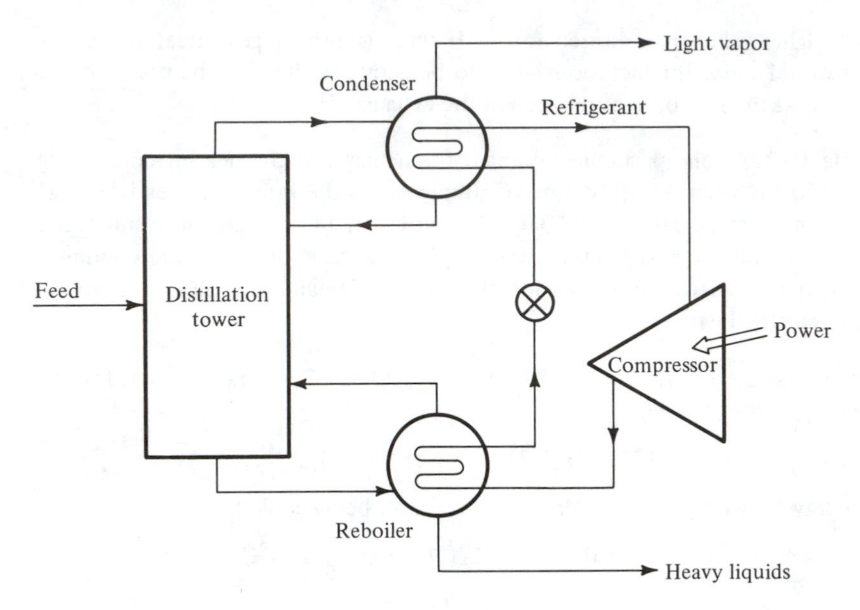

Figure 18-11 Heat pump serving a distillation tower.

temperature in order to preserve the flavor, enters the water vaporizer, which operates at subatmospheric pressure. The heat for the vaporization is provided by the condensation of refrigerant. The water vapor flows to the water condenser, which is the refrigerant evaporator. A pump elevates the pressure of the condensate so that the condensate can be rejected at atmospheric pressure. In the refrigerant circuit a water-cooled heat exchanger removes the heat of compression, because the rate of heat transfer in the water vaporizer must equal that of the water condenser.

Another example of an industrial heat pump is one that pumps heat from the condenser to the reboiler of a distillation column.[5] The condenser must be cooled at a low temperature, and the reboiler must be provided heat at a high temperature. Both these requirements can be met with the heat pump shown in Fig. 18-11. Compared with the conventional distillation tower, which rejects the condenser heat to the atmosphere and requires heat derived from a combustion process at the reboiler, the heat pump requires energy in the form of power to drive the compressor.

18-11 The future of the heat pump The heat pumps described in this chapter use compressors driven by electric motors, with the possible exception of the one in Fig. 18-11, which in a refinery or petrochemical plant could be driven by a steam or gas turbine. Natural gas may be used as the source of motive energy by using the gas to power an internal combustion engine[6] driving the compressor of the heat pump. The engine-driven heat pump has some advantages in heating applications because the assembly can recover heat from the exhaust gas and from the engine cooling fluid as well as heat from a source such as air or water. Another concept in non-electric-driven heat pumps is the absorption heat pump,[7] where both the generator heat and the heat absorbed at the evaporator from some external source can be supplied to a heating need.

While fuel-powered heat pumps will probably grow in importance, the main source of energy to drive heat pumps will be electricity, and thus the future of the heat pump is to a considerable extent linked to the fortunes of electricity serving as a source of heating energy.

PROBLEMS

18-1 An air-source heat pump uses a compressor with the performance characteristics shown in Fig. 18-4. The evaporator has an air-side area of 80 m^2 and a U value of 25 $W/m^2 \cdot K$. The airflow rate through the evaporator is 2 kg/s, and the condensing temperature is 40°C. Using the heat-rejection ratios of a hermetic compressor from Fig. 12-12, determine the heating capacity of the heat pump when the outdoor-air temperature is 0°C. *Ans.* 12.8 kW

18-2 The heat pump and structure whose characteristics are shown in Fig. 18-6 are in a region where the design outdoor temperature is -15°C. The compressor of the heat pump uses two cylinders to carry the base load and brings a third into service when needed. The third cylinder has a capacity equal to either of the other cylinders. How much supplementary resistance heat must be available at an outdoor temperature of -15°C? *Ans.* 5.8 kW

18-3 The air-source heat pump referred to in Figs. 18-4 and 18-5 operates 2500 h during the heating season, in which the average outdoor temperature is 5°C. The efficiency of the compressor motor is 80 percent, the motor for the outdoor air fan draws 0.7 kW, and the cost of electricity is 6 cents per kilowatthour. What is the heating cost for the season? *Ans.* $724

18-4 A decentralized heat pump serves a building whose air-distribution system is divided into one interior and one perimeter zone. The system uses a heat rejector, water heater, and storage tank (with a water capacity of 60 m^3) but no solar collector. The heat rejector comes into service when the temperature of the return-loop water reaches 32°C, and the boiler supplies supplementary heat when the return-loop water temperature drops to 15°C. Neither component operates when the loop water temperature is between 15 and 32°C. The heating and cooling loads of the different zones for two periods of a certain day are shown in Table 18-1. The loop water temperature is 15°C at the start of the day (7 A.M.). The decentralized heat pumps operate with a COP of 3.0. Determine the magnitude of (*a*) the total heat rejection at the heat rejector from 7 A.M. to 6 P.M. and (*b*) the supplementary heat provided from 6 P.M. to 7 A.M. *Ans.* (*a*) 11.6 GJ; (*b*) 3.84 GJ

Table 18-1 Heating and cooling loads in Prob. 18-4

	Interior zone		Perimeter zone	
	Heating, kW	Cooling, kW	Heating, kW	Cooling, kW
7 A.M. to 6 P.M.	....	260	....	40
6 P.M. to 7 A.M.	....	50	320	

18-5 The internal-source heat pump using the double-bundle heat pump shown in Fig. 18-9 is to satisfy a heating load of 335 kW when the outdoor temperature is −5°C, the return air temperature is 21°C, and the temperature of the cool supply air is 13°C. The minimum percentage of outdoor air specified for ventilation is 15 percent, and the flow rate of cool supply air is 40 kg/s. If the COP of the heat pump at this condition is 3.2, how much power must be provided by the supplementary heaters? *Ans.* 120 kW

REFERENCES

1. E. G. Granryd: Ground Source Heat Pump Systems in a Northern Climate, *15th Int. Congr. Refrig., Venice, 1979,* pap. E1–82.
2. H. P. Mogensen: The Ground as a Heat Source for Heat Pumps–Performance and Reactions, *15th Int. Congr. Refrig., Venice, 1979,* pap. E1-27.
3. Symposium on Solar-Assisted Heat Pumps, *ASHRAE Trans.* vol. 85, pt. 1, pp. 344–392, 1979.
4. Fruit Juice Concentrates, "ASHRAE Handbook and Product Director, Applications Volume," chap. 35, American Society of Heating, Refrigerating, and Air-Conditioning Engineers, Atlanta, Ga., 1978.
5. W. C. Petterson and T. A. Wells: Energy-Saving Schemes in Distillation, *Chem. Eng.*, vol. 84, no. 20, pp. 78–86, Sept. 26, 1977.
6. F. Steimle and J. Paul (eds.): "Antriebe fuer Waermepumpen," Vulkan-Verlag, Essen, 1978.
7. R. Lazzarin, M. Sovrano, R. Camporese, and E. Grinzato: Absorption Heat Pumps as Heating Systems, *15th Int. Cong. Refrig. Venice, 1979*, pap. E1–47.

CHAPTER

NINETEEN

COOLING TOWERS AND EVAPORATIVE CONDENSERS

19-1 Heat rejection to atmosphere Most refrigeration systems reject heat to the atmosphere. While there are applications where the rejected heat from the cycle is used for another purpose, as in certain heat pumps discussed in Chap. 18, and other applications where heat is rejected to a nearby body of water, most refrigeration systems reject heat to ambient air. One type of equipment for performing this heat exchange is the air-cooled condenser, discussed in Chap. 12, but another concept is to reject heat to ambient air through direct contact with water, in which a combined heat- and mass-transfer process takes place. The condensing temperature can usually be kept lower with one of the evaporative devices than with an air-cooled condenser because the condensing temperature in an ideal cooling tower or evaporative condenser approaches the *wet-bulb* temperature of the air in contrast to approaching the *dry-bulb* temperature of the air in an air-cooled condenser. Physical reasons also sometimes favor the choice of an evaporative condenser or a water-cooled condenser and cooling tower.

This chapter picks up principles of combined heat and mass transfer and the concept of enthalpy potential from Chap. 3 and uses them to explain the performance characteristics of the counterflow and crossflow cooling towers. The text also describes the construction of evaporative condensers and coolers and points out reasons for their use.

19-2 Cooling towers A cooling tower cools water by contacting it with air and evaporating some of the water. In most cooling towers serving refrigeration and air-conditioning systems one or more propeller or centrifugal fans move air vertically

Figure 19-1 A cooling-tower installation. *(The Marley Cooling Tower Company.)*

up or horizontally through the tower. A large surface area of water is provided by spraying the water through nozzles or splashing the water down the tower from one baffle to another. These baffles or fill materials have traditionally been wood but may also be made of plastic or ceramic materials. A cooling-tower configuration sometimes used for large-capacity power-plant applications is the hyperbolic shape, which resembles a chimney 50 to 100 m high in which the flow of air takes place by natural convection. Figure 19-1 shows a cooling tower where air is drawn in from the opposite two sides and is rejected out the top. Water enters the top through the rectangular boxes which distribute the water uniformly over the fill sections directly beneath the distributors.

Performance of cooling towers is often expressed in terms of *range* and *approach.* As shown in Fig. 19-2, the range is the reduction in temperature of the water through the cooling tower; the approach is the difference between the wet-bulb temperature of the entering air and temperature of the leaving water.

In the cooling tower a transfer takes place from the water to the unsaturated air. There are two driving forces for the transfer: the difference in dry-bulb temperatures and the difference in vapor pressures between the water surface and the air. These two driving forces combine to form the enthalpy potential, as explained in Secs. 3-8, 3-14, and 3-15.

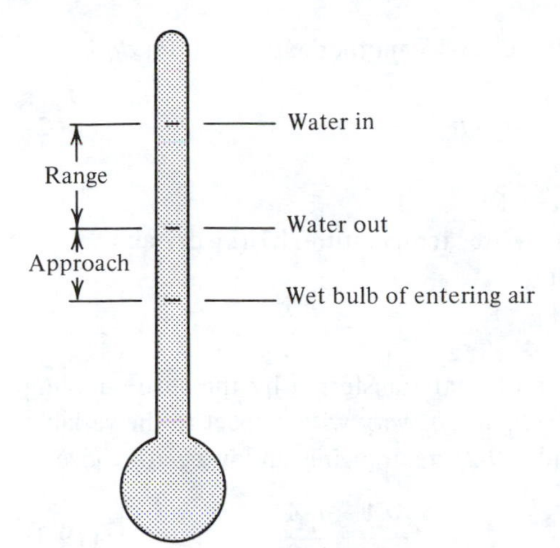

Figure 19-2 Range and approach in a cooling tower.

19-3 Analysis of a counterflow cooling tower One design of cooling tower is the counterflow type, in which air passes upward through a falling spray of water. Figure 19-3 shows a differential volume of a counterflow cooling tower with L kg/s of water entering from the top and G kg/s of air entering from the bottom. For simplicity, the small quantity of water which evaporates is neglected, so that both L and G remain constant throughout the tower.

Water enters the section at a temperature t°C and leaves at a slightly lower temperature $t - dt$. Air enters the section with an enthalpy of h_a kJ per kilogram of dry air and leaves with an enthalpy of $h_a + dh_a$. The total area of the wetted surface dA includes the surface area of the drops of water as well as the wetted slats or other fill material.

The rate of heat removed from the water dq is equal to the rate gained by the air:

$$dq = G\,dh_a = L(4.19\text{ kJ/kg}\cdot\text{K})\,dt \qquad \text{kW} \tag{19-1}$$

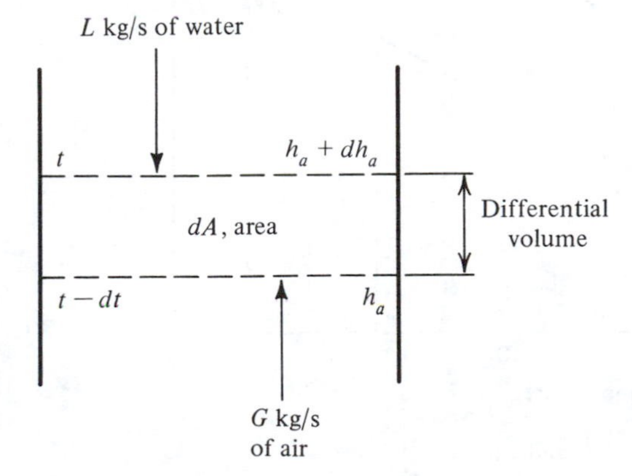

Figure 19-3 Exchange of energy in a differential volume of a counterflow cooling tower.

From the principles of enthalpy potential in Sec. 3-15 another expression for dq is

$$dq = \frac{h_c \, dA}{c_{pm}} (h_i - h_a) \tag{19-2}$$

where h_c = convection coefficient, $kW/m^2 \cdot K$
h_i = enthalpy of saturated air at the water temperature, kJ/(kg dry air)
h_a = enthalpy of air, kJ/(kg dry air)
c_{pm} = specific heat of moist air, kJ/kg · K

19-4 Stepwise integration To find the rate of heat transferred by the entire cooling tower, Eq. (19-2) must be integrated. Both h_i and h_a vary with respect to the variable of integration A. Combining Eqs. (19-1) and (19-2), rearranging, and integrating gives

$$4.19\, L \int_{t_{\text{out}}}^{t_{\text{in}}} \frac{dt}{h_i - h_a} = \int_0^A \frac{h_c \, dA}{c_{pm}} = \frac{h_c A}{c_{pm}} \tag{19-3}$$

where t_{in} and t_{out} are the water temperatures entering and leaving the tower, respectively.

A graphic visualization of the temperatures and enthalpies can be developed as in Fig. 19-4. Water enters the tower at t_{in} and leaves at t_{out}, and the enthalpies of satu-

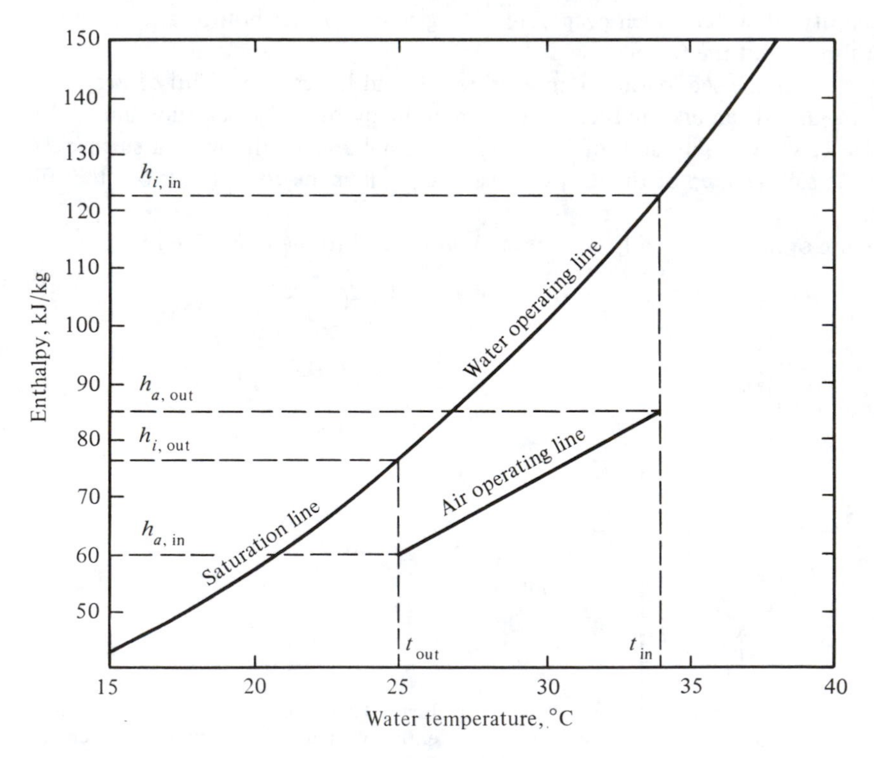

Figure 19-4 Enthalpy-temperature diagram of air and water.

rated air at these temperatures are $h_{i,\text{in}}$ and $h_{i,\text{out}}$, respectively. Designate the enthalpy of entering air as $h_{a,\text{in}}$ and that of the leaving air as $h_{a,\text{out}}$.

The saturation line in Fig. 19-4 represents the water temperature and enthalpy of the saturated air at this water temperature. Only the enthalpy coordinate applies to the air-operating line, however. The slope of the air-operating line is 4.19 *L*/*G*, which can be shown from Eq. (19-1). One traditional method[1] of performing the integration of Eq. (19-3) is a numerical process indicated by

$$\frac{h_c A}{c_{pm}} = 4.19L\ \Delta t \sum \frac{1}{(h_i - h_a)_m} \tag{19-4}$$

where $(h_i - h_a)_m$ is the arithmetic-mean enthalpy difference for an increment of volume. The procedure will be illustrated in Example 19-1.

Example 19-1 A counterflow cooling tower operates with a water flow rate of 18.8 kg/s and an airflow rate of 15.6 kg/s. When the wet-bulb temperature of entering air is 25°C and the entering water temperature is 34°C, the leaving water temperature is 29°C. Calculate $h_c A/c_{pm}$ for this cooling tower.

Solution The cooling tower will be imagined to be divided into 10 sections, as shown in Fig. 19-5, the water temperature dropping 0.5 K in each section.

As the water drops through the bottom section, for example, *t* decreases

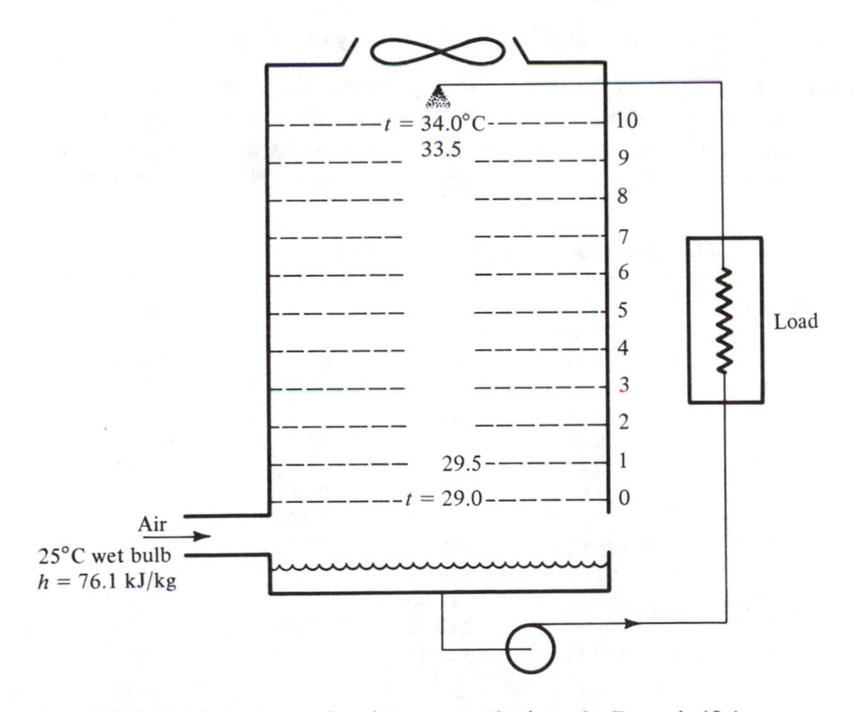

Figure 19-5 Division of tower into increments of volume for Example 19-1.

from 29.5 to 29.0°C. The wet-bulb temperature of the entering air almost precisely indicates the enthalpy of the air. If the air were saturated, its enthalpy would be 76.5 kJ/kg (from Table A-2), but if its relative humidity were of the order of 50 percent, the enthalpy would be approximately 76.1 kJ/kg (from the psychrometric chart of Fig. 3-1). The latter value will be used in the calculations.

The energy balance at the bottom section gives

$$h_{a,1} - h_{a,0} = \frac{L}{G} 4.19(0.5 \text{ K}) = 2.53 \text{ kJ/kg}$$

The enthalpy of the air leaving the bottom section $h_{a,1}$ is 76.1 + 2.53 = 78.63 kJ/kg, and the average enthalpy in this section is 77.36 kJ/kg.

The water has an average temperature of 29.25°C in the bottom section, and the enthalpy of saturated air at this temperature is 96.13 kJ/kg. The value of $(h_i - h_a)_m$ in this bottom section is 96.13 - 77.36 = 18.77 kJ/kg.

Moving up to the second section from the bottom, a similar procedure is followed to find $(h_i - h_a)_m$. The enthalpy of air entering the second section is the same as the enthalpy of air leaving the first section, 78.63 kJ/kg. Calculations for finding the summation of $1/(h_i - h_a)_m$ are shown in Table 19-1.

The value of $h_c A/c_{pm}$ can now be calculated from Eq. (19-4) as

$$\frac{h_c A}{c_{pm}} = (18.8 \text{ kg/s})(4.19)(0.5 \text{ K})(0.5097)$$

$$= 20.08 \text{ kW/(kJ/kg of enthalpy difference)}$$

The value of $h_c A/c_{pm}$ is a function of the dynamics of the airflow patterns and drop dynamics in the cooling tower,[2,3] but the magnitude remains essentially constant for a given cooling tower provided that the airflow rate and water flow rate remain constant, since they control h_c and the heat-transfer area A. The value of $h_c A/c_{pm}$

Table 19-1 Stepwise integration for solving Example 19-1

Section	Mean water temperature, °C	Average h_a, kJ/kg	Average h_i, kJ/kg	$(h_i - h_a)_m$	$\frac{1}{(h_i - h_a)_m}$
0-1	29.25	77.35	96.13	18.77	0.05328
1-2	29.75	79.89	98.70	18.81	0.05316
2-3	30.25	82.42	101.32	18.90	0.05291
3-4	30.75	84.95	104.00	19.05	0.05249
4-5	31.25	87.48	106.74	19.26	0.05192
5-6	31.75	90.01	109.54	19.53	0.05120
6-7	32.25	92.54	112.41	19.87	0.05033
7-8	32.75	95.07	115.35	20.28	0.04931
8-9	33.25	97.60	118.36	20.76	0.04817
9-10	33.75	100.13	121.43	21.30	0.04695
					0.5097

Table 19-2 Computer calculation of Example 19-1

Water temp., °C	Mean water temp., °C	Mean air enthalpy, kJ/kg	Mean h_i, kJ/kg	$(h_i - h_a)_m$	$\frac{1}{(h_i - h_a)_m}$
29.0–29.1	29.05	76.35	95.05	18.70	0.05348
29.1–29.2	29.15	76.86	95.56	18.70	0.05348
...	...	...	...	...	...
31.5–31.6	31.55	88.98	108.43	19.45	0.05141
...	...	...	...	...	...
33.9–34.0	33.95	101.09	122.77	21.67	0.04614
					2.5450

thus characterizes the cooling tower and is the basis for predicting its performance at other inlet water temperatures and other inlet wet-bulb temperatures.

The calculation shown in Example 19-1 conveniently lends itself to a computer solution through the application of an equation relating the enthalpy of saturated air to the water temperature

$$h_i = 4.7926 + 2.568t - 0.029834t^2 + 0.0016657t^3 \tag{19-5}$$

The equation represents the data with an error of approximately 0.1 percent between 11 and 40°C. Excerpts of the computer calculation of Example 19-1 using a temperature increment of 0.1 K are shown in Table 19-2. Applying Eq. (19-4) we get

$$\frac{h_c A}{c_{pm}} = (18.8 \text{ kg/s})(4.19)(0.1 \text{ K})(2.5450) = 20.047$$

which agrees with the hand calculation in Example 19-1.

The foregoing method has tacitly assumed that the temperature of the surface of the water droplets prevails through the droplet. Actually the interior of the droplet has a higher temperature than that of the surface, and heat flows by conduction to the surface where the heat- and mass-transfer process occurs. The experimentally determined value of $h_c A/c_{pm}$ includes the influence of this internal conduction.

Cooling-tower designers and manufacturers often use the *number of transfer units* (NTU) to refer to the term $h_c A/c_{pm}$. The higher the value of NTU, the closer the temperature of the water leaving the cooling tower will come to the wet-bulb temperature of the entering air.

19-5 Acceptance tests A manufacturer of a cooling tower may guarantee the tower to cool a specified flow rate of water from, say, 35 to 30°C when the wet-bulb temperature of entering air is 25°C. Quite likely when an acceptance test is run, the wet-bulb temperature of the air will not be 25°C and the water entering the tower will not be 35°C. The acceptance test is run, however, at the specified rates of water and airflow and with whatever water and air temperatures exist. The value of $h_c A/c_{pm}$ is calculated as in Example 19-1, and it should equal the value of $h_c A/c_{pm}$ which can be calculated from the performance data supplied by the manufacturer at the rated conditions.

19-6 Predicting outlet conditions from a tower When the value of $h_c A/c_{pm}$ is known and the entering air and water flow rates and conditions are known, it should be possible to predict the outlet water temperature. The procedure for making this prediction in a counterflow tower is not straightforward but requires iterative calculations. Since the temperature of the leaving water is initially unknown, a temperature can be assumed and the trial value of $h_c A/c_{pm}$ can be calculated as in Example 19-1. If the resulting $h_c A/c_{pm}$ is too high, the leaving water temperature should be raised for a new calculation.

19-7 State points of air through a cooling tower The stepwise integration as in Example 19-1 gives some information about the state of air as it passes through a counterflow cooling tower. The enthalpy of the air at the boundaries of each section are determined in the calculation and the state points of the air will lie somewhere on the lines of constant enthalpy, as shown in Fig. 19-6. The values of enthalpy are $h_{a,0} = 76.10$, $h_{a,1} = 78.63$, $h_{a,2} = 81.06$ kJ/kg, etc. In order to specify the conditions of the air completely, some other property in addition to the enthalpy must be calculated. A convenient property to determine is the temperature of the air. In order to calculate the dry-bulb temperature of the air t_a through the tower, the incoming temperature must be known.

A balance of the rate of sensible-heat transfer in any section of the cooling tower permits calculations of the outlet temperature of the air when the inlet temperature is known. For an arbitrary section n to $n + 1$ the sensible-heat balance is

$$Gc_{pm}(t_{a,n} - t_{a,n+1}) = h_c\ \Delta A \left(\frac{t_{a,n} + t_{a,n+1}}{2} - \frac{t_n + t_{n+1}}{2}\right)$$

where ΔA is the heat-transfer area in the n to $n + 1$ section.

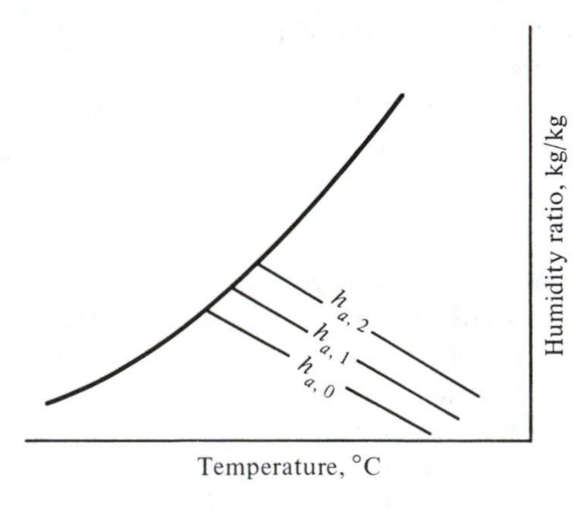

Figure 19-6 Enthalpy of air at various sections in the counterflow cooling tower of Example 19-1.

Solving for $t_{a,n+1}$ gives

$$t_{a,n+1} = \frac{t_{a,n} - \dfrac{h_c\,\Delta A}{2Gc_{pm}}(t_{a,n} - t_n - t_{n+1})}{1 + \dfrac{h_c\,\Delta A}{2Gc_{pm}}} \tag{19-6}$$

The magnitude of $h_c\,\Delta A/c_{pm}$ for the n to $n+1$ section is

$$\frac{h_c\,\Delta A}{c_{pm}} = 4.19L\,\Delta t\left(\frac{1}{\dfrac{h_{i,n}+h_{i,n+1}}{2} - \dfrac{h_{a,n}+h_{a,n+1}}{2}}\right) \tag{19-7}$$

The expression in the parentheses in Eq. (19-7) is $1/(h_i - h_a)_m$, a quantity calculated in the stepwise integration as in Table 19-1.

Example 19-2 Calculate the dry-bulb temperature of the air as it passes through the cooling tower of Example 19-1 if the air enters at 35°C.

Solution For section 0-1, $1/(h_i - h_a)_m = 0.05328$. Dividing Eq. (19-7) by 2G yields

$$\frac{h_c\,\Delta A}{2Gc_{pm}} = \frac{(4.19)18.8\,(0.5)\,(0.05328)}{2(15.6)} = 0.06726$$

From Eq. (19-6)

$$t_{a,1} = \frac{35.0 - (0.06726)\,(35.00 - 29.0 - 29.5)}{1 + 0.06726} = 34.28$$

A summary of the continued calculations is shown in Table 19-3.

Table 19-3 Dry-bulb temperatures through the cooling tower in Example 19-2

n	Section	$\dfrac{1}{(h_i - h_a)_m}$	$\dfrac{h_c\,\Delta A}{2Gc_{pm}}$	$t_{a,n+1}$
0	0-1	0.05328	0.06726	34.28
1	1-2	0.05316	0.06711	33.71
2	2-3	0.05291	0.06679	33.28
3	3-4	0.05249	0.06626	32.97
4	4-5	0.05192	0.06554	32.76
5	5-6	0.05120	0.06463	32.64
6	6-7	0.05033	0.06353	32.59
7	7-8	0.04931	0.06225	32.61
8	8-9	0.04817	0.06081	32.68
9	9-10	0.04695	0.05927	32.80

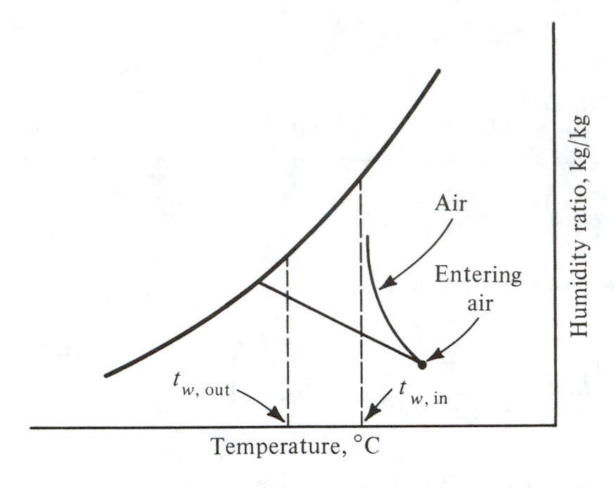

Figure 19-7 Path of air through a counterflow cooling tower.

The path of the air conditions will generally appear similar to that shown on the psychrometric chart of Fig. 19-7. The curve assumes a direction dictated by the straight-line law (Sec. 3-8) in that the curve drives toward the saturation line at the temperature of water t_w in contact with the air at that position. It is not uncommon for the temperature of the air first to drop as it passes up through the tower, reach a minimum, then increase from that point until it leaves the tower. Such a trend occurred in Example 19-2.

19-8 Crossflow cooling towers Although the counterflow cooling tower discussed so far is widely used in industrial service, another configuration is the crossflow tower, in which the air passes horizontally through the falling water sprays. Cooling towers used for air-conditioning systems are often located atop buildings, and the crossflow cooling tower usually has a lower profile, which lends itself better to architectural treatment.

The same principles of heat and mass transfer and balance of energy apply to a crossflow tower and to the counterflow type, but the geometric treatment of the crossflow tower is different.[4,5] Figure 19-8 shows a crossflow tower divided into 12 sections for purposes of analysis. Water enters the top at a temperature t_{in} while air enters from the left with an enthalpy h_{in}. In section 1 air enters at enthalpy h_{in} and leaves with enthalpy h_1. Also in section 1 water enters at temperature t_{in} and leaves at t_1. The enthalpy of air entering section 2 is h_1, and the water enters section 5 with a temperature t_1. The temperatures of water leaving sections 9, 10, 11, and 12 at the bottom of the tower are t_9, t_{10}, t_{11}, and t_{12}, respectively. These temperatures are all different, and the streams combine to form one stream of temperature t_{out}.

If the value of $h_c A/c_{pm}$ is known for the entire cooling tower, the outlet water temperature can be predicted when the inlet water temperature, inlet air enthalpy, and the flow rates of water and air are known. The tower can be divided into a number of small increments (12 increments, for example, in Fig. 19-8) and $(h_c A/c_{pm})/12$ assigned to each increment. Let

$$\frac{h_c\,\Delta A}{c_{pm}} = \frac{h_c A/c_{pm}}{12}$$

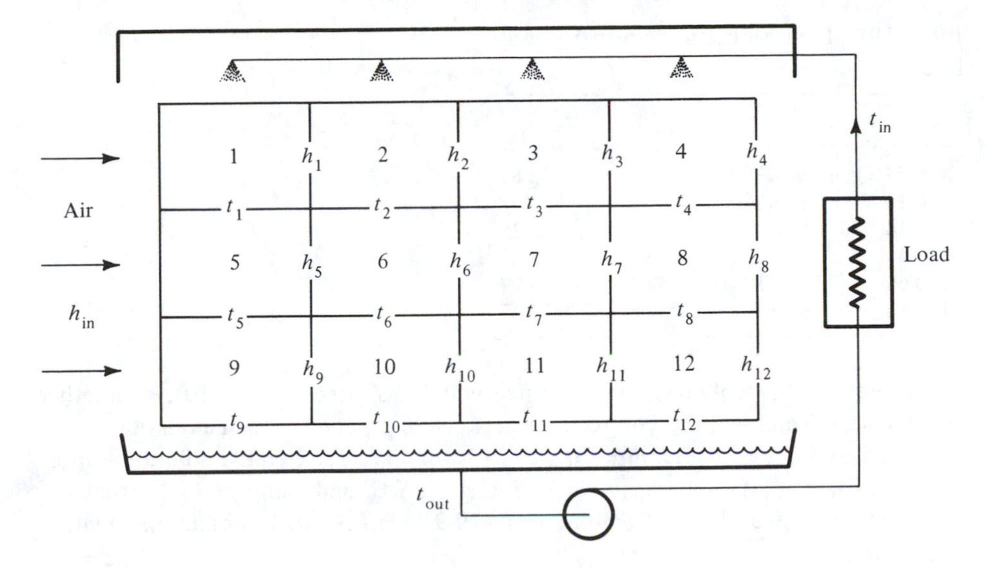

Figure 19-8 Analyzing a crossflow tower by dividing it into sections.

In Fig. 19-8, for example, the calculation can proceed in turn through section 1, 2, 3, and 4, calculating t_1 and h_1, t_2 and h_2, t_3 and h_3, and t_4. Drop next to the middle row of sections to calculate sections 5, 6, 7, and 8, and finally compute the sections in the lower row, sections 9, 10, 11, and 12, to obtain the values of t_9, t_{10}, t_{11}, and t_{12}.

The procedure for calculating outlet conditions for one section is illustrated in Example 19-3.

Example 19-3 In the cooling tower in Fig. 19-8, t_{in} = 37°C, h_{in} = 80.8, the water flow rate is 20 kg/s, the airflow is 18 kg/s, and $h_c A/c_{pm}$ = 21.5 kW/(kJ/kg of enthalpy difference). What are the values of t_1 and h_1 leaving section 1?

Solution For section 1, $L = 20/4 = 5$ kg/s, $G = 18/3 = 6$ kg/s, and $h_c\ \Delta A/c_{pm} = (h_c A/c_{pm})/12 = 1.79$ kW/(kJ/kg of enthalpy difference). The applicable equations are available from energy balances and the rate equation. Since the increment is small, it will be assumed that the arithmetic-mean difference of enthalpies will be sufficiently accurate in the rate equation. The equations are

$$q = (5 \text{ kg/s})(4.19)(t_{in} - t_1) \tag{19-8}$$

$$q = (6 \text{ kg/s})(h_1 - h_{in}) \tag{19-9}$$

$$q = \left(1.79 \frac{\text{kW}}{\text{kJ/kg}}\right)\left(\frac{h_{i,\text{in}} + h_{i,\text{out}}}{2} - \frac{h_{\text{in}} + h_1}{2}\right) \tag{19-10}$$

where q = rate of heat transfer, kW

$h_{i,\text{in}}$ = enthalpy of saturated air at t_{in} = h_i at 37°C = 143.24 kJ/kg from Table A-2

$h_{i,\text{out}}$ = enthalpy of saturated air at t_1

Table 19-4 Iterative calculation in Example 19-3

$h_{i,\text{out}}$	h_1	q	t_1
129.54	95.23	86.59	32.87
116.10	93.49	76.13	33.37
119.14	93.88	78.49	33.25
118.45	93.79	77.95	33.28
118.60	93.81	78.07	33.27
118.58	93.81	78.05	33.27

There are four unknowns in the three equations (19-8) to (19-10), so another equation is needed, i.e., the relation between $h_{i,\text{out}}$ and t_1. Such a relation is provided by Eq. (19-5). Because Eq. (19-5) is nonlinear, an iterative solution is most convenient. Choose a trial valve of t_1, say, 35°C, and compute $h_{i,\text{out}}$ from Eq. (19-5) as 129.54 kJ/kg. Combine Eqs. (19-9) and (19-10) to obtain an equation for h_1

$$h_1 = \frac{Gh_{\text{in}} + [(h_c\ \Delta A/c_{pm})/2]\ (143.24 + h_{i,\text{out}} - h_{\text{in}})}{(h_c A/c_{pm})/2 + G}$$

The value of h_1 thus computed is 95.23. From Eq. (19-9) $q = 86.59$, which when substituted into Eq. (19-8) yields an updated value of $t_1 = 32.87$. Table 19-4 shows a summary of the calculations that converge to $h_1 = 93.81$ kJ/kg and $t_1 = 33.27$°C.

19-9 Evaporative condensers and coolers The evaporative condenser combines the functions of the refrigerant condenser and the cooling tower. Figure 19-9 shows the elements of an evaporative condenser, and Fig. 19-10 is a cutaway photo of the coil

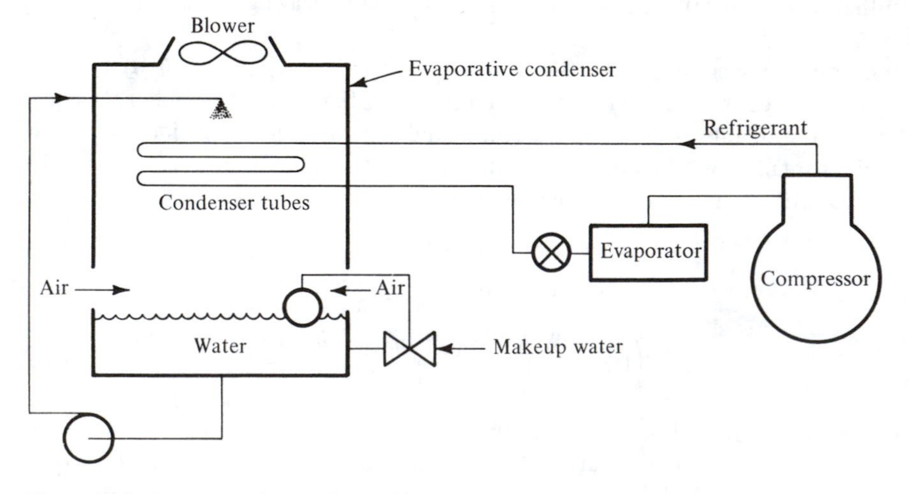

Figure 19-9 An evaporative condenser in a refrigeration system.

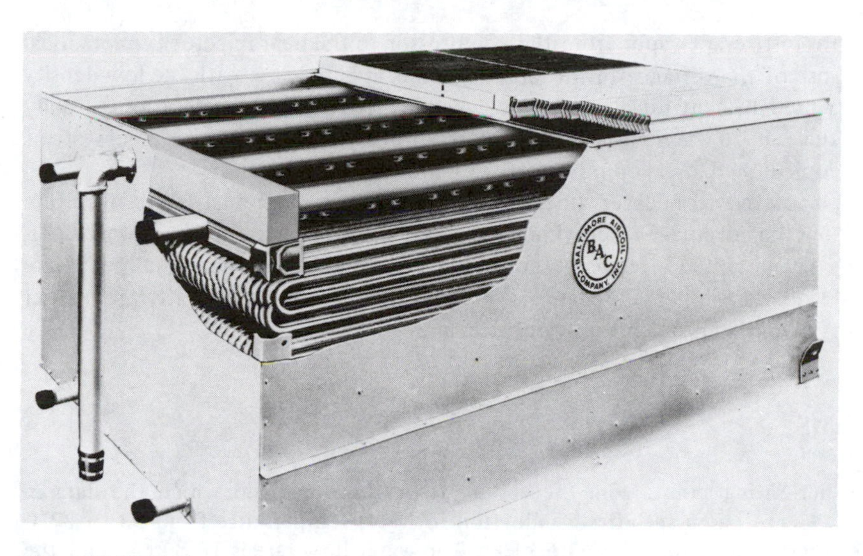

Figure 19-10 Coil section of an evaporative condenser. *(Baltimore Aircoil Company, Inc.)*

section showing the pipe coils, the water spray heads, and the eliminator plates on top that reduce the carryout of water droplets from the condenser. The discharge gas from the compressor condenses inside a bank of tubes over which water is spraying. The air flowing upward through the water spray eventually carries out the heat from the condensing refrigerant.

The sequence of the combined heat- and mass-transfer process in the evaporative condenser is (1) condensation of refrigerant vapor on the inside of the tubes, (2) conduction through the tube walls, (3) conduction and convection from the tube surface to the outside surface of the film of water that covers the tubes, and (4) simultaneous heat and mass transfer from the wetted surface to the airstream. Analyses of the heat-transfer process in the evaporative condenser are illustrated in Refs. 6 and 7.

A device closely related to the evaporative condenser is the evaporative cooler, where instead of condensing a refrigerant inside the tubes a liquid is cooled. One application for the evaporative cooler is the heat rejector in the decentralized heat pump described in Sec. 18-8.

19-10 When to use a cooling tower and evaporative condenser or cooler One performance advantage of the cooling tower and evaporative condenser is that the condensing temperature can drive toward the ambient wet-bulb temperature. Because the wet-bulb temperature is always equal to or lower than the dry-bulb temperature, the refrigeration system can operate with a lower condensing temperature and thus conserve energy compared with an air-cooled condenser. A disadvantage of the cooling tower is that maintenance costs are usually higher than for an air-cooled condenser, and attention must be paid to prevent the water from freezing in the cooling tower if operation is necessary during cold weather.

One situation that recommends the use of the cooling tower is where piping the high-pressure refrigerant vapor from the compressor to the heat rejector is impractical either because of the distance or the large-size-pipe, as is the case with the low-density refrigerants used in centrifugal-compressor systems. All these considerations usually influence the use of water-cooled condensers with cooling towers on large systems and of air-cooled condensers on moderate-sized and small systems.

The evaporative condenser finds widest acceptance in industrial refrigeration systems. It has the advantage of providing low condensing temperatures in hot weather, particularly in nonhumid regions. Industrial refrigeration systems usually operate year-round; in cold weather the water is often drained and the evaporative condenser operates dry, avoiding problems of water freezing.

PROBLEMS

19-1 Another rating point from the cooling-tower catalog from which the data in Example 19-1 are taken specifies a reduction in water temperature from 33 to 27°C when the entering-air enthalpy is 61.6 kJ/kg. The water flow rate is 18.8 kg/s, and the airflow rate is 15.6 kg/s. Using a stepwise integration with 0.5-K increments of change in water temperature, compute $h_c A/c_{pm}$ for the tower. *Ans.* 20.05 kW/(kJ/kg of enthalpy difference)

19-2 Solve Prob. 19-1 using a computer program and 0.1-K increments of change of water temperature. *Ans.* 20.06 kW/(kJ/kg of enthalpy difference)

19-3 If air enters the cooling tower in Prob. 19-1 with a dry-bulb temperature of 32°C, compute the dry-bulb temperatures as the air passes through the tower. For the stepwise calculation choose a change in water temperature of 0.5 K, for which the values of $1/(h_i - h_a)_m$ starting at the bottom section are, respectively, 0.04241, 0.04274, 0.04299, 0.04314, 0.04320, 0.04312, 0.04296, 0.04268, 0.04230, 0.04182, 0.04124, and 0.04055. *Ans.* Outlet air dry-bulb temperature = 31.0°C

19-4 A crossflow cooling tower operating with a water flow rate of 45 kg/s and an airflow rate of 40 kg/s has a value of $h_c A/c_{pm}$ = 48 kW/(kJ/kg of enthalpy difference). The enthalpy of the entering air is 80 kJ/kg, and the temperature of entering water is 36°C. Develop a computer program to predict the outlet water temperature when the tower is divided into 12 sections, as illustrated in Fig. 19-8. *Ans.* Water temperatures at t_9 to t_{12}: 29.11, 30.04, 30.85, and 31.56, respectively; combined outlet water temperature 30.4°C

REFERENCES

1. "ASHRAE Handbook, Fundamentals Volume," chap. 3, American Society of Heating, Refrigerating, and Air-Conditioning Engineers, Atlanta, Ga., 1981.
2. L. M. K. Boelter: Cooling Tower Performance Studies, *ASHVE Trans.*, vol. 45, pp. 615–638, 1939.
3. H. B. Nottage and L. M. K. Boelter: Dynamic and Thermal Behavior of Water Drops in Evaporative Cooling Processes, *ASHVE Trans.*, vol. 46, pp. 41–82, 1940.
4. H. Uchida: Graphical Analysis of a Cross-Flow Cooling Tower, *ASHRAE Trans.*, vol. 67, pp. 267–272, 1961.

5. N. Azmuner: A Method of Crossflow Cooling Tower Analysis and Design, *ASHRAE Trans.*, vol. 68, pp. 27-35, 1962.
6. W. Goodman: The Evaporative Condenser, *Heat., Piping, Air Cond.*, vol. 10, no. 5, p. 326, May 1938.
7. D. D. Wile: Evaporative Condensers, *Heat., Piping, Air Cond.*, vol. 30, no. 8, p. 153, August 1958.

CHAPTER

TWENTY

SOLAR ENERGY

20-1 Some fields of solar energy Since no life could exist on earth without the energy from the sun, the use of solar energy is as old as life. Even centuries ago living quarters were often built to take advantage of the warmth of the sun. The sun has also dried fruits and other foods and evaporated water from brine to leave salt. Earlier in the twentieth century solar collectors were used for heating domestic hot water. Since the precipitous increase in the cost of energy from fossil fuel in the mid-1970s and the prospect of their continued increase, solar energy has become the focus of much interest as one of the renewable resources.

Some of the areas of study and application of solar energy include direct conversion of solar energy into electricity, high-temperature collectors suitable for operating power generators, low-temperature flat-plate collectors, and design of buildings to use solar energy in a passive manner. Since the time of the day at which solar energy is available may not coincide perfectly with the demand for energy, thermal storage should be an integral component of the design of solar energy systems.

This chapter is an introduction to the study of solar energy and is intended to equip the engineer for making rough calculations of the potential for a solar system, for communication with the solar expert, and for elevating the awareness of the possibility of using solar energy in an installation. The key topics are solar geometry (the relationship of the sun's rays with a surface on earth) and solar intensity on these surfaces. Since glass is a key material in solar work, the response of glass to the sun's rays is explored. These topics provide insight for predicting the performance of a flat-plate fixed collector and the behavior of windows in a building using solar energy passively. Finally the chapter treats the integration of the solar collection equipment with storage and with the environmental system of the building.

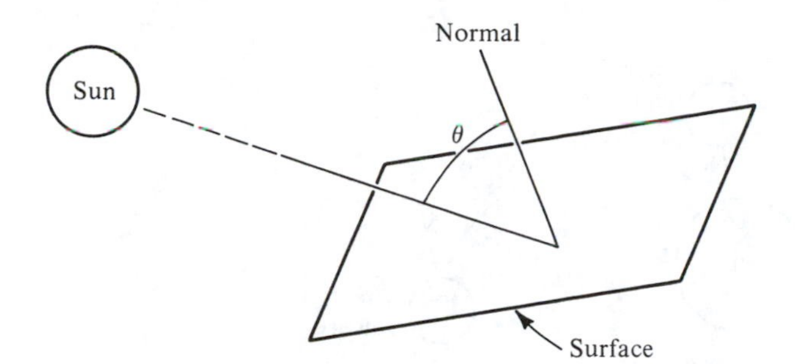

Figure 20-1 Incident angle θ.

20-2 Radiation intensity: an overview The approach to solar calculations to be taken in this chapter is that of arriving at a reasonably close estimate (±20 percent, for example) of the correct solar rates. Refined calculations are possible to reduce the error, and more convenient still are the charts and tables prepared to facilitate calculations. Through the approach taken in this chapter the dominant influences in solar calculations will be incorporated.

The usual objective in many solar calculations is to determine the solar *irradiation* of a given surface, i.e., the energy rate per unit area striking the surface. The key equation[1] for this calculation is

$$I_{i\theta} = I_{DN} \cos \theta + I_{d\theta} + I_r \tag{20-1}$$

where $I_{i\theta}$ = total solar irradiation of surface, W/m^2
I_{DN} = direct radiation from sun, W/m^2
θ = incident angle as shown in Fig. 20-1, deg
$I_{d\theta}$ = diffuse radiation component from sky, W/m^2
I_r = shortwave radiation reflected from other surfaces, W/m^2

The $I_{DN} \cos \theta$ term constitutes perhaps 85 percent of the total on clear days. The diffuse and reflected components $I_{d\theta}$ and I_r terms should not be neglected, however, because they persist on cloudy days and are thus sources of solar energy for flat-plate collectors.

To compute the major term in Eq. (20-1), the product of the direct radiation and $\cos \theta$, the task resolves into one of computing θ, which is a direct function of solar geometry, and I_{DN}, which is also influenced by solar geometry.

20-3 Solar geometry The following quantities must be known in order to compute the incident angle: latitude of the location in question, the hour angle, the sun's declination, the angle of the surface to horizontal, and the compass direction the surface faces. These terms will be discussed one at a time.

The latitude L is the angular distance of the point on the earth measured north (or south) of the equator. The hour angle H is the angle through which the earth must turn to bring the meridian of the point directly in line with the sun's rays. The hour angle expresses the time of day with respect to solar noon. One hour of time equals 360/24 or 15° degress of hour angle. There is a difference between clock time and solar time because the solar time varies at any instant depending on the east-west displacement. The clock time, on the other hand, is uniform throughout a time zone that may

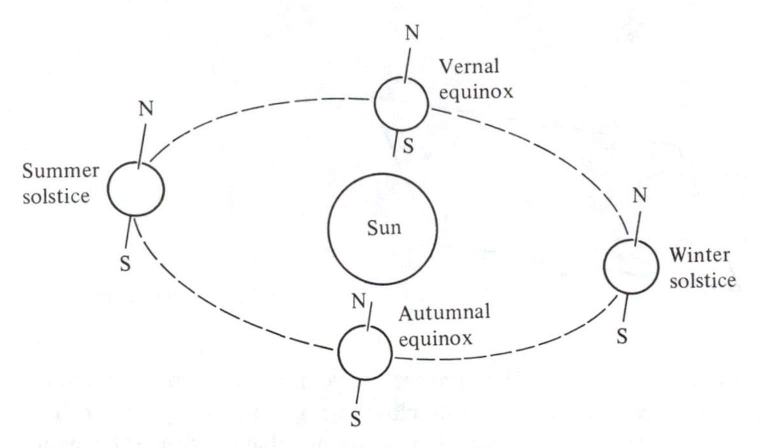

Figure 20-2 Earth's revolution about the sun.

span more than 1000 km in an east-west direction. When a time of day is used in this chapter, it will refer to solar time.

The sun's declination is the angular distance of the sun's rays north (or south) of the equator. In addition to rotating about its own axis, which causes night and day, the earth rotates once a year about the sun. The seasons are caused by the tilt of the earth's axis relative to the plane of this orbit around the sun. Figure 20-2 shows the earth's position at the start of each of the seasons. Figure 20-3 indicates the position of the earth at the winter solstice, showing that the rays of the sun strike the northern hemisphere less directly than they do the southern hemisphere.

The declination angle δ is zero at the autumnal and vernal equinoxes, is $+23.5^\circ$ at the summer solstice on June 21, and -23.5° at the winter solstice on December 21 in the northern hemisphere. At intervening periods of the year δ can be approximated by a sinusoidal variation

$$\delta = 23.47 \sin \frac{360(284 + N)}{365} \tag{20-2}$$

where N is the day of the year numbered from January 1.

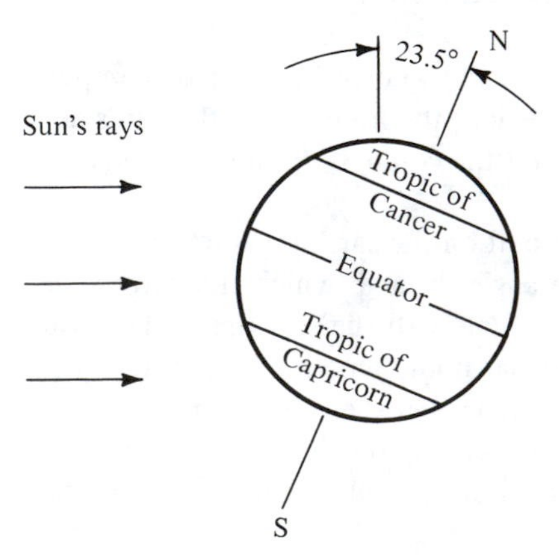

Figure 20-3 Relative position of sun's rays and axis at winter solstice.

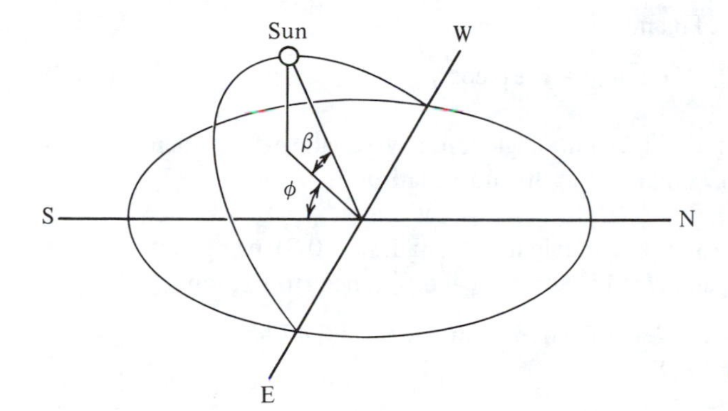

Figure 20-4 Solar altitude β and solar azimuth ϕ.

When calculating solar loads in Sec. 4-12 two solar angles were defined: the solar altitude β, which is the angle measured from a horizontal plane on earth up to the sun, and the solar azimuth angle ϕ, which is the angle between the sun's rays and the south. These angles are illustrated in Fig. 20-4. Equations for the computation of β and ϕ are

$$\sin \beta = \cos L \cos H \cos \delta + \sin L \sin \delta \tag{20-3}$$

$$\sin \phi = \frac{\cos \delta \sin H}{\cos \beta} \quad \text{for } \phi \leqslant 90^\circ \tag{20-4}$$

Example 20-1 What is the maximum solar altitude that occurs at any time during the year at 40° north latitude?

Solution The maximum altitude occurs at noon at the summer solstice, when $\delta = 23.5^\circ$

$$\begin{aligned} \sin \beta &= (\cos 40^\circ)(\cos 0^\circ)(\cos 23.5^\circ) + (\sin 40^\circ)(\sin 23.5^\circ) \\ &= 0.9588 \\ \beta &= 73.5^\circ \end{aligned}$$

which checks the value from Table 4-13.

In computing the incident angle θ of the sun's rays on an arbitrarily oriented surface (Fig. 20-1) the next step is to determine the incident angle of the sun's rays on horizontal and vertical surfaces. The angle of incidence is the angle between the sun's rays and a normal to the surface; so for the horizontal surface

$$\theta_{\text{hor}} = 90^\circ - \beta \tag{20-5}$$

For a vertical surface the orientation of the wall plays a role. In Sec. 4-12 an angle ψ, called the surface azimuth, was defined as the angle that a plane normal to the vertical surfaces makes with the south. Furthermore, the solar-surface azimuth γ is the angle between the solar azimuth and the surface azimuth

$$\gamma = \phi \pm \psi \tag{20-6}$$

The plus sign is chosen in Eq. (20-6) if ϕ and ψ are on opposite sides of south, and the negative sign chosen if ϕ and ψ are on the same side of south. The incident angle for a vertical surface is a function of β and γ,

$$\cos \theta_{ver} = \cos \beta \cos \gamma \tag{20-7}$$

Example 20-2 What is the incident angle on a vertical surface facing south-southwest at 11 A.M. on January 21 at 40° north latitude?

Solution The solar altitude β is obtainable from Eq. (20-3) using $L = 40°$, $\delta = -20.2°$ from Eq. (20-2), and $H = 15°$ for the 1-h difference from noon

$$\sin \beta = \cos L \cos H \cos \delta + \sin L \sin \delta = 0.4724$$
$$\beta = 28.2°$$

which checks the value in Table 4-13.

The solar azimuth ϕ can be determined from Eq. (20-4)

$$\phi = \sin^{-1} \frac{\cos \delta \sin H}{\cos \beta} = 16° \text{ east of south}$$

The surface azimuth ψ for a surface facing south-southwest is 22.5°; so

$$\gamma = 16° + 22.5° = 38.5°$$

The angle of incidence on the vertical surface is then calculated from Eq. (20-7)

$$\theta_{ver} = \cos^{-1} (\cos \beta \cos \gamma) = 46.4°$$

The final step to equip ourselves for computing an incident angle of the sun on an arbitrarily oriented surface is to introduce the tilt angle Σ, which is the angle of the surface tipped up from the horizontal from the south, as shown in Fig. 20-5. The general equation for the incident angle θ of the sun's rays on an arbitrarily oriented surface is

$$\cos \theta = \cos \beta \cos \gamma \sin \Sigma + \sin \beta \cos \Sigma \tag{20-8}$$

We now can compute the cos θ term in the major component of Eq. (20-1), the I_{DN} cos θ product. The next step will be to quantify I_{DN}.

20-4 Direct radiation from the sun I_{DN} The direct intensity of the sun's rays approaching the earth is relatively constant at 1353 W/m^2. Before these rays reach the earth, the intensity is attenuated because of absorption by the ozone, water vapor, and carbon dioxide in the atmosphere, even on cloudless days. The angle of passage through the atmosphere influences the degree of absorption too, so the solar altitude β affects

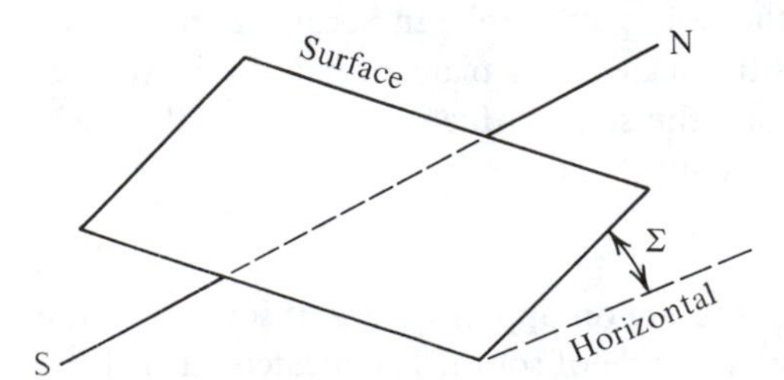

Figure 20-5 The tilt angle Σ of a surface.

I_{DN}. The equation for I_{DN} is

$$I_{DN} = \frac{A}{\exp(B/\sin\beta)} \tag{20-9}$$

where A = apparent solar irradiation, W/m^2
B = atmospheric extinction coefficient, dimensionless

Values of A and B depend on the month of the year and are tabulated in Ref. 1. The value of A is about 1230 W/m^2 in December and January and 1080 in midsummer. Values of B range from 0.14 in winter to 0.21 in summer. The maximum direct normal radiation at the earth's surface is of the order of 970 W/m^2.

20-5 Glazing characteristics We can now compute the intensity of direct irradiation, which is the intensity of the direct rays of the sun striking a surface. To use this energy it is necessary to pass it onto the inside of a solar collector or into a building. This transfer normally takes place through some glazing material, e.g., a glass or plastic, the transmittance of which determines the fraction of irradiation passing onto the collector or interior of the building. The incident angle influences the transmittance,[1] as shown in Fig. 20-6. The transmittance drops off rapidly for incident angles greater

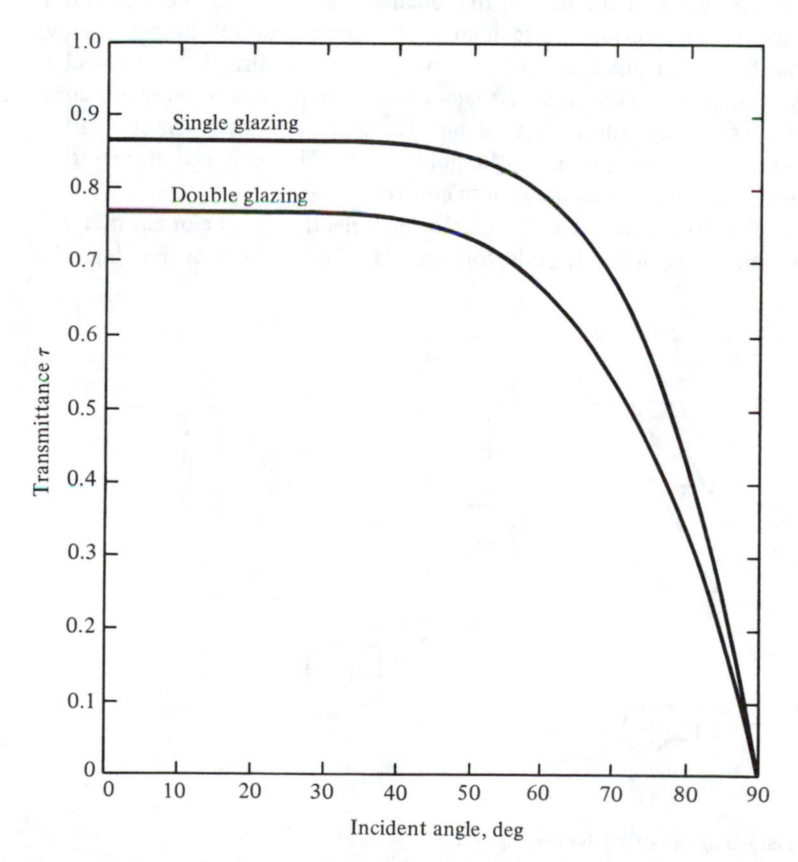

Figure 20-6 Transmittance of single and double glass as a function of the incident angle.

than 60°. Because the intensity of direct radiation passing into the collector or building is the product of the transmittance τ, I_{DN}, and cos θ, the intensity drops rapidly as θ increases beyond about 60°.

20-6 Solar collectors The basic purpose of any solar energy system is to collect solar radiation and convert it into useful thermal energy. System performance depends on several factors, including availability of solar energy, the ambient air temperature, the characteristics of the energy requirement, and especially the thermal characteristics of the solar system itself. Solar collection systems for heating or cooling are usually classified as passive or active. Passive systems collect and distribute solar energy without the use of an auxiliary energy source. They are dependent upon building design and the thermal characteristics of the materials used. Passive-system design will be presented in Sec. 20-9. Active systems, on the other hand, consist of components which are to a large extent independent of the building design and often require an auxiliary energy source for transporting the solar energy collected to its point of use. Active systems are more easily applied to existing buildings.

The major components of an active system are shown in Fig. 20-7. First the collector intercepts the sun's energy. A part of this energy is lost as it is absorbed by the cover glass or reflected back to the sky. Of the remainder absorbed by the collector, a small portion is lost by convection and reradiation, but most is useful thermal energy, which is then transferred via pipes or ducts to a storage mass or directly to the load as required. Energy storage is usually necessary since the need for energy may not coincide with the time when solar energy is available. Thermal energy is distributed either directly after collection or from storage to the point of use. The sequence of operation is managed by automatic and/or manual system controls.

Several types of solar collectors are available, and selection of one or another will depend on the intended application. Collectors are classified as fixed or tracking. The

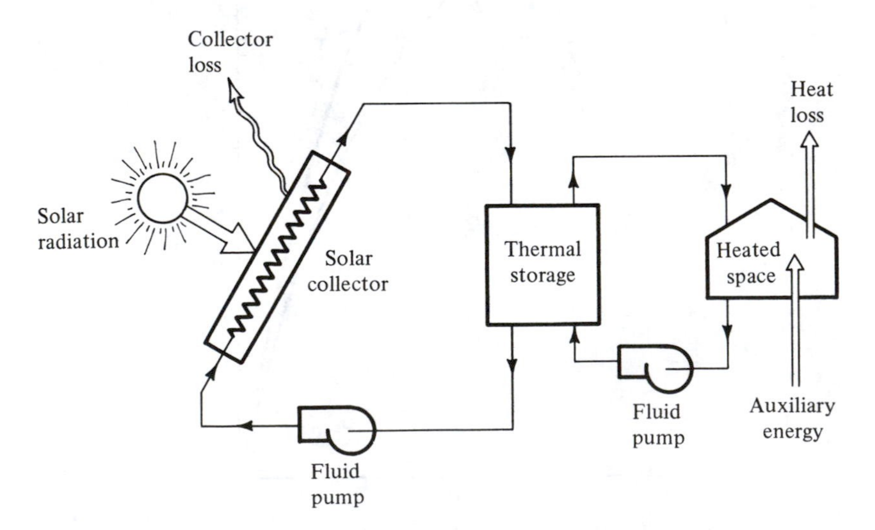

Figure 20-7 Schematic diagram of a solar heating system.

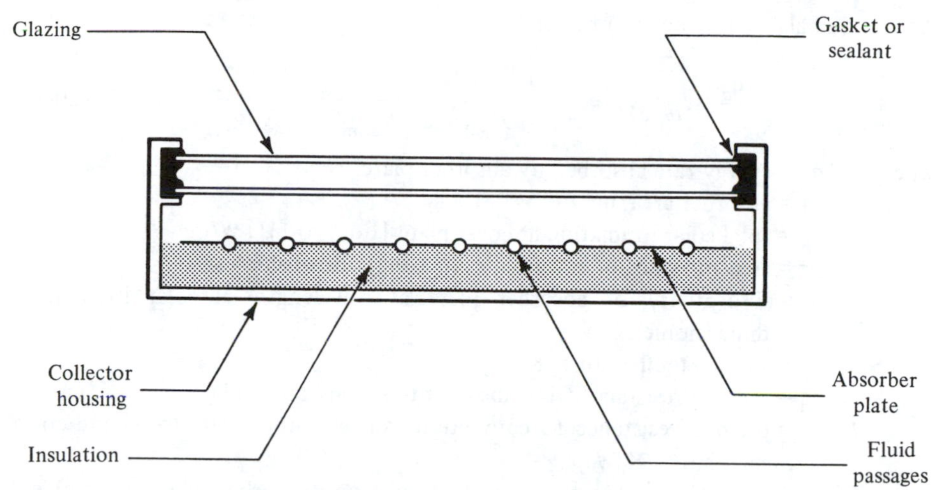

Figure 20-8 Cross section of a typical flat-plate collector.

tracking collectors are controlled to follow the sun throughout the day. Such systems are rather complicated and generally only used for special high-temperature applications. Fixed collectors are much simpler. Although their position or orientation may be adjusted on a seasonal basis, they remain "fixed" over a day's time. Fixed collectors are less efficient than tracking collectors; nevertheless they are generally preferred as they are less costly to buy and maintain.

Collectors may also be classified as flat-plate or concentrating. Concentrating collectors use mirrored surfaces or lenses to focus the collected solar energy on smaller areas to obtain higher working temperatures. Flat-plate collectors may be used for water heating and most space-heating applications. High-performance flat-plate or concentrating collectors are generally required for cooling applications since higher temperatures are needed to drive absorption-type cooling units.

The flat-plate collector consists of an absorber plate, cover glass, insulation, and housing (Fig. 20-8). The absorber plate is usually made of copper and coated to increase the absorption of solar radiation. The cover glass (or glasses) are used to reduce convection and reradiation losses from the absorber. The housing holds the absorber (insulated on the back and edges) and cover plates. The working fluid (water, ethylene glycol, air, etc.) is circulated in a serpentine fashion through the absorber plate to carry the solar energy to its point of use. The temperature of the working fluid in a flat-plate collector may range from 30 to 90°C, depending on the type of collector and the application. The collection efficiency of flat-plate collectors varies with design, orientation, time of day, and the temperature of the working fluid. How these factors influence the performance of a flat-plate collector can be illustrated by considering the collector in Fig. 20-8. The amount of solar irradiation reaching the top of the outside glazing will depend on the location, orientation, and tilt of the collector, as described in Secs. 20-2 to 20-4. The amount of useful energy collected will also depend on the optical properties (transmissivity and reflectivity), the properties of the absorber plate (absorptivity and emissivity), and losses by conduction, convection, and reradiation.

An energy balance for the absorber plate is

$$\frac{q_a}{A} = I_{i\theta}\tau_{c1}\tau_{c2}\alpha_a - \frac{T_a^4 - T_{c2}^4}{R_{\text{rad}}} - \frac{T_a - T_{c2}}{R_{\text{conv}}} - \frac{T_a - T_\infty}{R_{\text{cond}}} \tag{20-10}$$

where q_a = energy rate absorbed by absorber plate, W
A = absorber area, m^2
$I_{i\theta}$ = total solar irradiation at cover plate [Eq. (20-1)], W/m^2
τ = transmittance of cover plates (glazing), dimensionless
α = absorptivity of absorber plate at wavelength of solar irradiation, dimensionless
T = surface temperature, K
R_{rad} = thermal resistance from absorber to second cover plate, $m^2 \cdot K^4/W$
R_{conv} = thermal resistance to convection from absorber plate to second cover plate, $m^2 \cdot K/W$
R_{cond} = thermal resistance to conduction from absorber plate to ambient through the insulation, $m^2 \cdot K/W$

Subscripts
c_1 = first cover plate
c_2 = second cover plate
a = absorber
∞ = ambient

Equation (20-10) is somewhat unwieldy for frequent use; linearizing the radiation term, as in Sec. 2-16, and consolidating the driving force for all the losses as $t_{ai} - t_\infty$, leads to an approximation of Eq. (20-10) as

$$\frac{q_a}{A} = \left(I_{i\theta}\,\tau_{c1}\tau_{c2}\alpha_a - U(t_{ai} - t_\infty)\right) F_r \tag{20-11}$$

where t_{ai} = temperature of inlet fluid to absorber, °C
U = overall heat-transfer coefficient combining effects of radiation, convection, and conduction losses, $W/m^2 \cdot K$
F_r = empirically determined correction factor, dimensionless

The value of F_r is of the order of 0.9 for collectors using liquid. The values of U are also experimentally determined and have the typical range shown in Table 20-1.

Another important characterization of the collector is its efficiency η, which is defined as the energy rate transferred to the fluid divided by the solar irradiation on the cover plate,

$$\eta = \frac{q_a/A}{I_{i\theta}} = \left(\tau_{c1}\tau_{c2}\alpha_a - \frac{(t_{ai} - t_\infty)U}{I_{i\theta}}\right) F_r \tag{20-12}$$

Example 20-3 A 1- by 3-m flat-plate double-glazed collector is available for a solar-heating application. The transmittance of each of the two cover plates is 0.87, and the aluminum absorber plate has an $\alpha = 0.9$. Determine the collector efficiency when $I_{i\theta} = 800\ W/m^2$, $t_\infty = 10°C$, and $t_{ai} = 55°C$.

Table 20-1 Typical values[2] of U in Eq. (20-11)

Type of glazing	U, W/m^2 • K
Unglazed	13–15
Single	6–7
Double	3–4

Solution From Table 20-1 U is chosen as 3.5 W/m^2 • K; for a value of $F_r = 0.9$, Eq. (20-12) yields

$$\eta = \left(0.87(0.87)(0.9) - \frac{(55 - 10)(3.5)}{800}\right) 0.9 = 0.435$$

For the purpose of selecting collectors, designers often use a graph of collector efficiencies, as shown in Fig. 20-9. The trends shown in Fig. 20-9 are predictable from Eq. (20-12). The efficiency is a function of the optical and thermal properties of the cover plate and the absorber and also the term $(t_{ai} - t_\infty)/I_{i\theta}$. As the absorber tem-

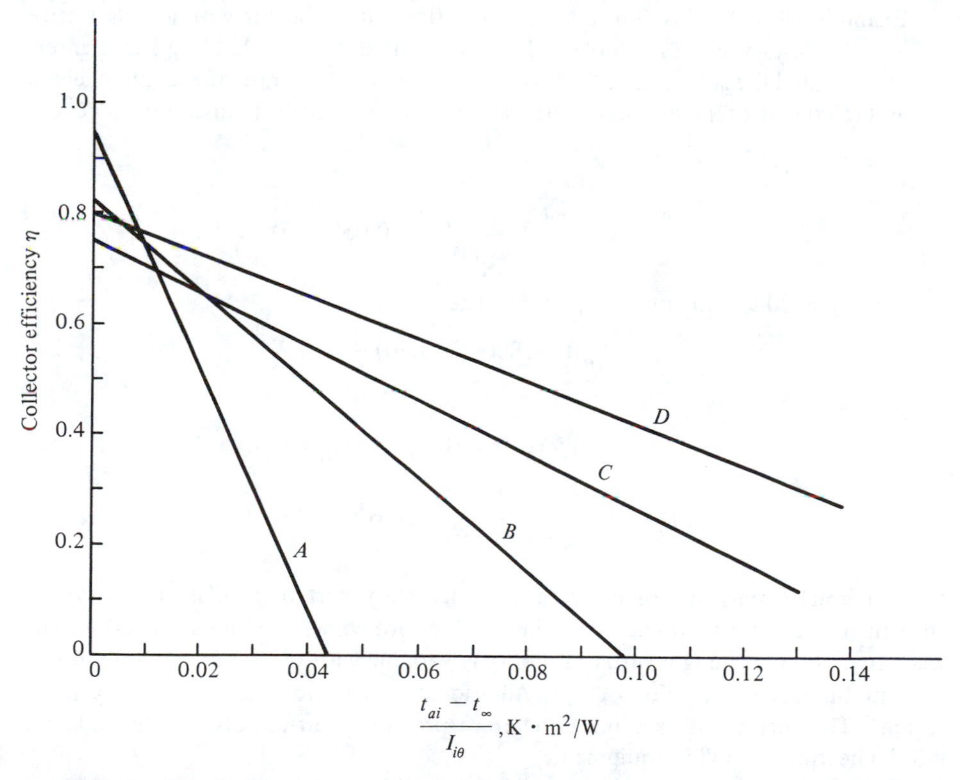

Figure 20-9 Collector efficiencies of typical flat-plate collectors. A = unglazed, nonselective absorber; B = single glass, nonselective absorber; C = double glass, nonselective absorber; D = double glass, selective absorber.

perature t_{ai} increases, the losses increase and the efficiency drops. Similarly, at low ambient temperatures the efficiency is low because of high losses. As the solar irradiation on the cover plate $I_{i\theta}$ increases, the efficiency increases because the loss from the collector $(t_{ai} - t_\infty)U$ is fairly constant for given absorber and ambient temperatures and becomes a smaller fraction as $I_{i\theta}$ increases.

Figure 20-9 also shows the effects of the cover plates. A collector with no cover plate or with a single cover plate is more efficient at low $t_{ai} - t_\infty$, where convective losses are small. A double-glazed collector is better at higher $t_{ai} - t_\infty$, where the convective losses would have been significantly larger than the additional transmission loss through the second cover plate.

The absorptivity and emissivity of a surface may vary with the wavelength of the incident radiation. Surface coatings for the absorber plate can be selected in such a way that the surface is highly absorbing at the short wavelength of solar radiation ($\alpha \approx 0.9$) but has a low emissivity ($\epsilon = 0.5$) at the longer wavelengths characteristic of a surface radiating at 100 to 200°C. Such surfaces are referred to as *selective surfaces*. The performance of a single-glazed collector can be upgraded by using a selective coating for the absorber surface without adding a second cover plate, as shown by the curve for collector D in Fig. 20-9.

Example 20-4 A 1- by 3-m double-glazed flat-plate collector with a nonselective absorber uses water as a coolant. If the coolant flows at 0.0333 kg/s and enters at 50°C and the solar flux is 800 W/m², determine (*a*) the rate of energy collected and (*b*) the exit temperature of the water when the ambient temperature is 10°C.

Solution (*a*)

$$\frac{t_{ai} - t_\infty}{I_{i\theta}} = \frac{50 - 10}{800} = 0.05$$

From Fig. 20-9 with curve C, $\eta = 0.50$. Then

$$q_a = I_{i\theta} A \eta = 800(3)\,(0.50) = 1.2 \text{ kW}$$

(*b*)

$$q_a = (w \text{ kg/s})\,(4.19 \text{ kJ/kg} \cdot \text{K})\,(t_o - t_{ai})$$

$$t_o = 50 + \frac{1.2}{0.033(4.19)} = 58.6°\text{C}$$

20-7 Thermal storage Thermal storage is a necessary part of a solar energy system since the demand for solar energy frequently does not coincide with its collection. The solar irradiation reaching a surface is not only variable with the season and the time of day but fluctuates with cloud cover. Additionally, demand is also frequently intermittent. The thermal storage must therefore provide a buffer between the collector and the heating or cooling equipment.

Solar energy can be stored as *sensible heat*, which involves only a change in temperature of the storage medium, or as *latent heat*, which involves a change in phase of the storage medium. Sensible-heat storage is more common, but there is active research in improved latent storage materials.[3]

Table 20-2 Specific heats of several common materials

Material	c, kJ/kg · K	ρc, kJ/m^3 · K
Water	4.19	390
Steel	0.46	250†
Pebbles, 20–40 mm	0.84	125†

† Assumes 70 percent packing density.

Any thermally and chemically stable solid material with a relatively high specific heat and high density may be used for sensible-heat storage. The high specific heat and high density are necessary to minimize the volume of the storage facility. The heat stored can be expressed as

$$Q_s = \rho V c(t_s - t_m) \tag{20-13}$$

where Q_s = heat stored, kJ
ρ = density, kg/m^3
V = volume, m^3
c = specific heat, kJ/kg · K
t_s = storage temperatre, °C
t_m = minimum useful temperature, °C

Table 20-2, giving values of c and ρc for some common materials, shows that water has 3 times the thermal-storage capacity of the same volume of pebbles, for example.

Latent-heat storage utilizes a change in the phase of the storage material. Two solid-liquid phase change processes can be used: the familiar melting and freezing process and a chemical reaction of the hydration and dehydration of salts. At temperatures below the hydration point the anhydrate becomes hydrated and crystalline with the evolution of heat. When the temperature is raised, the crystals dissolve in the water of hydration, absorbing heat. The advantage of latent-heat storage is that the heat of fusion is many times larger than the specific heat, which reduces the storage volume. A storage system using paraffin or wax requires only one-fourth the volume of a thermally equivalent water storage system. The disadvantage is that the performance of the latent storage materials now available degrades with time and they must be replaced in a few years; they are also considerably more expensive than sensible-heat storage systems using water or pebbles. Table 20-3 lists the melting points, heats of fusion, and densities of two common latent-heat storage materials.

One additional requirement for the use of latent-heat storage materials is the need

Table 20-3 Heat of fusion of latent-heat storage materials

Material	Melting point, °C	Heat of fusion, kJ/kg	ρ, kg/m^3	Heat of fusion per unit volume, kJ/m^3
Glauber salt	32	240	1100†	260,000
Amorphous paraffin wax	74	230	650†	150,000

† Assumes 30 percent of volume required for flow passages.

to find one with an appropriate melting point. The temperature at which the phase change takes place must fit the intended application and the capability of the available solar collector. Experimental data are available on a wide range of phase-change materials considered for thermal storage. Unfortunately, few materials have yet been brought to a stage of practical application.

The optimum storage capacity depends on the application. If the storage capacity is too small, some energy which might have been collected and used is, in effect, wasted. Collector area is then underused. If, on the other hand, the storage is oversized, costs increase and heat loss from the storage may become excessive.

Water-heating applications have a fairly regular daily demand, and storage of 1 to $1\frac{1}{2}$ times the daily demand is typical. Space-heating requirements are more irregular and likely to occur during periods when less solar energy is available for collection. For space-heating applications a slightly larger storage capacity can be justified—perhaps a 2-d requirement. A guideline often used for a water-storage system is 0.05 to 0.10 m^3 of storage capacity per square meter of collector area. When a system is designed for both heating and cooling, the storage capacity is generally based on heating requirements, although it is often advisable to provide some chilled-water storage to minimize cycling of the cooling unit.

Example 20-5 Select the size of the thermal-storage system for a building which has a heating requirement of 150 kJ/d, assuming that a 2-d storage is needed and that the solar energy system is sized to meet 70 percent of the heating load. Allow a 30°C temperature swing of the storage medium. Determine the volume when using (*a*) water and (*b*) latent-heat storage with Glauber salt. The specific heat of Glauber salt is 2.5 kJ/kg · K.

Solution (*a*)

$$\text{Volume} = \frac{\text{heat stored}}{\rho c\ \Delta t} = \frac{150{,}000(2)(0.7)}{1000(4.19)(30)}$$

$$= 1.67\ \text{m}^3$$

(*b*) When using Glauber salt

$$\text{Volume} = \frac{\text{heat stored}}{\rho(\text{heat of fusion} + c\ \Delta t)} = \frac{150{,}000(2)(0.7)}{1100[240 + (2.5)(30)]}$$

$$= 0.61\ \text{m}^3$$

20-8 Integration of solar and building systems Of the three applications for solar energy in buildings, water heating has developed most rapidly. Space-heating applications are becoming more frequent, but solar-energy cooling systems are still at a stage where their use is limited primarily to research and design development projects. The need for efficient high-temperature collection and for further improvement in the efficiency of heat-operated cooling units is presently the limiting factor in the development of cooling applications.

Figure 20-10 illustrates the components and controls required for a closed-loop hot-water system. Open-loop systems, which eliminate the heat exchanger in the

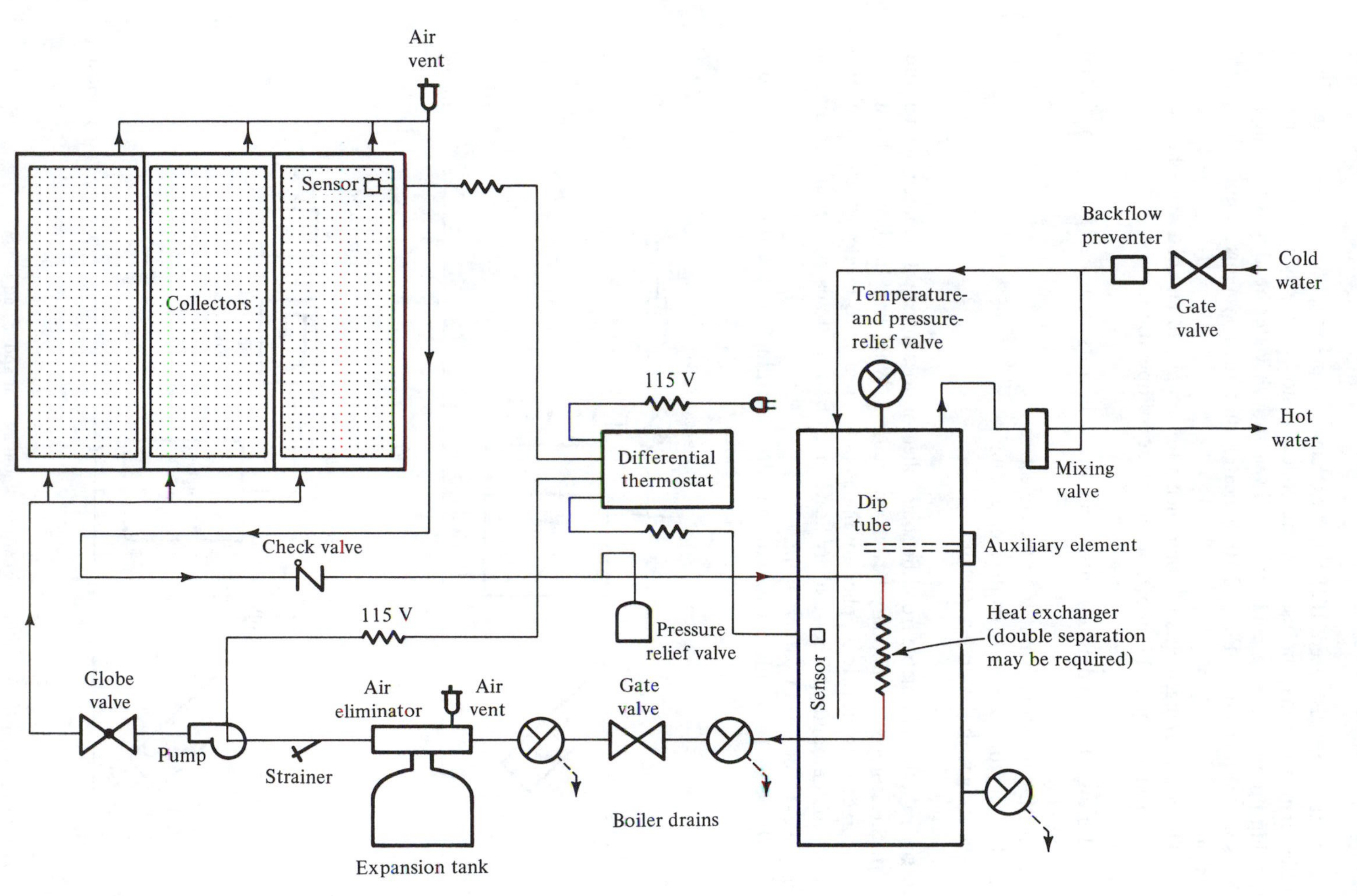

Figure 20-10 Closed-loop water-heating system.

storage tank, are also used. If the heat exchanger is present, however, a working fluid other than water may be used on the collector side of the system to eliminate the freezing problem during cold weather. Even though water heating is the simplest of the solar applications, Fig. 20-10 indicates that such systems require thoughtful design.

The basic arrangements of components for liquid and air solar space-heating systems, shown in Fig. 20-11, allow for four modes of operation:

1. Heating the building directly from the collectors (applicable only to the air system)
2. Heating the storage unit from the collectors
3. Heating the building from the storage unit
4. Heating with the auxiliary heater

Because the heat available from the solar collector does not always match the demand, both the storage and auxiliary heater are required. When heat has been depleted from the storage unit, heat is supplied by the auxiliary heater to meet the heating requirements of the building. The auxiliary heater must be sized to carry the full heating load in case extended bad weather occurs.

Figure 20-11 is only a schematic and does not illustrate the full complexity of a

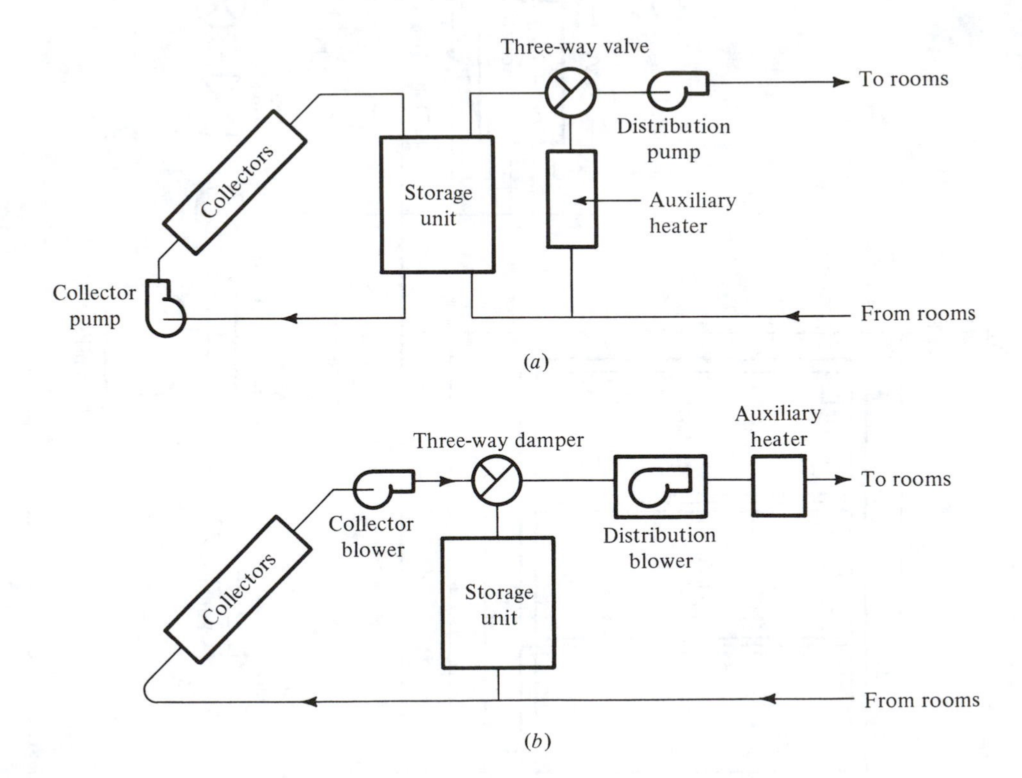

Figure 20-11 Solar space-heating systems: (*a*) liquid system and (*b*) air system.

solar heating system. At a minimum the auxiliary and control elements illustrated in Fig. 20-10 would be required.

Cooling systems are the most complex of the building applications of solar energy. They usually also incorporate water heating and space heating in the same system. The solar cooling system, such as shown in Fig. 20-12, requires a more sophisticated piping and control system than space-heating and water-heating applications. The components of the system common to both heating and cooling are:

1. Solar collectors
2. Heat storage tank
3. Auxiliary heater
4. Duct coil and air-supply ducts
5. Circulating pumps

The additional component required in the cooling mode is a refrigerating unit, an absorption chiller being the only device commercially available at present. The system often uses a cool storage tank.

In the cooling mode energy from the solar collector provides all or some of the heat required to drive the absorption chiller. The absorption unit produces chilled water, which is circulated through the coils in the air-supply duct to cool the space. A water-cooling tower is used to reject heat from both the absorber and condenser of the absorption unit. During the heating season solar-heated fluid is circulated directly through coils in the air-supply duct to provide heat in this configuration.

20-9 Passive solar design In contrast to active solar collectors and systems, the passive solar concept[4] uses only elements of the building to admit solar energy for heating and perhaps to store energy for later use. A strictly passive system uses no external energy to pump or circulate secondary fluids, although arrangements that are predominantly passive may use a fan to circulate building air.

The major objectives in passive solar design are to admit solar energy to the building during the heating season and store it, if possible, for use later in the day. A further objective is to reduce the energy flow into the building during the cooling season. The simple directions that are usually adequate for achieving favorable passive design are (1) to locate glass on the south side of the building and provide overhang if feasible, (2) to avoid to the extent possible west- and east-facing glass, and (3) to provide for thermal storage capacity to receive the direct rays of the sun.

South-facing glass is a key aperture for solar energy. Even with no overhang to shield the glass, south-facing glass naturally has some favorable characteristics. The intensity of direct radiation passing into the building I_T is

$$I_T = I_{DN}(\cos\theta)\tau \tag{20-14}$$

Table 20-4 shows values of I_T for several times of the day and months of the year. Because of the relatively small angle of incidence θ in the winter relative to summer and the influence of θ on τ and the cosine of θ, the magnitudes of transmission in December are 5 times that of June. In a natural way, then, the south-facing glass provides the high transmission when needed.

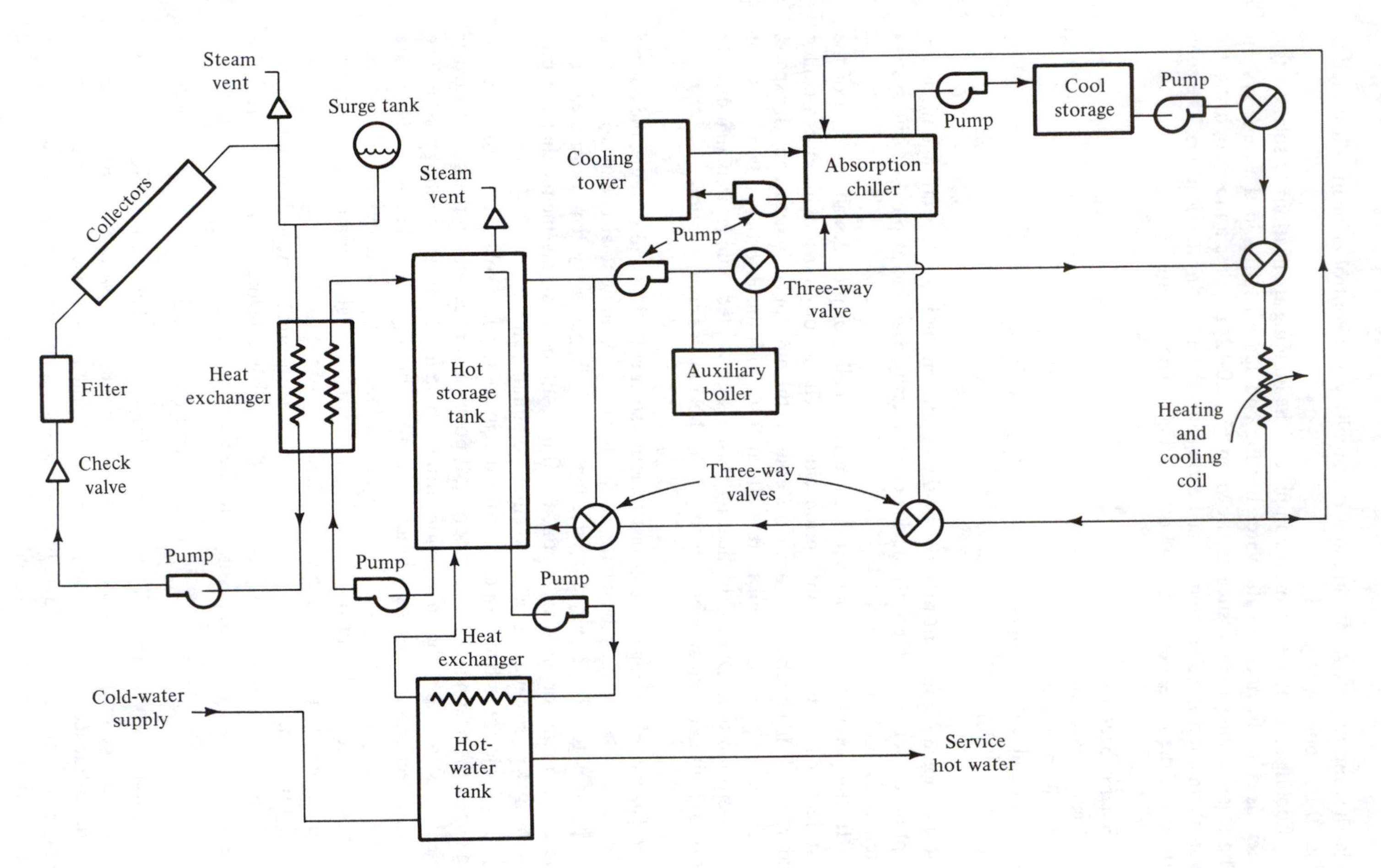

Figure 20-12 Solar heating and cooling system.

Table 20-4 Transmission into a building through south-facing double-pane glass at 40° north latitude

Month	Time of day	θ, deg	I_{DN}, W/m^2	τ	I_T, W/m^2
Dec	10 A.M.	35.2	826	0.76	512
Dec	12 noon	26.5	899	0.77	620
June	10 A.M.	78.1	847	0.37	65
June	12 noon	73.5	867	0.49	121

Admission of solar energy into the south windows during summer can be reduced further by using overhangs. The practice of constructing overhangs for south windows[5] was encouraged in the late 1940s, passed out of vogue during the decades of low-cost energy in the 1950s and 1960s, and became important again in the late 1970s. The idea is to place the window and construct the overhang so that a south window is shaded by the overhang from April through August and completely exposed to the sun in December, as shown in Fig. 20-13.

Example 20-6 Choose the dimensions x and y in Fig. 20-13 so that the south window at a latitude of 40° north is completely shaded at noon from April 21 to August 21 and is completely unshaded on December 21. The height of the window is 1.2 m.

Solution At noon, 40° north latitude, the solar altitudes β on April 21 (also Au-

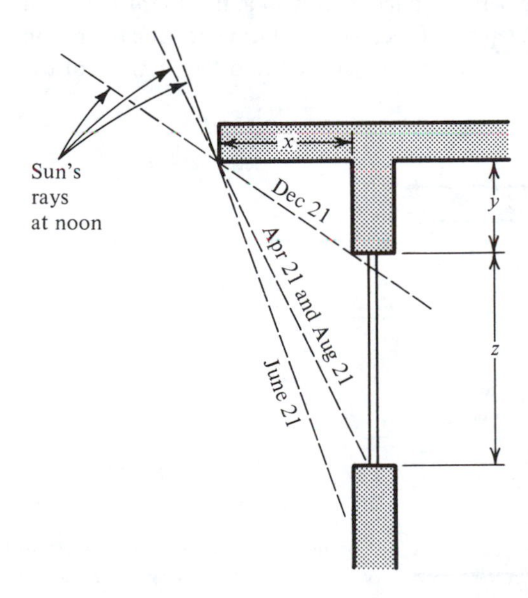

Figure 20-13 Use of overhang to control solar irradiation on a south window.

gust 21) and December 21 are, from Eq. (20-3),

$$\beta = \begin{cases} 61.5° & \text{Apr.} \\ 26.5° & \text{Dec.} \end{cases}$$

$$\frac{y}{x} = \tan 26.5° = 0.499$$

$$\frac{y+z}{x} = \frac{y+1.2}{x} = \tan 61.5° = 1.84$$

Solving gives

$$x = 0.895 \text{ m} \qquad \text{and} \qquad y = 0.447 \text{ m}$$

After the solar energy has been received in the building the next task is to use it effectively. In certain cases the energy may be needed instantaneously, but in others there may be a greater rate of solar energy flowing into rooms with south windows than can be used at the moment. Capability of thermal storage and/or the ability to transfer the heat to other sections of the building is useful. Thermal storage occurs inherently because the sun's rays first warm the floor and perhaps the walls and furnishings. The energy stored in these masses flows slowly by convection to the air in the space. More effective but more elaborate thermal storage is possible, still within the framework of passive design, by the using such devices as the Trombe-type wall shown in Fig. 20-14. An opaque wall is placed inside the window so that the sun's rays fall directly on it. Some of the heat that enters the wall conducts to the space, but much of it is first absorbed in the wall and slowly passes to room air as the air flows by natural convection upward across the outer surface of the wall.

Another practice that combines the passive concept with an active thermal system is the application of a decentralized heat pump (Sec. 18-8). During heating operation the excess heat flowing into southern exposures is transferred to other sections of the building where heating is needed.

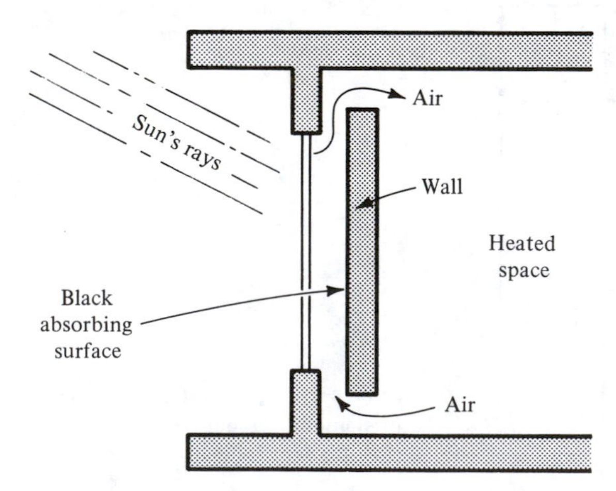

Figure 20-14 Trombe-type wall to enhance thermal storage in a passive solar system.

20-10 Economics of solar installations The acceptance of solar energy for heating and cooling has been somewhat slow, but if the costs of energy from fossil fuel continue to increase at a rapid rate, the economic prospects of solar energy are likely to improve. From the standpoint of cost effectiveness, passive solar systems are usually considered the most attractive, followed by water heating, space heating, and finally space cooling.

PROBLEMS

20-1 Using Eq. (20-3), compute the hour of sunrise on the shortest day of the year at 40° north latitude. *Ans.* 7:26 A.M.

20-2 Compute the solar azimuth angle at 32° north latitude at 9 A.M. on February 21. *Ans.* See Table 4-13

20-3 (*a*) What is the angle of incidence of the sun's rays with a south-facing roof that is sloped at 45° with the horizontal at 8 A.M. on June 21 at a latitude of 40° north? (*b*) What is the compass direction of the sun at this time? *Ans.* (*a*) 64°

20-4 As an approach to selecting the tilt angle Σ of a solar collector a designer chooses the sum of $I_{DN} \cos \theta$ at 10 A.M. and 12 noon on January 21 as the criterion on which to optimize the angle. At 40° north latitude, with values of $A = 1230$ W/m^2 and $B = 0.14$ in Eq. (20-9), what is this optimum tilt angle? *Ans.* 61.5°

20-5 Plot the efficiency of the collector described in Example 20-3 versus temperature of fluid entering the absorber over the range of 30 to 140°C fluid temperatures. The ambient temperature is 10°C. If the collector is being irradiated at 750 W/m^2, determine the rate of collection at entering fluid temperatures of (*a*) 50°C and (*b*) 100°C.

20-6 A 1.25- by 2.5-m flat-plate collector receives solar irradiation at a rate of 900 W/m^2. It has a single cover plate with $\tau = 0.9$, and the absorber has an absorptivity $\alpha_a = 0.9$. Experimentally determined values are $F_r = 0.9$ and $U = 6.5$ W/m^2 · K. The cooling fluid is water. If the ambient temperature is 32°C and the fluid temperature is 60°C entering the absorber, what are (*a*) the collector efficiency, (*b*) the fluid out-

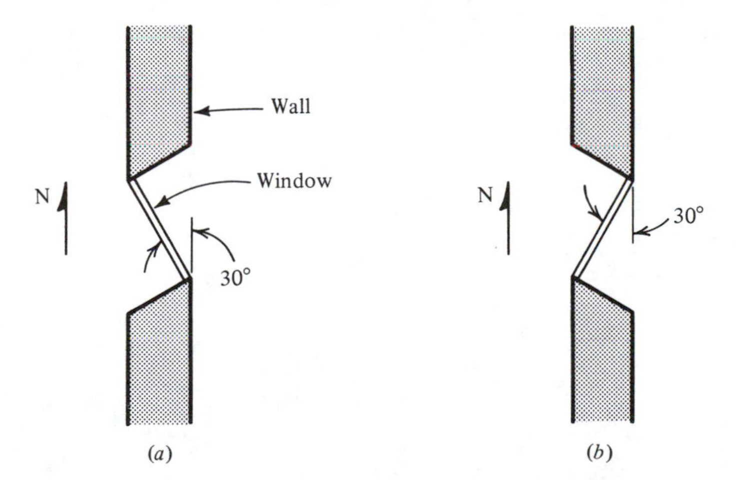

Figure 20-15 West-window orientations in Prob. 20-7.

let temperature for a flow rate of 25 kg/h, and (*c*) the inlet temperature to the absorber at which the output drops to zero?

20-7 Two architects have different notions of how to orient windows on the west side of a building in order to be most effective from a solar standpoint–summer and winter. The windows are double-glazed. The two designs are shown in Fig. 20-15. Compute at 40° north latitude the values of I_T from Eq. (20-14) for June 21 at 2 and 6 P.M. and January 21 at 2 P.M. and then evaluate the pros and cons of the two orientations.

REFERENCES

1. "ASHRAE Handbook and Product Directory, Applications Volume," chap. 58, American Society of Heating, Refrigerating, and Air-Conditioning Engineers, Atlanta, Ga., 1978.
2. J. R. Howell, R. B. Bannerot, and G. C. Vliet: "Solar/Thermal Energy Systems; Analysis and Design," McGraw-Hill, New York, 1982.
3. Thermal Storage Application, ASHRAE Symp., *ASHRAE Trans.*, vol. 85, pt. 1, pp. 480–524, 1979.
4. Passive Solar Systems, ASHRAE Symp., *ASHRAE Trans.*, vol. 85, pt. 1, pp. 443–477, 1979.
5. Solar Orientation in Home Design, *Univ. Ill. Small Homes Counc. Circ.* C 3.2, Urbana, 1945; reissued 1977.
6. F. Trombe et al.: "Some Characteristics of the CNRS Solar House Collectors," CNRS Solar Laboratory, Font Romeu/Odeillo, France, 1976.
7. J. A. Duffie and W. A. Beckman: "Solar Engineering of Thermal Processes," Wiley, New York, 1980.
8. F. Kreith and J. F. Kreider: "Principles of Solar Engineering," Hemisphere, Washington, 1978.
9. J. L. Threlkeld: "Thermal Environmental Engineering," Prentice-Hall, Englewood Cliffs, N.J., 1970.

CHAPTER

TWENTY-ONE

ACOUSTICS AND NOISE CONTROL

21-1 The study of sound and acoustics The subject of acoustics is broad enough to challenge specialists in many branches of the field. Engineers and scientists may concentrate on the acoustic design of concert halls and theaters, designing quieter mechanical equipment, reducing aircraft noise, or such specialized fields as underwater sound and medical applications of ultrasonics. One of the most important and extensive applications of acoustics and noise control relates to mechanical systems in buildings. Here concern must be focused on ensuring such things as adequate speech privacy, acceptable sleeping habitats, and little or no annoyance from equipment noise.

A basic understanding of acoustics is important for designers of air-conditioning systems because the system is a primary generator of sound in a building. Noise results from the operation of compressors, fans, and pumps and from flow when air or water moves through ducts and pipes or, especially, on abrupt drops in pressure. The air-conditioning system is also linked with sound and noise concerns not only because it is a generator of noise but also because pipes or ducts transmit noises from conversations or office equipment from one room to another.

This chapter seeks to provide a foundation from which to proceed in deciding how to control sound in a building through proper design or the installation of sound-absorbing devices and materials. Following an introduction to the physics of sound waves, this chapter explains the distinction between sound power level and sound pressure level. A generator of sound emits a certain magnitude of sound power, but this magnitude normally cannot be measured directly. Instead, instruments are available to measure the sound pressure level. The sound power level of the generator and the sound pressure level measured some distance from the source are related through the acoustic characteristics of the enclosure. With this understanding of power level and pressure level, the text attempts to provide an appreciation of how sound absorption affects room characteristics and how sound is transmitted and attenuated (reduced) through ducts to provide the background for further study of sound and noise control in buildings.

21-2 One-dimensional sound waves A sound wave consists of a rapid oscillation of air pressure. A sound generator has a vibrating surface that alternately compresses and expands the air in contact with it. These alternate compressions and rarefactions move through the air at sonic velocity and reach the receiver (perhaps a human ear) still retaining their alternate compressions and rarefactions. A speaker placed at the end of a tube as in Fig. 21-1, vibrating with a sinusoidal displacement at a frequency f Hz causes a pressure wave to propagate down the tube, which can be represented by

$$p = p_0 \sin \left[2\pi f \left(t + \frac{x}{c}\right)\right] \tag{21-1}$$

where p = instantaneous sound pressure amplitude above and below atmospheric, Pa
p_0 = maximum amplitude of pressure fluctuation, Pa
f = frequency, Hz
t = time, s
x = distance, m
c = sonic velocity = 344 m/s at standard air conditions

At a given time t the distance from a position x to a point on the next wave where the pressure is the same as at x requires that

$$p = p_0 \sin \left[2\pi f \left(t + \frac{x}{c}\right)\right] = p_0 \sin \left[2\pi f \left(t + \frac{x + \lambda}{c}\right)\right] \tag{21-2}$$

where λ is the wavelength in meters. For the two sine functions in Eq. (21-2) to be equal, the arguments must differ by 2π or a multiple of 2π, so

$$2\pi f \left(t + \frac{x}{c}\right) + 2\pi = 2\pi f \left(t + \frac{x + \lambda}{c}\right) \tag{21-3}$$

and

$$\lambda = \frac{c}{f}$$

The order of magnitude of wavelengths in the audible range is therefore between $\frac{1}{2}$ cm at 10 kHz and several meters at low frequencies.

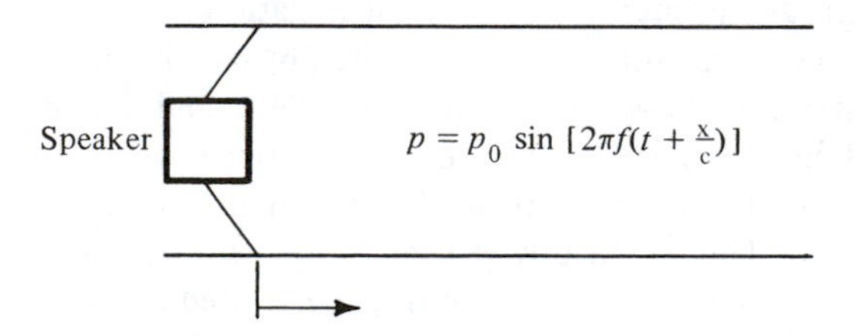

Figure 21-1 Sound-pressure variation in a tube.

21-3 Standing waves If a sound reflector is installed in the tube of Fig. 21-1 as shown in Fig. 21-2 so that its distance from the speaker is equal to one wavelength of the tone emitted by the speaker, a standing wave will develop. At position x the pressure is the algebraic sum of the primary pressure wave just coming from the speaker and the reflected wave. The distance the reflected wave has traveled is $\lambda + (\lambda - x)$; so the combination of the two pressure contributions is

$$p = p_0 \sin\left[2\pi f\left(t + \frac{x}{c}\right)\right] + p_0 \sin\left[2\pi f\left(t + \frac{\lambda + \lambda - x}{c}\right)\right]$$

Since the presence of the two λ's in the second term only adds 4π to the argument of the sine term and does not change its value,

$$p = p_0 \sin\left[2\pi f\left(t + \frac{x}{c}\right)\right] + p_0 \sin\left[2\pi f\left(t - \frac{x}{c}\right)\right] \tag{21-4}$$

By using the properties of the sine of the sum and difference of angles, Eq. (21-4) can be revised to

$$p = 2p_0 \sin 2\pi ft \cos \frac{2\pi fx}{c} \tag{21-5}$$

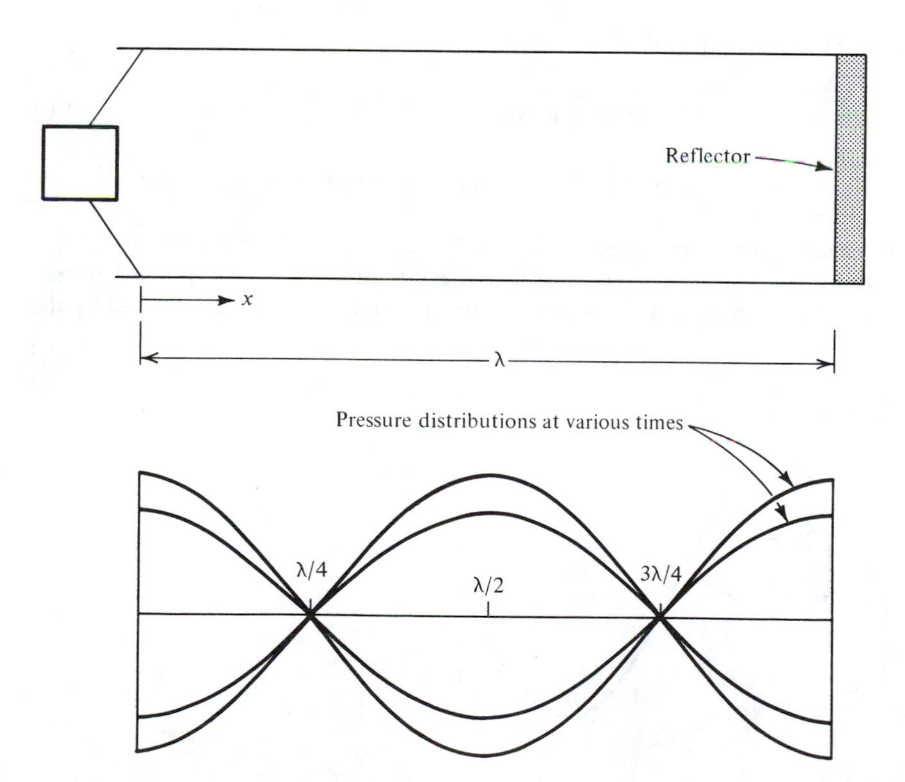

Figure 21-2 Standing waves.

The distribution of pressure expressed by Eq. (21-5) is shown in Fig. 21-2. When x = 0 and $\lambda/2$, the magnitude of the cosine term is 1.0 and the pressure varies between $+2p_0$ and $-2p_0$. When $x = \lambda/4$ and $3\lambda/4$, the cosine term is zero, indicating that the pressure is zero at all times.

Standing waves can develop in rooms, particularly those with highly reflective surfaces when some pure tones are present. Sometimes these standing waves can be detected even with the ear by noticing the variations in sound intensity in moving from one position to another. Standing waves will occur not only when the length of the tube equals the wavelength but also when the length is an integer multiple of the wavelength.

21-4 Energy in a sound wave The next several sections lay the groundwork for quantifying the measurement of sound. A wave, as in Fig. 21-3, possesses energy due to two sources. Energy was required to distort the pressures from the equilibrium atmospheric pressure and equals the integral of $p\,dV$ over the wave. Kinetic energy is also present because of the motion of the air and can be represented by the integral of $(u^2/2)(\rho\,dV)$. The sum of these energies for a wave of area A m^2 results from the integration

$$\text{Energy in wave} = \int_0^\lambda p\rho A\,dx + \int_0^\lambda \frac{u^2}{2}\rho A\,dx \qquad \text{J}$$

The result of this integration[1] is

$$\text{Energy in wave} = \frac{Ap_0^2}{2c\rho f} \qquad \text{J} \tag{21-6}$$

The energy in a wave is proportional to the square of the pressure amplitude.

21-5 Intensity, power, and pressure Equation (21-6) provides the key for relating sound power and sound pressure through the intensity. This relationship between sound power and pressure is important because a sound source is characterized by the

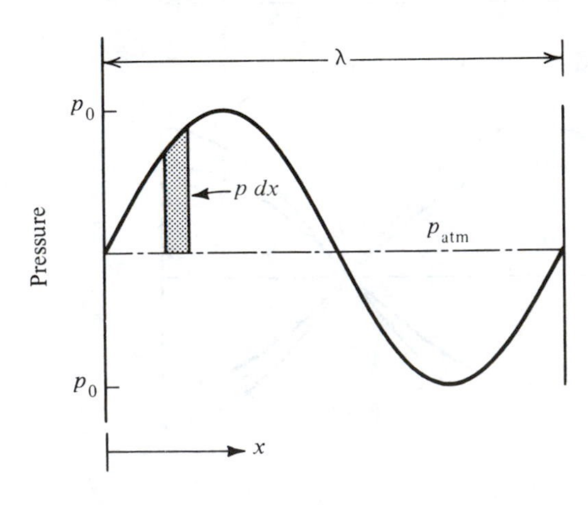

Figure 21-3 Energy in a wave.

power it generates, but this power cannot be measured directly. The sound source causes sound pressure waves in the surrounding air whose magnitudes can be measured. Furthermore, the extent to which the human ear perceives the sound is directly related to the sound pressure level.

The intensity I is defined as the rate at which sound energy passes a point per unit area. In a sound field, the intensity is the number of joules passing through each square meter per second. The intensity thus has units of watts per square meter. One means of calculating the intensity is to multiply the energy expressed in Eq. (21-6) by the rate at which waves flow past a point and then divide by the area. Since this rate of wave passage is the frequency

$$I = \frac{(A\, p_0^2/2\rho c f)f}{A} = \frac{p_0^2}{2\rho c} \quad \text{W/m}^2 \tag{21-7}$$

Since for a sine wave the root-mean square of the pressure p_{rms} equals $p_0/\sqrt{2}$, an alternate expression for Eq. (21-7) is

$$I = \frac{p_{\text{rms}}^2}{\rho c} \tag{21-8}$$

If a source of sound exists within a room, this source emits a certain *acoustic power* in watts. At any distance from the source the statement cannot be made that a certain value of power prevails; instead, a certain intensity exists. If, as in Fig. 21-4, a source of power of E W radiates uniformly in all directions, the intensity at a distance r from the source, considering only direct sound radiation from the source and no reflections, will be the power divided by the area of the sphere at a radius r

$$I = \frac{E}{4\pi r^2} \tag{21-9}$$

Equations (21-7) and (21-8) relate the sound intensity to the amplitude of the pressure fluctuation, while Eq. (21-9) relates the intensity to the sound power, at least for direct radiation from a nondirectional source. The importance of relating these quantities is that sound-measuring instruments are capable of measuring p_{rms}^2, which is proportional to intensity. Conversion from an intensity measurement to power also demands knowledge of the acoustic characteristics of the room, discussed later in this chapter.

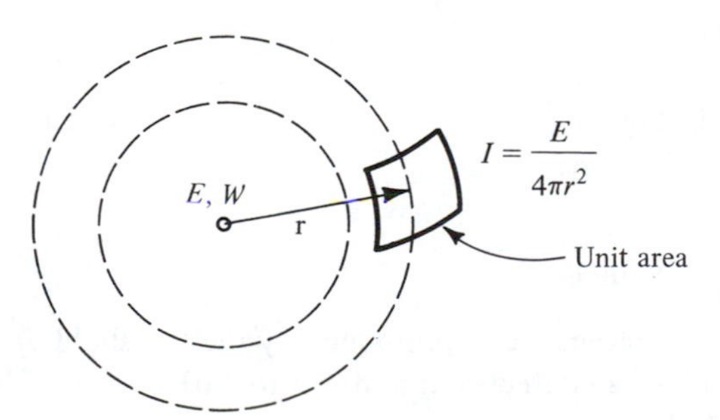

Figure 21-4 Uniform radiation from a sound source.

21-6 Sound power level While it would be possible to express the power emitted by a source directly in watts, the standard form of presentation is the *sound power level*

$$\text{PWL} = 10 \log \frac{E}{E_0} \tag{21-10}$$

where PWL = sound power level, dB
E = power emitted by source, W
E_0 = reference level, W

The expression $\log (E/E_0)$ has units of bels, and the multiplying factor of 10 converts the units into decibels (dB), in which the magnitudes are somewhat more convenient. The reference level E_0 can be chosen arbitrarily and is usually 1 pW.

Example 21-1 Calculate the sound power level of (*a*) a whisper that emits a power of 1 nW and (*b*) a rocket engine that emits 10 MW.

Solution
(*a*)

$$\text{PWL} = 10 \log \frac{10^{-9}}{10^{-12}} = 30 \text{ dB}$$

(*b*)

$$\text{PWL} = 10 \log \frac{10^{7}}{10^{-12}} = 190 \text{ dB}$$

21-7 Intensity level and sound pressure level The sound intensity level IL is defined as

$$\text{IL} = 10 \log \frac{I}{I_0} \tag{21-11}$$

where IL = intensity level, dB
I_0 = reference intensity level, = 1 pW/m^2 arbitrarily

The intensity level is not used in routine acoustic work, but it will be useful at several points in this chapter in developing relationships between sound power level and sound pressure level.

The sound pressure level is defined as

$$\text{SPL} = 10 \log \frac{p_{\text{rms}}^2}{p_{\text{ref}}^2} \tag{21-12}$$

where SPL = sound pressure level, dB

$$p_{\text{ref}} = 20 \ \mu\text{Pa}$$

The reference pressure of 20 μPa was chosen because it is approximately the threshold of hearing. A person, then, with good hearing can detect sounds down to 0 dB.

The SPL and IL are approximately equal. When the expression for the intensity from Eq. (21-8) is substituted into the definition for the intensity level, we have

$$IL = 10 \log \frac{p_{rms}^2/\rho c}{10^{-12}}$$

Using $\rho = 1.18\ kg/m^3$ and $c = 344\ m/s$ gives

$$IL = 10 \log \frac{p_{rms}^2}{0.0000202^2} \approx SPL$$

21-8 Sound spectrum While the knowledge of the overall SPL may be adequate in some instances, it is often useful and even necessary to know the frequency distribution of the sound. For example, it may be necessary to know whether most of the sound intensity occurs in the low-, medium-, or high-frequency range. In analyzing machine noises, the information that the principal contributor to the total SPL occurs at a certain frequency may be helpful in pinpointing the offending member of the machine. If undesirable sound is being transmitted through an air duct and is to be reduced by placing an absorber in the duct, the absorber should be selected with its most effective absorption in the frequency range of the noise. Since the human ear is not equally sensitive to all frequencies, a high SPL in a frequency range where the ear is insensitive may not be objectionable.

The audible range is from about 20 to 20,000 Hz, and the standard division into octave bands covering most of this range is shown in Table 21-1. An octave band is the range through which the frequency doubles. A standard accessory for sound-level meters is and octave-band analyzer, which filters out all but the desired octave band so that the pressure level of the band in question can be determined separately.

21-9 Combination of sound sources Since more than one source of sound often contributes to the total, some method must be developed for determining the total SPL when the individual SPLs are known. In Fig. 21-5, which shows two sources and a receiver, if SPL_1 is the sound pressure level at the receiver when only source 1 is active and SPL_2 when source 2 alone is active, when they are both active,

$$SPL \neq SPL_1 + SPL_2$$

Table 21-1 Octave bands

Octave bands, Hz	Midfrequency, Hz
45–90	63
90–180	125
180–355	250
355–710	500
710–1400	1000
1400–2800	2000
2800–5600	4000
5600–11,200	8000

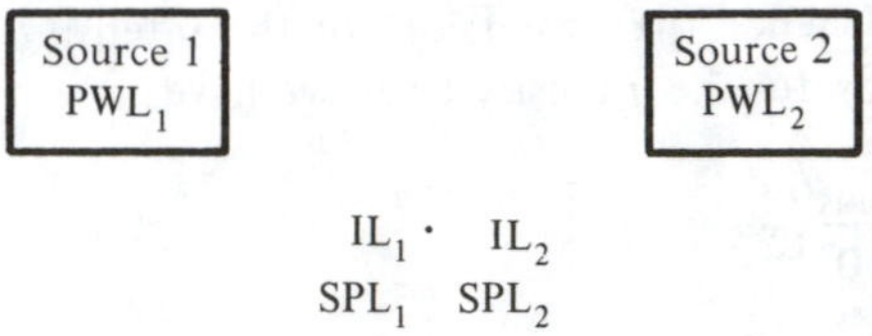

Source 1 PWL_1		Source 2 PWL_2

$IL_1 \cdot \quad IL_2$

$SPL_1 \quad SPL_2$

Figure 21-5 Combination of two random sound sources.

The SPL values in decibels do not add to give the combined SPL. The intensity I W/m^2 at the receiver, however, is the sum of the intensities contributed by each source. The foregoing statement is true only for random noise, and not, for example, if sources 1 and 2 are pure tones of the same frequency. In the latter case the waves may combine to partially reinforce or partially cancel each other. If the sources were of slightly different frequencies, periodic beats would develop.

Using the relation that $I = I_1 + I_2$ and assigning the subscript 1 to the source contributing the largest IL at the receiver gives

$$\text{SPL} = \text{IL} = 10 \log \frac{I}{10^{-12}} = 10 \log \frac{I_1 + I_2}{10^{-12}}$$

and

$$\text{SPL} = 10 \log \frac{I_1(1 + I_2/I_1)}{10^{-12}}$$

$$\text{SPL} = 10 \log \frac{I_1}{10^{-12}} + 10 \log \left(1 + \frac{I_2}{I_1}\right) \tag{21-13}$$

In Eq. (21-13)

$$10 \log \frac{I_1}{10^{-12}} = \text{IL}_1 = \text{SPL}_1 \tag{21-14}$$

The antilog of Eq. (21-11) is

$$I = 10^{-12}(10^{\text{IL}/10})$$

which applied to I_1 and I_2 yields

$$I_1 = 10^{-12}(10^{\text{IL}_1/10}) \quad \text{and} \quad I_2 = 10^{-12}(10^{\text{IL}_2/10}) \tag{21-15}$$

Substituting Eq. (21-15) into Eq. (21-13) gives

$$\text{SPL} = \text{SPL}_1 + 10 \log (1 + 10^{-(\text{IL}_1 - \text{IL}_2)/10}) \tag{21-16}$$

The combined SPL is therefore the SPL due to the source contributing the higher SPL of the two plus a quantity dependent upon the difference $\text{SPL}_1 - \text{SPL}_2$. The quantity to be added to SPL_1 is plotted in Fig. 21-6 from values calculated from Eq. (21-16). The graph shows that when two sources of equal intensity are combined, the total is 3 dB higher than either one separately. If a second source has an SPL of 10 dB lower than the first, the total SPL is only 0.5 dB higher than the first.

Example 21-2 Three sound sources provide equal SPL readings at a receiver when active individually. How much higher is the combined SPL when all three sources are active than when only one individual source is active?

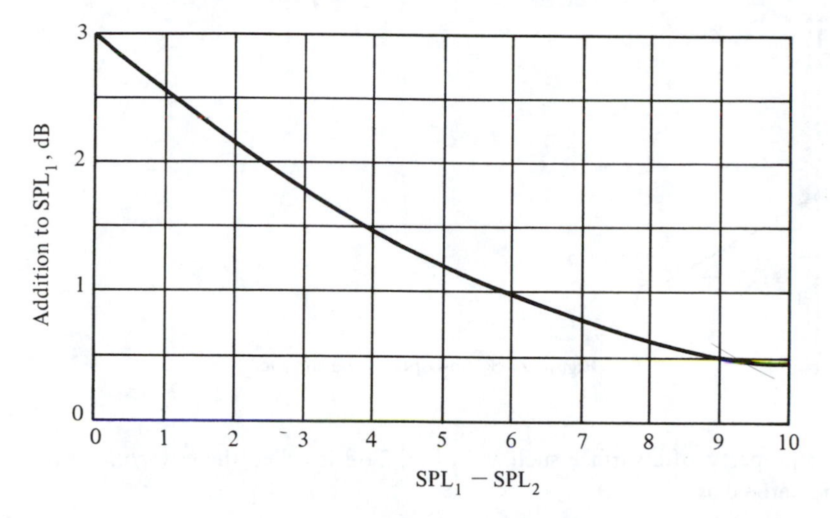

Figure 21-6 Total SPL when combining two sound sources.

Solution The sources are combined schematically in Fig. 21-7 showing that the combination of two sources gives a total of SPL_1 + 3 dB. Combining the result of that pair with the third source adds another 1.7 dB to SPL_1 + 3, giving a total of SPL_1 + 4.7 dB.

A point illustrated by Example 21-2 is that if there are several sound sources, all contributing approximately the same SPL, removal of just one of them does not reduce the overall SPL appreciably. In Example 21-2 if one source were removed, the reduction in total SPL would be only 1.7 dB.

21-10 Absorptivity In the situations discussed so far in this chapter it has been assumed that only direct radiation of sound from a source reaches the receiver. This situation is rare since it would occur only out of doors or in a room where all the surfaces absorb all the sound that strikes them. The usual case is where the enclosure has surfaces that are neither perfectly absorptive nor perfectly reflective. The measure of

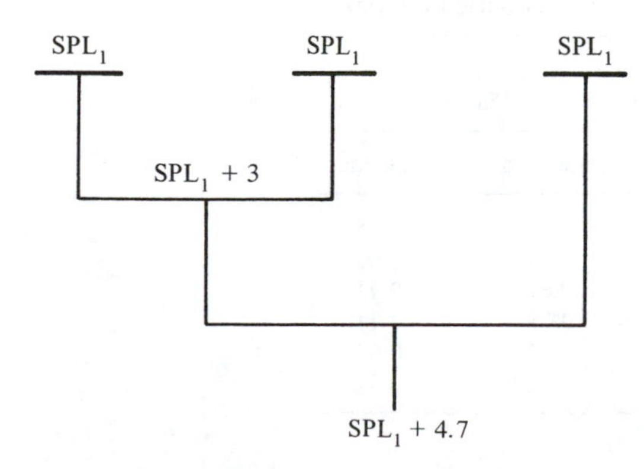

Figure 21-7 Schematic combination of sources in Example 21-2.

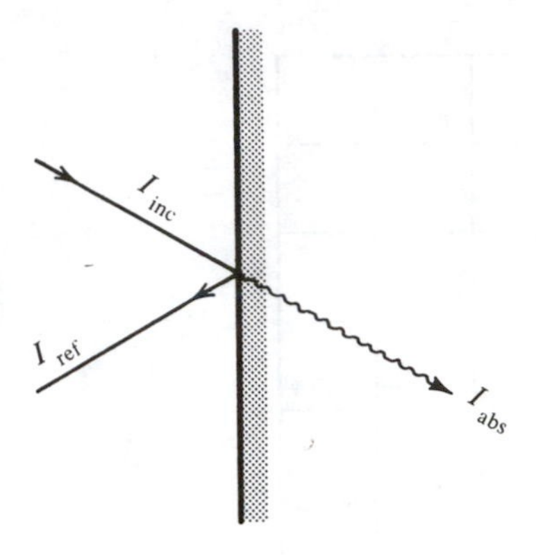

Figure 21-8 Absorption at a surface.

the absorptive property of a surface such as in Fig. 21-8 is called the *absorption coefficient* α and defined as

$$\alpha = \frac{I_{abs}}{I_{inc}}$$

where I_{inc} = sound intensity striking surface, W/m^2
I_{abs} = intensity absorbed, W/m^2
The order of magnitude of α might be from 0.01 to 0.05 for concrete and from 0.2 to 0.8 for an acoustical material.

The absorption coefficient of a material usually varies with the frequency. Table 21-2 shows the absorption coefficients of several commercial acoustic materials commonly installed as ceiling tile or cemented to walls. Two trends are discernible from Table 21-2: (1) it is usually more difficult to absorb low-frequency sound, and (2) the acoustic material has a higher absorption coefficient spaced away from a wall or ceiling than cemented directly to it. The explanation for the latter property is that sound-absorbing materials function by converting the air motion into heat due to friction. At the wall or ceiling this air motion is small, while away from the wall it is much higher. The maximum air motion will, in fact, occur at a distance of $\lambda/4$ from a hard reflective surface.

Table 21-2 Absorption coefficients of several acoustic materials

Frequency Hz	Johns-Manville Comet		Armstrong Classic	
	Cemented	Suspended	Cemented	Suspended
125	0.10	0.49	0.10	0.35
250	0.25	0.32	0.19	0.62
500	0.61	0.51	0.64	0.71
1000	0.58	0.64	0.78	0.71
2000	0.47	0.57	0.72	0.68
4000	0.38	0.41	0.52	0.52

When a room consists of a number of different surfaces, the mean absorption coefficient $\bar{\alpha}$ is defined as

$$\bar{\alpha} = \frac{S_1\alpha_1 + S_2\alpha_2 + \cdots + S_n\alpha_n}{S_1 + S_2 + \cdots + S_n} \tag{21-17}$$

where S_1, S_2, etc., are areas in square meters having absorption coefficients α_1, α_2, etc., respectively.

21-11 Room characteristics A source of sound can be characterized by the power it emits. To say that a source has a certain pressure level has no meaning. At a position removed from the source the sound pressure level can be measured, but any statement about the sound power at this position is ambiguous. The sound power emitted by the source and the pressure level measured at a point away from the source are related. The relation depends upon the distance between the two but is also a function of the characteristics of the enclosure. The relationship between the PWL of the source, the SPL measured at some distance from the source, and the characteristics of the room will now be discussed. The rooms considered will be so-called *large enclosures*, defined as those having dimensions much larger than the wavelengths of interest.

The graph in Fig. 21-9 will be the form chosen to present the relationships, and it shows the difference between the SPL and the PWL as a function of the distance between the source and receiver. The parameter of the family of curves is the room constant R, which is defined as

$$R = \frac{S\bar{\alpha}}{1 - \bar{\alpha}} \tag{21-18}$$

where R = room constant, m^2
S = total surface area of room, m^2

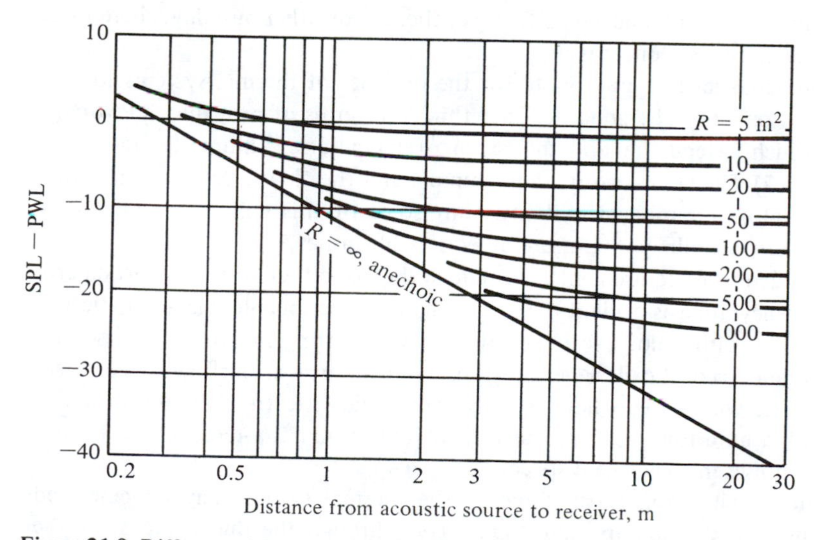

Figure 21-9 Difference between SPL and PWL for rooms of various characteristics with nondirectional source.

The two extreme room conditions within whose boundaries all rooms lie are the *anechoic* and the *reverberant*. In the anechoic room all the surfaces are perfectly absorptive; in the reverberant room all surfaces are perfectly reflective. The line in Fig. 21-9 for the anechoic room can be determined simply because the SPL is that due only to direct radiation

$$\text{SPL} - \text{PWL} = \text{IL} - \text{PWL}$$

$$\text{SPL} - \text{PWL} = 10 \log \frac{E/4\pi r^2}{1 \text{ pW/m}^2} - 10 \log \frac{E}{1 \text{ pW}}$$

then

$$\text{SPL} - \text{PWL} = -10 \log 4\pi r^2 \tag{21-19}$$

Equation (21-19) plotted on Fig. 21-9 is the straight line applicable to $R = \infty$ for $\bar{\alpha} = 1.0$.

The reverberant room cannot be plotted on Fig. 21-9 because if there is no absorptivity a continuous introduction of energy into the room would progressively build up the intensity and SPL would continue to increase with time. Of course, no completely reverberant room exists.

For actual rooms the room constant R lies somewhere between zero and infinity; so both the direct and reflected intensity must be considered in determining the value of SPL - PWL.

21-12 Acoustic design in buildings Now that some definitions and a few basic principles of acoustics have been presented, it is appropriate to pause for a broad view of what is important from an acoustical standpoint in a building. What, for example, would be a standard review process for the designer of the heating and air-conditioning system of a building to execute in order to include acoustic considerations in the design process? The scope to be described will envision an occupied workplace, such as an office building or school, and not a factory, theater, or other building where there are special sound and noise concerns.

Sound and acoustics are appropriate for the designer of thermal systems to consider for two reasons: (1) The good designer thinks in terms of providing a favorable environment which extends beyond the realm of temperature, humidity, radiation, and air motion. (2) The compressors, fans, pumps, and the flow of air are sometimes the major generators of noise in buildings, but some of this same equipment may offer the means of controlling other more objectionable noises.

The typical acoustic requirement is to reduce the magnitude of noise, particularly if a single-frequency tone is evident. Another highly objectionable characteristic is a repeated change of either noise level or frequency occurring in intervals of several seconds. Sometimes acoustic designers purposely introduce an airflow noise, a so-called *white noise,* whose frequency extends over a wide spectrum. White noise can be used to mask a pure tone, such as the 120-Hz hum from a fluorescent light fixture or conversation noises in an adjacent space.

Low-frequency vibrations attributable to the thermal system may be generated by a compressor, fan, or pump, and transmitted through the ducts, pipes, or the structural members of the building. Control of this type of noise is usually provided by

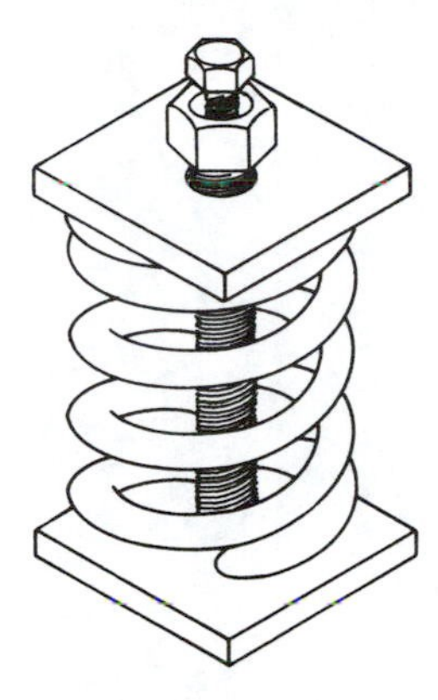

Figure 21-10 A vibration isolator.

isolating the machine from the transmitting medium.[2] The machine should be installed on a resilient mount, one type of which is shown in Fig. 21-10, that uses a spring to damp out the low frequencies. The low-frequency mount is sometimes fastened in turn to a load-supporting flexible material that damps high frequencies. Frequently the machine is mounted on an *inertia block* of concrete, which is then supported by the springs or flexible pads. Care must be exercised in this application to ensure that the resulting resonant frequency of the isolated equipment is not near a frequency of a vibrating element of the equipment.

To inhibit the structural transmission of machine noise through ducts and pipes, flexible connectors are available. The sides of large rectangular ducts sometimes flex in and out in an objectionable "tin-canning" action that can be corrected by stiffening the sheet metal with a piece of angle iron.

21-13 Fan- and air-noise transmission in ducts A frequently performed acoustic analysis of the air-conditioning system in a building is based on the following proposition of noise generation and transmission:[3]

1. The source of noise is the fan (perhaps both the return-air and supply-air fans but especially the supply-air fan, which generates high air pressures). The sound power divides, half passing in the same and half in the opposite direction of airflow.
2. In the duct system the sound power passes through straight sections of duct, elbows, branches, and finally to the outlet diffuser. In those elements there may be a net decrease or a net increase (due to flow-induced noise) of the sound power.
3. Calculation of the sound power at each element along the duct provides the magnitude of the sound power at the air outlet to the room. Figure 21-9 can then be used to compute the sound pressure level at one or more positions in the room. If the sound pressure level is too high at the design position in the room, its magnitude can be reduced by the appropriate combinations of increasing the R value of the room through the application of sound-absorbing material or by installing a sound attenuator in the duct, such as the one shown in Fig. 21-11. Other possibilities include selecting lower-noise equipment or duct components or reducing the flow velocities.

Figure 21-11 Sound attenuator for a duct. *(Environmental Elements Corporation, Subsidiary of Koppers, Co., Inc.)*

21-14 Conclusions The field of acoustics and noise control is extensive enough to command the full-time efforts of many specialists. The designer of air-conditioning components and systems normally cannot concentrate on acoustics alone and must be satisfied with knowledge of only the most important topics. This knowledge should include an understanding of the definition of terms, a qualitative understanding of design features that result in desirable acoustic conditions, and, conversely, what design practices to avoid. The designer should also be able to analyze an acoustic problem in an existing installation and know in which direction to proceed to cure the problem. As the engineer proceeds through the design of the air-conditioning system, reduction of noise generation, prevention of its transmission through ducts, pipes, and structural members, and the absorption of noise that is generated should be kept in mind. A practice followed by some design firms is to submit tentative designs to an acoustic specialist for review of potential noise problems.

PROBLEMS

21-1 A tube 1.5 m long has a speaker at one end and a reflecting plug at the other. The frequency of a pure-tone generator driving the speaker is to be set so that standing waves will develop in the tube. What frequency is required? *Ans.* 229 Hz, 458 Hz, etc.

21-2 The sound power emitted by a certain rocket engine is 10^7 W, which is radiated uniformly in all directions.

(*a*) Calculate the amplitude of the sound pressure fluctuation 10 m removed from the source. *Ans.* 2540 Pa

(*b*) What percentage is this amplitude of the standard atmospheric pressure? *Ans.* 2.54%

21-3 At a distance of 3 m from a sound source of 100 W that radiates uniformly in all directions what is the SPL due to direct radiation from this source? *Ans.* 119.5 dB

21-4 An octave-band measurement resulted in the following SPL measurements in decibels for the eight octave bands listed in Table 21-1: 65.4, 67.3, 71.0, 74.2, 72.6, 70.9, 67.8, and 56.0, respectively. What is the expected overall SPL reading? *Ans.* 79.2 dB

21-5 A room has a ceiling area of 25 m^2 with acoustic material that has an absorption coefficient of 0.55; the walls and floor have a total area of 95 m^2 with an absorption coefficient of 0.12. A sound source located in the center of the room emits a sound power level of 70 dB. What is the SPL at a location 3 m from the source? *Ans.* 62 dB

21-6 In computing the transmission of sound power through a duct, the standard calculation procedure for a branch takeoff is to assume that the sound power in watts divides in a ratio of the areas of the two branches. If a PWL of 78 dB exists before the branch, what is the distribution of power in the two branches if the areas of the branches (*a*) are equal and (*b*) are in a ratio of 4:1? *Ans.* (*a*) 75 dB; (*b*) 71, 77 dB

REFERENCES

1. L. L. Beranek: "Acoustics," McGraw-Hill, New York, 1954.
2. L. F. Yerges: "Sound, Noise and Vibration Control," Van Nostrand Reinhold, New York, 1969.
3. "ASHRAE Handbook and Product Directory, Systems Volume," American Society of Heating, Refrigerating, and Air-Conditioning Engineers, Atlanta, Ga., 1980.
4. "ASHRAE Handbook, Fundamentals Volume," American Society of Heating, Refrigerating, and Air-Conditioning Engineers, Atlanta, Ga., 1981.
5. L. L. Beranek: "Noise and Vibration Control," McGraw-Hill, New York, 1971.
6. L. E. Kinser and A. R. Frey: "Fundamentals of Acoustics," Wiley, New York, 1967.
7. L. L. Faulkner: "Handbook of Industrial Noise Control," Industrial, New York, 1976.
8. C. M. Harris: "Handbook of Noise Control," McGraw-Hill, New York, 1957.

APPENDIX

Table A-1 Water: properties of liquid and saturated vapor

t, °C	Saturation pressure, kPa	Specific volume, m^3/kg		Enthalpy, kJ/kg		Entropy, kJ/kg · K	
		Liquid	Vapor	Liquid	Vapor	Liquid	Vapor
0	0.6108	0.0010002	206.3	-0.04	2501.6	-0.0002	9.1577
2	0.7055	0.0010001	179.9	8.39	2505.2	0.0306	9.1047
4	0.8129	0.0010000	157.3	16.80	2508.9	0.0611	9.0526
6	0.9345	0.0010000	137.8	25.21	2512.6	0.0913	9.0015
8	1.0720	0.0010001	121.0	33.60	2516.2	0.1213	8.9513
10	1.2270	0.0010003	106.4	41.99	2519.9	0.1510	8.9020
12	1.4014	0.0010004	93.84	50.38	2523.6	0.1805	8.8536
14	1.5973	0.0010007	82.90	58.75	2527.2	0.2098	8.8060
16	1.8168	0.0010010	73.38	67.13	2530.9	0.2388	8.7593
18	2.062	0.0010013	65.09	75.50	2534.5	0.2677	8.7135
20	2.337	0.0010017	57.84	83.86	2538.2	0.2963	8.6684
22	2.642	0.0010022	51.49	92.23	2541.8	0.3247	8.6241
24	2.982	0.0010026	45.93	100.59	2545.5	0.3530	8.5806
26	3.360	0.0010032	41.03	108.95	2549.1	0.3810	8.5379
28	3.778	0.0010037	36.73	117.31	2552.7	0.4088	8.4959
30	4.241	0.0010043	32.93	125.66	2556.4	0.4365	8.4546
32	4.753	0.0010049	29.57	134.02	2560.0	0.4640	8.4140
34	5.318	0.0010056	26.60	142.38	2563.6	0.4913	8.3740
36	5.940	0.0010063	23.97	150.74	2567.2	0.5184	8.3348
38	6.624	0.0010070	21.63	159.09	2570.8	0.5453	8.2962
40	7.375	0.0010078	19.55	167.45	2574.4	0.5721	8.2583
42	8.198	0.0010086	17.69	175.31	2577.9	0.5987	8.2209
44	9.100	0.0010094	16.04	184.17	2581.5	0.6252	8.1842
46	10.086	0.0010103	14.56	192.53	2585.1	0.6514	8.1481

Table A-1 (continued)

t, °C	Saturation pressure, kPa	Specific volume, m^3/kg Liquid	Specific volume, m^3/kg Vapor	Enthalpy, kJ/kg Liquid	Enthalpy, kJ/kg Vapor	Entropy, kJ/kg · K Liquid	Entropy, kJ/kg · K Vapor
48	11.162	0.0010112	13.23	200.89	2588.6	0.6776	8.1125
50	12.335	0.0010121	12.05	209.26	2592.2	0.7035	8.0776
52	13.613	0.0010131	10.98	217.62	2595.7	0.7293	8.0432
54	15.002	0.0010140	10.02	225.98	2599.2	0.7550	8.0093
56	16.511	0.0010150	9.159	234.35	2602.7	0.7804	7.9759
58	18.147	0.0010161	8.381	242.72	2606.2	0.8058	7.9431
60	19.920	0.0010171	7.679	251.09	2609.7	0.8310	7.9108
62	21.84	0.0010182	7.044	259.46	2613.2	0.8560	7.8790
64	23.91	0.0010193	6.469	267.84	2616.6	0.8809	7.8477
66	26.15	0.0010205	5.948	276.21	2620.1	0.9057	7.8168
68	28.56	0.0010217	5.476	284.59	2623.5	0.9303	7.7864
70	31.16	0.0010228	5.046	292.97	2626.9	0.9548	7.7565
72	33.96	0.0010241	4.646	301.35	2630.3	0.9792	7.7270
74	36.96	0.0010253	4.300	309.74	2633.7	1.0034	7.6979
76	40.19	0.0010266	3.976	318.13	2637.1	1.0275	7.6693
78	43.65	0.0010279	3.680	326.52	2640.4	1.0514	7.6410
80	47.36	0.0010292	3.409	334.92	2643.8	1.0753	7.6132
82	51.33	0.0010305	3.162	343.31	2647.1	1.0990	7.5850
84	55.57	0.0010319	2.935	351.71	2650.4	1.1225	7.5588
86	60.11	0.0010333	2.727	360.12	2653.6	1.1460	7.5321
88	64.95	0.0010347	2.536	368.53	2656.9	1.1693	7.5058
90	70.11	0.0010361	2.361	376.94	2660.1	1.1925	7.4799
92	75.61	0.0010376	2.200	385.36	2663.4	1.2156	7.4543
94	81.46	0.0010391	2.052	393.78	2666.6	1.2386	7.4291
96	87.69	0.0010406	1.915	402.20	2669.7	1.2615	7.4042
98	94.30	0.0010421	1.789	410.63	2672.9	1.2842	7.3796
100	101.33	0.0010437	1.673	419.06	2676.0	1.3069	7.3554
102	108.78	0.0010453	1.566	427.50	2679.1	1.3294	7.3315
104	116.68	0.0010469	1.466	435.95	2682.2	1.3518	7.3078
106	125.04	0.0010485	1.374	444.40	2685.3	1.3742	7.2845
108	133.90	0.0010502	1.289	452.85	2688.3	1.3964	7.2615
110	143.26	0.0010519	1.210	461.32	2691.3	1.4185	7.2388
112	153.16	0.0010536	1.137	469.78	2694.3	1.4405	7.2164
114	163.62	0.0010553	1.069	478.26	2697.2	1.4624	7.1942
116	174.65	0.0010571	1.005	486.74	2700.2	1.4842	7.1723
118	186.28	0.0010588	0.9463	495.23	2703.1	1.5060	7.1507
120	198.54	0.0010606	0.8915	503.72	2706.0	1.5276	7.1293

Source: Abstracted by permission from Ref. 1.

Table A-2 Moist air:[2] thermodynamic properties of saturated air at atmospheric pressure of 101.325 kPa

t, °C	Vapor pressure, kPa	Humidity ratio, kg/kg	Specific volume, m^3kg	Enthalpy, kJ/kg
-40	0.01283	0.000079	0.6597	-40.041
-35	0.02233	0.000138	0.6740	-34.868
-30	0.03798	0.000234	0.6884	-29.600
-25	0.06324	0.000390	0.7028	-24.187
-20	0.10318	0.000637	0.7173	-18.546
-18	0.12482	0.000771	0.7231	-16.203
-16	0.15056	0.000930	0.7290	-13.795
-14	0.18107	0.001119	0.7349	-11.314
-12	0.21716	0.001342	0.7409	-8.745
-10	0.25971	0.001606	0.7469	-6.073
-8	0.30975	0.001916	0.7529	-3.285
-6	0.36846	0.002280	0.7591	-0.360
-4	0.43716	0.002707	0.7653	2.724
-2	0.51735	0.003206	0.7716	5.991
0	0.61072	0.003788	0.7781	9.470
1	0.6566	0.00407	0.7813	11.200
2	0.7055	0.00438	0.7845	12.978
3	0.7575	0.00471	0.7878	14.807
4	0.8130	0.00505	0.7911	16.692
5	0.8719	0.00542	0.7944	18.634
6	0.9347	0.00582	0.7978	20.639
7	1.0013	0.00624	0.8012	22.708
8	1.0722	0.00668	0.8046	24.848
9	1.1474	0.00716	0.8081	27.059
10	1.2272	0.00766	0.8116	29.348
11	1.3119	0.00820	0.8152	31.716
12	1.4017	0.00876	0.8188	34.172
13	1.4969	0.00937	0.8225	36.719
14	1.5977	0.01001	0.8262	39.362
15	1.7044	0.01069	0.8300	42.105
16	1.8173	0.01141	0.8338	44.955
17	1.9367	0.01218	0.8377	47.918
18	2.0630	0.01299	0.8417	50.998
19	2.1964	0.01384	0.8457	54.205
20	2.3373	0.01475	0.8498	57.544
21	2.4861	0.01572	0.8540	61.021
22	2.6431	0.01674	0.8583	64.646
23	2.8086	0.01781	0.8626	68.425
24	2.9832	0.01896	0.8671	72.366
25	3.1671	0.02016	0.8716	76.481
26	3.3609	0.02144	0.8763	80.777
27	3.5649	0.02279	0.8811	85.263
28	3.7797	0.02422	0.8860	89.952
29	4.0055	0.02572	0.8910	94.851
30	4.2431	0.02732	0.8961	99.977
31	4.4928	0.02900	0.9014	105.337
32	4.7552	0.03078	0.9068	110.946

Table A-2 (continued)

t, °C	Vapor pressure, kPa	Humidity ratio, kg/kg	Specific volume, m^3kg	Enthalpy, kJ/kg
33	5.0308	0.03266	0.9124	116.819
34	5.3201	0.03464	0.9182	122.968
35	5.6237	0.03674	0.9241	129.411
36	5.9423	0.03895	0.9302	136.161
37	6.2764	0.04129	0.9365	143.239
38	6.6265	0.04376	0.9430	150.660
39	6.9935	0.04636	0.9497	158.445
40	7.3778	0.04911	0.9567	166.615
40	7.3778	0.04911	0.9567	166.615
41	7.7803	0.05202	0.9639	175.192
42	8.2016	0.05509	0.9713	184.200
43	8.6424	0.05833	0.9790	193.662
44	9.1036	0.06176	0.9871	203.610
45	9.5856	0.06537	0.9954	214.067
46	10.0896	0.06920	1.0040	225.068
47	10.6161	0.07324	1.0130	236.643
48	11.1659	0.07751	1.0224	248.828
49	11.7402	0.08202	1.0322	261.667
50	12.3397	0.08680	1.0424	275.198
52	13.6176	0.09720	1.0641	304.512
54	15.0072	0.10887	1.0879	337.182
56	16.5163	0.12198	1.1141	373.679
58	18.1531	0.13674	1.1429	414.572
60	19.9263	0.15341	1.1749	460.536
62	21.8447	0.17228	1.2105	512.391
64	23.9184	0.19375	1.2504	571.144
66	26.1565	0.21825	1.2953	638.003
68	28.5701	0.24638	1.3462	714.531
70	31.1693	0.27884	1.4043	802.643
75	38.5562	0.38587	1.5925	1092.010
80	47.3670	0.55201	1.8792	1539.414
85	57.8096	0.83634	2.3633	2302.878
90	70.1140	1.41604	3.3412	3856.547

Table A-3 Ammonia: properties of liquid and saturated vapor[3]

		Enthalpy, kJ/kg		Entropy, kJ/kg · K		Specific volume, L/kg	
t, °C	P, kPa	h_f	h_g	s_f	s_g	v_f	v_g
-60	21.99	-69.5330	1373.19	-0.10909	6.6592	1.4010	4685.08
-55	30.29	-47.5062	1382.01	-0.00717	6.5454	1.4126	3474.22
-50	41.03	-25.4342	1390.64	0.09264	6.4382	1.4245	2616.51
-45	54.74	-3.3020	1399.07	0.19049	6.3369	1.4367	1998.91
-40	72.01	18.9024	1407.26	0.28651	6.2410	1.4493	1547.36
-35	93.49	41.1883	1415.20	0.38082	6.1501	1.4623	1212.49
-30	119.90	63.5629	1422.86	0.47351	6.0636	1.4757	960.867
-28	132.02	72.5387	1425.84	0.51015	6.0302	1.4811	878.100
-26	145.11	81.5300	1428.76	0.54655	5.9974	1.4867	803.761
-24	159.22	90.5370	1431.64	0.58272	5.9652	1.4923	736.868
-22	174.41	99.5600	1434.46	0.61865	5.9336	1.4980	676.570
-20	190.74	108.599	1437.23	0.65436	5.9025	1.5037	622.122
-18	208.26	117.656	1439.94	0.68984	5.8720	1.5096	572.875
-16	227.04	126.729	1442.60	0.72511	5.8420	1.5155	528.257
-14	247.14	135.820	1445.20	0.76016	5.8125	1.5215	487.769
-12	268.63	144.929	1447.74	0.79501	5.7835	1.5276	450.971
-10	291.57	154.056	1450.22	0.82965	5.7550	1.5338	417.477
-9	303.60	158.628	1451.44	0.84690	5.7409	1.5369	401.860
-8	316.02	163.204	1452.64	0.86410	5.7269	1.5400	386.944
-7	328.84	167.785	1453.83	0.88125	5.7131	1.5432	372.692
-6	342.07	172.371	1455.00	0.89835	5.6993	1.5464	359.071
-5	355.71	176.962	1456.15	0.91541	5.6856	1.5496	346.046
-4	369.77	181.559	1457.29	0.93242	5.6721	1.5528	333.589
-3	384.26	186.161	1458.42	0.94938	5.6586	1.5561	321.670
-2	399.20	190.768	1459.53	0.96630	5.6453	1.5594	310.263
-1	414.58	195.381	1460.62	0.98317	5.6320	1.5627	299.340
0	430.43	200.000	1461.70	1.00000	5.6189	1.5660	288.880
1	446.74	204.625	1462.76	1.01679	5.6058	1.5694	278.858
2	463.53	209.256	1463.80	1.03354	5.5929	1.5727	269.253
3	480.81	213.892	1464.83	1.05024	5.5800	1.5762	260.046
4	498.59	218.535	1465.84	1.06691	5.5672	1.5796	251.216
5	516.87	223.185	1466.84	1.08353	5.5545	1.5831	242.745
6	535.67	227.841	1467.82	1.10012	5.5419	1.5866	234.618
7	555.00	232.503	1468.78	1.11667	5.5294	1.5901	226.817
8	574.87	237.172	1469.72	1.13317	5.5170	1.5936	219.326
9	595.28	241.848	1470.64	1.14964	5.5046	1.5972	212.132
10	616.25	246.531	1471.57	1.16607	5.4924	1.6008	205.221
11	637.78	251.221	1472.46	1.18246	5.4802	1.6045	198.580
12	659.89	255.918	1473.34	1.19882	5.4681	1.6081	192.196
13	682.59	260.622	1474.20	1.21515	5.4561	1.6118	186.058
14	705.88	265.334	1475.05	1.23144	5.4441	1.6156	180.154
15	729.79	270.053	1475.88	1.24769	5.4322	1.6193	174.475
16	754.31	274.779	1476.69	1.26391	5.4204	1.6231	169.009
17	779.46	279.513	1477.48	1.28010	1.4087	1.6269	163.748
18	805.25	284.255	1478.25	1.29626	5.3971	1.6308	158.683
19	831.69	289.005	1479.01	1.31238	5.3855	1.6347	153.804
20	858.79	293.762	1479.75	1.32847	5.3740	1.6386	149.106

Table A-3 (continued)

		Enthalpy, kJ/kg		Entropy, kJ/kg · K		Specific volume, L/kg	
t, °C	P, kPa	h_f	h_g	s_f	s_g	v_f	v_g
21	886.57	298.527	1480.48	1.34452	5.3626	1.6426	144.578
22	915.03	303.300	1481.18	1.36055	5.3512	1.6466	140.214
23	944.18	308.081	1481.87	1.37654	5.3399	1.6507	136.006
24	974.03	312.870	1482.53	1.39250	5.3286	1.6547	131.950
25	1004.6	317.667	1483.18	1.40843	5.3175	1.6588	128.037
26	1035.9	322.471	1483.81	1.42433	5.3063	1.6630	124.261
27	1068.0	327.284	1484.42	1.44020	5.2953	1.6672	120.619
28	1100.7	332.104	1485.01	1.45604	5.2843	1.6714	117.103
29	1134.3	336.933	1485.59	1.47185	5.2733	1.6757	113.708
30	1168.6	341.769	1486.14	1.48762	5.2624	1.6800	110.430
31	1203.7	346.614	1486.67	1.50337	5.2516	1.6844	107.263
32	1239.6	351.466	1487.18	1.51908	5.2408	1.6888	104.205
33	1276.3	356.326	1487.66	1.53477	5.2300	1.6932	101.248
34	1313.9	361.195	1488.13	1.55042	5.2193	1.6977	98.3913
35	1352.2	366.072	1488.57	1.56605	5.2086	1.7023	95.6290
36	1391.5	370.957	1488.99	1.58165	5.1980	1.7069	92.9579
37	1431.5	375.851	1489.39	1.59722	5.1874	1.7115	90.3743
38	1472.4	380.754	1489.76	1.61276	5.1768	1.7162	87.8748
39	1514.3	385.666	1490.10	1.62828	5.1663	1.7209	85.4561
40	1557.0	390.587	1490.42	1.64377	5.1558	1.7257	83.1150
41	1600.6	395.519	1490.71	1.65924	5.1453	1.7305	80.8484
42	1645.1	400.462	1490.98	1.67470	5.1349	1.7354	78.6536
43	1690.6	405.416	1491.21	1.69013	5.1244	1.7404	76.5276
44	1737.0	410.382	1491.41	1.70554	5.1140	1.7454	74.4678
45	1784.3	415.362	1491.58	1.72095	5.1036	1.7504	72.4716
46	1832.6	420.358	1491.72	1.73635	5.0932	1.7555	70.5365
47	1881.9	425.369	1491.83	1.75174	5.0827	1.7607	68.6602
48	1932.2	430.399	1491.88	1.76714	5.0723	1.7659	66.8403
49	1983.5	435.450	1491.91	1.78255	5.0618	1.7712	65.0746
50	2035.9	440.523	1491.89	1.79798	5.0514	1.7766	63.3608
51	2089.2	445.623	1491.83	1.81343	5.0409	1.7820	61.6971
52	2143.6	450.751	1491.73	1.82891	5.0303	1.7875	60.0813
53	2199.1	455.913	1491.58	1.84445	5.0198	1.7931	58.5114
54	2255.6	461.112	1491.38	1.86004	5.0092	1.7987	56.9855
55	2313.2	466.353	1491.12	1.87571	4.9985	1.8044	55.5019

Table A-4 Refrigerant 11: properties of liquid and saturated vapor[4]

		Enthalpy, kJ/kg		Entropy, kJ/kg · K		Specific volume, L/kg	
t, °C	P, kPa	h_f	h_g	s_f	s_g	v_f	v_g
-30	9.24	174.25	373.57	0.90099	1.72074	0.62466	1581.77
-25	12.15	178.53	376.11	0.91824	1.71447	0.62894	1225.53
-20	15.78	182.81	378.66	0.93517	1.70885	0.63331	960.954
-15	20.25	187.09	381.22	0.95179	1.70377	0.63777	761.949
-10	25.71	191.39	383.77	0.96813	1.69922	0.64234	610.466
-8	28.20	193.11	384.80	0.97459	1.69753	0.64419	560.196
-6	30.88	194.83	385.82	0.98100	1.69592	0.64606	514.840
-4	33.76	196.55	386.84	0.98738	1.69438	0.64795	473.883
-2	36.86	198.27	387.86	0.99371	1.69291	0.64985	436.764
0	40.18	200.00	388.89	1.00000	1.69150	0.65178	403.130
1	41.92	200.86	389.40	1.00313	1.69082	0.65275	387.493
2	43.73	201.73	389.91	1.00625	1.69016	0.65372	372.593
3	45.60	202.59	390.42	1.00936	1.68951	0.65470	358.366
4	47.54	203.46	390.93	1.01246	1.68888	0.65568	344.792
5	49.53	204.32	391.44	1.01555	1.68826	0.65667	331.859
6	51.60	205.19	391.95	1.01863	1.68766	0.65766	319.500
7	53.73	206.05	392.46	1.02170	1.68707	0.65866	307.698
8	55.93	206.92	392.97	1.02476	1.68650	0.65966	296.427
9	58.21	207.79	393.47	1.02782	1.68594	0.66067	285.648
10	60.55	208.65	393.98	1.03086	1.68539	0.66168	275.347
11	62.97	209.52	394.49	1.03389	1.68486	0.66270	265.483
12	65.47	210.39	395.00	1.03692	1.68434	0.66327	256.063
13	68.04	211.26	395.51	1.03994	1.68383	0.66475	247.037
14	70.70	212.13	396.02	1.04294	1.68333	0.66578	238.396
15	73.43	213.00	396.52	1.04594	1.68285	0.66682	230.130
16	76.25	213.87	397.03	1.04893	1.68238	0.66786	222.205
17	79.15	214.74	397.54	1.05191	1.68193	0.66891	214.614
18	82.14	215.61	398.04	1.05488	1.68148	0.66997	207.332
19	85.21	216.48	398.55	1.05785	1.68105	0.67102	200.361
20	88.38	217.35	399.05	1.06080	1.68062	0.67209	193.665
21	91.64	218.22	399.56	1.06375	1.68021	0.67316	187.245
22	94.99	219.10	400.06	1.06669	1.67982	0.67424	181.089
23	98.44	219.97	400.57	1.06961	1.67942	0.67532	175.166
24	101.98	220.84	401.07	1.07254	1.67905	0.67641	169.485
25	105.62	221.72	401.57	1.07545	1.67868	0.67750	164.034
26	109.37	222.59	402.07	1.07838	1.67832	0.67860	158.786
27	113.21	223.47	402.57	1.08125	1.67798	0.67971	153.754
28	117.16	224.34	403.08	1.08414	1.67764	0.68082	148.903
29	121.22	225.22	403.58	1.08702	1.67731	0.68194	144.246
30	125.38	226.10	404.08	1.08989	1.67699	0.68307	139.768
32	134.05	227.85	405.07	1.09561	1.67638	0.68533	131.305
34	143.18	229.61	406.07	1.10130	1.67581	0.68763	123.462
36	152.78	231.37	407.06	1.10696	1.67527	0.68995	116.135
38	162.87	233.13	408.05	1.11259	1.67476	0.69230	109.430
40	173.46	234.90	409.04	1.11819	1.67429	0.69468	103.151

Table A-4 (continued)

		Enthalpy, kJ/kg		Entropy, kJ/kg · K		Specific volume, L/kg	
t, °C	P, kPa	h_f	h_g	s_f	s_g	v_f	v_g
45	202.28	239.32	411.49	1.13206	1.67324	0.70074	89.2884
50	234.64	243.75	413.93	1.14576	1.67237	0.70700	77.6428
55	270.83	248.21	416.34	1.15929	1.67165	0.71346	67.8040
60	311.10	252.68	418.73	1.17267	1.67109	0.72014	59.4543
70	405.15	261.68	423.42	1.19898	1.67031	0.73421	46.2114
80	519.21	270.79	427.98	1.22479	1.66992	0.74937	36.3872

Table A-5 Refrigerant 12: properties of liquid and saturated vapor[5]

		Enthalpy, kJ/kg		Entropy, kJ/kg · K		Specific volume, L/kg	
t, °C	P, kPa	h_f	h_g	s_f	s_g	v_f	v_g
-60	22.62	146.463	324.236	0.77977	1.61373	0.63689	637.911
-55	29.98	150.808	326.567	0.79990	1.60552	0.64226	491.000
-50	39.15	155.169	328.897	0.81964	1.59810	0.64782	383.105
-45	50.44	159.549	331.223	0.83901	1.59142	0.65355	302.683
-40	64.17	163.948	333.541	0.85805	1.58539	0.65949	241.910
-35	80.71	168.369	335.849	0.86776	1.57996	0.66563	195.398
-30	100.41	172.810	338.143	0.89516	1.57507	0.67200	159.375
-28	109.27	174.593	339.057	0.90244	1.57326	0.67461	147.275
-26	118.72	176.380	339.968	0.90967	1.57152	0.67726	136.284
-24	128.80	178.171	340.876	0.91686	1.56985	0.67996	126.282
-22	139.53	179.965	341.780	0.92400	1.56825	0.68269	117.167
-20	150.93	181.764	342.682	0.93110	1.56672	0.68547	108.847
-18	163.04	183.567	343.580	0.93816	1.56526	0.68829	101.242
-16	175.89	185.374	344.474	0.94518	1.56385	0.69115	94.2788
-14	189.50	187.185	345.365	0.95216	1.56250	0.69407	87.8951
-12	203.90	189.001	346.252	0.95910	1.56121	0.69703	82.0344
-10	219.12	190.822	347.134	0.96601	1.55997	0.70004	76.6464
-9	227.04	191.734	347.574	0.96945	1.55938	0.70157	74.1155
-8	235.19	192.647	348.012	0.97287	1.55897	0.70310	71.6864
-7	243.55	193.562	348.450	0.97629	1.55822	0.70465	69.3543
-6	252.14	194.477	348.886	0.97971	1.55765	0.70622	67.1146
-5	260.96	195.395	349.321	0.98311	1.55710	0.70780	64.9629
-4	270.01	196.313	349.755	0.98650	1.55657	0.70939	62.8952
-3	279.30	197.233	350.187	0.98989	1.55604	0.71099	60.9075
-2	288.82	198.154	350.619	0.99327	1.55552	0.71261	58.9963
-1	298.59	199.076	351.049	0.99664	1.55502	0.71425	57.1579
0	308.61	200.000	351.477	1.00000	1.55452	0.71590	55.3892
1	318.88	200.925	351.905	1.00335	1.55404	0.71756	53.6869
2	329.40	201.852	352.331	1.00670	1.55356	0.71924	52.0481
3	340.19	202.780	352.755	1.01004	1.55310	0.72094	50.4700
4	351.24	203.710	353.179	1.01337	1.55264	0.72265	48.9499
5	263.55	204.642	353.600	1.01670	1.55220	0.72438	47.4853
6	374.14	205.575	354.020	1.02001	1.55176	0.72612	46.0737
7	386.01	206.509	354.439	1.02333	1.55133	0.72788	44.7129
8	398.15	207.445	354.856	1.02663	1.55091	0.72966	43.4006
9	410.58	208.383	355.272	1.02993	1.55050	0.73146	42.1349
10	423.30	209.323	355.686	1.03322	1.55010	0.73326	40.9137
11	436.31	210.264	356.098	1.03650	1.54970	0.73510	39.7352
12	449.62	211.207	356.509	1.03978	1.54931	0.73695	38.5975
13	463.23	212.152	356.918	1.04305	1.54893	0.73882	37.4991
14	477.14	213.099	357.325	1.04632	1.54856	0.74071	36.4382
15	491.37	214.048	357.730	1.04958	1.54819	0.74262	35.4133
16	505.91	214.998	358.134	1.05284	1.54783	0.74455	34.4230
17	520.76	215.951	358.535	1.05609	1.54748	0.74649	33.4658
18	535.94	216.906	358.935	1.05933	1.54713	0.74846	32.5405
19	551.45	217.863	359.333	1.06258	1.54679	0.75045	31.6457
20	567.29	218.821	359.729	1.06581	1.54645	0.75246	30.7802

Table A-5 (continued)

t, °C	P, kPa	Enthalpy, kJ/kg		Entropy, kJ/kg · K		Specific volume, L/kg	
		h_f	h_g	s_f	s_g	v_f	v_g
21	583.47	219.783	360.122	1.06904	1.54612	0.75449	29.9429
22	599.98	220.746	360.514	1.07227	1.54579	0.75655	29.1327
23	616.84	221.712	360.904	1.07549	1.54547	0.75863	28.3485
24	634.05	222.680	361.291	1.07871	1.54515	0.76073	27.5894
25	651.62	223.650	361.676	1.08193	1.54484	0.76286	26.8542
26	669.54	224.623	362.059	1.08514	1.54453	0.76501	26.1422
27	687.82	225.598	362.439	1.08835	1.54423	0.76718	25.4524
28	706.47	226.576	362.817	1.09155	1.54393	0.76938	24.7840
29	725.50	227.557	363.193	1.09475	1.54363	0.77161	24.1362
30	744.90	228.540	363.566	1.09795	1.54334	0.77386	23.5082
31	764.68	229.526	363.937	1.10115	1.54305	0.77614	22.8993
32	784.85	230.515	364.305	1.10434	1.54276	0.77845	22.3088
33	805.41	231.506	364.670	1.10753	1.54247	0.78079	21.7359
34	826.36	232.501	365.033	1.11072	1.54219	0.78316	21.1802
35	847.72	233.498	365.392	1.11391	1.54191	0.78556	20.6408
36	869.48	234.499	365.749	1.11710	1.54163	0.78799	20.1173
37	891.64	235.503	366.103	1.12028	1.54135	0.79045	19.6091
38	914.23	236.510	366.454	1.12347	1.54107	0.79294	19.1156
39	937.23	237.521	366.802	1.12665	1.54079	0.79546	18.6362
40	960.65	238.535	367.146	1.12984	1.54051	0.79802	18.1706
41	984.51	239.552	367.487	1.13302	1.54024	0.80062	17.7182
42	1008.8	240.574	367.825	1.13620	1.53996	0.80325	17.2785
43	1033.5	241.598	368.160	1.13938	1.53968	0.80592	16.8511
44	1058.7	242.627	368.491	1.14257	1.53941	0.80863	16.4356
45	1084.3	243.659	368.818	1.14575	1.53913	0.81137	16.0316
46	1110.4	244.696	369.141	1.14894	1.53885	0.81416	15.6386
47	1136.9	245.736	369.461	1.15213	1.53856	0.81698	15.2563
48	1163.9	246.781	369.777	1.15532	1.53828	0.81985	14.8844
49	1191.4	247.830	370.088	1.15851	1.53799	0.82277	14.5224
50	1219.3	248.884	370.396	1.16170	1.53770	0.82573	14.1701
52	1276.6	251.004	370.997	1.16810	1.53712	0.83179	13.4931
54	1335.9	253.144	371.581	1.17451	1.53651	0.83804	12.8509
56	1397.2	255.304	372.145	1.18093	1.53589	0.84451	12.2412
58	1460.5	257.486	372.688	1.18738	1.53524	0.85121	11.6620
60	1525.9	259.690	373.210	1.19384	1.53457	0.85814	11.1113
62	1593.5	261.918	373.707	1.20034	1.53387	0.86534	10.5872
64	1663.2	264.172	374.180	1.20686	1.53313	0.87282	10.0881
66	1735.1	266.452	374.625	1.21342	1.53235	0.88059	9.61234
68	1809.3	268.762	375.042	1.22001	1.53153	0.88870	9.15844
70	1885.8	271.102	375.427	1.22665	1.53066	0.89716	8.72502
75	2087.5	277.100	376.234	1.24347	1.52821	0.92009	7.72258
80	2304.6	283.341	376.777	1.26069	1.52526	0.94612	6.82143
85	2538.0	289.879	376.985	1.27845	1.52164	0.97621	6.00494
90	2788.5	296.788	376.748	1.29691	1.51708	1.01190	5.25759
95	3056.9	304.181	375.887	1.31637	1.51113	1.05581	4.56341
100	3344.1	312.261	374.070	1.33732	1.50296	1.11311	3.90280

Table A-6 Refrigerant 22: properties of liquid and saturated vapor[6]

		Enthalpy, kJ/kg		Entropy, kJ/kg • K		Specific volume, L/kg	
t, °C	P, kPa	h_f	h_g	s_f	s_g	v_f	v_g
-60	37.48	134.763	379.114	0.73254	1.87886	0.68208	537.152
-55	49.47	139.830	381.529	0.75599	1.86389	0.68856	414.827
-50	64.39	144.959	383.921	0.77919	1.85000	0.69526	324.557
-45	82.71	150.153	386.282	0.80216	1.83708	0.70219	256.990
-40	104.95	155.414	388.609	0.82490	1.82504	0.70936	205.745
-35	131.68	160.742	390.896	0.84743	1.81380	0.71680	166.400
-30	163.48	166.140	393.138	0.86976	1.80329	0.72452	135.844
-28	177.76	168.318	394.021	0.87864	1.79927	0.72769	125.563
-26	192.99	170.507	394.896	0.88748	1.79535	0.73092	116.214
-24	209.22	172.708	395.762	0.89630	1.79152	0.73420	107.701
-22	226.48	174.919	396.619	0.90509	1.78779	0.73753	99.9362
-20	244.83	177.142	397.467	0.91386	1.78415	0.74091	92.8432
-18	264.29	179.376	398.305	0.92259	1.78059	0.74436	86.3546
-16	284.93	181.622	399.133	0.93129	1.77711	0.74786	80.4103
-14	306.78	183.878	399.951	0.93997	1.77371	0.75143	74.9572
-12	329.89	186.147	400.759	0.94862	1.77039	0.75506	69.9478
-10	354.30	188.426	401.555	0.95725	1.76713	0.75876	65.3399
-9	367.01	189.571	401.949	0.96155	1.76553	0.76063	63.1746
-8	380.06	190.718	402.341	0.06585	1.76394	0.76253	61.0958
-7	393.47	191.868	402.729	0.97014	1.76237	0.76444	59.0996
-6	407.23	193.021	403.114	0.97442	1.76082	0.76636	57.1820
-5	421.35	194.176	403.496	0.97870	1.75928	0.76831	55.3394
-4	435.84	195.335	403.876	0.98297	1.75775	0.77028	53.5682
-3	450.70	196.497	404.252	0.98724	1.75624	0.77226	51.8653
-2	465.94	197.662	404.626	0.99150	1.75475	0.77427	50.2274
-1	481.57	198.828	404.994	0.99575	1.75326	0.77629	48.6517
0	497.59	200.000	405.361	1.00000	1.75279	0.77834	47.1354
1	514.01	201.174	405.724	1.00424	1.75034	0.78041	45.6757
2	530.83	202.351	406.084	1.00848	1.74889	0.78249	44.2702
3	548.06	203.530	406.440	1.01271	1.74746	0.78460	42.9166
4	565.71	204.713	406.793	1.01694	1.74604	0.78673	41.6124
5	583.78	205.899	407.143	1.02116	1.74463	0.78889	40.3556
6	602.28	207.089	407.489	1.02537	1.74324	0.79107	39.1441
7	621.22	208.281	407.831	1.02958	1.74185	0.79327	37.9759
8	640.59	209.477	408.169	1.03379	1.74047	0.79549	36.8493
9	660.42	210.675	408.504	1.03799	1.73911	0.79775	35.7624
10	680.70	211.877	408.835	1.04218	1.73775	0.80002	34.7136
11	701.44	213.083	409.162	1.04637	1.73640	0.80232	33.7013
12	722.65	214.291	409.485	1.05056	1.73506	0.80465	32.7239
13	744.33	215.503	409.804	1.05474	1.73373	0.80701	31.7801
14	766.50	216.719	410.119	1.05892	1.73241	0.80939	30.8683
15	789.15	217.937	410.430	1.06309	1.73109	0.81180	29.9874
16	812.29	219.160	410.736	1.06726	1.72978	0.81424	29.1361
17	835.93	220.386	411.038	1.07142	1.72848	0.81671	28.3131
18	860.08	221.615	411.336	1.07559	1.72719	0.81922	27.5173
19	884.75	222.848	411.629	1.07974	1.72590	0.82175	26.7477
20	909.93	224.084	411.918	1.08390	1.72462	0.82431	26.0032

Table A-6 (continued)

		Enthalpy, kJ/kg		Entropy, kJ/kg • K		Specific volume, L/kg	
t, °C	P, kPa	h_f	h_g	s_f	s_g	v_f	v_g
21	935.64	225.324	412.202	1.08805	1.72334	0.82691	25.2829
22	961.89	226.568	412.481	1.09220	1.72206	0.82954	24.5857
23	988.67	227.816	412.755	1.09634	1.72080	0.83221	23.9107
24	1016.0	229.068	413.025	1.10048	1.71953	0.83491	23.2572
25	1043.9	230.324	413.289	1.10462	1.71827	0.83765	22.6242
26	1072.3	231.583	413.548	1.10876	1.71701	0.84043	22.0111
27	1101.4	232.847	413.802	1.11290	1.71576	0.84324	21.4169
28	1130.9	234.115	414.050	1.11703	1.71450	0.84610	20.8411
29	1161.1	235.387	414.293	1.12116	1.71325	0.84899	20.2829
30	1191.9	236.664	414.530	1.12530	1.71200	0.85193	19.7417
31	1223.2	237.944	414.762	1.12943	1.71075	0.85491	19.2168
32	1255.2	239.230	414.987	1.13355	1.70950	0.85793	18.7076
33	1287.8	240.520	415.207	1.13768	1.70826	0.86101	18.2135
34	1321.0	241.814	415.420	1.14181	1.70701	0.86412	17.7341
35	1354.8	243.114	415.627	1.14594	1.70576	0.86729	17.2686
36	1389.2	244.418	415.828	1.15007	1.70450	0.87051	16.8168
37	1424.3	245.727	416.021	1.15420	1.70325	0.87378	16.3779
38	1460.1	247.041	416.208	1.15833	1.70199	0.87710	15.9517
39	1496.5	248.361	416.388	1.16246	1.70073	0.88048	15.5375
40	1533.5	249.686	416.561	1.16659	1.69946	0.88392	15.1351
41	1571.2	251.016	416.726	1.17073	1.69819	0.88741	14.7439
42	1609.6	252.352	416.883	1.17486	1.69692	0.89097	14.3636
43	1648.7	253.694	417.033	1.17900	1.69564	0.89459	13.9938
44	1688.5	255.042	417.174	1.18315	1.69435	0.89828	13.6341
45	1729.0	256.396	417.308	1.18730	1.69305	0.90203	13.2841
46	1770.2	257.756	417.432	1.19145	1.69174	0.90586	12.9436
47	1812.1	259.123	417.548	1.19560	1.69043	0.90976	12.6122
48	1854.8	260.497	417.655	1.19977	1.68911	0.91374	12.2895
49	1898.2	261.877	417.752	1.20393	1.68777	0.91779	11.9753
50	1942.3	263.264	417.838	1.20811	1.68643	0.92193	11.6693
52	2032.8	266.062	417.983	1.21648	1.68370	0.93047	11.0806
54	2126.5	268.891	418.083	1.22489	1.68091	0.93939	10.5214
56	2223.2	271.754	418.137	1.23333	1.67805	0.94872	9.98952
58	2323.2	274.654	418.141	1.24183	1.67511	0.95850	9.48319
60	2426.6	277.594	418.089	1.25038	1.67208	0.96878	9.00062
62	2533.3	280.577	417.978	1.25899	1.66895	0.97960	8.54016
64	2643.5	283.607	417.802	1.26768	1.66570	0.99104	8.10023
66	2757.3	286.690	417.553	1.27647	1.66231	1.00317	7.67934
68	2874.7	289.832	417.226	1.28535	1.65876	1.01608	7.27605
70	2995.9	293.038	416.809	1.29436	1.65504	1.02987	6.88899
75	3316.1	301.399	415.299	1.31758	1.64472	1.06916	5.98334
80	3662.3	310.424	412.898	1.34223	1.63239	1.11810	5.14862
85	4036.8	320.505	409.101	1.36936	1.61673	1.18328	4.35815
90	4442.5	332.616	402.653	1.40155	1.59440	1.28230	3.56440
95	4883.5	351.767	386.708	1.45222	1.54712	1.52064	2.55133

Table A-7 Refrigerant 22: properties of superheated vapor[6]

t, °C	v, L/kg	h, kJ/kg	s, kJ/kg · K	v, L/kg	h, kJ/kg	s, kJ/kg · K	v, L/kg	h, kJ/kg	s, kJ/kg · K
	Saturation temperature, −20°C			Saturation temperature, −10°C			Saturation temperature, 0°C		
−20	92.8432	397.467	1.7841						
−15	95.1474	400.737	1.7969						
−10	97.4256	404.017	1.8095	65.3399	401.555	1.7671			
−5	99.6808	407.307	1.8219	67.0081	404.983	1.7800			
0	101.915	410.610	1.8341	68.6524	408.412	1.7927	47.1354	405.361	1.7518
5	104.130	413.926	1.8461	70.2751	411.845	1.8052	48.3899	408.969	1.7649
10	106.328	417.258	1.8580	71.8785	415.283	1.8174	49.6215	412.567	1.7777
15	108.510	420.606	1.8697	73.4644	418.730	1.8295	50.8328	416.159	1.7903
20	110.678	423.970	1.8813	75.0346	422.186	1.8414	52.0259	419.649	1.8026
25	112.832	426.353	1.8928	76.5904	425.653	1.8531	53.2028	423.339	1.8148
	Saturation temperature, 5°C			Saturation temperature, 10°C			Saturation temperature, 15°C		
5	40.3556	407.143	1.7446						
10	41.4580	410.851	1.7578	34.7136	408.835	1.7377			
15	42.5379	414.542	1.7708	35.6907	412.651	1.7511	29.9874	410.430	1.7311
20	43.5979	418.222	1.7834	36.6454	416.442	1.7642	30.8606	414.362	1.7556
25	44.6401	421.894	1.7958	37.5804	420.215	1.7769	31.7114	418.260	1.7578
30	45.6665	425.562	1.8080	38.4981	423.974	1.7894	32.5427	422.133	1.7707
35	46.6786	429.229	1.8200	39.4002	427.724	1.8017	33.3568	425.985	1.7833
40	47.6779	432.897	1.8319	40.2884	431.469	1.8137	34.1556	429.823	1.7956
45	48.6656	436.569	1.8435	41.1642	435.211	1.8256	34.9409	433.650	1.8078
50	49.6427	440.247	1.8550	42.0286	438.954	1.8373	35.7139	437.470	1.8197

Table A-7 (continued)

	Saturation temperature, 20°C			Saturation temperature, 25°C			Saturation temperature, 30°C		
20	26.0032	411.918	1.7246						
25	26.7900	415.977	1.7383	22.6242	413.289	1.7183			
30	27.5542	419.991	1.7517	23.3389	417.487	1.7322	19.7417	414.530	1.7120
35	28.2989	423.970	1.7646	24.0306	421.627	1.7458	20.3962	418.881	1.7262
40	29.0264	427.922	1.7774	24.7027	425.721	1.7590	21.0272	423.159	1.7400
45	29.7389	431.852	1.7899	25.3575	429.779	1.7718	21.6381	427.378	1.7534
50	30.4379	435.766	1.8021	25.9974	433.807	1.7844	22.2316	431.549	1.7664
55	31.1250	439.668	1.8141	26.6239	437.813	1.7967	22.8101	435.683	1.7791
60	31.8012	443.561	1.8258	27.2386	441.801	1.8087	23.3733	439.787	1.7915
65	32.4678	447.450	1.8374	27.8427	445.777	1.8206	23.9288	443.867	1.8036

	Saturation temperature, 32°C			Saturation temperature, 34°C			Saturation temperature, 36°C		
35	19.0907	417.648	1.7182	17.8590	416.325	1.7099			
40	19.7093	422.014	1.7322	18.4675	420.792	1.7243	17.2953	419.483	1.7162
45	20.3062	426.310	1.7458	19.0526	425.174	1.7382	17.8708	423.961	1.7304
50	20.8847	430.549	1.7591	19.6178	429.487	1.7517	18.4247	428.358	1.7442
55	21.4471	434.743	1.7719	20.1660	433.747	1.7647	18.9603	432.690	1.7575
60	21.9956	438.900	1.7845	20.6994	437.963	1.7775	19.4802	436.970	1.7704
65	22.5318	443.028	1.7968	21.2199	442.143	1.7899	19.9865	441.207	1.7830
70	23.0571	447.133	1.8089	21.7289	446.294	1.8021	20.4807	445.410	1.7954
75	23.5726	451.219	1.8207	22.2278	450.424	1.8141	20.9643	449.586	1.8074
80	24.0794	455.292	1.8323	22.7176	454.535	1.8258	21.4385	453.739	1.8193

Table A-7 (continued)

t, °C	v, L/kg	h, kJ/kg	s, kJ/kg · K	v, L/kg	h, kJ/kg	s, kJ/kg · K	v, L/kg	h, kJ/kg	s, kJ/kg · K
	Saturation temperature, 38°C			Saturation temperature, 40°C			Saturation temperature, 42°C		
40	16.1865	418.076	1.7080	15.1350	416.561	1.6995			
45	16.7545	422.664	1.7225	15.6982	421.274	1.7144	14.6964	419.779	1.7061
50	17.2991	427.155	1.7365	16.2355	425.871	1.7287	15.2286	424.496	1.7208
55	17.8240	431.568	1.7501	16.7514	430.374	1.7426	15.7373	429.101	1.7349
60	18.3320	435.918	1.7632	17.2491	434.803	1.7560	16.2264	433.617	1.7486
65	18.8255	440.218	1.7760	17.7313	439.171	1.7690	16.6987	438.062	1.7618
70	19.3063	444.477	1.7885	18.2001	443.491	1.7817	17.1568	442.449	1.7747
75	19.7760	448.703	1.8008	18.6571	447.771	1.7940	17.6024	446.788	1.7872
80	20.2358	452.901	1.8127	19.1038	452.019	1.8061	18.0371	451.090	1.7995
85				19.5412	456.241	1.8180	18.4622	455.360	1.8115
	Saturation temperature, 45°C			Saturation temperature, 50°C					
45	13.2841	417.308	1.6931						
50	13.8136	422.241	1.7084	11.6693	417.839	1.6864			
55	14.3154	427.025	1.7231	12.1721	423.028	1.7024			
60	14.7946	431.693	1.7372	12.6447	428.026	1.7175			
65	15.2550	436.268	1.7509	13.0932	432.877	1.7319			
70	15.6995	440.769	1.7641	13.5219	437.613	1.7458			
75	16.1303	445.209	1.7769	13.9342	442.258	1.7593			
80	16.5492	449.599	1.7895	14.3325	446.828	1.7723			
85	16.9578	453.950	1.8017	14.7187	451.337	1.7850			
90	17.3571	458.267	1.8137	15.0943	455.796	1.7973			

Table A-8 Refrigerant 502: Properties of liquid and saturated vapor[7]

t, °C	Pressure P, kPa	Enthalpy, kJ/kg h_f	h_g	Entropy, kJ/kg · K s_f	s_g	Specific volume, L/kg v_f	v_g
-40	129.64	158.085	328.147	0.83570	1.56512	0.68307	127.687
-30	197.86	167.883	333.027	0.87665	1.55583	0.69890	85.7699
-25	241.00	172.959	335.415	0.89719	1.55187	0.70733	71.1552
-20	291.01	178.149	337.762	0.91775	1.54826	0.71615	59.4614
-15	348.55	183.452	340.063	0.93833	1.54500	0.72538	50.0230
-10	414.30	188.864	342.313	0.95891	1.54203	0.73509	42.3423
-8	443.04	191.058	343.197	0.96714	1.54092	0.73911	39.6747
-6	473.26	193.269	344.071	0.97536	1.53985	0.74323	37.2074
-4	504.98	195.497	344.936	0.98358	1.53881	0.74743	34.9228
-2	538.26	197.740	345.791	0.99179	1.53780	0.75172	32.8049
0	573.13	200.000	346.634	1.00000	1.53683	0.75612	30.8393
1	591.18	201.136	347.052	1.00410	1.53635	0.75836	29.9095
2	609.65	202.275	347.467	1.00820	1.53588	0.76062	29.0131
3	628.54	203.419	347.879	1.01229	1.53542	0.76291	28.1485
4	647.86	204.566	348.288	1.01639	1.53496	0.76523	27.3145
5	667.61	205.717	348.693	1.02048	1.53451	0.76758	26.5097
6	687.80	206.872	349.096	1.02457	1.53406	0.76996	25.7330
7	708.43	208.031	349.496	1.02866	1.53362	0.77237	24.9831
8	729.51	209.193	349.892	1.03274	1.53318	0.77481	24.2589
9	751.05	210.359	350.285	1.03682	1.53275	0.77728	23.5593
10	773.05	211.529	350.675	1.04090	1.53232	0.77978	22.8835
11	795.52	212.703	351.062	1.04497	1.53190	0.78232	22.2303
12	818.46	213.880	351.444	1.04905	1.53147	0.78489	21.5989
13	841.87	215.061	351.824	1.05311	1.53106	0.78750	20.9883
14	865.78	216.245	352.199	1.05718	1.53064	0.79014	20.3979
15	890.17	217.433	352.571	1.06124	1.53023	0.79282	19.8266
16	915.06	218.624	352.939	1.06530	1.52982	0.79555	19.2739
17	940.45	219.820	353.303	1.06936	1.52941	0.79831	18.7389
18	966.35	221.018	353.663	1.07341	1.52900	0.80111	18.2210
19	992.76	222.220	354.019	1.07746	1.52859	0.80395	17.7194
20	1019.7	223.426	354.370	1.08151	1.52819	0.80684	17.2336
21	1047.1	224.635	354.717	1.08555	1.52778	0.80978	16.7630
22	1075.1	225.858	355.060	1.08959	1.52737	0.81276	16.3069
23	1103.7	227.064	355.398	1.09362	1.52697	0.81579	15.8649
24	1132.7	228.284	355.732	1.09766	1.52656	0.81887	15.4363
25	1162.3	229.506	356.061	1.10168	1.52615	0.82200	15.0207
26	1192.5	230.734	356.385	1.10571	1.52573	0.82518	14.6175
27	1223.2	231.964	356.703	1.10973	1.52532	0.82842	14.2263
28	1254.6	233.198	357.017	1.11375	1.52490	0.83171	13.8468
29	1286.4	234.436	357.325	1.11776	1.52448	0.83507	13.4783
30	1318.9	235.677	357.628	1.12177	1.52405	0.83848	13.1205
32	1385.6	238.170	358.216	1.12978	1.52318	0.84551	12.4356
34	1454.7	240.677	358.780	1.13778	1.52229	0.85282	11.7889
36	1526.2	243.200	359.318	1.14577	1.52137	0.86042	11.1778
38	1600.3	245.739	359.828	1.15375	1.52042	0.86834	10.5996
40	1677.0	248.295	360.309	1.16172	1.51943	0.87662	10.0521

Table A-8 (continued)

		Enthalpy, kJ/kg		Entropy, kJ/kg · K		Specific volume, L/kg	
t, °C	Pressure P, kPa	h_f	h_g	s_f	s_g	v_f	v_g
45	1880.3	254.762	361.367	1.18164	1.51672	0.89908	8.80325
50	2101.3	261.361	362.180	1.20159	1.51358	0.92465	7.70220
55	2341.1	268.128	362.684	1.22168	1.50983	0.95430	6.72295
60	2601.4	275.130	362.780	1.24209	1.50518	0.98962	5.84240
70	3191.8	290.465	360.952	1.28562	1.49103	1.09069	4.28602
80	3900.4	312.822	350.672	1.34730	1.45448	1.34203	2.70616

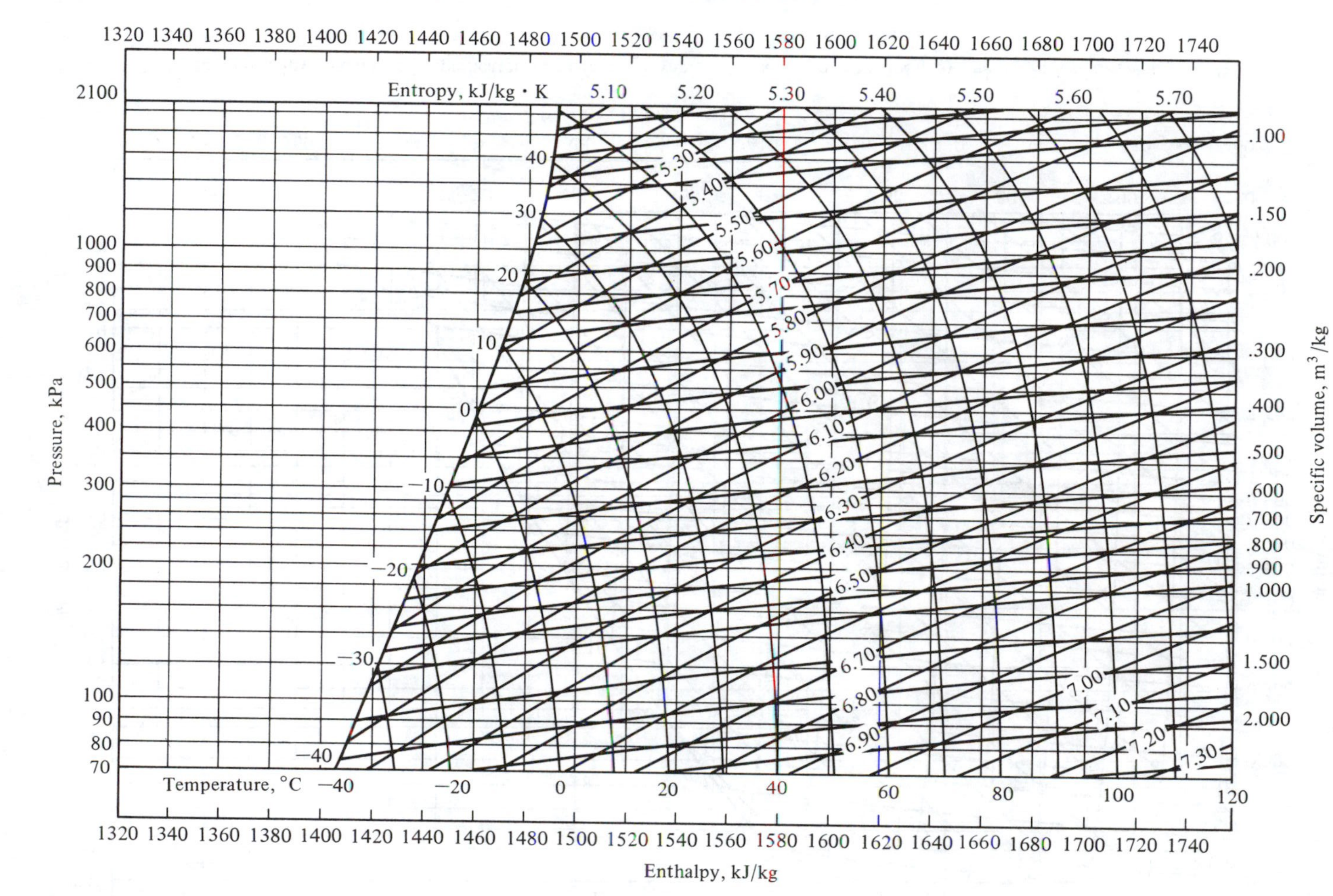

Figure A-1 Pressure-enthalpy diagram of superheated ammonia vapor. *(Prepared for this book by the Technical University of Denmark from Data in Ref. 8.)*

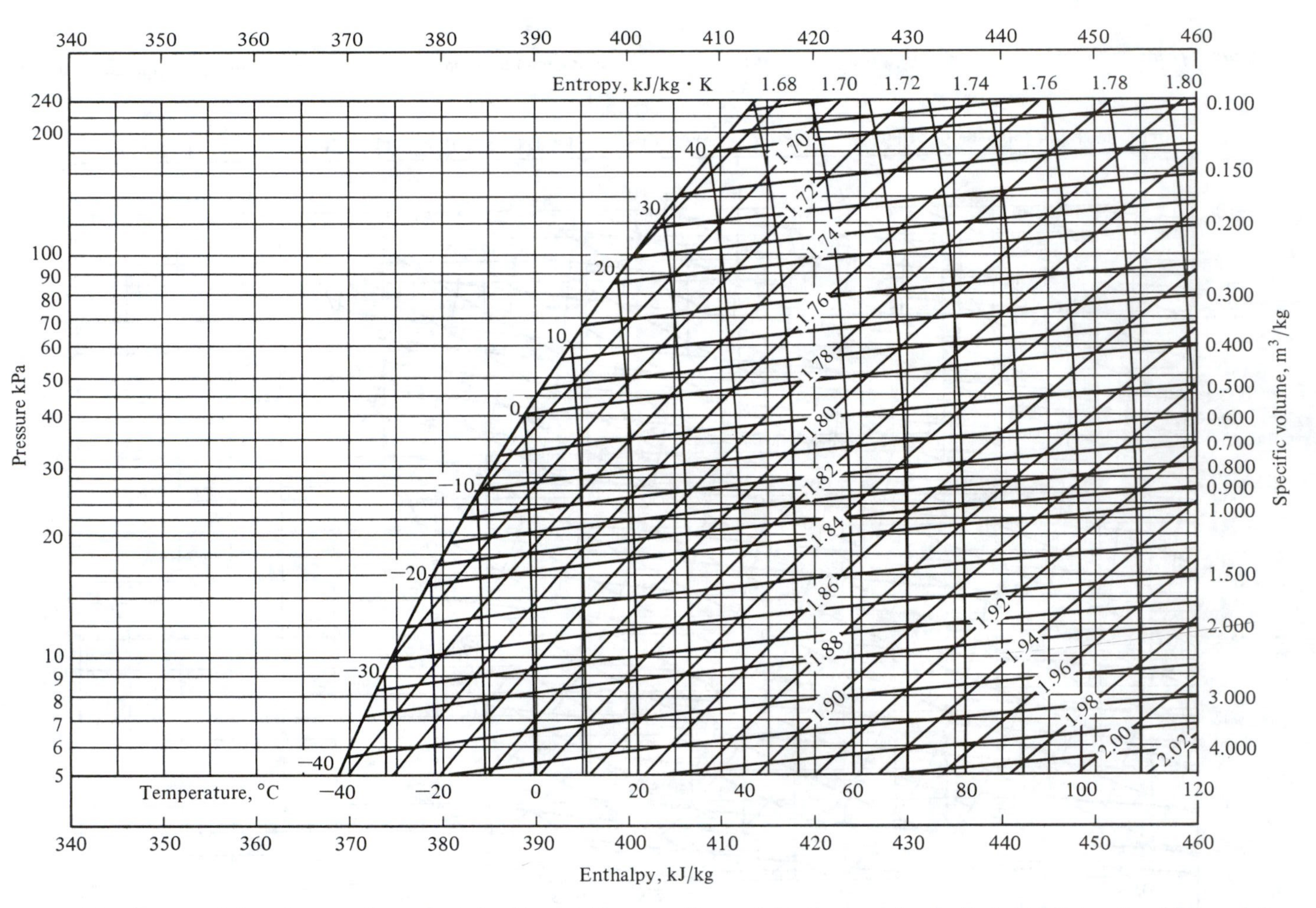

Figure A-2 Pressure-enthalpy diagram of superheated refrigerant 11 vapor. *(Prepared for this book by the Technical University of Denmark from data in Ref. 9.)*

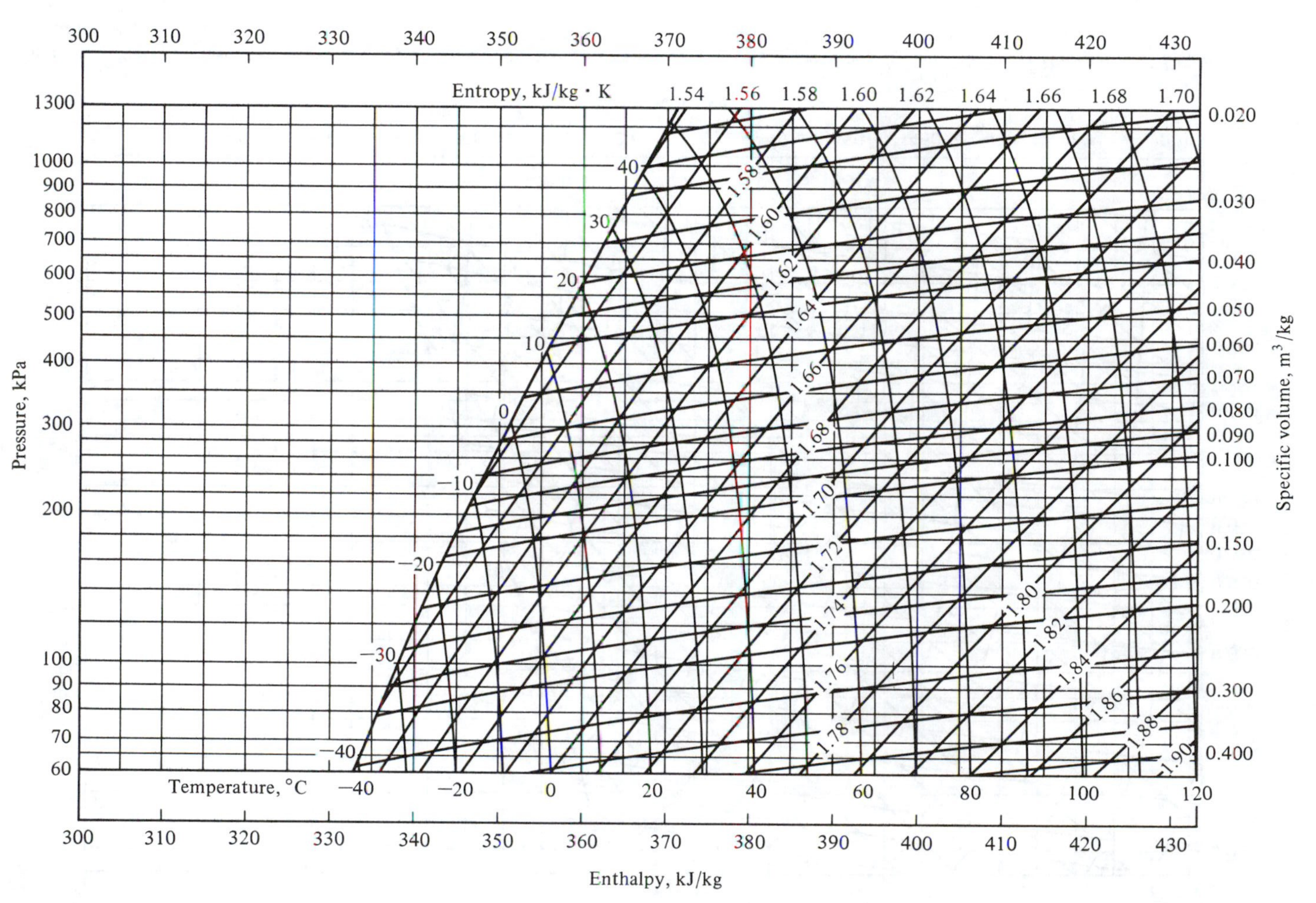

Figure A-3 Pressure-enthalpy diagram of superheated refrigerant 12 vapor. *(Prepared for this book by the Technical University of Denmark from data in Ref. 9.)*

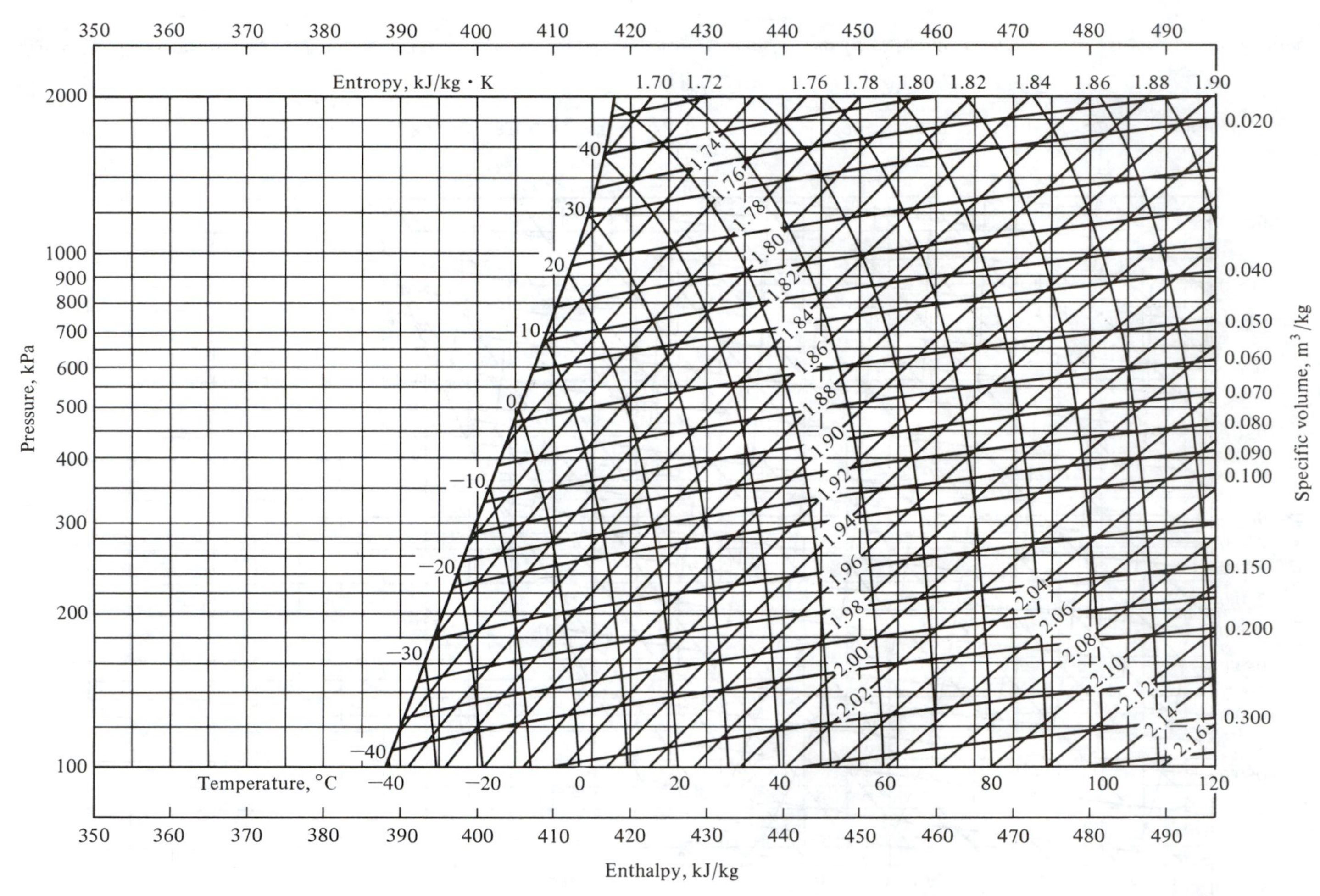

Figure A-4 Pressure-enthalpy diagram of superheated refrigerant 22 vapor. *(Prepared for this book by the Technical University of Denmark from data in Ref. 9.)*

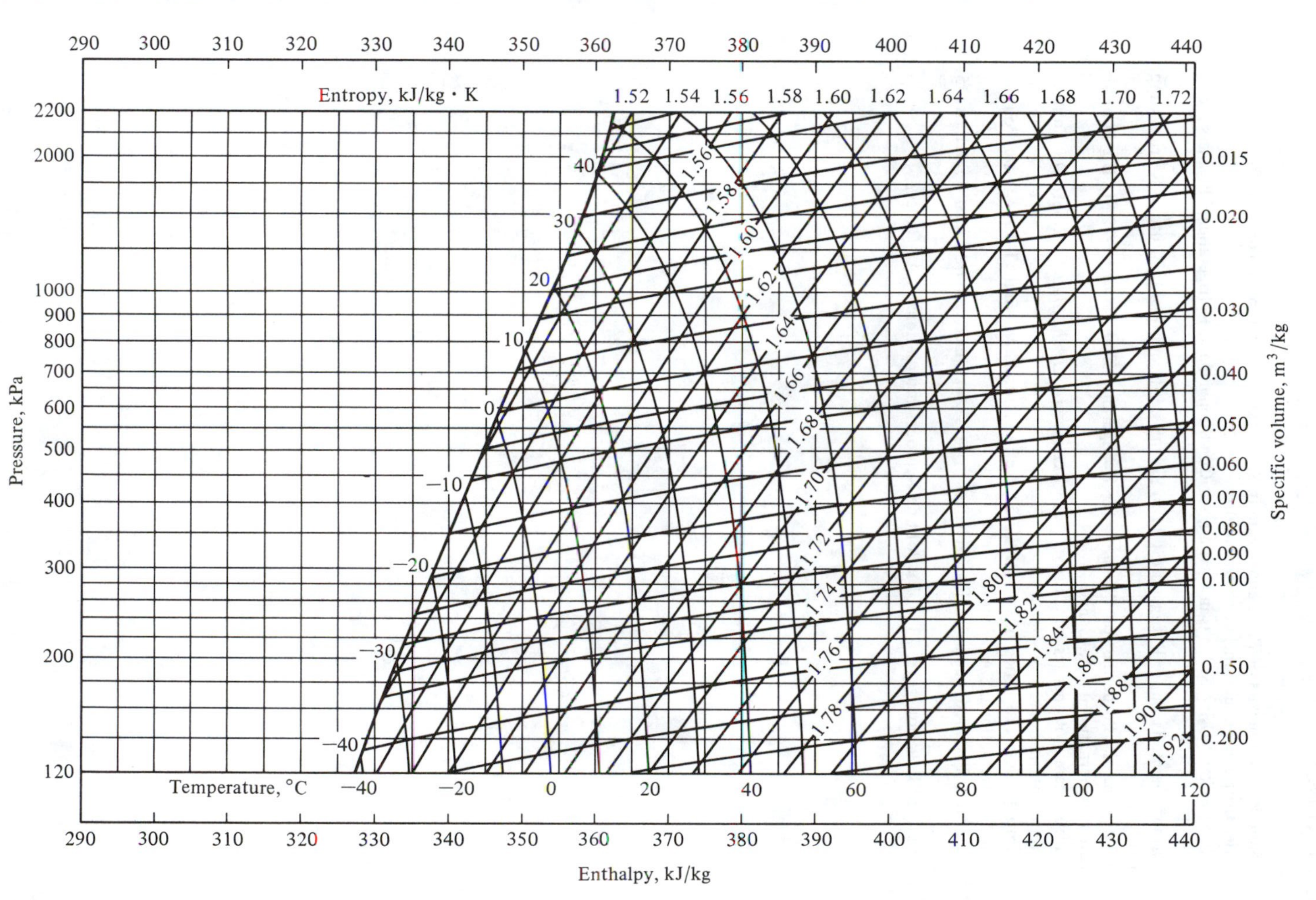

Figure A-5 Pressure-enthalpy diagram of superheated refrigerant 502 vapor. *(Prepared for this book by the Technical University of Denmark from data in Ref. 9.)*

REFERENCES

1. E. Schmidt: "Properties of Water and Steam in SI-Units," Springer, New York, 1969.
2. Carrier Corporation, personal communication.
3. W. F. Stoecker, "Using SI Units in Heating, Air Conditioning, and Refrigeration," Business News, Troy, Mich., 1977.
4. Thermodynamic Table for Refrigerant R 11 in SI-Units, International Institute of Refrigeration, Paris.
5. Thermodynamic Properties of "Freon" 12 Refrigerant, *Tech. Bull.* T-12-SI, Du Pont de Nemours International S.A., Geneva.
6. Thermodynamic Properties of "Freon" 22 Refrigerant, *Tech. Bull.* T-22-SI, Du Pont de Nemours International S.A., Geneva.
7. Thermodynamic Properties of "Freon" 502 Refrigerant, *Tech. Bull.* T-502-SI, Du Pont de Nemours International S.A., Geneva.
8. R. Döring, Thermodynamische Eigenschaften von Ammoniak, *Kaelte-Klima Ing.* Extra 5, 1978.
9. Refrigerant Equations, *Pap.* 2313, Du Pont de Nemours International S.A., Geneva.

INDEX